Human Genetics

A Problem-Based Approach

Human Genetics

A Problem-Based Approach

Bruce R. Korf, MD, PhD

Associate Professor of Neurology (Pediatrics)
Harvard Medical School
Director, Clinical Genetics Program
Children's Hospital
Boston, Massachusetts

Blackwell
Science

Blackwell Science
Editorial offices:
238 Main Street, Cambridge, Massachusetts 02142, USA
Osney Mead, Oxford OX2 0E1, England
25 John Street, London WC1N 2BL, England
23 Ainslie Place, Edinburgh EH3 6AJ, Scotland
54 University Street, Carlton, Victoria 3053, Australia

Other Editorial Offices:
Arnette Blackwell SA, 224, Boulevard Saint Germain, 75007 Paris, France
Blackwell Wissenschafts-Verlag GmbH Kurfürstendamm 57, 10707 Berlin, Germany
Zehetnergasse 6, A-1140 Vienna, Austria

Distributors:
USA
Blackwell Science Inc.
238 Main Street
Cambridge, Massachusetts 02142
(Telephone orders: 800-215-1000 or 617-876-7000; fax orders: 617-492-5263)
Canada
Copp Clark, Ltd.
2775 Matheson Blvd. East
Mississauga, Ontario
Canada, L4W 4P7
(Telephone orders: 800-263-4374 or 905-238-6074)
Australia
Blackwell Science Pty., Ltd.
54 University Street
Carlton, Victoria 3053
(Telephone orders: 03-9347-0300; fax orders 03-9349 3016)
Outside North America and Australia
Blackwell Science, Ltd.
c/o Marston Book Services, Ltd.
P.O. Box 269
Abingdon
Oxon OX14 4YN
England
(Telephone orders: 44-01235-465500; fax orders 44-01235-465555)

Acquisitions: Joy Ferris Denomme
Development: Kathleen Broderick
Production: Karen Feeney
Manufacturing: Lisa Flanagan
Typeset by Publication Services
Printed and bound by Quebecor

Printed in the United States of America

97 98 99 5 4 3

Dedication

To Shelley, Jessica, and Katie

Contents

Preface

Both medicine and genetics are undergoing revolutionary change, one in response to out-of-control costs and the other driven by unprecedented technical advance. New models of medical practice are evolving, with increasing attention to primary care and preventative medicine. Genetics will play a central role in this new approach, particularly as the genetic bases for common disease come to light. Today's medical students therefore face the dual challenge of mastering an ever complex body of knowledge along with the need to develop a set of problem solving skills that will foster career-long learning.

It has become clear that the traditional lecture-based medical curriculum is not well suited to these needs, opening the door to new approaches to medical education. The problem-based approach has been the mainstay of curriculum change at many schools, including here at Harvard Medical School. The first two years at Harvard Medical School now emphasize student-driven learning using a combination of lectures and problem-based tutorial exercises. Courses are taught in 6–12 week blocks, of which one is Genetics, Embryology, and Reproduction; this book is based on the genetics component of that course.

The initial impetus to write this book was to provide a set of clinical problems in genetics, but as the book evolved it became clear that there was another opportunity to change the way genetics is taught. If genetics is taught at all to medical students, it is usually done in an historical context, covering Mendelian genetics, cytogenetics, molecular genetics, etc. The power of modern molecular analysis, however, is that the mysteries of clinical experience are now being explained in precise molecular terms. Powerful concepts about the structure and function of the genome are emerging, and it is time to synthesize these into a new didactic model. Techniques such as pedigree analysis or gene cloning become the means; understanding basic ideas such as the mechanisms underlying inborn errors of metabolism are the ends.

I am indebted to students, teachers, and colleagues who have helped shape this work, some deliberately, others unwittingly. A series of first year Harvard Medical School classes have taught me what works and what does not. I have learned the culture of problem-based learning along with colleagues on the Genetics, Embryology, and Reproduction planning committee, including Philip Leder, Daniel Federman, Elizabeth Hay, Betsy Williams, Sohela Garib, and Elizabeth Armstrong. Innumerable colleagues have reviewed portions of the text and have provided illustrative material. The latter are acknowledged in figure legends; the former have included Alan Beggs, Peter Byers, Antonio Cao, Jonathan Fletcher, Uta Francke, Donald Johns, Mark Korson, Harvey Levy, Cynthia Morton, Richard Parad, Reed Pyeritz, Cliff Tabin, David Whiteman, Mary Ellen Wohl, and Joseph Wolfsdorf. Their many suggestions and corrections do not render the chapters immune from my further tampering, but have contributed enormously to maintaining accuracy and readability. Several editors at Blackwell Science deserve credit for stimulating me to undertake this project, providing vital feedback, and being patient. These include Victoria Reeders, Coleen Traynor, Kathleen Broderick, and Joy Ferris Denomme.

The processes of teaching and learning are inseparable. If the approach used in this book is successful, it will spawn a process of further evolution driven by feedback from those who use it.

B.R.K.

Introduction

The science of genetics is concerned with the mechanisms whereby biological traits are passed from generation to generation and expressed in the individual. Although its roots can be traced to the experiments of Mendel in the 1860s and the subsequent birth of cell biology, modern genetics is almost entirely a creation of the twentieth century. Genetics has a long tradition of merging clinical observation with basic laboratory investigation. This has culminated in the emergence of molecular biology, in which biological processes are understood from the level of the gene through proteins that direct the development and function of the organism. Recently, recombinant DNA technology has provided tools of enormous power, able to reveal the structure and function of genes. The entire human genome has been targeted for study, promising a revolution in understanding human development and physiology. Genetics is therefore playing an increasingly central role in medical science, and no physician can remain ignorant of its fundamental principles and approaches.

The teaching of genetics has focused traditionally on a number of distinct subject areas, such as mendelian genetics, cytogenics, molecular genetics, and population genetics. Although this approach may mirror the sequence in which our knowledge has emerged, a major opportunity to integrate concepts across disciplinary lines is lost. Indeed, the power of the modern molecular approach is best demonstrated by its ability to reveal the mechanisms underlying sometimes mysterious observations made in classic genetics.

Consequently, this book will use a very different approach. Each of the 10 chapters herein focuses on a fundamental concept in genetics. The most basic, that a gene may direct the synthesis of an enzyme and that mutation can lead to deficiency of the enzyme and consequent disruption of a metabolic pathway, comprises the theme for Chapter 1. This chapter introduces the concept of autosomal recessive inheritance and the molecular basis for gene mutation. The mechanisms of dominant inheritance of mutations in genes that encode structural proteins are explored in Chapter 2. Chapter 3 then describes a new method for understanding the structure of genes based on knowledge of their location in the genome. X-linked genetic transmission is discussed next, in Chapter 4, along with the concept of X chromosome inactivation. This is followed, in Chapter 5, by a review of chromosomes and chromosomal anomalies. The inheritance of complex multifactorial genetic traits is considered in Chapter 6. Although this is a relatively unexplored area, it promises to be a source of major insight into the mechanisms of relatively common disorders such as heart disease and diabetes mellitus. A newly emerging field, inheritance through the mitochondrial genome, also is considered (Chapter 7). The study of mitochondrial DNA has led to the discovery of a new class of disorders of energy metabolism. The genetics of cancer and the notion of somatic genetic change are the subject of Chapter 8. Chapter 9 addresses the field of developmental genetics and, finally, Chapter 10 deals with population genetics.

Each chapter begins with a clinical case history, and details of the case unfold as the chapter proceeds. The cases will provide a driving force to introduce new concepts and maintain a link with their clinical application. Chapters end with another case that will continue the theme and introduce new ideas. Part of the challenge in teaching genetics is to provide a basis for mastery of new developments in a rapidly evolving field. Cases at the ends of chapters are structured to encourage self-directed learning.

Three basic themes may be discerned in the genetic approach to medicine, and these will compose the framework for each chapter. The first concerns the means by which genetic traits flow through families. This is most easily understood by the study of rare genetic variants resulting from changes in single genes. Increasingly, however, more complex and common traits are being scrutinized, leading to insights that may influence the health of large populations. The second theme concerns the flow of genetic information in the cell from DNA through RNA to protein. This provides the link from a heritable trait to a physiological event, and it is here that great advances are being made in diagnosis and management of disease. The third theme involves the ethical context for human genetics. To the extent that the purview of genetics is now all of medicine, ethical issues in genetics are those of medicine itself. Yet genetics adds a dimension by taking into consideration health across generations. Attention to ethical issues must be integral to all aspects of medical education and is particularly important in the teaching of genetics.

1 Inborn Errors of Metabolism

OVERVIEW

Elucidation of the mechanism by which genes control biologic processes has been a major scientific advance in this century. The basic tenet—that genes consist of DNA and function to encode the structure of proteins—was established by the early 1950s. Since that time, attention has focused on the molecular mechanisms through which genes exert their effects. The flow of information from genes to proteins first came to light from the study of enzymes, proteins that catalyze biochemical reactions. In the 1940s, Beadle and Tatum demonstrated that strains of the fungus *Neurospora* that were unable to grow on media lacking specific nutrients were genetically unable to make the enzymes needed to synthesize those nutrients. Even before that, during the first decade of the twentieth century, Garrod had recognized genetic traits in families that he attributed to "inborn errors of metabolism," an inherited deficiency in the ability to carry out an essential biochemical reaction.

This chapter will explore the path from genes to enzymes. We will begin with the story of a child with albinism, one of the inborn errors of metabolism originally described by Garrod. After briefly describing the clinical entity and its biochemical basis, we will see how this disorder is genetically transmitted as an autosomal recessive trait. This will be the first of the basic modes of mendelian genetic transmission to be considered. Returning to the biochemical basis of albinism, it will become apparent why genetic metabolic disorders tend to be inherited as recessive traits. Biochemistry, however, provides little insight into the mechanisms by which gene mutations occur and lead to deficient enzyme function. To understand this, it is necessary to isolate and purify the gene that encodes an enzyme—in this case, tyrosinase, the enzyme that is deficient in one form of albinism. Genes are isolated by a process of *cloning*, the basic approach to which will be described, as will the means by which a cloned gene is characterized.

How does this help us to better understand a disorder such as albinism? We will see how gene mutations are identified in affected individuals and how this knowledge provides the basis for genetic testing of affected individuals and their family members. Although such testing is of limited value in families with albinism, it is of great importance in caring for those with other inborn errors, some of which have devastating effects on health. From diagnosis we will turn to treatment, looking at medical treatments, both established and experimental, and at the prospect of genetic therapy. A personal statement by an individual affected with albinism is provided in the Perspective. This chapter will close with the story of another inborn error of metabolism, which considers some of the public health issues raised by newborn screening.

1.1 Oculocutaneous Albinism

Ben and Linda are concerned that their 1-year-old daughter Katie is having a problem with her eyes. Katie seems to have difficulty focusing on objects, and her eyes tend to jiggle back and forth when she moves them. Also, although she responds to light and has learned to smile responsively, Katie does not seem to recognize her parents from across the room. Ben and Linda bring Katie to her pediatrician, who refers her to an ophthalmologist. The ophthalmologist notes that Katie has blue eyes and that there is no pigment in her retina. A dermatologist then is asked to see Katie, and the dermatologist is impressed by her very fair skin and white hair (Figure 1-1). It has been a source of amusement to her parents that Katie has a much fairer complexion and lighter hair than anyone else in the family, but they have assumed that her hair would darken as she gets older. The dermatologist plucks a few scalp hairs and sends them for assay of tyrosinase activity (Figure 1-2). No enzyme activity is found, confirming a clinical diagnosis of oculocutaneous albinism.

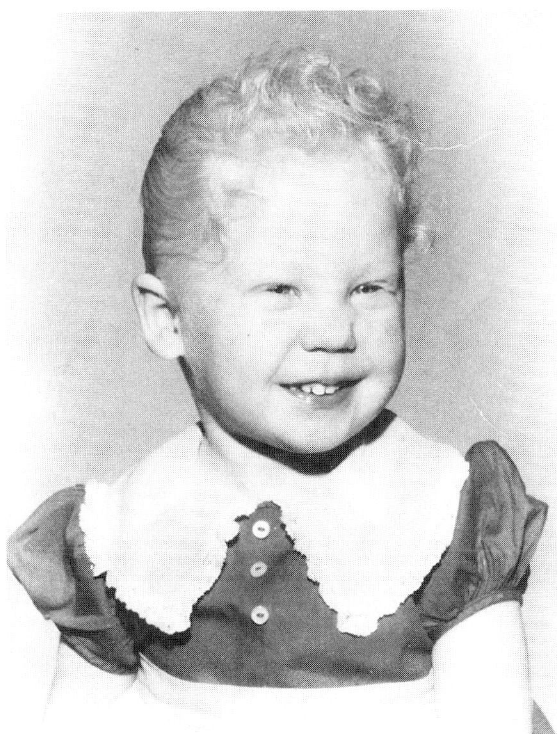

FIGURE 1-1 Child with oculocutaneous albinism, with white hair, fair skin, and eyes that are sensitive to light.

Oculocutaneous albinism (OCA) is characterized by total or nearly total absence of pigmentation. The major pigment in the body is melanin, a complex molecule synthesized by melanocytes in organelles called *melanosomes*. The biochemical pathway leading to melanin begins with the amino acid tyrosine (Figure 1-3). This is first hydroxylated to dihydroxyphenylalanine (DOPA), and then DOPA is oxidized to DOPA quinone. Both reactions are catalyzed by the enzyme tyrosinase. Deficiency of tyrosinase activity is responsible for the most common form of OCA. Affected individuals cannot form DOPA or DOPA quinone and therefore cannot synthesize melanin. They have normal numbers of melanocytes, but the melanocytes are not pigmented (Figure 1-4). The standard tyrosinase assay is performed by incubating freshly plucked hair bulbs in the presence of tyrosine. Normally this results in generation of melanin, but no melanin is produced in the absence of tyrosinase activity.

Melanin, the major pigment of hair and skin, provides protection from the damaging effects of ultraviolet light. Persons with albinism are extremely sensitive to sunburn and have a high risk of developing skin cancer. They must avoid exposure to direct sunlight by wearing a hat and long clothing and using sunscreen.

Ben and Linda have heard of albinism and can now understand why Katie has fair skin and white hair. They assume that this will not cause medical problems. They have difficulty understanding why Katie's vision is abnormal, however.

Melanin is also present in the eye. Iris pigmentation gives rise to eye color, so persons with albinism have blue or gray irises. Pigment is normally present in the retina and is important for vision. Lack of this pigment leads to very poor visual acuity (20/200 or worse) and extreme sensitivity to light. Poor visual acuity makes it difficult to focus on objects at a distance, causing the eyes to jiggle back and forth (a condition referred to as *nystagmus*) as they try to achieve visual fixation. For unknown reasons, absence of tyrosinase activity also results in aberrant migration of nerve fibers along the optic pathways during embryonic development. This leads to defective stereoscopic vision in persons with albinism.

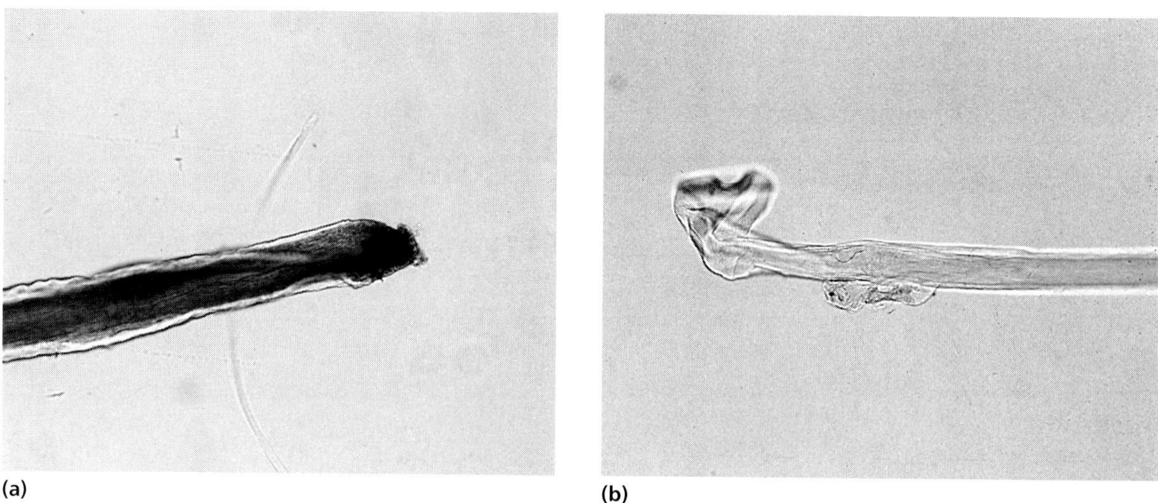

FIGURE 1-2 Hair bulb test for oculocutaneous albinism. Hair bulbs are incubated in presence of tyrosine; melanin is produced by normal hair (**a**) but not hair from a person with oculocutaneous albinism due to tyrosinase deficiency (**b**). (Photograph courtesy of Dr. William Oetting, University of Minnesota.)

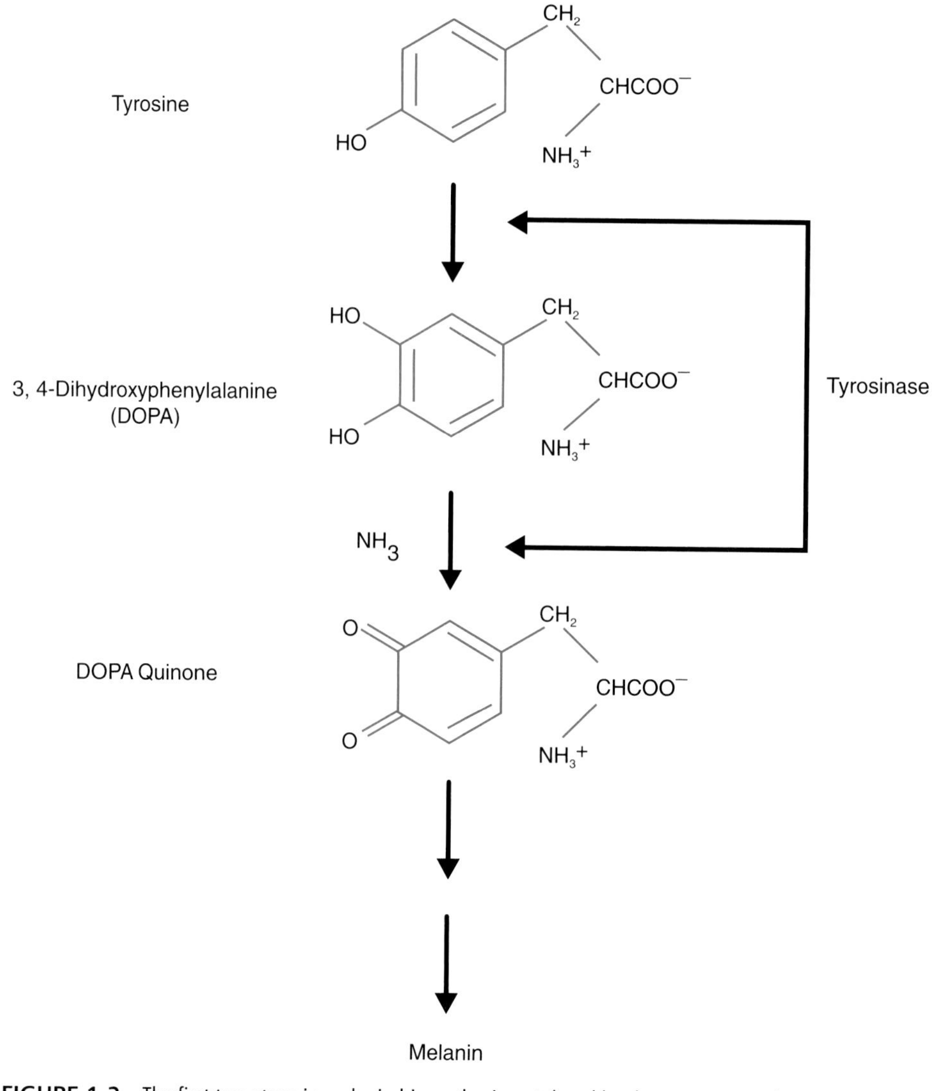

Tyrosine — HO, CH_2, $CHCOO^-$, NH_3^+

↓

3, 4-Dihydroxyphenylalanine (DOPA) — HO, HO, CH_2, $CHCOO^-$, NH_3^+

Tyrosinase

NH_3 ↓

DOPA Quinone — O, O, CH_2, $CHCOO^-$, NH_3^+

↓

↓

Melanin

FIGURE 1-3 The first two steps in melanin biosynthesis, catalyzed by the enzyme tyrosinase.

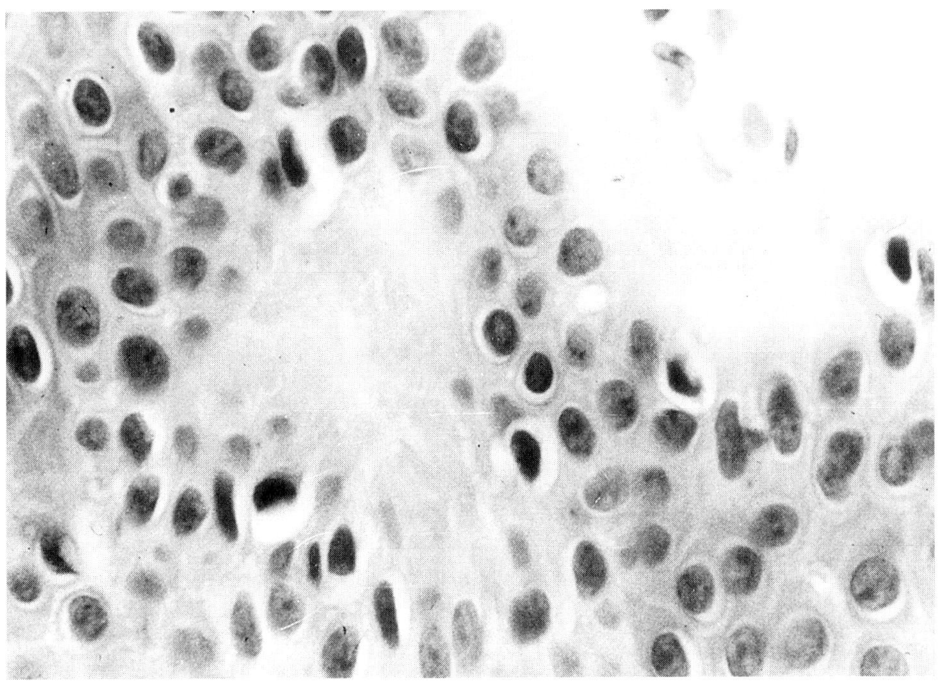

FIGURE 1-4 Photomicrograph of skin from individual with oculocutaneous albinism. No pigment granules are seen in melanocytes (cells with halolike region surrounding nucleus). (Courtesy of Dr. Cynthia Magro, Pathology Services, Boston.)

Oculocutaneous Albinism

- Lack of pigmentation
- Fair skin and hair
- Decreased visual acuity
- Lack of stereoscopic vision

1.2 Autosomal Recessive Genetic Transmission

Ben and Linda are referred to a geneticist, which surprises them because neither of them has albinism and there is no known history of albinism in either of their families. They have two other children, Jamie, a 3-year-old boy, and Susan, a 5-year-old girl (Figure 1-5). Both have light brown hair and normal vision. The geneticist takes a complete family history. Ben and Linda are a bit put off when asked if they might be related to one another, which they are not (Ben's ancestry is Irish and German, and Linda's is Italian).

OCA is inherited as an **autosomal recessive** trait. An affected child such as Katie is homozygous for a mutation in the gene encoding tyrosinase. Her

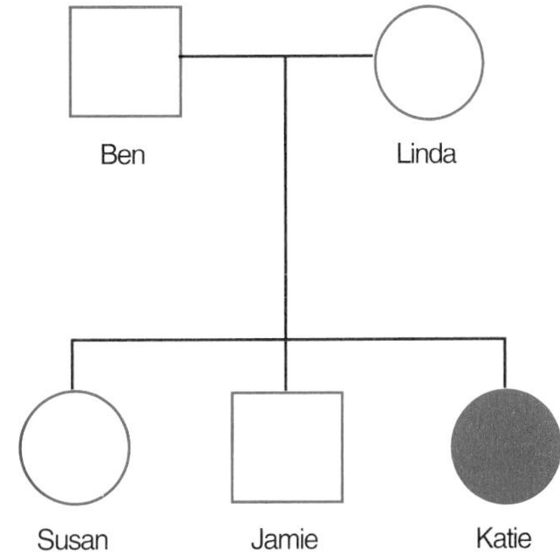

FIGURE 1-5 Family pedigree for Ben and Linda.

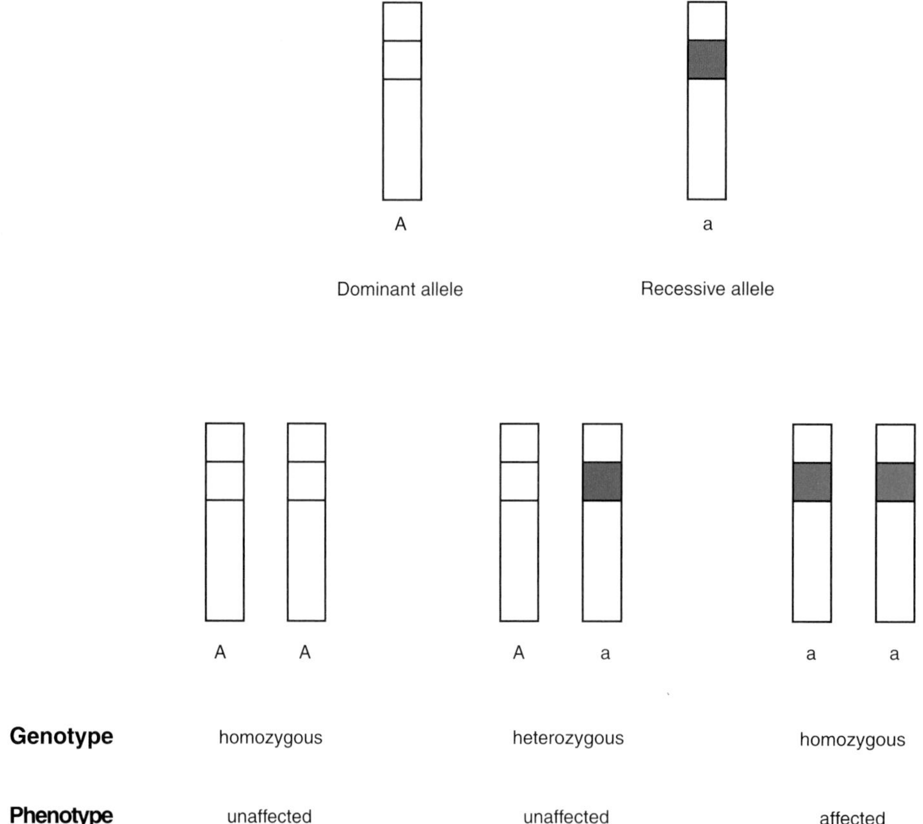

FIGURE 1-6 Genotype and phenotype associated with gene locus with dominant allele *A* and recessive allele *a*.

parents are heterozygous carriers for the mutation but do not manifest any clinical signs of OCA. It is not expected that there would be a family history of OCA in prior generations, although it is likely that Ben and Linda have many relatives who are also OCA carriers. Because tyrosinase mutant alleles are relatively rare, it is unlikely that other members of the family who might be carriers also have partners who are carriers.

The foundation for our modern understanding of genetic segregation is that a diploid organism contains two copies of every gene (excepting those carried on the sex chromosomes). One copy of each gene is inherited from each parent. These gene copies separate during the formation of haploid germ cells and are reunited at the time of fertilization. The individual copies of a particular gene are called *alleles*. The genetic constitution of an individual with respect to a particular trait is referred to as **genotype,** whereas the corresponding physical manifestation of the trait is the **phenotype**. In human genetics, traits can be transmitted as **autosomal** or **sex-linked, dominant** or **recessive**. Gender is determined by the sex chromosomes: A man has an X and a Y chromosome, and a woman has two X chromosomes. Genes located on a sex chromosome are said to be *sex-linked*, whereas the non-sex-linked genes are referred to as *autosomal*.

For an autosomal recessive trait to be expressed, both copies of a gene pair must be present in an altered (**mutant**) form. A person who expresses the phenotype of an autosomal recessive trait is said to be **homozygous** for the mutant gene (Figure 1-6). Autosomal recessive traits arise in children of phenotypically normal parents when both parents are carriers—that is, each parent has one normal (**wild-type**) allele and one mutant allele. The parents are said to be **heterozygous**. Each parent has a 50% chance of transmitting the wild-type or the mutant allele to a germ cell, and therefore each child has a one in four chance of inheriting the gene mutation from both parents and being affected by the disorder (Figure 1-7). Heterozygous carriers do not manifest features of the genetic trait.

The wild-type allele is said to be *dominant* to the mutant; the mutant allele is *recessive.* Recessive traits tend to occur among children in a sibship but not in other members of a family (Figure 1-8). The parents are heterozygous but do not manifest the mutant phenotype. Other members of the parents' families (e.g., their siblings, parents) may also be carriers but, if the mutant allele is rare in the population, it is unlikely that other relatives will be homozygous.

The chances that both members of a couple carry the same rare mutant allele are increased if they are related to one another (i.e., they are **consanguineous**) and hence have inherited the same rare allele from a common ancestor (Figure 1-9). That is why Ben and Linda were questioned about the possibility of their being relatives. The frequency of consanguinity tends to be increased in families with rare autosomal recessive disorders, but it is important to remember

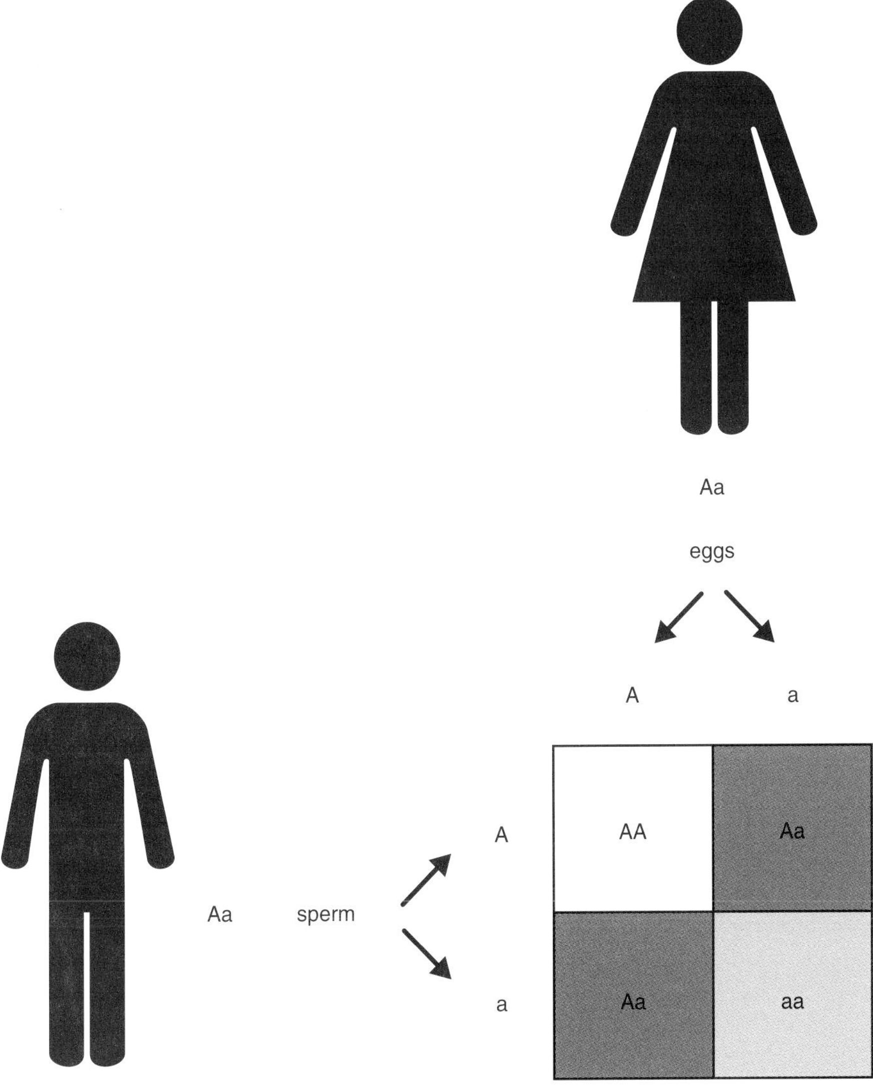

FIGURE 1-7 Segregation of dominant and recessive alleles in family in which both parents are heterozygous.

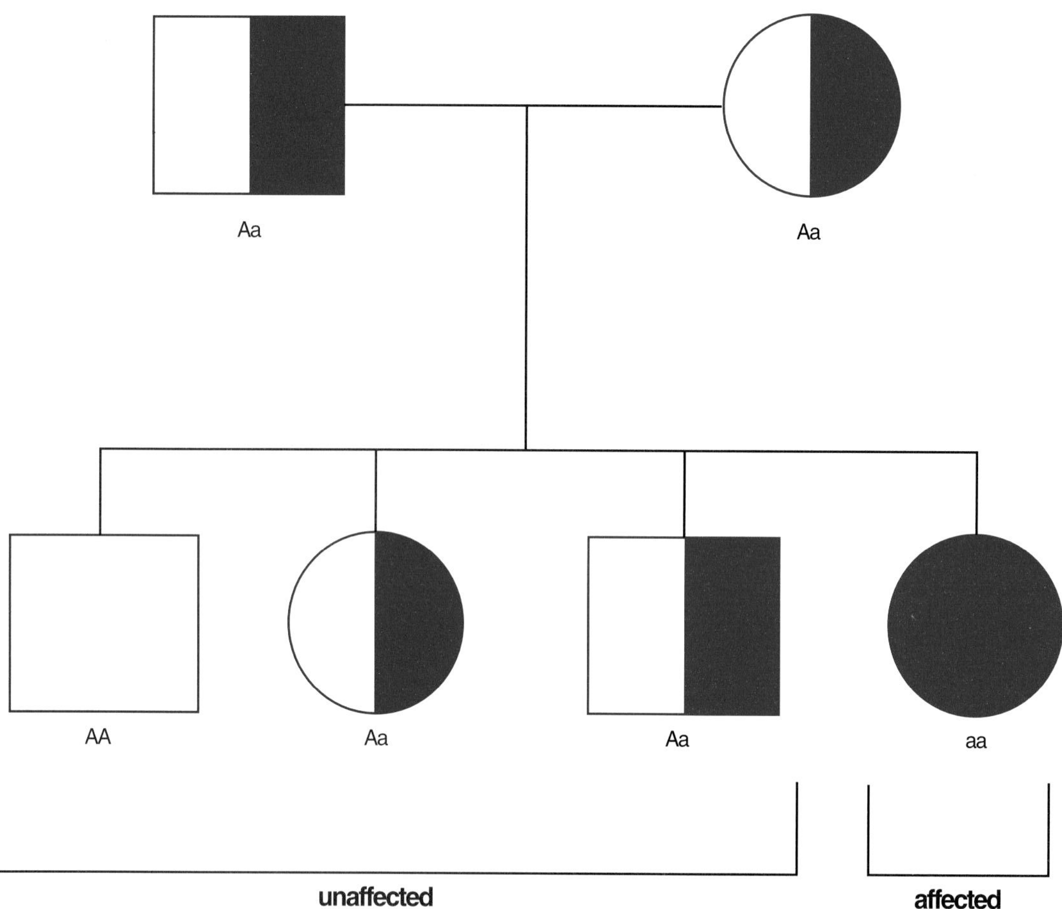

FIGURE 1-8 Pedigree illustrating segregation of autosomal recessive trait. Allele *A* is dominant, *a* is recessive.

that in most populations consanguinity is relatively infrequent, even in families with autosomal recessive traits. Symbols used to draw pedigrees are shown in Figure 1-10.

Autosomal Recessive Transmission

- Parents are heterozygous carriers
- Affected children are homozygous for mutant gene
- Carrier couple has one in four chance of having affected offspring
- Affects members of sibship

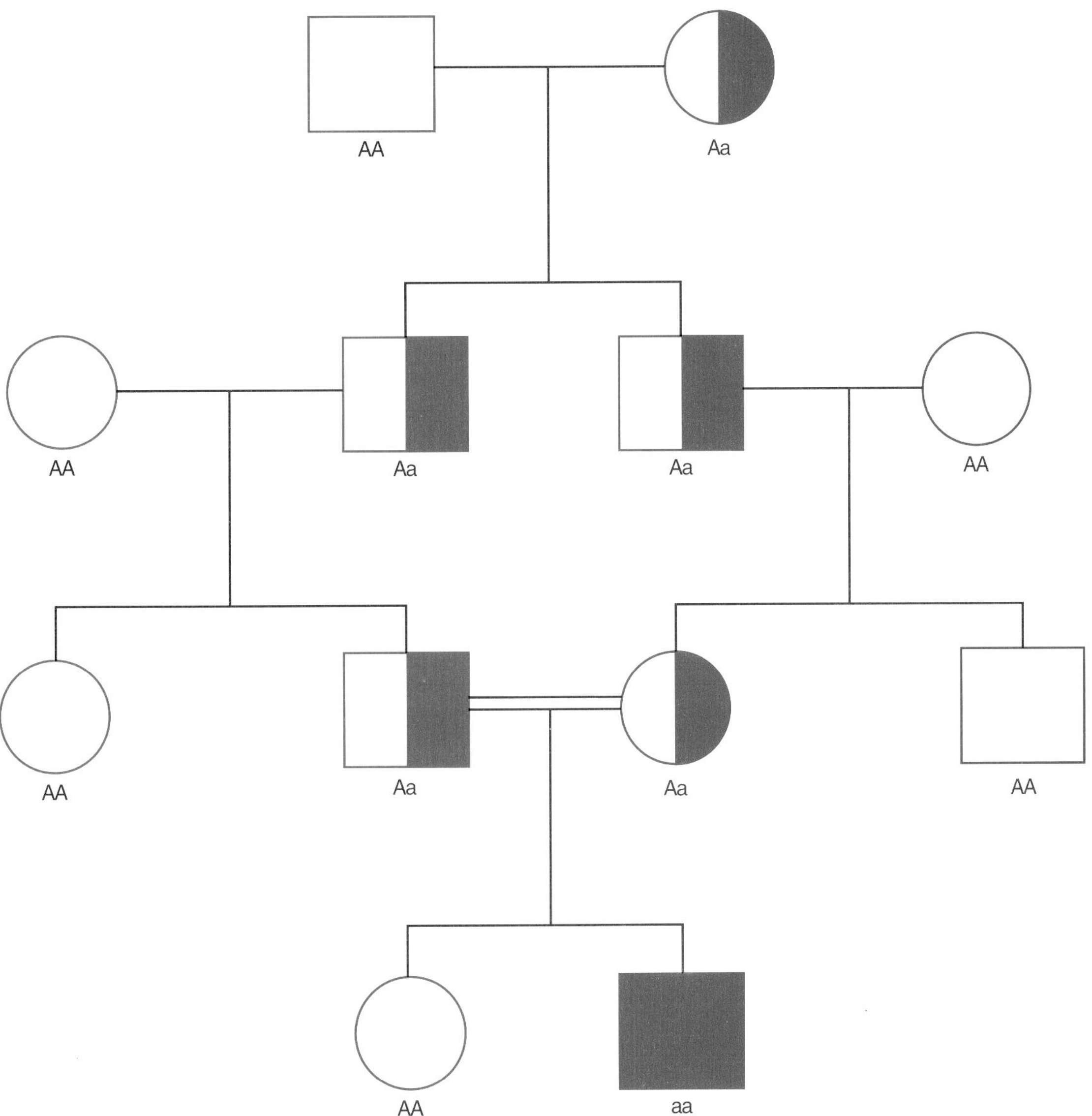

FIGURE 1-9 The individual who is homozygous for *aa* has inherited the *a* allele from his great-grandmother, transmitted through both parents.

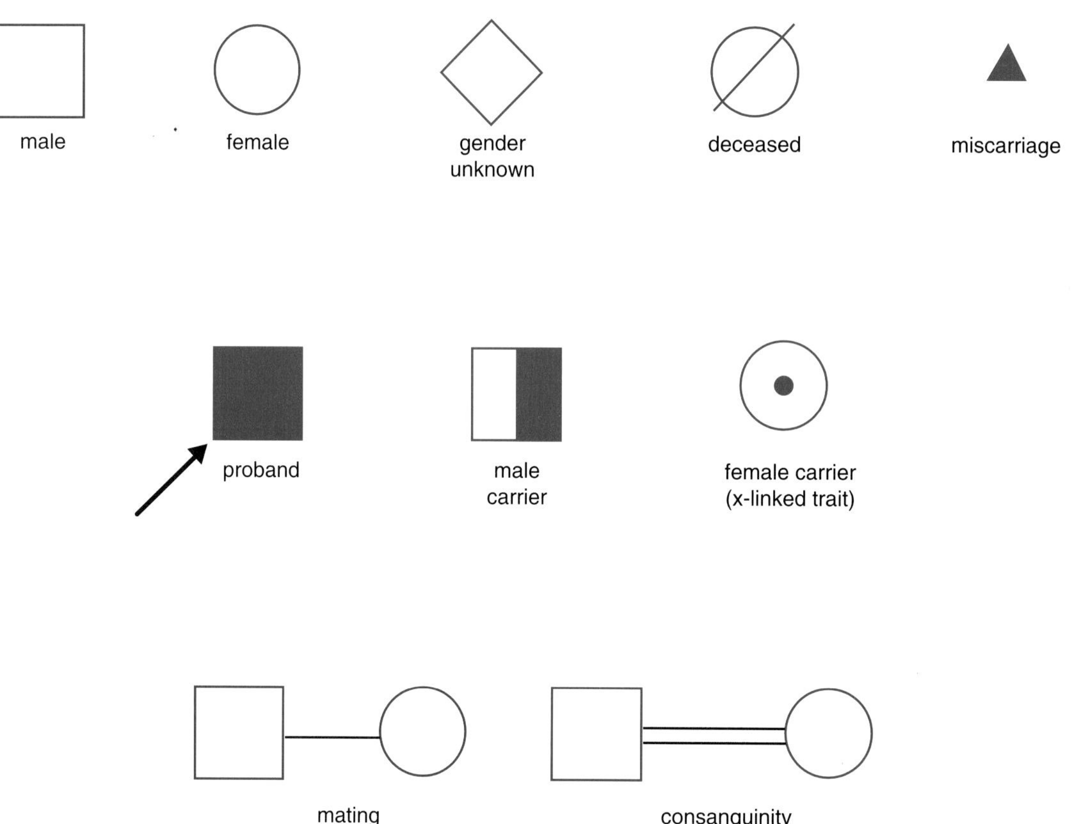

FIGURE 1-10 Symbols used to draw pedigrees. A filled in symbol indicates a phenotypically affected individual.

1.3 Inborn Errors of Metabolism

OCA is one of a class of disorders referred to as *inborn errors of metabolism*. This term was coined in 1908 by the British physician Sir Archibald Garrod. Working in the early years after Mendel's laws of inheritance had been "rediscovered," Garrod recognized several disorders that tend to run in families. He postulated that each was due to a hereditary block in the body's ability to carry out some essential chemical reaction. Although this hypothesis predated by more than 40 years the elucidation of the chemical basis of heredity, the formulation is essentially correct.

A landmark series of experiments done in the 1940s by Beadle and Tatum helped establish the relationship between genes and proteins. These investigators induced mutations in the mold *Neurospora* by irradiation of spores. Wild-type *Neurospora* organisms grow on media including certain salts, a sugar such as sucrose, and the vitamin biotin. Molds grown from irradiated spores had, in some cases, lost the ability to grow on this "minimal" medium. Mutants were found that required a specific nutrient—for example, an amino acid—for growth. From study of these inborn errors of metabolism in *Neurospora*, the concept "one gene, one enzyme" was developed (Figure 1-11). This postulates that for each protein there is a separate and distinct gene to encode its production.

Many inborn errors are now known in humans. Two described by Garrod (albinism and alcaptonuria) involve enzymes in the pathway of tyrosine metabolism (Figure 1-12). Phenylketonuria (PKU) is one of the most common and best-known inborn errors of metabolism. The enzyme missing in this disorder is phenylalanine hydroxylase, which is required to convert phenylalanine to tyrosine. In the absence of this enzyme, phenylalanine builds up in the blood to toxic concentrations. Excess phenylalanine is converted to phenylpyruvic acid, also a toxic

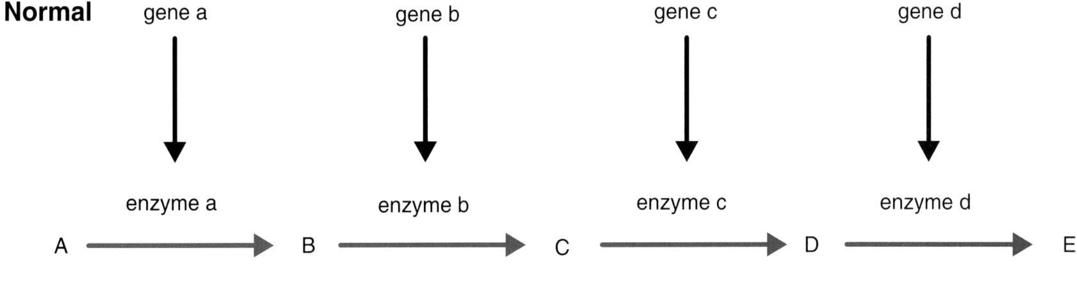

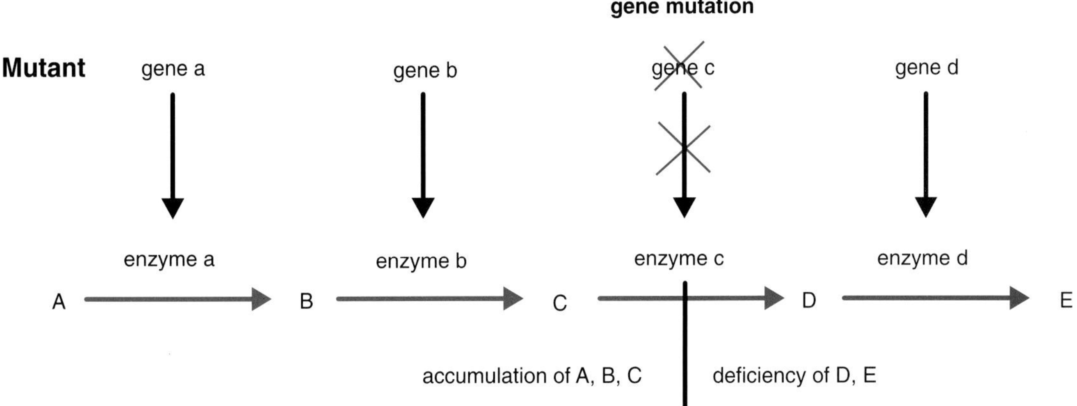

FIGURE 1-11 "One gene, one enzyme" concept. Normally, a specific gene directs the synthesis of each individual enzyme. Mutation of a gene leads to deficiency of the enzyme and consequent accumulation of substrate and deficiency of product.

compound. The phenotype is a progressive neurologic disorder with seizures and mental retardation. Treatment consists of restriction of phenylalanine intake by using a phenylalanine-free protein extract as the main source of amino acids. Adherence to this diet leads to essentially normal neurologic function. PKU was one of the most common causes of mental retardation before dietary therapy was devised. Now newborns throughout the developed world are screened for PKU, so that the disorder can be diagnosed and treatment begun before neurologic damage occurs.

Another inborn error of metabolism associated with buildup of a toxic substance is carbamyl phosphate synthetase (CPS) deficiency. CPS is the enzyme through which ammonia gains entry into the urea cycle (Figure 1-13). The urea cycle leads to the formation of urea, allowing toxic ammonia to be excreted safely from the body. In the absence of CPS activity, ammonia cannot be converted to carbamyl phosphate and therefore does not enter the urea cycle. Instead it accumulates in the blood, where it is highly toxic, causing brain edema and eventual death of neurons. Symptoms begin after the first protein feeding and consist of vomiting, lethargy and, eventually, coma and death. Emergency treatment consists of renal dialysis to remove excessive ammonia. If an infant survives the neonatal period, long-term management includes restriction of protein intake just to levels required for growth and administration of sodium phenylacetate and sodium benzoate, which complex with ammonia to form compounds able to be excreted safely in the urine.

The biochemical basis for the dominance of wild-type alleles over mutant alleles in inborn errors of metabolism can be understood by considering how enzymes function (Figure 1-14). Enzymes are proteins that catalyze chemical reactions. An enzyme is not consumed during the reaction, so only small

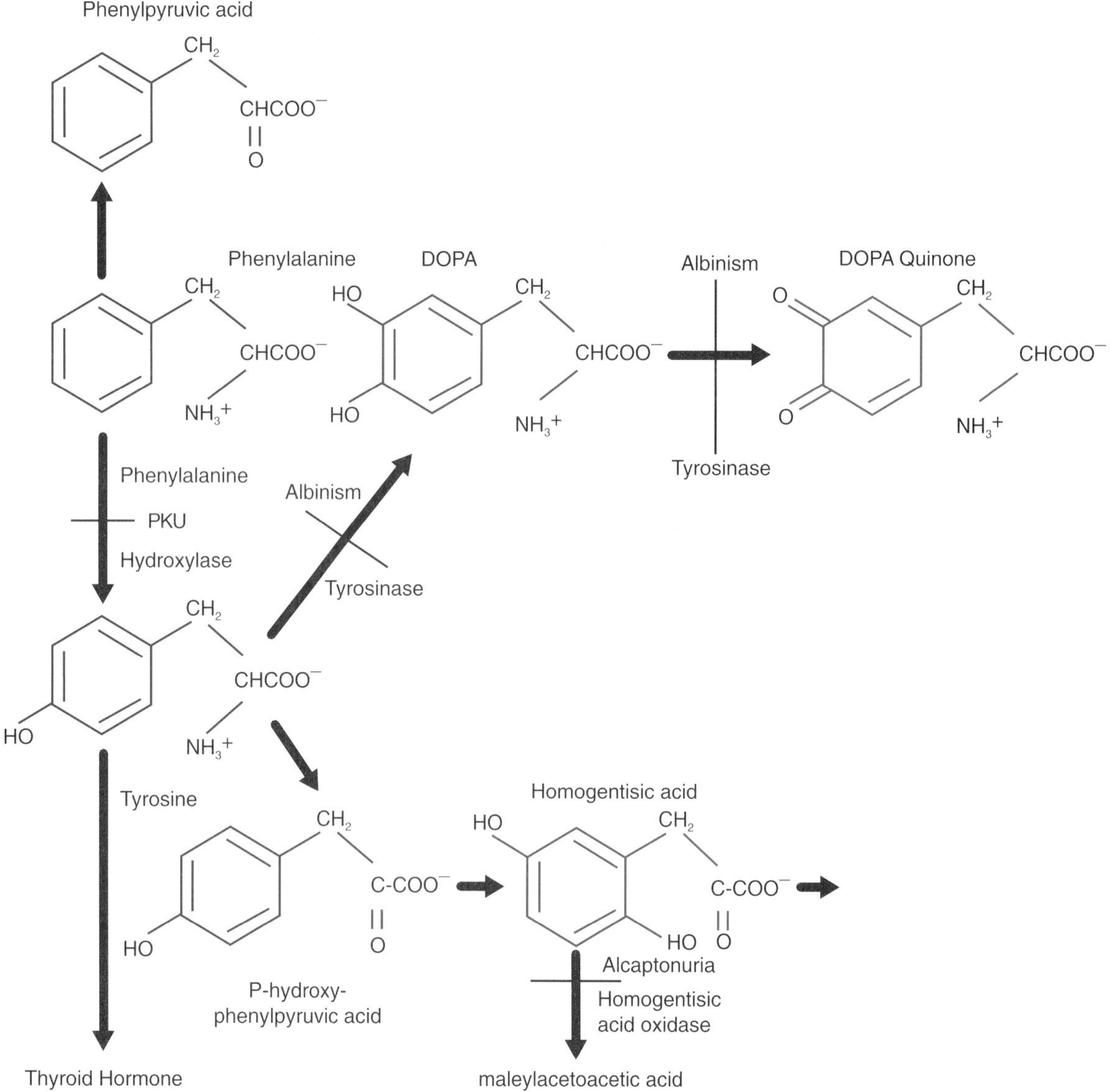

FIGURE 1-12 Metabolic pathways involving tyrosine. Phenylalanine is converted to tyrosine by phenylalanine hydroxylase, the enzyme blocked in phenylketonuria. Lack of homogentisic acid oxidase leads to alkaptonuria, and deficiency of tyrosinase to albinism.

quantities are required for a reaction to be carried out. In a person homozygous for a mutation in the gene encoding an enzyme, little or no enzyme activity is present, so he or she will manifest the abnormal phenotype. A heterozygous individual expresses at least 50% of the normal level of enzyme activity due to expression of the wild-type allele. This is usually sufficient to prevent phenotypic expression.

Individuals with OCA lack activity of tyrosinase owing to mutation of the gene encoding this enzyme. Heterozygous carriers have approximately 50% of the normal amount of tyrosinase activity, sufficient to produce adequate quantities of melanin. A child whose parents both are carriers for a tyrosinase mutation has a 50% chance of inheriting the mutant form of the tyrosinase gene from each parent, and hence a 25% risk of developing albinism.

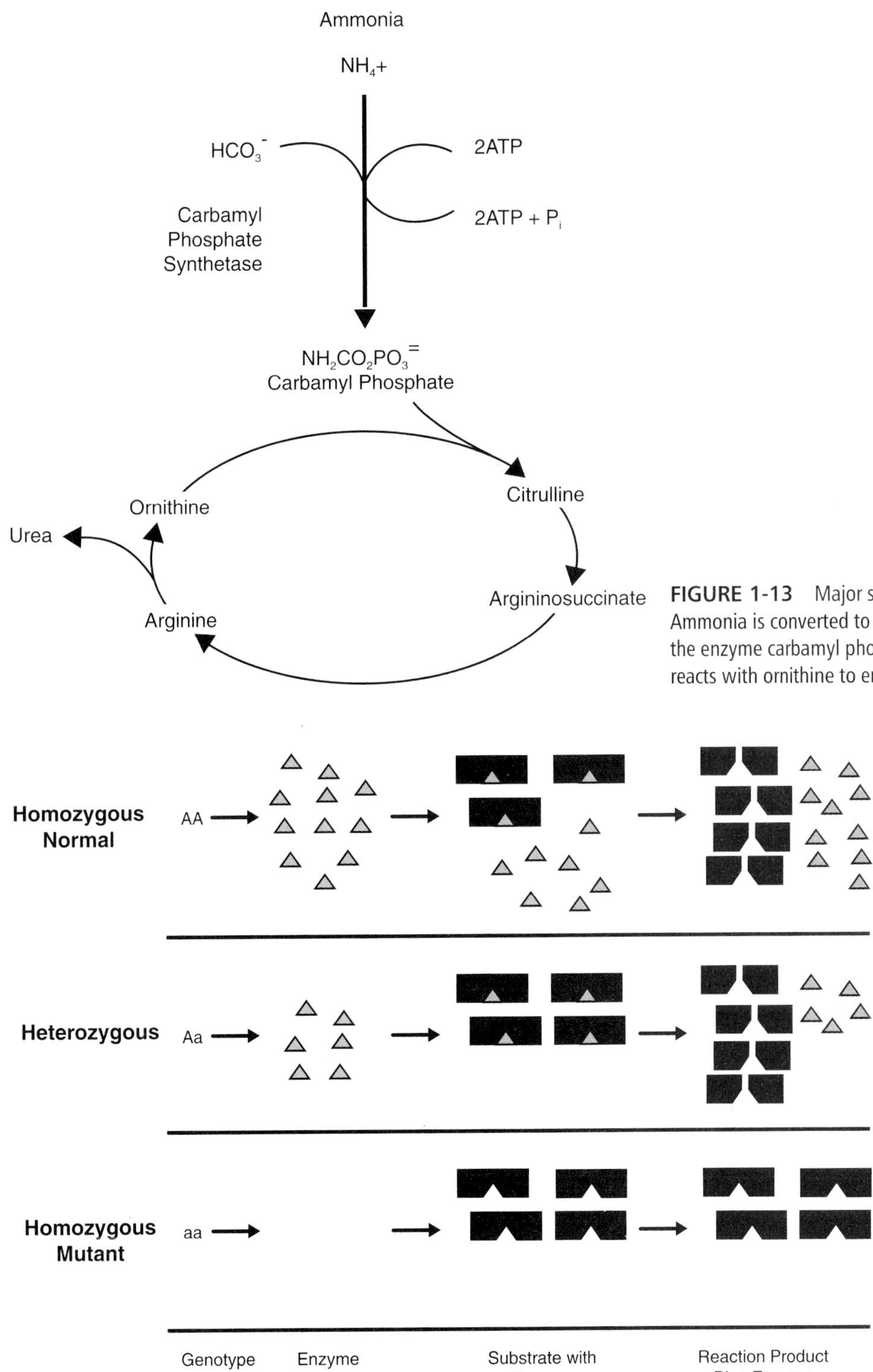

FIGURE 1-13 Major steps of the urea cycle. Ammonia is converted to carbamyl phosphate by the enzyme carbamyl phosphate synthetase, which reacts with ornithine to enter the urea cycle.

FIGURE 1-14 Model explaining recessive transmission of most enzyme deficiencies. Normally, more than sufficient enzyme is synthesized to carry out a reaction. A heterozygote still has sufficient enzyme, but a homozygote for a mutation does not make enough enzyme to complete the reaction.

So far, two pathogenetic mechanisms of inborn errors of metabolism have been described: deficiency of the product of an enzyme reaction (as in albinism) or toxic buildup of a substrate (as in PKU or CPS deficiency). Another mechanism is the gradual accumulation of substances within the cell due to hereditary deficiency of enzymes required for their breakdown. This is the case with lysosomal storage diseases, in which one or another of the enzymes required to metabolize cell membrane components in the lysosome is missing. A hallmark of these disorders is the progressive buildup of membrane debris in the lysosome, leading to progressive loss of cell function.

An example of a lysosomal storage disorder is Tay-Sachs disease, in which the enzyme hexosaminidase A is missing (Figure 1-15). This enzyme removes *N*-acetylhexosamine from a variety of substrates, especially a complex component of the cell membrane of neurons, GM_2-ganglioside. Deficiency of hexosaminidase A leads to a gradual loss of nerve cells, paralleled clinically by loss of developmental abilities. An afflicted child achieves normal development for several months and then starts to lose ground, ultimately becoming paralyzed and usually

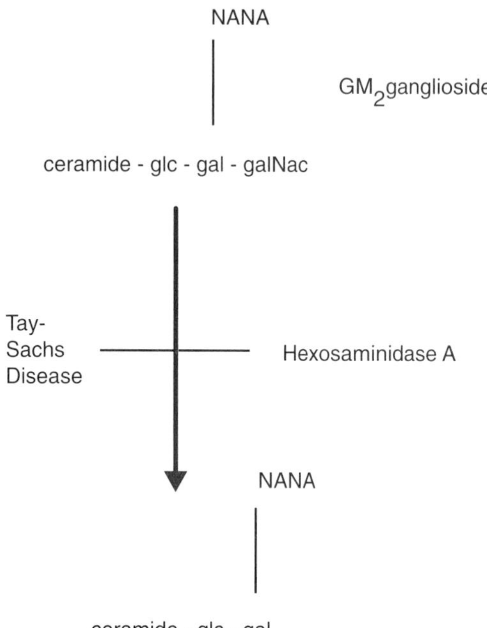

FIGURE 1-15 Metabolism of GM_2-ganglioside by hexosaminidase A, the enzyme missing in Tay-Sachs disease. The enzyme cleaves an *N*-acetyl-galactosamine residue from GM_2-ganglioside.

dying before age 3. Engorged nerve cells can be dramatically visualized in the eye, where swollen retinal ganglion cells render the normally transparent retina cloudy, obscuring the red color imparted by the rich retinal vascular supply. One area, the macula, is spared this effect owing to the lack of ganglion cells. This is the origin of the "cherry-red spot" (Figure 1-16), a clinical sign seen in Tay-Sachs disease and in several other lysosomal storage diseases. Tay-Sachs disease is inherited in an autosomal recessive manner. It is found worldwide but is especially prevalent in the Ashkenazi Jewish population. There is, sadly, no cure for Tay-Sachs disease.

Not all inborn errors result from deficient production of enzyme responsible for a particular metabolic pathway. Many reactions depend on coenzymes that function alongside the enzyme; deficiency of a coenzyme can also result in a metabolic block. For example, methylmalonic acid is an intermediate in the breakdown of branched-chain amino acids. One step in this pathway involves conversion of L-methylmalonyl CoA to succinyl coenzyme A (CoA) by the enzyme methylmalonyl CoA mutase (Figure 1-17). Failure of this reaction leads to a buildup of toxic methylmalonic acid, leading to acidosis and clinical signs of vomiting, lethargy and, eventually, coma. In some instances, the reaction is blocked by mutation of the gene encoding the enzyme itself. The reaction also requires the cofactor adenosylcobalamin. Some persons with methylmalonic acidemia have a mutation in one of the genes involved in cobalamin synthesis rather than in the mutase enzyme.

As can be seen from these examples, the consequences of enzyme deficiency can be far-reaching. This is illustrated by OCA (Figure 1-18), in which absence of melanin resulting from the tyrosinase mutation leads to susceptibility to sunburn and skin cancer, as well as decreased visual acuity and abnormal stereoscopic vision. The tendency for a single gene mutation to lead to a variety of seemingly disconnected physiologic effects is referred to as **pleiotropy**.

The clinical features of an inborn error of metabolism result from deficient enzyme activity, but how does a change in a gene cause such deficiency? Unraveling this problem has been the focus of research in genetics during the second half of the twentieth century. To follow this story, we must turn our attention from the biochemistry of the cell to the biochemistry of DNA.

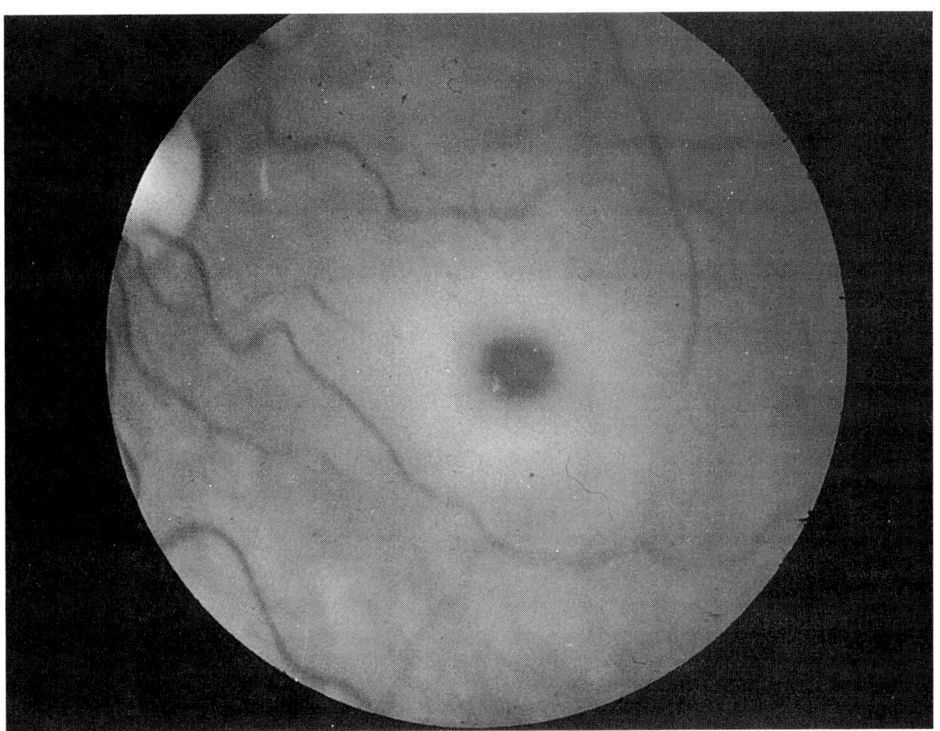

FIGURE 1-16 Cherry-red spot from an infant with Tay-Sachs disease. (Courtesy of Dr. Robert Peterson, Children's Hospital, Boston.)

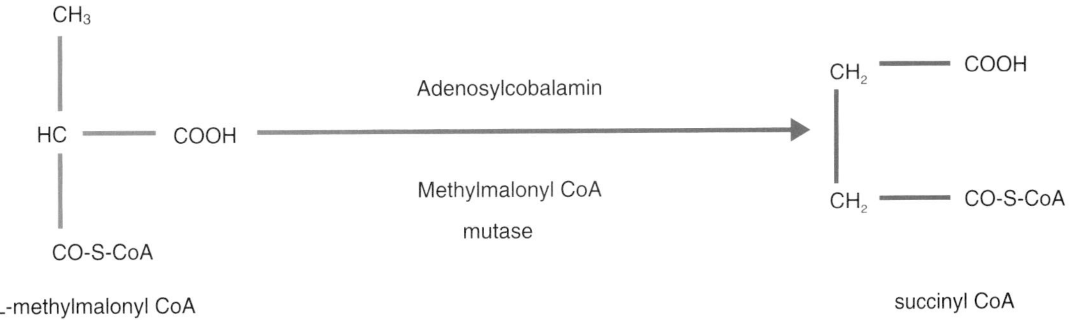

FIGURE 1-17 Conversion of L-methylmalonyl CoA to succinyl CoA is catalyzed by methylmalonyl CoA mutase and requires the cofactor adenosylcobalamin.

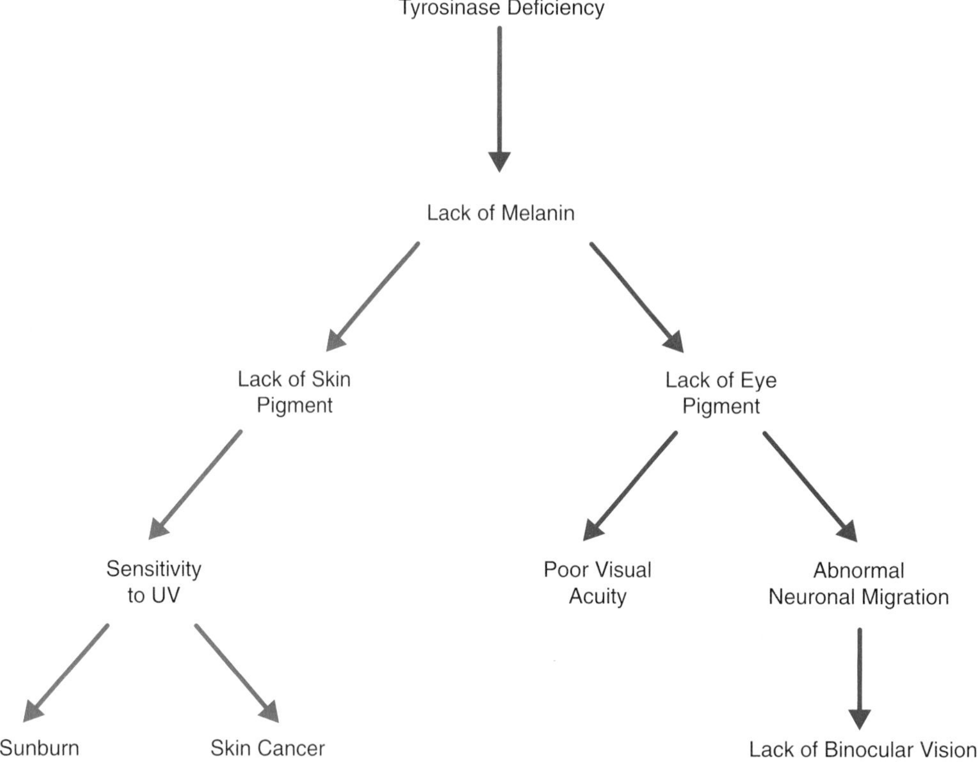

FIGURE 1-18 Diagram of pleiotropic effects resulting from tyrosinase deficiency in albinism.

1.4 The Chemical Basis of Heredity

The chemical basis for genetic transmission came to light following the elucidation of the structure of DNA by Watson and Crick. DNA consists of a pair of strands of a sugar-phosphate backbone attached to a set of pyrimidine and purine bases (Figure 1-19). The strands are bound together by hydrogen bonds between adenine and thymine bases and between guanosine and cytosine bases. Together these strands form a double helix. The strands separate during DNA replication, and the base sequence of the newly synthesized strand is dictated by the complementarity of adenine with thymine and guanosine with cytosine. DNA therefore contains within its structure the information necessary for its replication.

The sequence of bases in DNA also provides the code that determines the structure of proteins (Figure 1-20). Proteins consist of chains of amino acids; the specific ordering of amino acids determines the unique properties of each protein. The amino acid sequence of a protein is determined by the sequence of bases in the stretch of DNA that encodes the protein. Each amino acid is represented in DNA by a triplet of bases. This genetic code is more or less universal to all organisms. The base sequence of one strand of DNA is copied into a complementary RNA, which is in turn translated on the ribosome into protein.

The first human mutation to be elucidated was the one responsible for sickle cell anemia. Sickle cell is a severe anemia that occurs predominantly in blacks. The primary physiologic defect is a tendency for red blood cells to assume a sickle shape under conditions of reduced oxygen tension (Figure 1-21). Sickle cells tend to obstruct small capillaries, leading to episodes of painful infarction of tissues such as bone. Infarction of the spleen leads to gradual loss of spleen tissue and, in turn, increased vulnerability to bacterial infection.

The predominant protein of the red blood cell is hemoglobin, the major form of which consists of a

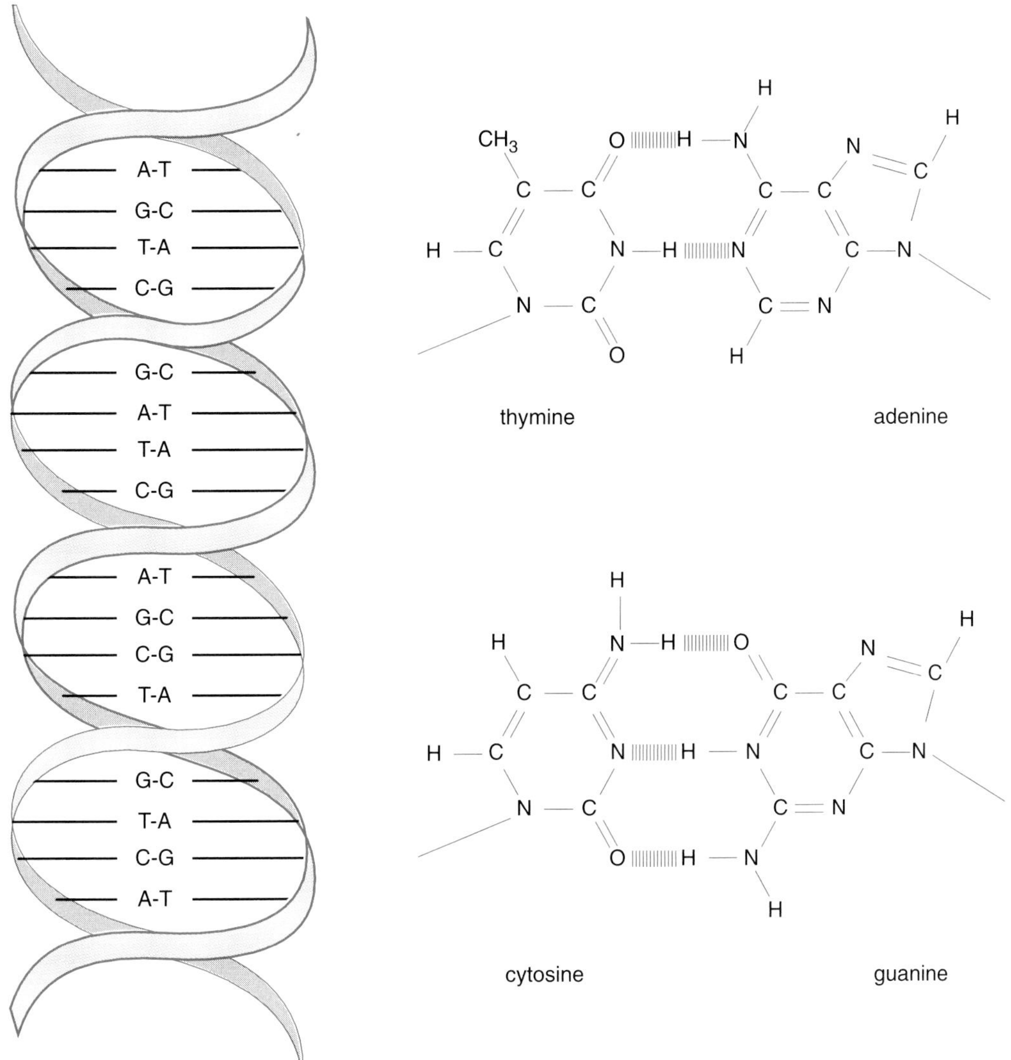

FIGURE 1-19 Double-helical DNA molecule (left), the strands of which are held together by hydrogen bonding between adenine and thymine base pairs, or guanine and cytosine.

tetramer of two alpha-chains and two beta-chains. The globin tetramer binds four molecules of heme, which serve as the oxygen-carrying substance in blood. In 1949, Linus Pauling and his colleagues demonstrated by protein electrophoresis that sickle cell hemoglobin migrates differently in an electric field than wild-type hemoglobin, indicating a difference in net charge between the two molecules. Ingram subsequently found that valine was substituted for glutamic acid in the sickle cell beta-chain. The genetic code for glutamic acid (GAG) differs by only one base from that for valine (GTG), suggesting that a single base pair change in the beta-globin DNA is responsible for the disorder (Figure 1-22). The discovery of other mutations in genes responsible for human disease has proved to be more difficult, as few proteins can be studied as easily as globin. It was not until the advent of gene-cloning technologies in the 1970s that molecular genetic analysis became routinely possible.

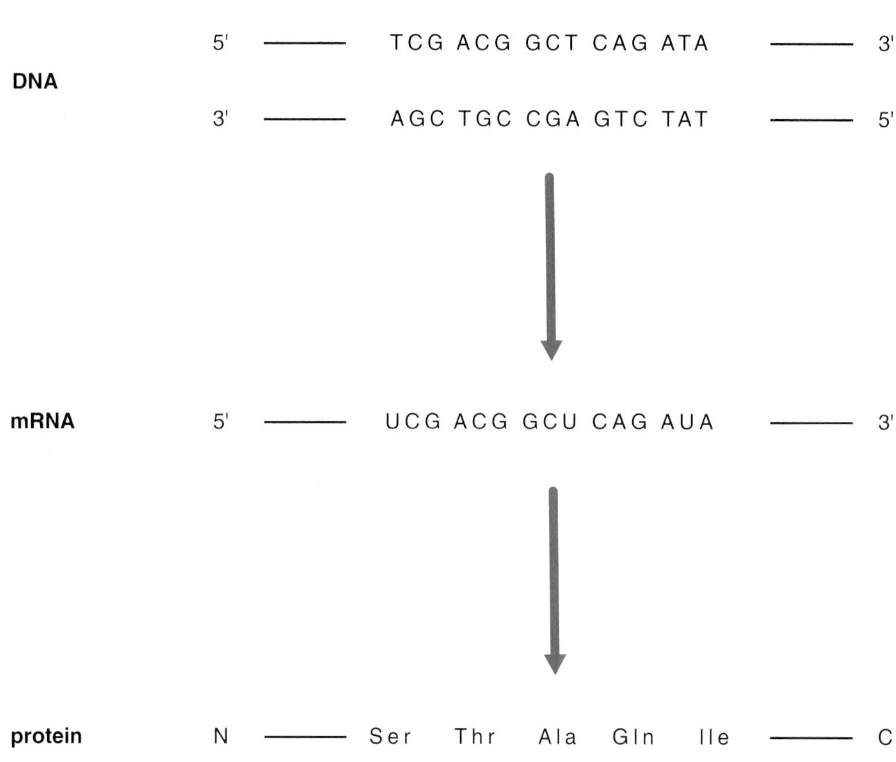

FIGURE 1-20 The base sequence of DNA is copied into a messenger RNA (mRNA), which is, in turn, translated into protein.

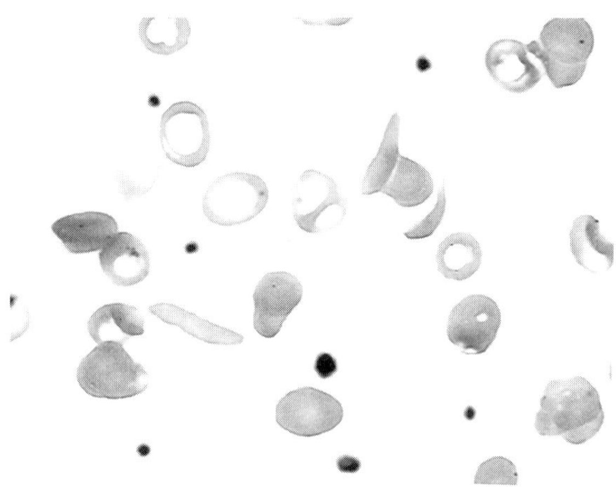

FIGURE 1-21 Photomicrograph of sickle cells. (Photograph courtesy of Dr. Orah Platt, Children's Hospital, Boston.)

- Genes consist of DNA.
- Information flows from DNA to RNA to protein.
- A mutation consists of a change of DNA sequence, leading to aberrant or absent expression of the corresponding protein.

1.5 Gene Cloning

The human genome consists of approximately 3 billion base pairs of DNA, encoding 100,000 or more proteins. Individual genes range in size from a few thousand to more than a million base pairs. The problem posed in isolating a single gene for analysis is thus formidable. Key developments that made this possible were the discovery of restriction endonucleases and the invention of cloning vectors.

Restriction endonucleases are enzymes derived from bacteria that cleave DNA at specific sites. In bacteria, they function to cut the DNA of viruses,

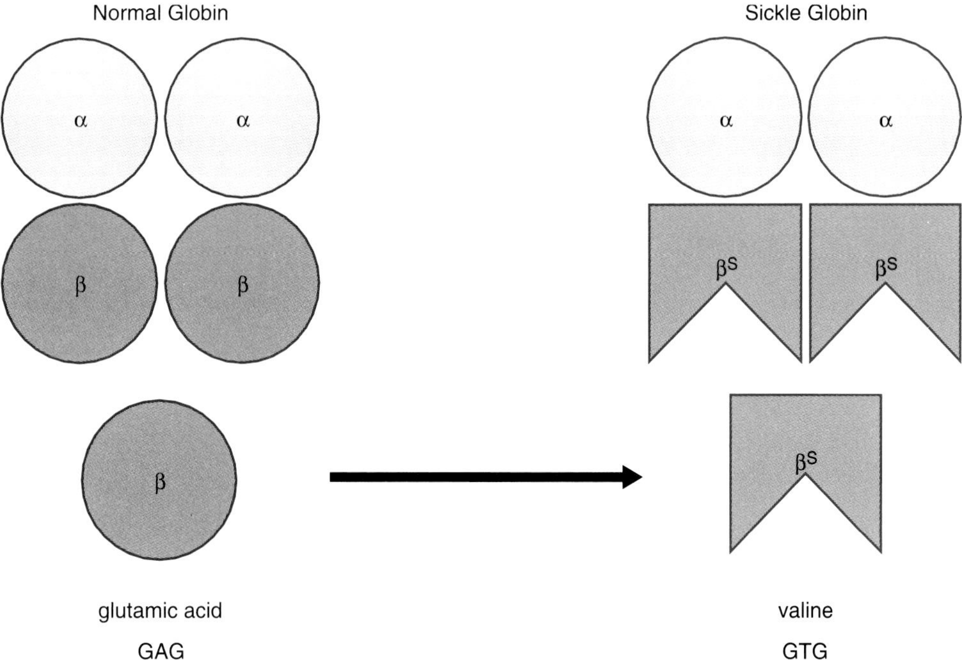

FIGURE 1-22 Adult hemoglobin is normally a tetramer of alpha- and beta-globin chains. The chains are mutated in sickle globin. The mutation consists of substitution of valine for glutamic acid in the protein owing to a base change of A to T in the codon for this amino acid.

thereby restricting the viruses from growth in the bacteria (hence the name *restriction enzyme*). In this way, they function as a kind of primitive immune system. Each restriction enzyme recognizes a particular base sequence, usually four to eight bases, often having axial symmetry (Figure 1-23). The recognition specificity is very high: Even a single base change in the enzyme recognition sequence will cause the enzyme not to cut. Cuts of the two strands of DNA may be staggered, leaving single-stranded, non–base paired overhangs. Alternatively, some enzymes cut the two strands at the same site, leaving "blunt" ends. Restriction enzymes are named after the bacteria from which they are derived—for example, *EcoRI* from *Escherichia coli*.

A cloning vector is a DNA sequence that can incorporate foreign DNA and replicate independently within a prokaryotic cell. The simplest example is the bacterial plasmid. This is a double-stranded circular DNA molecule found naturally in many bacteria. Plasmids contain a site from which DNA synthesis can begin, as well as one or more antibiotic resistance genes (Figure 1-24). The latter are a nuisance to clinicians but are a very helpful feature for

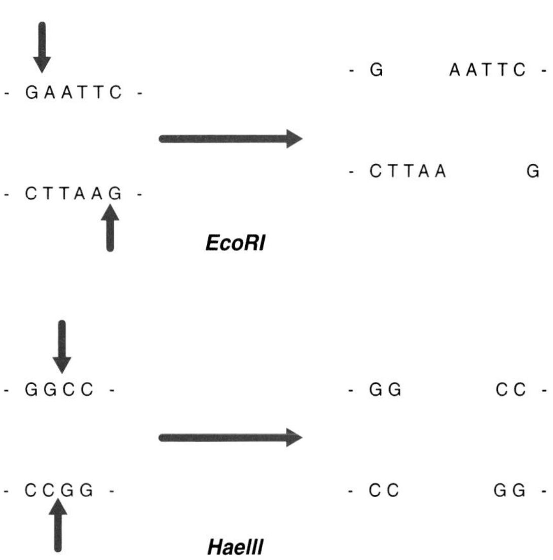

FIGURE 1-23 The restriction enzyme *EcoRI* recognizes the palindromic hexamer GAATTC and cleaves asymmetrically to result in non–base paired ends. The enzyme *HaeIII* cuts at the center of the four-base palindrome GGCC to leave blunt ends.

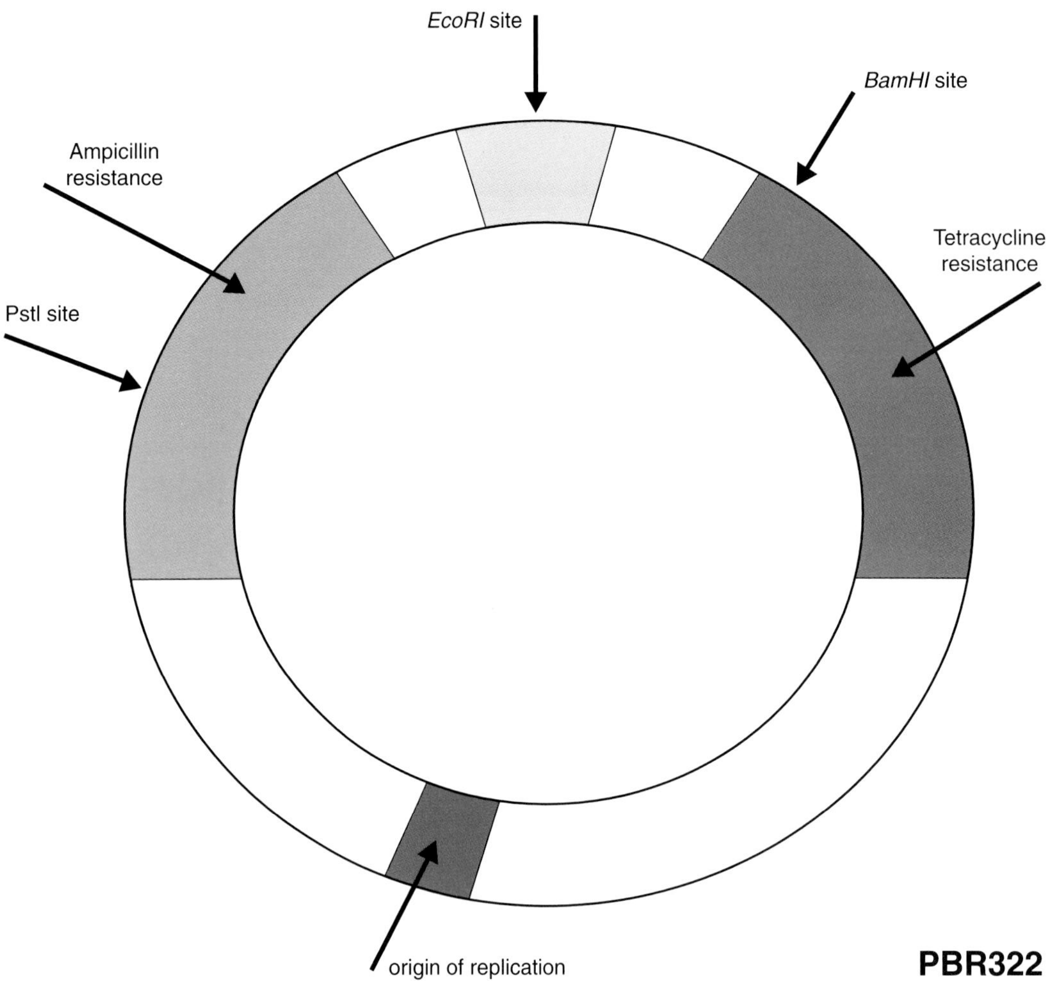

FIGURE 1-24 Diagram of cloning plasmid pBR322. There is a single *EcoRI* enzyme site, as well as a single *PstI* site (the latter located in an ampicillin resistance gene). Two antibiotic resistance genes are present, and there is an origin of replication. Any of the unique restriction enzyme recognition sites can be used as a site to insert foreign DNA.

gene cloning. Plasmids commonly used as cloning vectors have been modified in the laboratory to make them more convenient to use. An important modification is the insertion of a set of restriction enzyme recognition sites in one region, usually within an antibiotic resistance gene. The sites are chosen so that they are unique in the plasmid, and therefore cleavage of the plasmid with one of these enzymes yields a single break in the circular DNA, resulting in a linear molecule. If an enzyme that produces staggered cuts is used, there will be unpaired bases at each end. A basic cloning experiment employing a plasmid is outlined in Figure 1-25.

The usefulness of plasmids as cloning vectors is limited by their ability to accommodate only up to approximately 10,000 base pairs (10 Kb) of DNA. Larger sequences can be accommodated in bacteriophages, viruses that infect bacteria. The prototype is **phage lambda** (Figure 1-26). A cloning vector has been created from lambda by replacing genes normally used for the lysogenic process with a cloning site containing the recognition sequences for one or more restriction enzymes. Foreign DNA cut with an appropriate restriction enzyme can be inserted into this site just as was done with the plasmid DNA. The resulting recombinant phage can then be used to infect bacteria. Once inside, the phage replicates and eventually lyses the cell, giving rise to a plaque of lysed bacteria from which phage DNA can be isolated, and the DNA inserted in the phage is released with the restriction enzyme. The size limitation for phage cloning is approximately 23 Kb.

Other cloning vectors in common use are **cosmids** and **yeast artificial chromosomes (YACs)**. Cosmids can accommodate up to approximately 50 Kb of DNA, and YACs can be used to clone hundreds of kilobases or even more than a million base pairs. These vectors are useful for cloning whole genes or groups of genes. Details of their use will be described later. Table 1-1 presents a summary of cloning vectors.

TABLE 1-1 Major cloning vectors and quantity of inserted DNA that can be accommodated

Vector	Capacity
Plasmid	10 Kb
Phage	23 Kb
Cosmid	50 Kb
Yeast artificial chromosome	>100 Kb

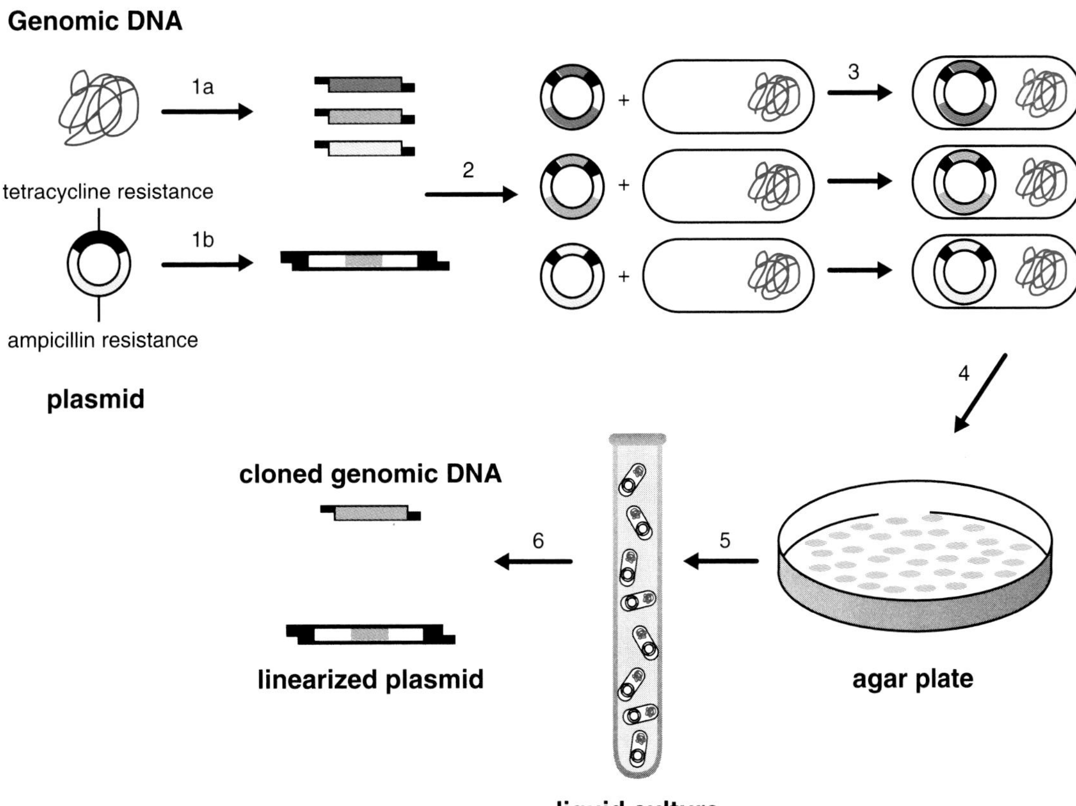

FIGURE 1-25 Steps in plasmid cloning. Both genomic DNA and the plasmid are cut with the same restriction enzyme, resulting in non–base paired ends (Steps *1a* and *1b*). In this example, the plasmid is cut within the tetracycline resistance gene. Individual DNA fragments attach to plasmid DNA through their complementary non–base paired ends and are covalently annealed with an enzyme known as a ligase (step *2*). Recombinant plasmids then are taken up by bacterial cells, in which they multiply (step *3*). The bacteria are grown in single-cell-derived colonies on an agar plate (step *4*). Bacteria with recombinant plasmids will grow on ampicillin but not tetracycline because the inserted DNA disrupts the tetracycline resistance gene. Bacteria are picked from a single colony and grown in liquid culture (step *5*). Plasmids isolated from this colony (step *6*) can be treated with restriction enzyme to release their DNA insert.

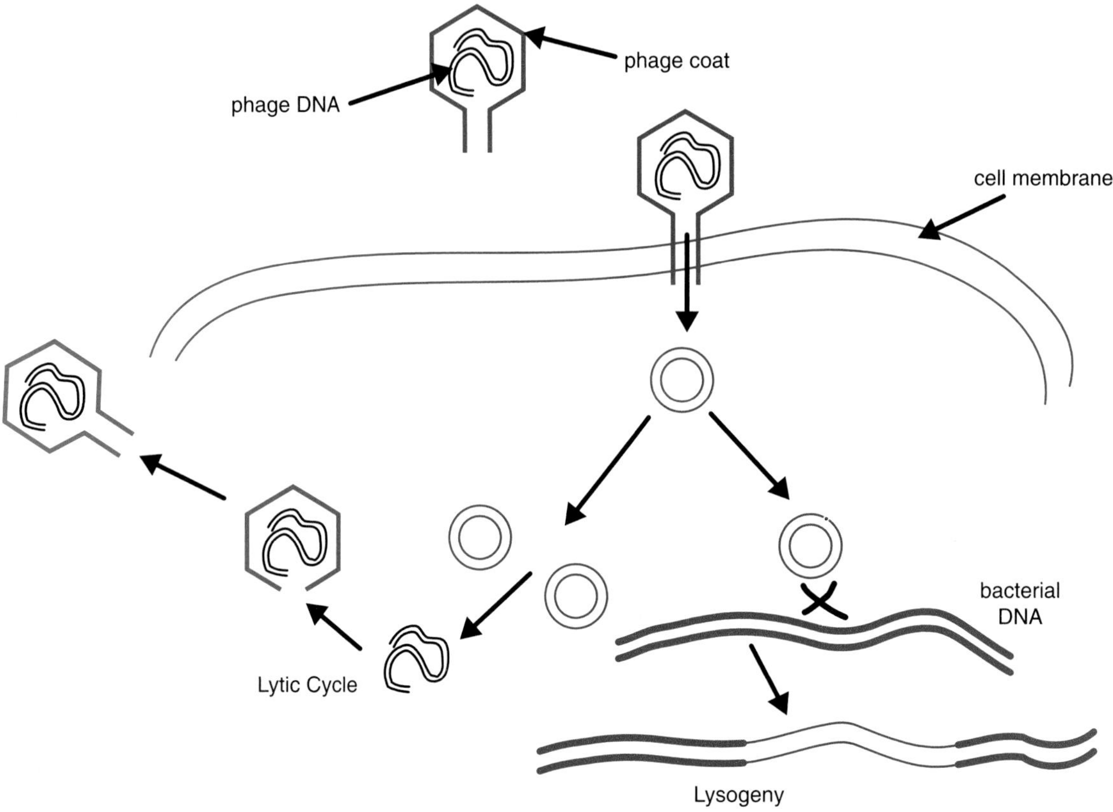

FIGURE 1-26 Life cycle of bacteriophage lambda. The phage injects its DNA into the bacterial cell, where it adopts a circular configuration. The phage DNA either can integrate stably into the bacterial chromosome by recombination (lysogeny) or can replicate itself, direct the synthesis of viral protein, and then lyse the cell.

1.6 Cloning the Gene for Tyrosinase

Further understanding of the nature of gene mutations responsible for albinism requires knowledge of the structure of the tyrosinase gene. The key to any gene-cloning experiment is to identify the DNA sequence of interest from the many thousands of gene fragments that may be isolated. In the case of tyrosinase, this was accomplished by starting with RNA from cells known to synthesize tyrosinase, copying the RNA into DNA, and cloning the DNA in lambda phage. A DNA copy of RNA was made using the reverse transcriptase obtained from single-stranded RNA retroviruses, which, when they infect cells, use this enzyme to copy their RNA genome into DNA. The DNA copy is referred to as *cDNA*. The phage vector was designed so that cloned sequences would be transcribed and translated into protein in *E. coli*. Bacteria infected with a phage containing the tyrosinase cDNA were identified using an antibody to tyrosinase protein. Details about the approach to cloning tyrosinase are provided in Figure 1-27.

Having cloned cDNA from tyrosinase-producing cells, it remained to establish that the inserts in clones that reacted with tyrosinase antibody actually corresponded with tyrosinase transcript. First, it was shown that RNA complementary to the cloned DNA was transcribed only in cells that express tyrosinase (Figure 1-28). Evidence of transcription was seen using RNA from melanocytes or melanocytic cell lines, but not from human and mouse fibroblasts or lymphocytes.

Although this demonstrated that the cloned cDNA was homologous to RNA molecules produced only in cells known to transcribe tyrosinase, it still was possible that the clones corresponded with some other gene also expressed in these cell types. Further

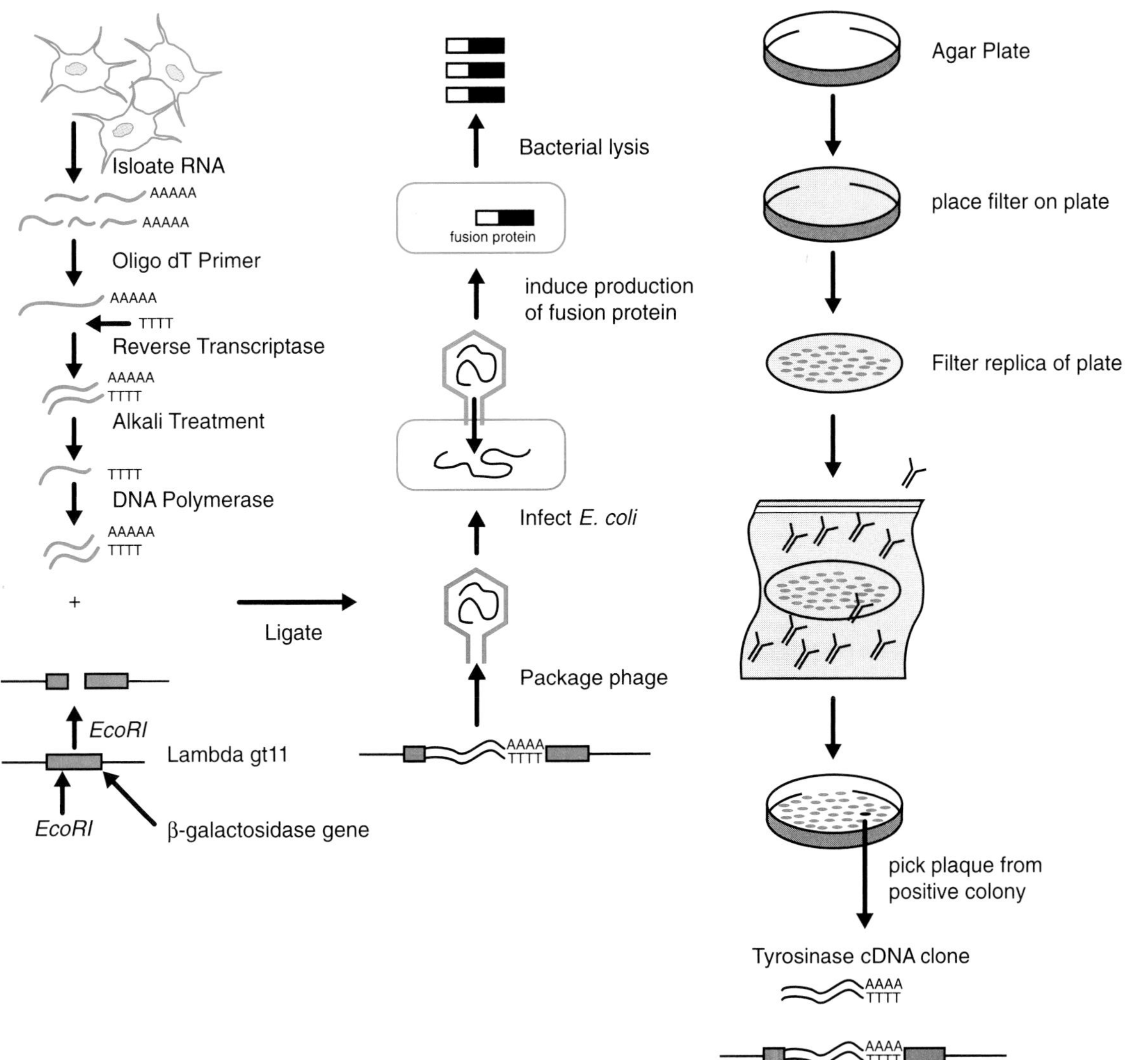

FIGURE 1-27 Cloning of tyrosinase cDNA. RNA was isolated from cells that express tyrosinase. Most mRNA molecules contain a "tail" of 100 to 300 adenines at their 3′ ends ("polyA tail"). A synthetic oligo dT sequence (a string of thymidines) was bound to these polyA tails and served as a starting point (primer) for synthesis of a DNA copy of the RNA (cDNA) using the enzyme reverse transcriptase. The RNA then was digested away with alkali treatment and a second DNA copy made with enzyme DNA polymerase, rendering the cDNA double-stranded. The phage vector lambda gt11 was digested with *EcoRI*, and cDNA molecules were ligated with the phage arms. These then were packaged into phage particles and infected into *E. coli* cells. Lambda gt11 contains a sequence for synthesis of the protein beta-galactosidase, which is located at the site into which foreign DNA is cloned. Stimulation of transcription of beta-galactosidase results in synthesis of a fusion protein consisting of part of beta-galactosidase and part of the protein encoded by the cloned sequence. After stimulation of transcription, bacteria were plated on agar and plaques of lysed bacteria were absorbed onto filters. The filters were incubated in a bag with tyrosinase antibody. The antibody became bound to tyrosinase on the filter, and the binding was detected with an immunologic staining reaction. This identified the clones containing tyrosinase cDNA, which then were isolated and used as a source of tyrosinase cDNA. (Data from Kwon BS, Haq AK, Pomerantz SH, Halaban R. Isolation and sequence of a cDNA clone from human tyrosinase that maps at the mouse c-albino locus. Proc Natl Acad Sci USA 1987;84:7473–7477.)

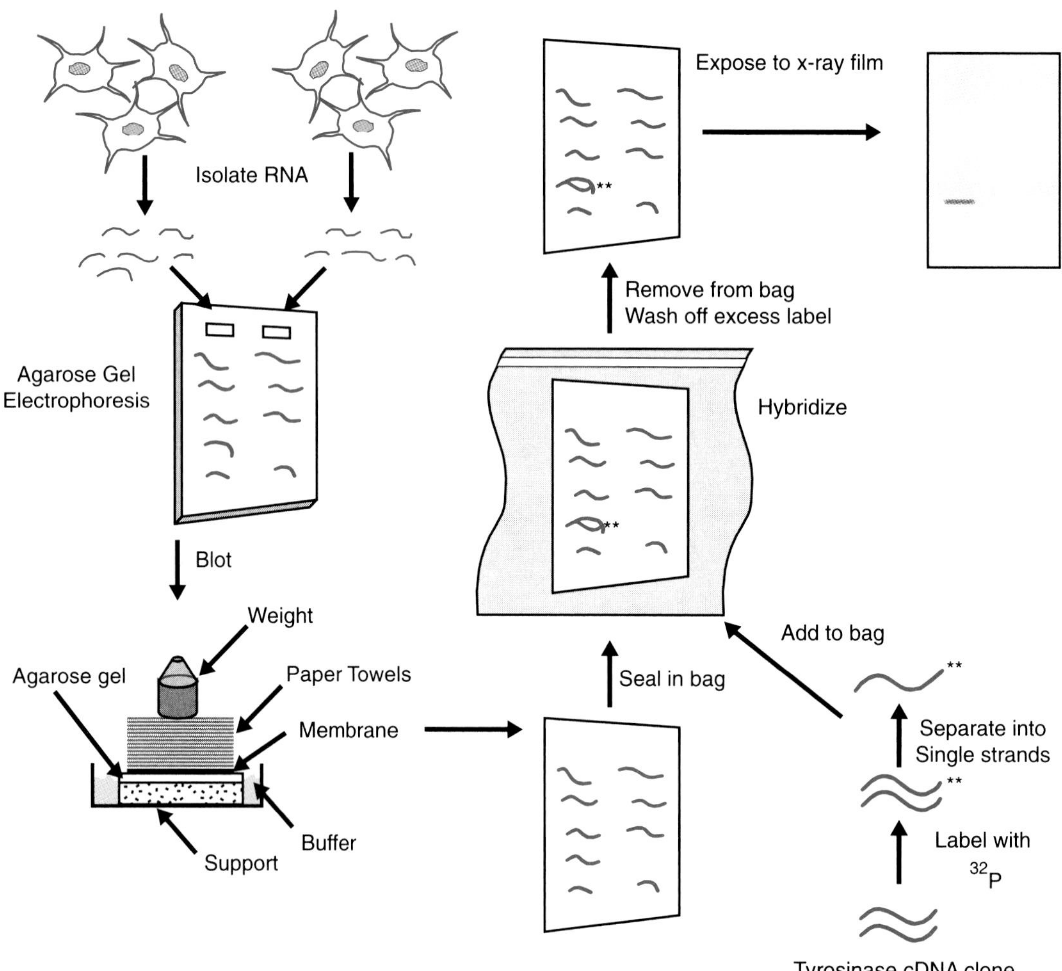

FIGURE 1-28 Analysis of tyrosinase RNA. RNA was isolated from cells that either do (left) or do not (right) synthesize tyrosinase. The RNA was subjected to electrophoresis in an agarose gel, separating fragments by molecular weight. The agarose gel then was blotted onto a filter that was placed in a bag containing radioactively labeled single-stranded tyrosinase cDNA. This probe DNA bound to homologous tyrosinase RNA, resulting in a band corresponding to the position in the blot where this RNA was located when the filter was washed and exposed to x-ray film. The formation of a double helix from complementary molecules of different origin is referred to as *hybridization*.

evidence that cloned cDNA actually represented tyrosinase was provided by analysis of albino mice with mutant tyrosinase alleles produced by irradiation (Figure 1-29). DNA corresponding with the putative tyrosinase cDNA was found in wild-type mice but not in mice with radiation-induced albinism. It was concluded that the tyrosinase gene was missing in the albino mice.

Blots of single-stranded genomic DNA are known as **Southern blots**, named for Dr. E.M. Southern, who devised the technique. RNA blots are referred to as **northern blots**, a play on the previously coined term *Southern blot*. In both instances, the single-stranded nucleic acid to be analyzed is immobilized on a membrane and is exposed to a *probe*, consisting of a single-stranded radioactively labeled piece of DNA or RNA. The formation of a double helix from two complementary nucleic acids from different sources is referred to as *hybridization*. Having one of the molecules immobilized on a membrane facilitates

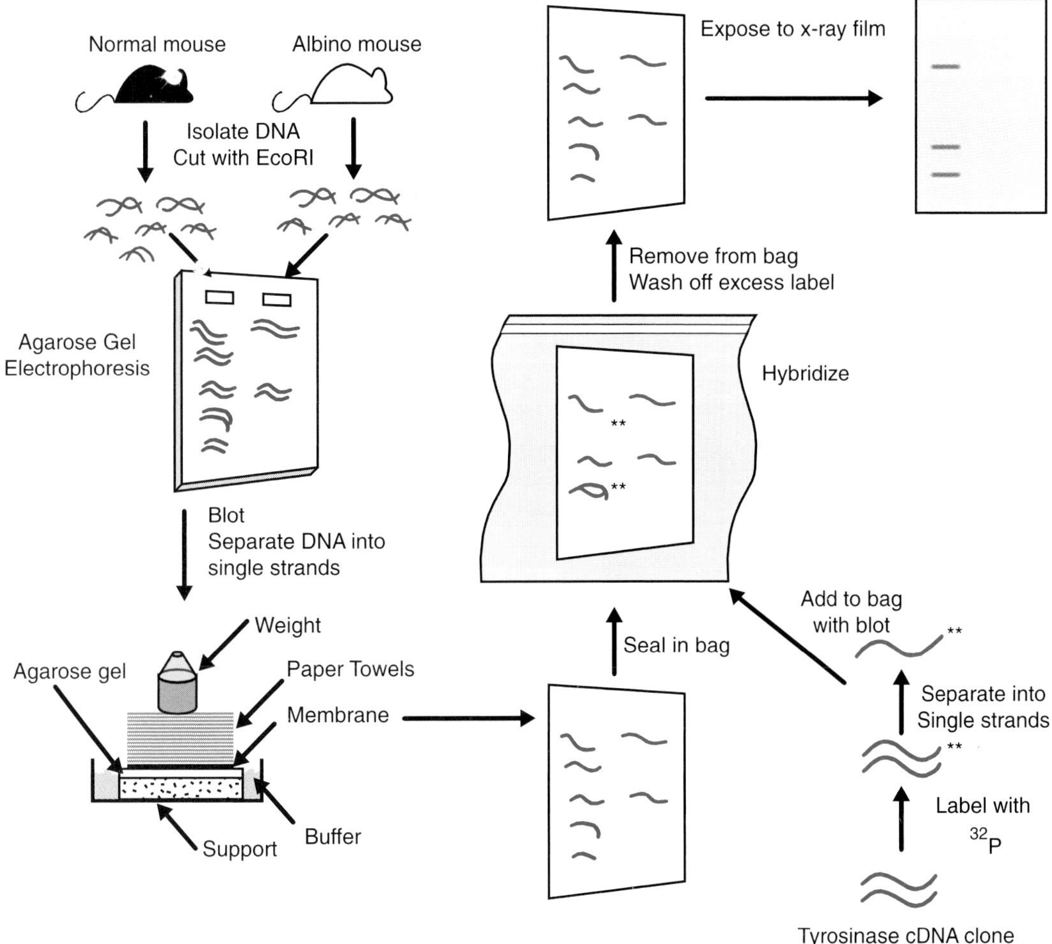

FIGURE 1-29 Analysis of DNA from albino and nonalbino mice. Genomic DNA was isolated and cut into fragments with *EcoRI*. The fragments then were subjected to electrophoresis in an agarose gel. The DNA in the gel was separated into single strands by alkali treatment and was blotted onto a filter. This then was hybridized with labeled tyrosinase cDNA. Exposure of the filter to x-ray film revealed three bands in the lane containing DNA from the normal mouse but no hybridization in the lane with DNA from the albino mouse. (Data from Kwon BS, Haq AK, Pomerantz SH, Halaban R. Isolation and sequence of a cDNA clone from human tyrosinase that maps at the mouse c-albino locus. Proc Natl Acad Sci USA 1987;84:7473–7477.)

visualization of the result of hybridization. The hybrid molecules are rendered visible by placing the blot on an x-ray film, which is exposed by the radioactivity of the probe DNA. The location of the target DNA in the blot indicates its molecular weight, as the blot is prepared after separation of fragments by electrophoresis.

Why were three DNA fragments of different sizes seen in the Southern blot of mouse DNA hybridized with tyrosinase cDNA? To understand this, it is necessary to consider an important feature of gene structure (Figure 1-30). Most genes consist of a much larger amount of DNA than would be required to encode a protein. Segments of DNA-encoding portions of protein are separated from one another by segments of non-coding DNA. The coding sequences are referred to as **exons** and the non-coding sequences as **introns**. The entire gene is transcribed but, before the transcript is exported from the nucleus to the cytoplasm, introns are cut out and exons spliced together to form the mature mRNA. A very substantial proportion of a gene may consist of

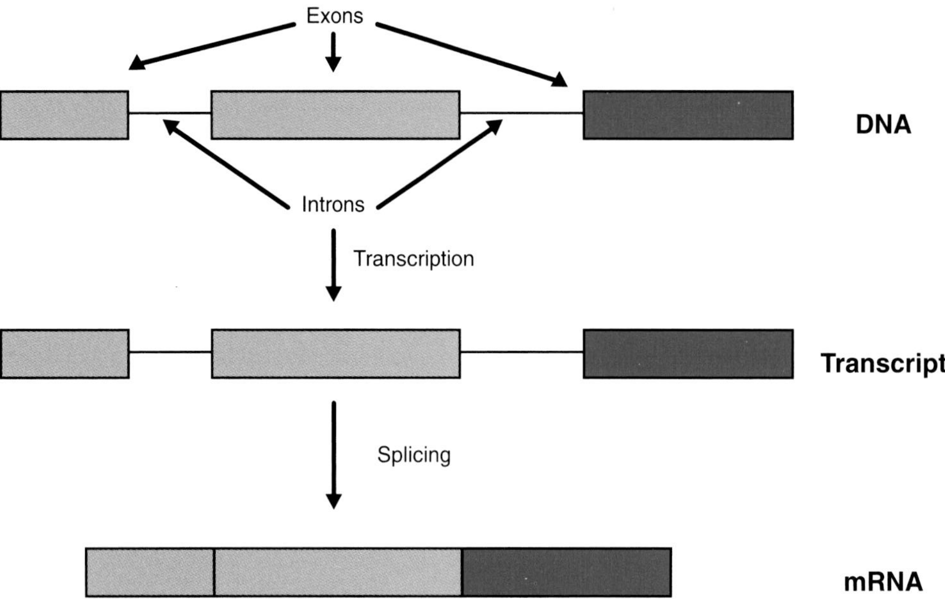

FIGURE 1-30 A genomic DNA sequence includes both introns and exons. The entire sequence is transcribed, but the introns are spliced out of the transcript to produce the mature mRNA.

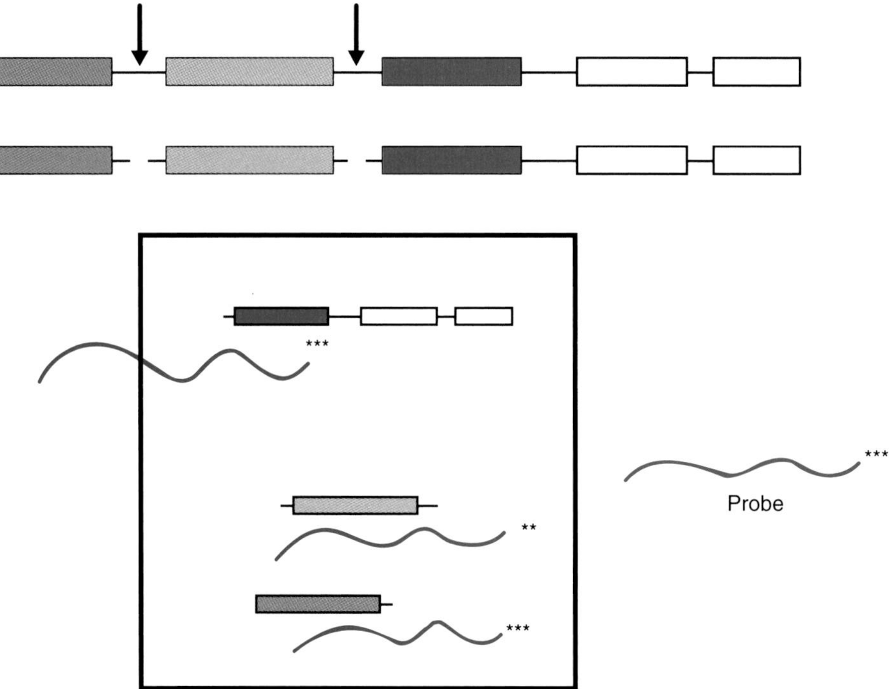

FIGURE 1-31 Tyrosinase consists of five exons. Possible *EcoRI* digestion sites are indicated by the arrows. Two cutting sites cleave the genomic DNA into three fragments, each of which hybridizes to a different region of the probe tyrosinase cDNA, resulting in three bands on the Southern blot.

introns, but the function of introns is unknown. cDNA clones are copies of fully processed mRNA and therefore include only exons, which may be separated by considerable distance in genomic DNA. It is likely that there are several *EcoRI* sites between the various exons in genomic DNA, and therefore the regions of tyrosinase DNA homologous to the cloned cDNA were cut into several fragments by the enzyme, resulting in multiple DNA fragments of different sizes on the Southern blot (Figure 1-31).

Having determined that the cDNA clones most likely corresponded with the tyrosinase gene, the next step in characterization of the gene was to determine the nucleic acid sequence (Figure 1-32). The nucleotide sequence of tyrosinase is shown in Figure 1-33. It should be noted that the entire transcript was not represented in any one cDNA clone. The complete cDNA sequence was reconstructed from three overlapping clones. It is common for cDNA clones to contain partial sequences, probably as a result of inefficiency of the reverse transcriptase reaction used to form the cDNA. Prior to sequencing the tyrosinase cDNA, the amino acid sequence of the protein was unknown. Protein sequencing is a very laborious process; it is generally easier to clone a gene, determine its nucleotide sequence, and infer the amino acid sequence, than it is to sequence the protein

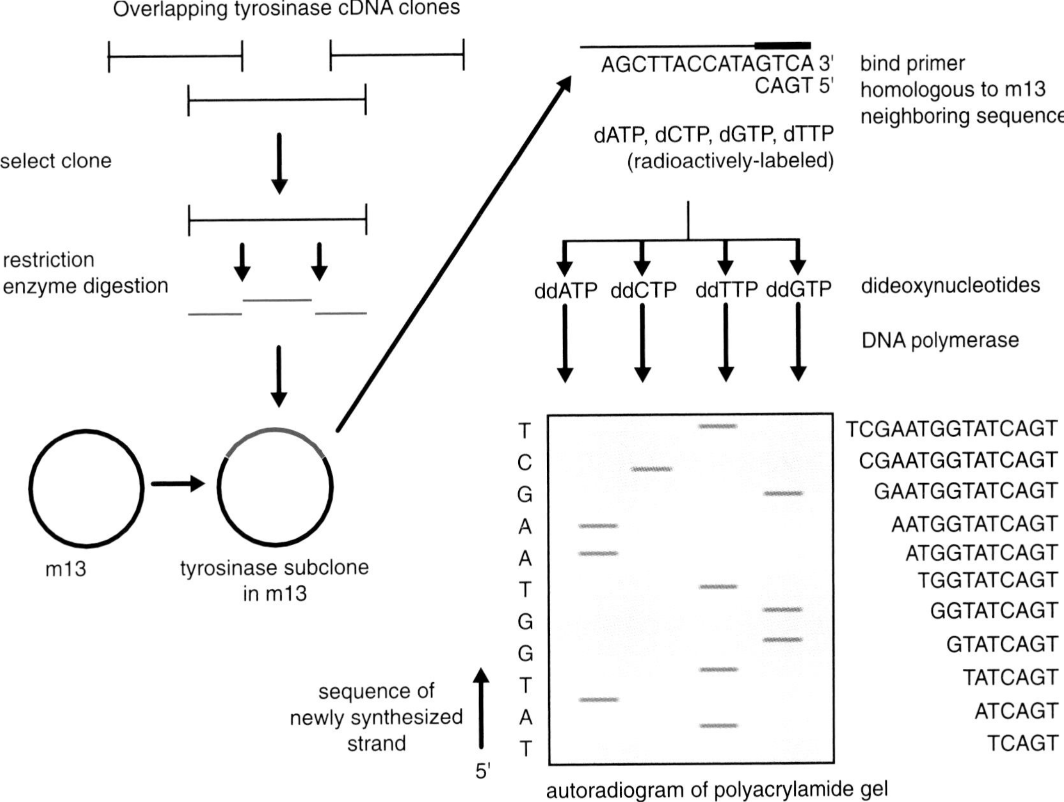

FIGURE 1-32 Dideoxy sequencing of tyrosinase cDNA. Overlapping cDNA clones were inserted in the single-stranded bacteriophage vector **m13**. These subclones then were sequenced. A primer homologous to m13 DNA adajcent to the cloning site is used as a point of initiation of DNA synthesis. Four separate reactions are performed, each of which contains a proportion of one nucleotide represented as a dideoxynucleotide. The dideoxynucleotide or the normal nucleotide can be incorporated into the growing strand, but incorporation of the dideoxynucleotide results in termination of DNA synthesis. Thus, each reaction results in a population of differently sized fragments that can be separated on a polyacrylamide gel and visualized by autoradiography. The sequence homologous to the template is read from bottom to top, 5′ to 3′, by noting the lane in which a band appears.

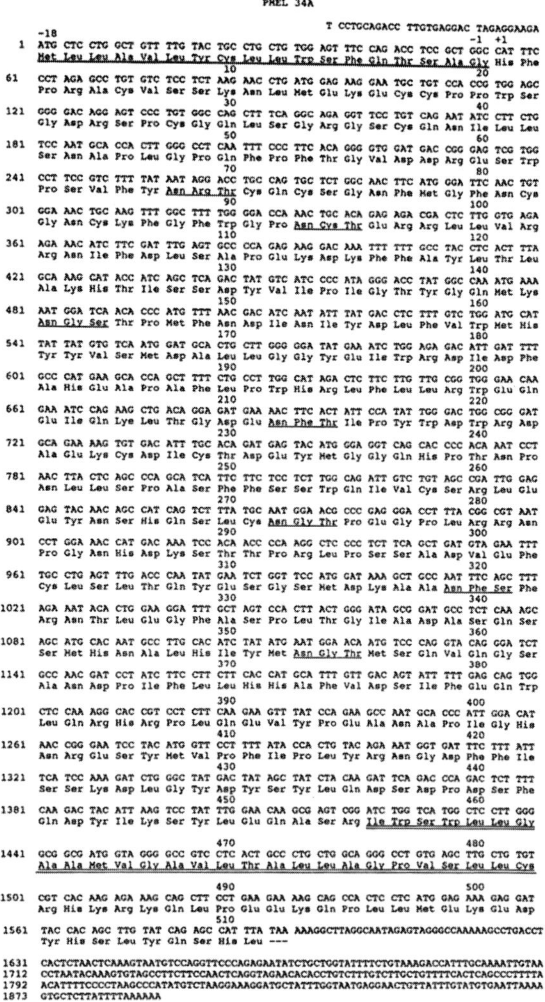

FIGURE 1-33 Tyrosinase sequence. Triplet codons are shown, with corresponding amino acids indicated. A stop codon is encountered at base 1588. (Reproduced by permission. Kwon BS, Haq AK, Pomerantz SH, Halaban R. Isolation and sequence of a cDNA clone from human tyrosinase that maps at the mouse c-albino locus. Proc Natl Acad Sci USA 1987;84:7473–7477.)

directly. To infer the amino acid sequence from the DNA sequence, it is necessary to decide where to start reading the triplets of bases, referred to as establishing a *reading frame*. The reading frame for tyrosinase was determined by searching for a reading frame that did not result in rapidly encountering a stop codon. The reading frame shown in Figure 1-33 does not lead to a stop codon until position 1645.

Several features of the protein can be predicted from analysis of the amino acid sequence, which is based on comparison of the sequence with that of other proteins of known structure. The first 12 amino acids include 10 hydrophobic residues, a common characteristic of proteins inserted into the cell membrane or secreted from the cell. Possible sites of glycosylation and copper binding were identified, as well as a region that is likely to span the cell membrane.

With the tyrosinase cDNA in hand, it was possible to clone the entire tyrosinase gene (Figure 1-34). A set of clones of random fragments of genomic DNA is known as a **genomic library**. A human genomic DNA library in lambda phage was screened by hybridizing DNA from bacterial plaques with radioactively labeled tyrosinase cDNA. The plaques identified as containing DNA homologous to tyrosinase were used to isolate recombinant phages that were grown in large quantities. Human DNA isolated from these phages contained a portion of the tyrosinase gene, both exons and introns. Through study of overlapping clones, the sequence of the entire gene was determined. Tyrosinase was found to consist of five exons spanning a distance of more than 50 Kb.

1.7 Tyrosinase Mutations

Katie is now 4 years old. She is in preschool and doing very well. Her parents have received a call from the dermatologist responsible for the original diagnosis of albinism in Katie. The dermatologist, having learned that researchers are interested in receiving blood samples from individuals with OCA, to test for tyrosinase mutations, requests that the family permit their blood samples to be submitted. It is explained that this will not provide a clinical benefit to Katie in the immediate future but will help to advance research on albinism. The family agrees, and blood is drawn from Katie and from both her parents.

The cloning of tyrosinase enabled identification of mutations in affected individuals. Mutations responsible for albinism in one family were described by Spritz and colleagues. A Southern blot of restriction enzyme–digested genomic DNA hybridized with tyrosinase cDNA gave a normal pattern. This was in contrast to the results described previously

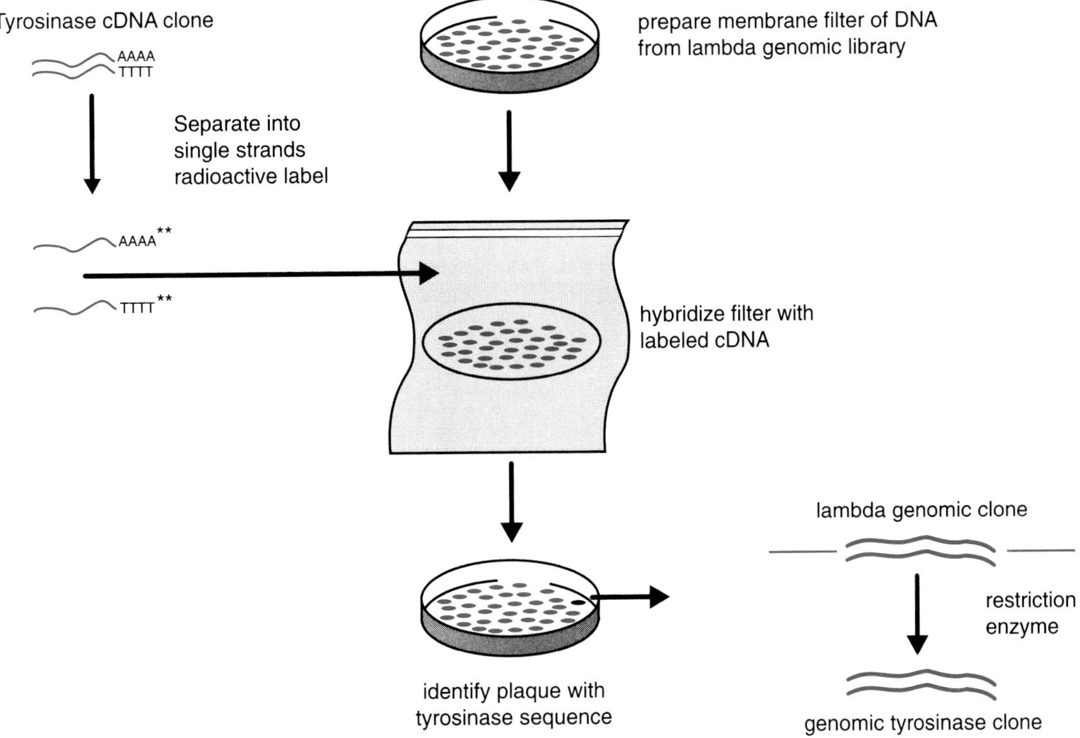

FIGURE 1-34 Cloning tyrosinase genomic DNA. A membrane filter was prepared from a lambda phage library containing random fragments of genomic DNA. DNA in the filter was hybridized with labeled tyrosinase cDNA, identifying the clones containing homologous DNA. These were then isolated and used to prepare the genomic clone. (Data from Giebel LB, Strunk KM, Spritz RA. Organization and nucleotide sequences of the human tyrosinase gene and a truncated tyrosinase-related segment. Genomics 1991;9:435–445.)

with DNA from albino mice and indicates that albinism in this family was not due to deletion of the tyrosinase gene. In this instance, the mutation was found by sequencing the gene, accomplished with a strategy based on the *polymerase chain reaction* (PCR).

PCR is an extraordinarily powerful technique widely used in genetic analysis. The approach, outlined in Figure 1-35, is based on amplification of a specific DNA sequence, so that from one copy of the target sequence millions of duplicates are made that can be subjected to further study. Application of PCR requires knowledge of the sequence of segments of DNA flanking the target region. Oligonucleotides, usually 20 to 30 bases long, are synthesized that are complementary to opposite DNA strands on either side of the target. The DNA to be studied is separated into single strands by heating, and then the oligonucleotides are allowed to bind to their complements when the reaction mixture is cooled. These double-stranded regions serve as primers for a DNA synthesis reaction catalyzed by DNA polymerase. The double-stranded reaction products are again melted at high temperature, and the primers are allowed to bind to their complements by cooling the mixture. DNA synthesis is allowed to proceed once more, but this time the substrates include not only the original target strands but also the newly synthesized DNA strands. This process is repeated cyclically, the number of copies of the target sequence doubling with each cycle. After 20 to 30 cycles, millions of copies of the target are obtained.

To identify tyrosinase mutations (Figure 1-36), oligonucleotide primers were made corresponding to the first and last 19 or 20 bases of each of the five exons, or within adjacent intron sequences. PCR

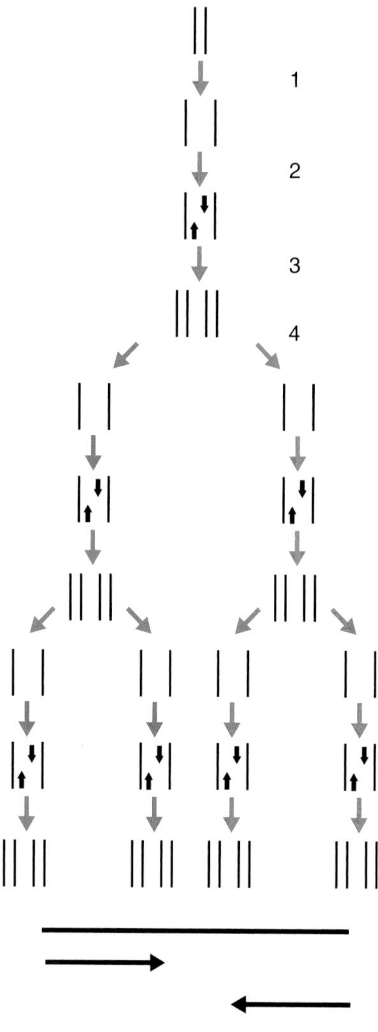

FIGURE 1-35 Diagram of polymerase chain reaction (PCR). Strands of target DNA are separated (*1*) and primers allowed to bind to opposite strands (*2*). A DNA synthesis reaction then is allowed to proceed, copying the target DNA (*3*). This process is repeated cyclically (*4*), resulting in exponential amplification of the target sequence.

products were subjected to electrophoresis, and the DNA corresponding to each reaction product was cut out of the gel and purified. This DNA then was used in a dideoxy sequencing reaction. Sequence analysis from one individual in a family with albinism revealed three base changes from the wild-type sequence. One, at position 174, substituted a tyrosine (codon TAT) for a serine (codon TCT). A change of base sequence in a gene that results in a change of amino acid in the corresponding protein is referred to as a **missense mutation.** Previous studies had determined the position 174 mutation to be a common base change found in many members of the general population. The substitution of tyrosine for serine at this site apparently does not interfere with enzyme activity. Such mutations are considered to be **conservative.** The individual with albinism was also heterozygous for two other tyrosinase mutations, one substituting lysine for threonine at position 355 (ACA to AAA) and the other substituting asparagine for aspartic acid at position 365 (GAT to AAT). Sequencing of PCR products obtained from the patient's parents revealed that the mother was heterozygous for the mutation at codon 355 and the father was heterozygous for the codon 365 mutation.

To facilitate the study of other family members and other individuals with albinism, an assay based on allele-specific oligonucleotide hybridization was devised (Figure 1-37). The assay revealed the presence of both normal and mutant DNA in the proband, indicating that the patient was heterozygous for both mutations. Her father had the normal codon 355 sequence but the mutant codon 365 sequence, and the opposite was seen in the mother. Three other children with albinism in the family gave results similar to the proband. Examination of a number of unrelated individuals with albinism revealed one who was heterozygous for the codon 355 mutation but none for the codon 365 mutation.

Both the codon 355 and 365 mutations change the net charge of tyrosinase and hence alter the physical properties of the enzyme. Moreover, they both occur near a copper-binding site important for normal enzyme function. The codon 355 mutation also disrupts a potential *N*-glycosylation site. These missense mutations are therefore very likely to be the cause of albinism in the proband.

The results of studies of DNA obtained from Katie and her parents are shown in Figure 1-38. Katie is found to have two different tyrosinase mutations. She has a C-to-A change at codon 373, inherited from her mother, and a one-base deletion leading to frameshift at codon 225, inherited from her father.

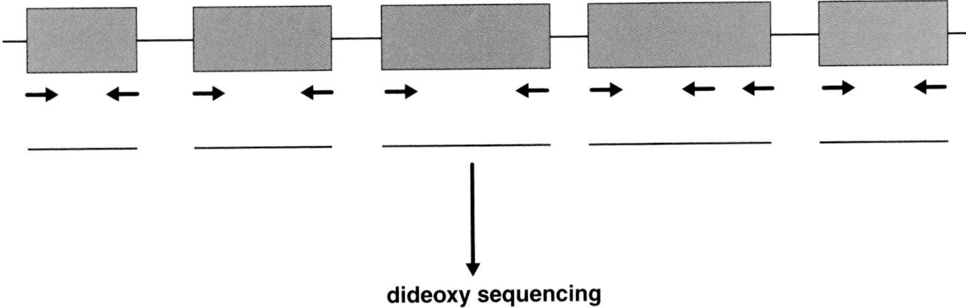

FIGURE 1-36 Scheme for detection of tyrosinase mutations. PCR primers were prepared to amplify each of the exons of the gene. These products were purified, and the amplified fragments were sequenced by a dideoxy reaction.

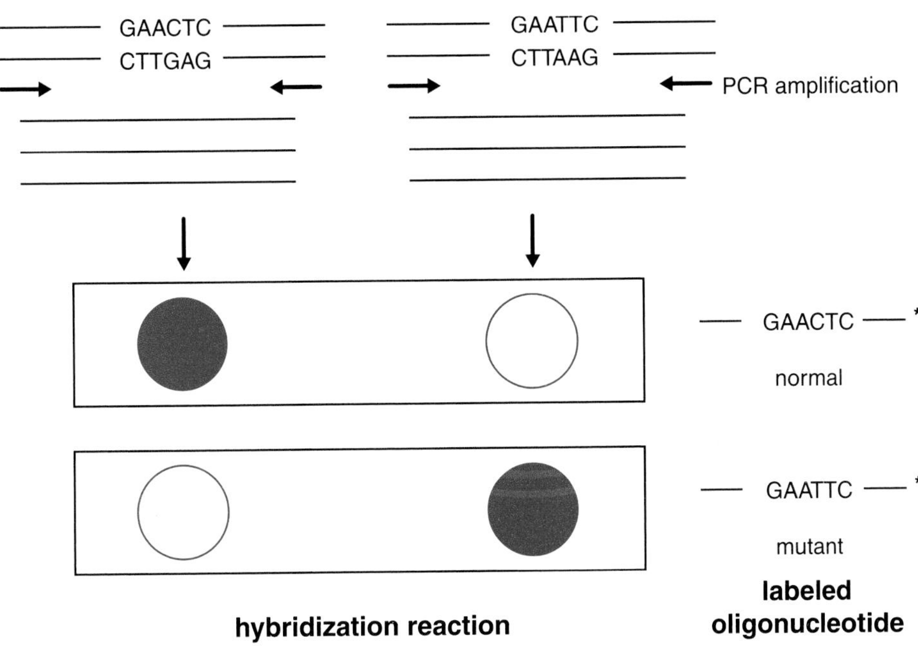

FIGURE 1-37 Allele-specific oligonucleotide mutation detection scheme. Oligonucleotides (19 bases in length) corresponding to the base sequence of the wild-type (left) and mutant (right) sequences surrounding codons 355 and 365 were synthesized and labeled with ^{32}P. PCR products from amplification of exon 3 (in which these codons are located) were applied to nylon membranes, denatured into single strands, and hybridized with either the normal or the mutant oligonucleotides. Conditions of hybridization were such that only a perfect match of base sequence would result in a stable double helix between the PCR product in the filter and the labeled oligonucleotide. Exposure to x-ray film resulted in a dark spot only if the sequence exactly corresponding to the oligonucleotide was present in the PCR reaction. Sequences shown are hypothetical and do not represent the actual tyrosinase sequence.

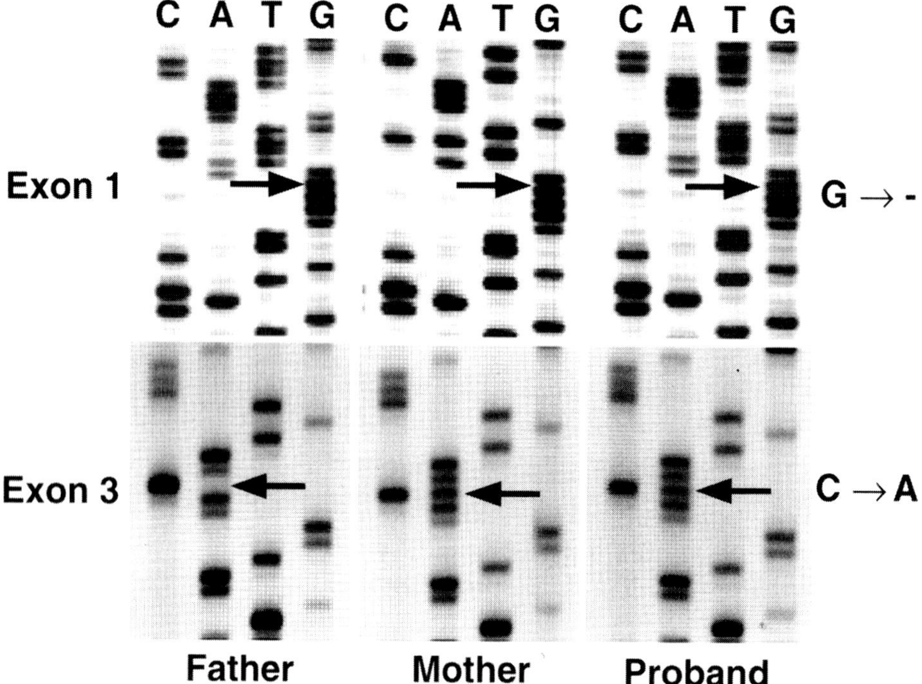

FIGURE 1-38 Identification of tyrosinase mutations by DNA sequencing. The proband has a C-to-A change in exon 3, inherited from mother, and a deletion of one G in exon 1, inherited from father. Mother and proband are heterozygous for the C-to-A change and therefore have bands at both C and A on the gel (indicated by arrow). Father and proband are heterozygous for deletion of one G. As a result, they have two sequencing patterns superimposed above the site of the arrow, displaced by one band. (Courtesy of Dr. William Oetting, University of Minnesota.)

1.8 Genetic Heterogeneity in Albinism

An individual who has two alleles with different mutations is said to be a **compound heterozygote** (Figure 1-39). Although the alleles are different, both result in dysfunctional protein, and hence enzyme activity is low. The parents, in contrast, are heterozygous for one mutation each and can be presumed to have approximately 50% of the normal tyrosinase activity. The unrelated person with albinism who was homozygous for the codon 355 mutation would have albinism on the basis of true homozygosity for a single mutation.

Point mutations can differ widely in their impact on enzyme function. At one extreme is the substitution of tyrosine for serine at position 174. This conservative change has little or no impact on the function of the enzyme. Consequently, whichever is considered the normal or wild-type sequence, there is no selection against individuals carrying the variant form, and so both forms are common in the general population. This is referred to as a genetic **polymorphism**, defined as a genetic variation in which the frequency of two or more alleles is at least 1% in the population. Polymorphisms may exist for a variety of reasons, which will be explored later in this book. They are particularly common in situations in which there is no selective pressure to maintain a particular DNA sequence, such as in introns or in regions of DNA between genes. It is estimated that the DNA sequence of any two individuals is likely to differ by approximately 1 in every 200 bases. Some of this genetic variation accounts for phenotypic differences between individuals, but much of it is of no genetic consequence. As will be seen in later chapters, genetic polymorphisms are key tools in efforts to map the human genome.

It is important to remember that finding a base sequence variant in a gene from a person with a genetic disorder does not necessarily constitute

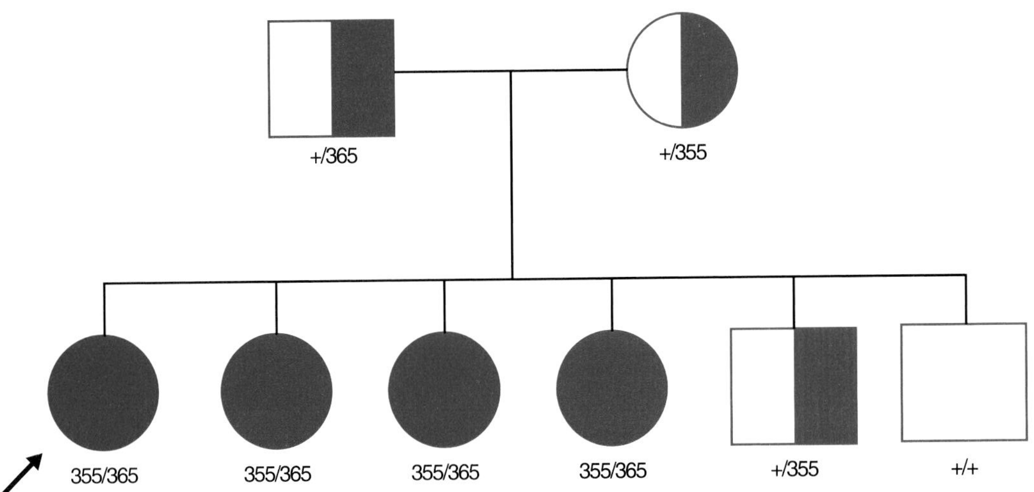

FIGURE 1-39 Pedigree of family studied for albinism mutation. The proband (arrow) is a compound heterozygote for the codon 355/365 mutations. The same is true for her three affected sisters. Her father is heterozygous for the 365 mutation, and her mother is heterozygous for the 355 mutation, as is one brother. The other brother has two wild-type (+) alleles.

finding a disease-causing mutation. A variant responsible for a genetic disorder will be present only in persons with the disorder and not in unaffected individuals. It also should be possible to infer how the variant would disrupt function of the protein. The codon 355 and 365 tyrosinase mutations are not seen in the general population and lead to nonconservative substitutions at a critical region of the enzyme. Often, however, it can be difficult to prove the etiologic effect of an apparent mutation. This will be explored later in this book.

Many different tyrosinase missense mutations have been identified in various individuals with OCA. One mutation, a substitution of leucine for proline at codon 81 of tyrosinase, accounts for 20% of OCA alleles in a group of white individuals in the United States. This allele was not found in any unaffected persons, suggesting that it is not a polymorphism. OCA affects approximately 1 of 39,000 whites and 1 of 28,000 blacks. A cysteine-to-arginine (TGC to CGC) substitution at codon 89 has been found in one American black with OCA. This individual was homozygous for the mutation, suggesting that this allele, which has not been found in a nonblack albino, may be more common in the black population. Rare mutations tend to remain within specific populations if most matings occur within that population, a phenomenon known as the **founder effect.** The codon 89 mutation would occur in a white individual only if there were a rare new mutation or if that individual had a black ancestor who happened to carry the codon 89 mutation.

Analysis of point mutations responsible for OCA has revealed that the mutations are not randomly distributed along the gene (Figure 1-40). Four clusters of mutation sites have been found. Two involve binding sites for copper apparently necessary for enzyme activity. The other clusters are in exons 1 and 4. The functional significance of these sites is not known. It is hoped that further study of the distribution of mutations in tyrosinase will help elucidate structure-function relationships for this enzyme.

Tyrosinase mutations described thus far have been missense mutations, in which a single base change in DNA results in aberrant incorporation of an amino acid in the protein. Another class of mutations found in association with albinism are frameshifts (Figure 1-41). When a protein is synthesized, translation begins at the first triplet encoding the first amino acid (usually AUG, encoding methionine) and proceeds to read triplets until a stop codon is reached. There are three stop codons—UAA, UAG, and UGA. A deletion or insertion of one or two bases in an exon will result in a shift of reading frame. From that point, incorrect amino acids will be inserted into the growing peptide until a stop codon is

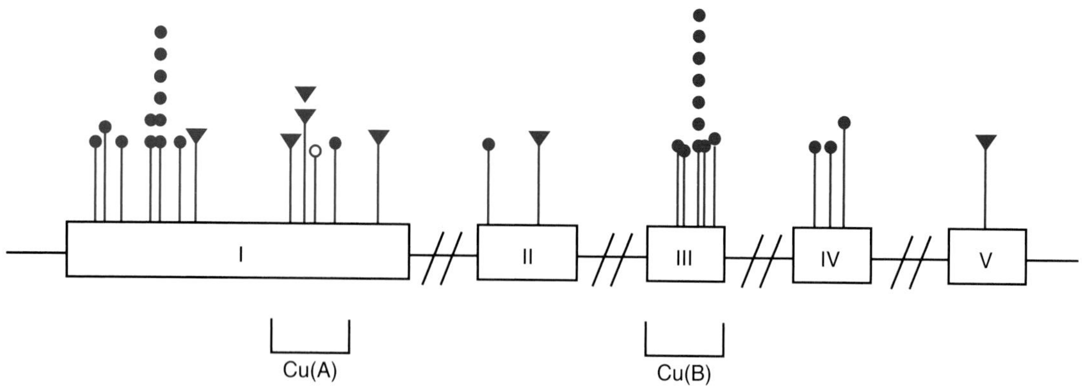

FIGURE 1-40 Map of tyrosinase mutations. The five exons are numbered, and the two copper binding sites are indicated. *Closed circles* indicate point mutations; *triangles* are chain termination mutations; and the *open circle* is a polymorphism. (Modified from King RA, Mentink MM, Oetting WS. Mol Biol Med 1991;8:19–29.)

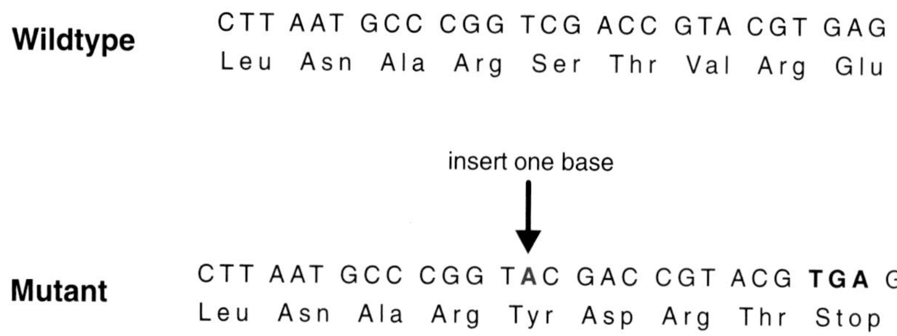

FIGURE 1-41 Frameshift mutation. The wild-type DNA sequence and corresponding amino acid sequence are shown at top. Insertion of a single A causes a shift of reading frame. From that point on, incorrect amino acids are incorporated into the protein, and eventually a stop codon is encountered, leading to termination of translation and truncation of the protein.

reached. Usually a stop codon occurs soon, and a truncated protein results that may be unstable and is degraded in the cell. Both insertions and deletions have been described in tyrosinase, resulting in alleles for albinism.

The existence of a variety of different mutations in different individuals with the same disorder is referred to as *genetic heterogeneity*. Heterogeneity has proved to be the rule rather than the exception, as more and more genetic disorders have been studied at the molecular level. Mutations at many sites in a gene can have a similar ability to disrupt the function of a protein. In some instances, individuals with different mutations may have an identical phenotype. This would be the case if each mutation resulted in complete abolition of protein function. Other times, the mutations may result in similar, but not identical, phenotypes. This is the case for some persons with albinism, as will be described in the next section.

1.9 Genotype-Phenotype Correlations in OCA

The classic form of OCA, in which tyrosinase activity is absent, is designated OCA type IA. Several other forms have been identified clinically (Table 1-2). OCA type II is defined by the presence of normal tyrosinase activity. (It sometimes is referred to as *tyrosinase-positive albinism*.) Individuals with OCA type II usually make a small amount of pigment and have light yellow hair. They also lack visual pigmentation, but the visual defect usually is milder than in OCA type I. The basis for OCA type II currently is being studied actively. The major gene involved is referred to as the *P gene*. Mutations of the P gene are responsible for the pink-eyed dilution phenotype (abbreviated *p*) in mice and type II OCA in humans. The P gene product is thought to be involved in transport of tyrosine into the cell.

TABLE 1-2 Clinical characteristics of different types of oculocutaneous albinism (OCA)

Feature	Type IA OCA	Type IB OCA	Type II OCA
Skin color	Pink/white	White (tans slightly)	White
Hair color	White	Yellow	Yellow
Eye color	Blue/gray	Blue	Blue

There are also clinically distinct forms of type I OCA, including types IA (classic OCA) and type IB (yellow mutant). Type IB involves partial deficiency of tyrosinase activity. In one inbred Amish family, individuals affected with OCA type IB were found to be homozygous for a substitution of leucine for proline at codon 406. The Amish population is relatively inbred. Homozygosity for rare alleles in this population is another example of the founder effect. Other individuals with OCA type IB have been found to be compound heterozygotes, having two different tyrosinase mutations. These are often missense mutations that result in partial deficiency of tyrosinase activity.

A particularly interesting type of albinism has been reported in a few individuals who have a temperature-sensitive mutation of tyrosinase. The phenotype is white hair in warmer areas of the body, such as the axillae and scalp, but progressive darkening of hair color in the cooler extremities. One such individual has been found to have a tyrosinase mutation in which glutamine is substituted for arginine at codon 422. This renders the protein temperature-sensitive: enzyme activity is reduced at normal body temperature but is higher at cooler temperatures. The phenotype is similar to that which accounts for the pigmentation of the Siamese cat.

1.10 Molecular Diagnosis of Inborn Errors of Metabolism

Ben and Linda are told that knowledge of the tyrosinase mutations responsible for OCA in Katie makes it possible to offer prenatal diagnosis for OCA. This would be done by sampling fetal tissue at 10 to 12 weeks of pregnancy by chorionic villus sampling or at 16 to 18 weeks by amniocentesis. Fetal DNA would then be analyzed for tyrosinase mutations. Ben and Linda decide that they are not interested in having prenatal diagnosis of albinism, although they would like to have additional children.

Knowledge of the molecular basis for inborn errors of metabolism has opened new avenues of diagnosis as well as promising new means of treatment. Prenatal diagnosis can be offered to couples who already have a child with a recessive disorder and therefore are at 25% risk of having another affected child. The decision to undergo prenatal testing is complex. The motivation, in many instances, is to make a diagnosis early in pregnancy and interrupt the pregnancy if the fetus is affected. This raises controversial issues regarding termination of pregnancy. Attitudes of couples with a family history of an inherited disorder differ widely. Factors include feelings about abortion and experience with and attitude toward the medical burden imposed by the disorder. It should be noted, though, that many couples who choose prenatal testing do not necessarily intend to terminate an affected pregnancy. Some undergo prenatal diagnosis in the hope of being reassured that the fetus will not be affected (which is 75% likely for an autosomal recessive disorder). At the time of testing, they may not yet have faced the question of whether they would terminate an affected fetus. Some are clear that they would not terminate the pregnancy but would like to know whether the fetus is affected so that they may plan for the medical needs of the child.

Attitudes toward prenatal diagnosis of genetic disorders vary from individual to individual and may vary from culture to culture. In albinism, for example, there has been relatively little interest in prenatal diagnosis in North America and Europe. In the Middle East and parts of Asia, however, prenatal diagnosis of the condition has been in greater demand. The reasons for this difference probably are complex

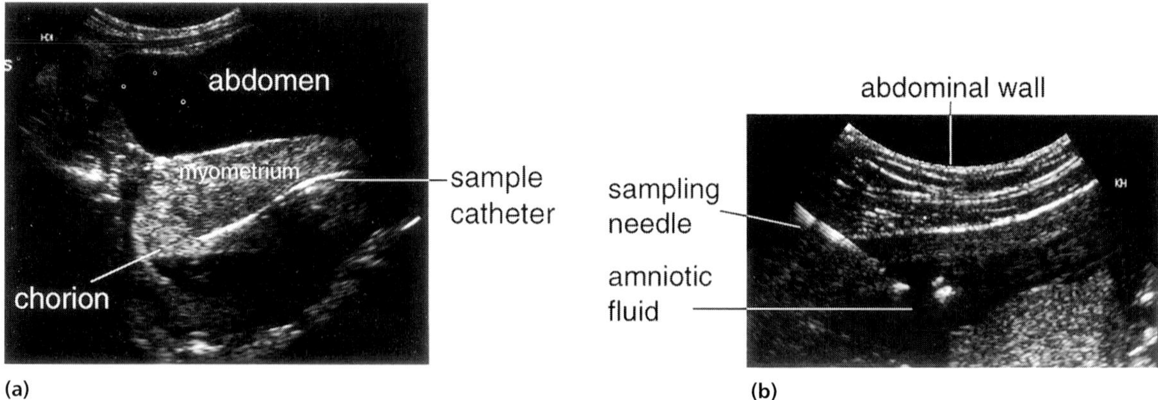

FIGURE 1-42 Ultrasound photographs illustrating chorionic villus biopsy (**a**) and amniocentesis (**b**). (Courtesy of Drs. Jodi Abbott and Deborah Levine, Beth Israel Hospital, Boston.)

but have to do, in part, with the perception of the severity of the phenotype in different environments and cultures.

Prenatal diagnosis of genetic disorders requires obtaining fetal tissue for analysis. There are two major ways of obtaining this tissue. At 10 to 12 weeks of gestation, a sample of the fetal placenta can be obtained with a biopsy catheter inserted either through the cervix or transabdominally—in either case, under ultrasonographic guidance (Figure 1-42a). This is referred to as *chorionic villus sampling* (CVS). The tissue can be used to isolate DNA or can be grown in culture. At 16 to 18 weeks of pregnancy (and, in some centers, at 11 to 12 weeks), a sample of amniotic fluid can be withdrawn with a needle inserted transabdominally (*amniocentesis*) (Figure 1-42b). This fluid contains fetal cells, mainly derived from skin and bladder (amniotic fluid is largely derived from fetal urine), that can be grown in culture.

Until recently, the only approach to prenatal diagnosis of inborn errors was assay of enzyme activity in cultured fetal cells. This limited diagnosis to conditions associated with enzymes that are expressed in cultured chorionic villus or amniotic fluid cells. Many metabolic disorders involve enzymes that are expressed in a limited set of tissues, such as brain or liver, and therefore are not amenable to prenatal diagnosis. The advent of molecular genetic analysis has changed all this, because fetal DNA can be analyzed for any gene mutation, regardless of the tissue in which the gene is expressed.

Application of molecular prenatal diagnosis to inborn errors of metabolism requires prior knowledge of the mutation carried by the parents. Usually, this is determined from study of an affected child. Because genetic heterogeneity is common, searching for mutations can be time-consuming and expensive. Also, disease-causing mutations must be distinguished from polymorphisms; a test based on a sequence variant that is a polymorphism rather than a true mutation would lead to an erroneous diagnosis.

Molecular analysis has a role in diagnostic testing aside from prenatal diagnosis. For example, DNA-

TABLE 1-3 Summary of common mutations causing Tay-Sachs disease in the Ashkenazi Jewish population

Mutation	Phenotype
Exon 11 four base insertion	Infantile Tay-Sachs disease
Intron 12 splice donor	Infantile Tay-Sachs disease
Codon 269 G to A (gly to ser)	Adult Tay-Sachs disease

Source: Triggs-Raine BL, Feigenbaum ASJ, Natowicz M, et al. Screening for carriers of Tay-Sachs disease among Ashkenazi Jews. N Engl J Med 1990;323:6–12.

based methods are increasingly playing a role in screening for carriers of Tay-Sachs disease. As already noted, Tay-Sachs is a devastating neurologic disease. The carrier frequency in the Ashkenazi Jewish population is approximately 1 in 30, high enough to justify screening for carrier status. The goal is to identify couples at risk of having an affected child and to offer them prenatal diagnosis. Owing to the unequivocally tragic nature of the illness, this mode of screening has been widely accepted by the Jewish community. Screening is done by assay of the hexosaminidase A enzyme in white blood cells. False negative results are rare, but there are some individuals whose carrier status is indeterminate.

The gene for hexosaminidase A has been cloned, and it has been found that three mutations account for 98% of the mutations in the Ashkenazi Jewish population (Table 1-3). One occurs at the 5′ border of intron 12, and changes a G to a C (Figure 1-43). The 5′ border of most introns consists of the sequence GU (the splice "donor") and the 3′ borders end in AG nucleotides (the splice "acceptor"). These

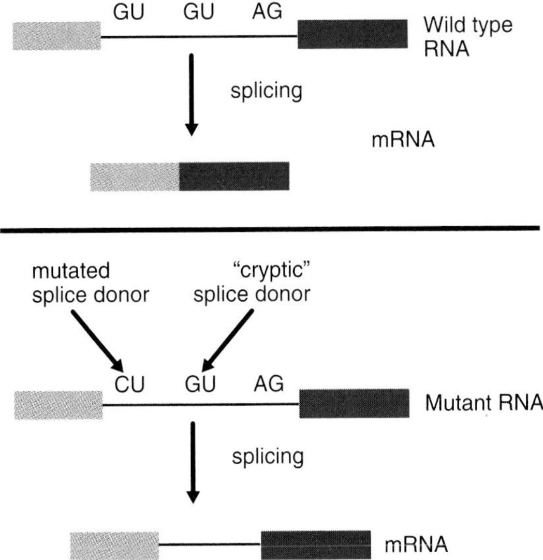

FIGURE 1-43 RNA splicing recognizes a GU base pair at the 5′ end of an intron and an AG at the 3′ end. A mutation of the GU ("donor") to CU in this case causes the 5′ splice not to occur at a normal site. Instead, the splice may occur at another GU within the intron. The result is that some intron material is retained in the final mRNA.

nucleotides mark the sites where introns begin and end and are recognized by the splicing machinery in the nucleus. Mutation in the splice donor leads to failure of splicing at the correct site. The splicing machinery scans further until a suitable donor sequence is found (often within the same intron, there is a sequence that can function as a donor, which is referred to as a *cryptic donor*). The consequence is that a splice is initiated downstream of the normal 5′ splice site, and some intron material is included in the mature mRNA. Translation of the intron sequence leads to an abnormal segment of protein and, usually, a stop codon soon is encountered, truncating the protein. In some instances, splice donor or acceptor site mutations lead to complete failure of splicing of an exon, with the result that the entire exon is spliced out (referred to as *exon skipping*).

A second hexosaminidase A mutation in the Jewish population consists of a four–base pair insertion in exon 11. This leads to a frameshift and consequent truncation of the protein. Together, these mutations account for almost all cases of infantile Tay-Sachs disease in the Jewish population. A third mutation has been found in individuals with adult-onset Tay-Sachs disease. This condition begins in the second or third decade of life. It is associated with progressive loss of motor function and psychosis. The mutation is a substitution of G to A, leading to insertion of serine for glycine at codon 269. It is notable that the two mutations responsible for severe infantile Tay-Sachs disease cause complete loss of enzyme function, whereas the later-onset form is a single amino acid substitution, a less drastic change in the protein.

PCR assays have been developed for all three mutations (Figure 1-44). The four-base insertion in exon 11 is detected by PCR amplification of the region including the mutation, cleavage with *HaeIII*, and polyacrylamide gel electrophoresis. Normally, a 43–base pair fragment is seen, but the insertion changes the fragment size to 47 base pairs. Moreover, in individuals heterozygous for this mutation, undigested PCR products form heteroduplex molecules that migrate very slowly. The heteroduplexes consist of a wild-type strand base-paired with a mutant strand and form during the final PCR cycle. Two types of heteroduplex molecules form between one or the other mutant and wild-type DNA strands. Base pair mismatch at the four-base insertion site creates a bubble, and this leads to slow migration in the gel.

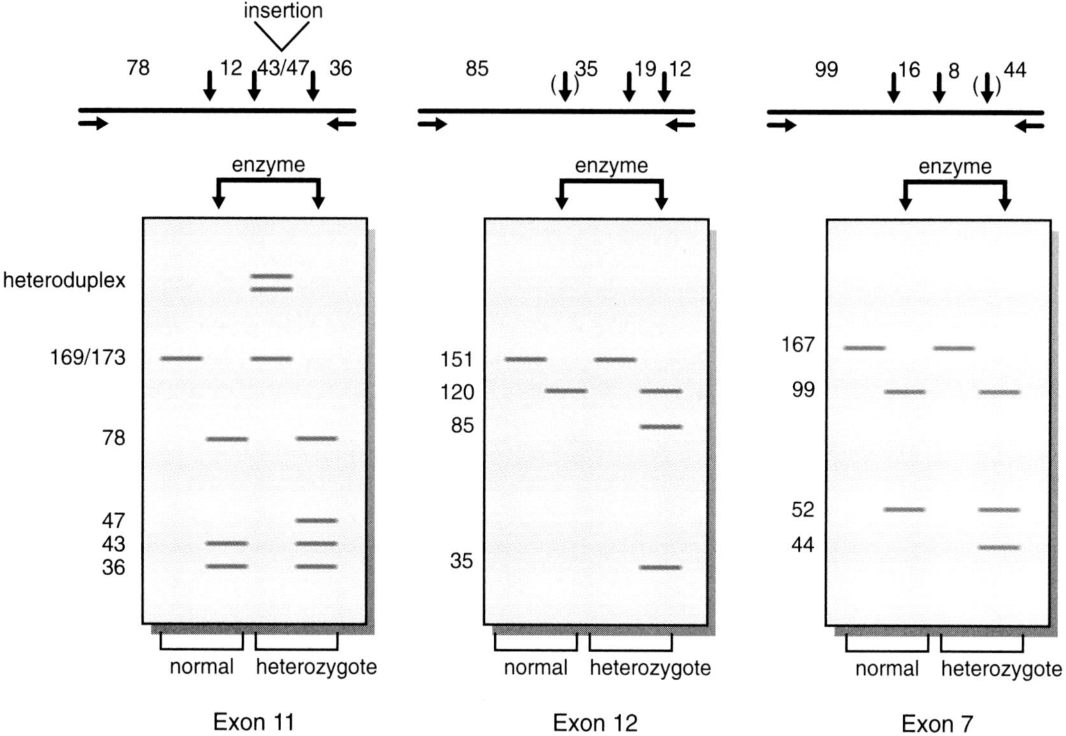

FIGURE 1-44 Identification of Tay-Sachs mutants by PCR. The diagrams at the top show the location of PCR primers (*horizontal arrows*) and restriction enzyme recognition sites (*vertical arrows*). The numbers indicate the sizes of restriction digest fragments. In the electrophoresis gel diagrams, the lanes marked at the top contain samples of PCR products digested with the appropriate restriction enzyme, whereas the other lanes contain nondigested DNA after PCR. The molecular weights of fragments in the gels are indicated by numbers to the left of the gels. Fragments smaller than 20 bases cannot be seen in these gels. (Data from Triggs-Raine BL, Feigenbaum ASJ, Natowicz M, et al. Screening for carriers of Tay-Sachs disease among Ashkenazi Jews. N Engl J Med 1990;323:6–12.)

The intron 12 mutation creates a new recognition site for the enzyme *DdeI*, cutting a fragment of 120 bases into two fragments of 85 and 35 bases each. The exon 7 point mutation creates a new *EcoRII* site, reducing a 52-base band to 44 bases. Using these three simple PCR assays, approximately 98% of Ashkenzi Jewish Tay-Sachs carriers can be detected. The assay is inexpensive to perform and is much less prone to false positive results than is the biochemical assay. Currently, it is used to clarify equivocal results of biochemical screening. Although molecular analysis has a negligible false positive rate, a false negative rate of 2% limits its use as a primary means of population screening.

1.11 Treatment of Inborn Errors of Metabolism

Katie is now 7 years old. She is in second grade and is doing very well in school. She loves to read, although she must hold a book close to her eyes. She wears glasses and sits near the front of the classroom. Katie has learned to avoid exposure to direct sunlight, and she is very conscientious about wearing sunscreen. Her general health is excellent.

Treatment of OCA is aimed at amelioration of symptoms. There is currently no way to replace the missing enzyme in all melanocytes or to replace the gene. With avoidance of sun exposure, however, life expectancy can be normal. The greatest physical challenge faced by individuals with albinism is visual impairment. Here, too, treatment is available to help deal with the problem.

Other instances of treatment of inborn errors of metabolism have already been mentioned. A prime example is dietary treatment of PKU. Restriction of phenylalanine intake in children with PKU has essentially eradicated the mental retardation associated with this disorder. The basis for treatment of PKU is avoidance of a substrate that cannot be metabolized owing to a hereditary lack of the necessary enzyme. Many other metabolic disorders are subject to similar modes of treatment, and newborns are screened for some of the more common ones.

Organ transplantation has been used to treat some metabolic disorders. Liver transplantation has been performed for some children with blocks in the urea cycle, for example. Advances in surgical technique and immunosuppression have vastly improved the safety and success of liver transplantation. A major obstacle currently is the relative scarcity of livers available for transplant.

Likewise, bone marrow transplantation has been used to treat some lysosomal storage diseases. The rationale is that macrophages from transplanted bone marrow function as scavengers to engulf and digest membrane components that cannot be digested by the enzyme-deficient host cells. This has been tried with a set of disorders known as *mucopolysaccharidoses*. These disorders lead to buildup of proteoglycans (protein molecules with attached sugars) and are manifested clinically as progressive neurologic deterioration, growth failure, skeletal deformity, coarse facial features, and heart failure. Reduction in circulating mucopolysaccharide has been demonstrated in transplant recipients, but it has been difficult to document major clinical improvement. One problem with this mode of treatment is that transplanted reticuloendothelial cells have limited access to the brain, which is the site of the most significant pathology. The procedure remains experimental but is the subject of intense study.

Because the pathogenesis of inborn errors of metabolism is based on deficiency of an enzyme, enzyme replacement would seem a reasonable therapeutic approach. Most attempts have been unsuccessful, however. Administration of a foreign protein is difficult as it must be provided parenterally (a protein would be digested if taken orally). The half-life of an injected protein usually is short, and there is a risk of stimulating an immune reaction. Finally, it is difficult to target an injected foreign enzyme to the tissue or cells in which it is normally active.

Substantial progress toward enzyme replacement has been made in Gaucher disease, a lysosomal storage disorder in which the enzyme glucocerebrosidase is absent. Macrophages become swollen with storage material and cause enlargement of the liver and spleen. In addition, presence of engorged macrophages in bones leads to bone and joint pain. The missing enzyme can be purified in quantity from placenta, but initial efforts at treatment of patients by enzyme infusion were disappointing. It then was found that there are mannose receptors in the macrophage's cell membrane that are responsible for targeting mannose-containing substances for the lysosome. Chemical modification of purified glucocerebrosidase to expose mannose residues leads to efficient transport of the enzyme into the lysosomes of macrophages. Infusion of this modified enzyme, referred to as *alglucerase*, has been demonstrated to be highly effective in reducing the size of the liver and spleen and reducing bone pain.

The development of alglucerase represents an important technical achievement but also tells a story about the economics of new drug development. The product was developed—and is produced—by one company, Genzyme, Inc. The dosage recommended initially was 60 units/kg of body weight administered every week to severely affected patients and every other week to others. The drug is sold at $3.50 per unit, so the cost of treating a 30-kg child every other week is $163,800 per year. The cost would increase as the child grows, and treatment would need to continue for the patient's entire life. This results in major problems of access to the medication, particularly for those who may not have excellent medical insurance. In part, the high cost reflects the drug company's need to recover its very high costs of developing the new medication. The potential market for alglucerase is small because Gaucher disease is rare, and therefore the cost per dose is high. Recent studies have indicated that smaller doses can be given more often than once weekly and can result in reduction of the cost per patient.

Perspective

Living with Albinism

LINDA MARUCCI

I grew up in a family where I was the second child to be born with albinism. My brother, who is 3 years older, also has the condition. We both have white hair and poor vision, and we both hid in the house on hot, sunny summer days.

I didn't feel different or alone at home, and my parents had already dealt with their own grief about having an albino child by the time I was born. We also seemed to be doing much better than many of the doctors had predicted when we were very young. My parents had encountered many doctors who gave them a grim picture of what life would be like for a child with albinism. Some even made them feel we might be less intelligent than other children. Most of these doctors knew little about albinism, and I still find today that there are doctors who provide incorrect information or who are afraid to say that they just don't know. This is much less the case today, but it can be devastating for patients and families when it does happen.

Another type of misconception we had to deal with as children was the kind generated by superstition, such as "albinos can see in the dark," "albinos have pink eyes," and "albinos are good or bad luck." Probably the most distressing was "albinos die young." I didn't know what to believe as a child. Even when my parents would reassure me, I thought they were just protecting me.

Although things could be a struggle, I had real friends and people who loved me, and I did the same things my friends did most of the time. My parents had to hold their breath when my brother or I learned to ride a bicycle, played softball, or crossed a busy street, but we all made it through. The more obvious problems were poor vision and repeated sunburn. There were many visits to eye and skin specialists, and my brother and I hated these appointments. Most of these doctors were very kind and wanted to be helpful, but it was difficult to get small children to participate in eye exams. Eye doctors would also become disappointed at how little our vision could be corrected, and that made us feel sad. The vision I had was all I had ever known, so I didn't feel a great loss until someone started measuring it. Reading glasses and sunglasses were important, though. We worked it out so that glasses and visual aids would maximize the vision we had and learned not to expect dramatic improvements. This is still the approach I use when coping with my visual limitations. I'm always open to new ideas about improving my sight, but my main concern is to read a book more comfortably or to read the bathroom scale myself.

Then there were the dermatology appointments. When I was about 7 years old, we were invited to participate in the testing of a new sunscreen (PABA). We were very eager to try anything that would really work as protection from sunburn, since until this time nothing had. We either dressed in long sleeves and long pants or avoided the sun during the hottest part of the day. We went for many appointments to the dermatology department, and they took samples of our hair and skin. We didn't like this, but it was explained that it was important. Then groups of residents were brought in to examine us. They stood around talking about the type of albinism we have and looked into our eyes with penlights but never acknowledged us as people. Nobody spoke *to* me, only *about* me. The whole process made me feel anxious and uncomfortable, and now it makes me angry. Still, I am grateful to these people because, without their dedication, PABA would never have been available.

As an adult, I saw a geneticist who helped me to understand albinism and how genes are passed through families. This information made me understand that the albinism gene was in my family for many generations and was not due to some mistake of mine or my parents. I had known that this was true rationally but still had felt guilty for who I was. I also learned about the risk of having children affected with albinism, and this gave me the courage to have my wonderful and beautiful daughter, who is unaffected.

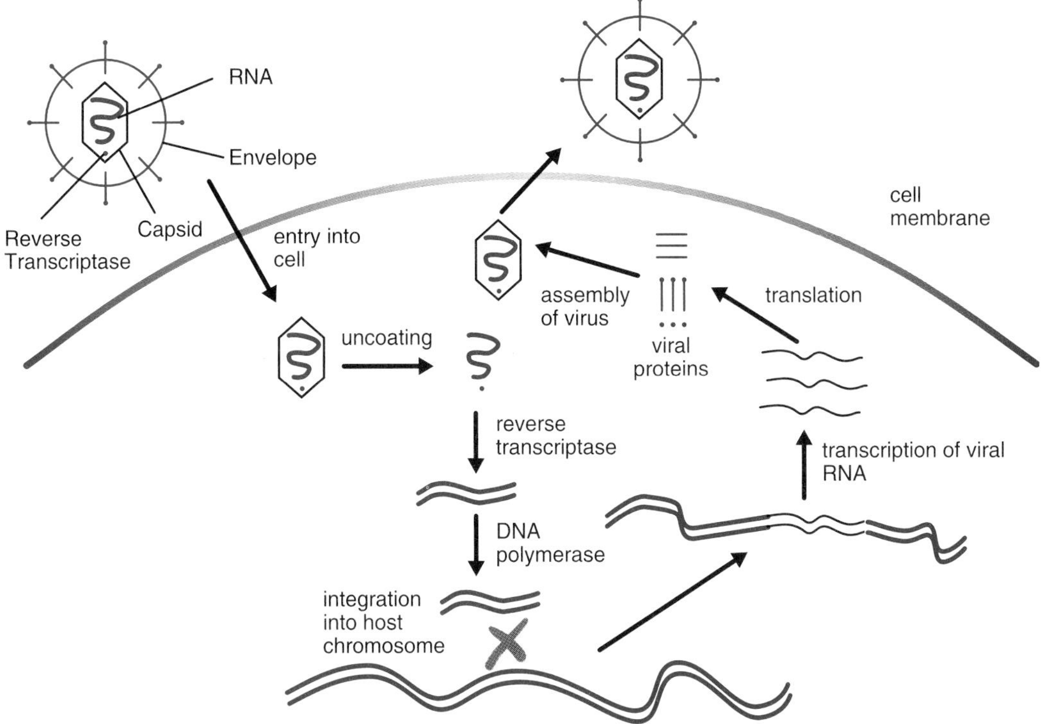

FIGURE 1-45 Life cycle of a retrovirus. The virus sheds its coat as it enters a cell. The viral RNA is made into DNA with reverse transcriptase found inside the viral particle. The viral cDNA is inserted into the host DNA, where it is both replicated and transcribed. The transcripts direct the synthesis of new viral proteins, and the viral RNA then is packaged into new particles that are released from the cell.

The high cost and relative inconvenience of alglucerase infusion has stimulated interest in gene replacement therapy for Gaucher disease. The cDNA for glucosylcerebrosidase has been cloned, and the sequence has been inserted into a retroviral vector. Retroviruses are single-stranded RNA viruses that infect eukaryotic cells (Figure 1-45). After gaining entry into the cell, the viral RNA is transcribed into DNA by reverse transcriptase. The viral DNA then is integrated into the host chromosome, where the viral genes are transcribed. Moreover, the integration is stable, so the viral genome is replicated each time the host cell replicates.

The glucocerebrosidase vector was created from a retroviral sequence from which viral genes were excised and the glucocerebrosidase cDNA was inserted along with a gene that confers resistance to the antibiotic G418 ("neo" gene) (Figure 1-46). Cells were infected with recombinant virus. Because the recombinant virus is devoid of many of the genes necessary for viral replication, the cells also were infected with a *helper virus*, a retrovirus that can synthesize proteins necessary for viral replication but that contains a mutation rendering it unable to package its own genome into new virus. Recombinant virus produced by these cells was used to infect macrophages from patients with Gaucher disease. Infected macrophages were selected by growth in the presence of G418. The transduced cells have been shown to make essentially normal quantities of glucocerebrosidase. The experiment has been done successfully with both mature macrophages and with bone marrow stem cells.

Some technical obstacles still must be overcome before gene transfer can be tested in treating a patient with Gaucher disease. These include improvement of the efficiency of gene transfer and prevention of overgrowth of the marrow by nontransfected cells. Nonetheless, the approach remains a very promising hope for the future, both for Gaucher disease and perhaps for other inborn errors of metabolism.

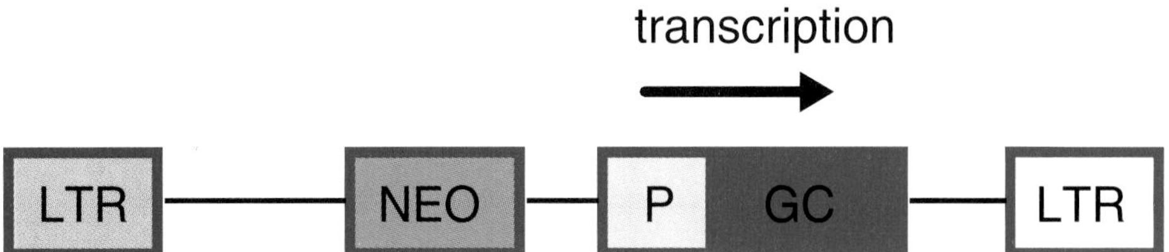

FIGURE 1-46 Recombinant retrovirus containing the glucocerebrosidase gene (GC) adjacent to a promoter (*P*). The LTR sequences are normally part of the viral genome. (Data from Fink JK, Gorrell PH, Perry LK, et al. Correction of glucocerebrosidase deficiency after retroviral-mediated gene transfer into hematopoietic progenitor cells from patients with Gaucher disease. Proc Natl Acad Sci USA 1990;87:2334–2338 and Nolta JA, Yu XJ, Bahner I, Kohn DB. Retroviral-mediated transfer of the human glucocerebrosidase gene into cultured Gaucher bone marrow. J Clin Invest 1992;90:342–348.)

Case Study

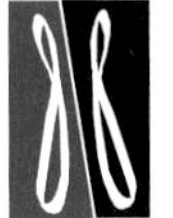

Part I **December 1965**

Jocelyn is born after a full-term, uncomplicated pregnancy. Her birth weight is 3100 g, and she is a healthy and vigorous baby. She is being breast-fed and seems to be feeding well. A few drops of blood are taken from her heel on day 3 of life, just prior to her being discharged from the nursery with her mother (Figure 1-47); the blood is blotted onto a card and sent to the state laboratory. Her parents are told that this is a routine test done for all newborns, and so they put it out of their minds. They take Jocelyn home, and she continues to feed well and seems to be thriving. After 1 week, however, they receive a call from their pediatrician, who explains that an abnormal laboratory result has come back (Figure 1-48). At first Jocelyn's parents are mystified, unaware that any laboratory test had been done. The pediatrician then reminds them about the heel stick. He has arranged for the family to take Jocelyn to a special clinic at the nearby children's hospital.

Screening of newborns for several inborn errors of metabolism is standard throughout the United States and in most of the developed world. The rationale is that identification of infants with inborn errors can lead to institution of treatment prior to the onset of clinical signs of the disorder. This can prevent otherwise irreversible neurologic damage and other medical problems.

Screening generally is performed by obtaining a blood sample by heel stick just before the baby is discharged from the nursery. During fetal life, the disorder is masked because the buildup of toxic compounds due to enzyme block is prevented by clearance through the placenta. PKU will be evident by biochemical testing soon after birth, however.

TESTING LAB
☐ Female ☐ Male
181588
Baby's Last Name
Ethnicity/Race:
Premature: ☐ Yes ☐ No
Antibiotics: ☐ Yes ☐ No
Transfusion: ☐ Yes ☐ No
Birth Date — Date — Time — A.M. P.M.
Specimen Collected → Date — Time
Mother's Name — Mother's Age
Address
City/State/Zip — County
SUBMITTING HEALTH CARE PROVIDER
(Results and Bills For Unpaid Charges Will Be Sent To Submitter)
Name
Address
City/State/Zip — County
Phone
Name of Physician if not indicated above
NEWBORN SCREENING RESULTS
(Tests for diseases listed below)
☐ All Tests Within Normal Limits
(normal values on reverse side)
Phenylketonuria
Homocystinuria
Maple Syrup Urine Dis.
Galactosemia
Hypothyroidism
Hemoglobinopathy
☐ Unsatisfactory
COMPLETELY FILL ALL CIRCLES WITH BLOOD
SOAK THRU FROM ONE SIDE ONLY
DO NOT USE CAPILLARY TUBES

FIGURE 1-47 Spots of blood on filter paper card to be sent to state laboratory.

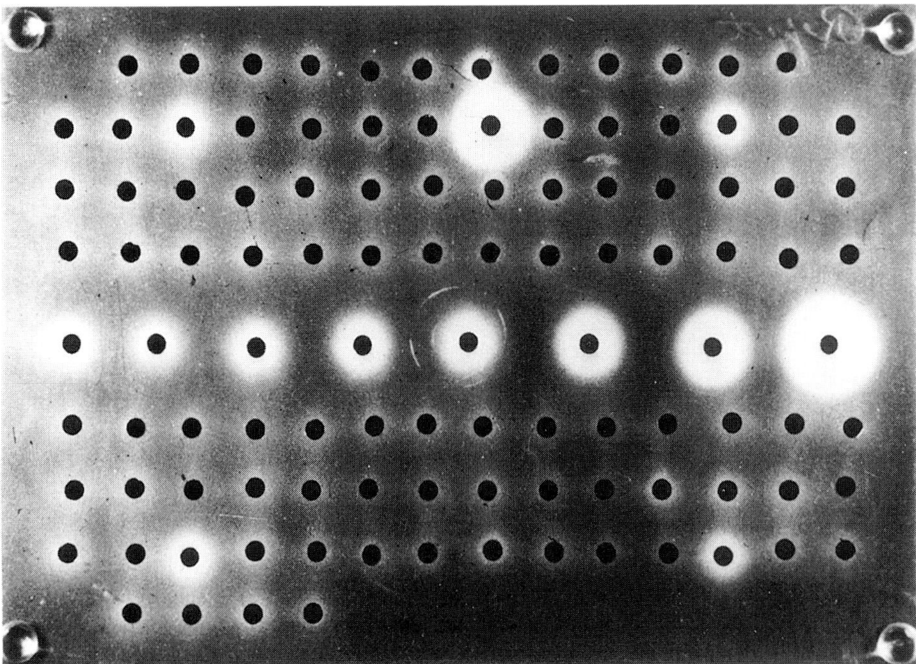

FIGURE 1-48 Guthrie plate for screening phenylketonuria. Each dark circle represents a disc of a blood spot from a newborn. The fifth row contains controls, with increasing quantities of phenylalanine from left to right. This results in a zone of bacterial growth of increasing size. The eighth sample in the second row is from a newborn with phenylketonuria, showing a large zone of bacterial growth. (Courtesy of Dr. Harvey Levy, Children's Hospital, Boston, and Massachusetts State Laboratory.)

Metabolic disorders commonly included in screening tests are PKU, maple syrup urine disease (disorder of branched-chain amino acid catabolism), galactosemia (disorder of galactose catabolism), and homocystinuria. In some areas, newborns also are screened for biotinidase deficiency (lack of enzyme required to recycle the cofactor biotin), sickle cell disease, congenital hypothyroidism, congenital adrenal hyperplasia, and congenital toxoplasmosis. The criteria for screening for a particular disorder are as follows: 1) The disorder produces irreversible damage if untreated early in life; 2) treatment prevents the damage but only if begun in the newborn period when the infant is asymptomatic; and 3) the natural history of the disease is known.

The PKU screening test commonly is done using a bacteriologic assay. Discs are punched out from the filter papers on which blood samples are collected and are placed in wells in an agar plate onto which a confluent lawn of bacteria has been inoculated and which includes an analog of phenylalanine that inhibits bacterial growth. Phenylalanine present in the blood samples diffuses out of the filter paper into the agar and, by competing with the analog, stimulates a zone of bacterial growth. The higher the concentration of phenylalanine, the larger is the zone of growth. The concentration in a particular blood sample can be estimated by comparing the size of the zone to a standard of known concentration on the same plate.

When a sample is found to have a high concentration of phenylalanine, the baby's physician is notified. The child is referred to a specialty clinic, where a sample of blood is drawn for quantification of phenylalanine. A value greater than 20 mg/dl is indicative of classic PKU. Some newborns have phenylalanine values in an intermediate range (7–20 mg/dL), which is indicative of atypical or mild PKU. Either form of PKU is treated by reducing phenylalanine intake. Elevated levels of less than 7 mg/dL correspond with non-PKU benign hyperphenylalaninemia and require no treatment. Benign hyperphenylalaninemia mutations are missense mutations. PKU mutations also are usually missense mutations but might be mutations that lead to deficient production of protein (e.g., stop codons, splicing mutations, deletions).

Part II **December 1965**

The next day, Jocelyn is taken to the metabolism clinic. Her parents are told that the laboratory result indicates that Jocelyn probably has phenylketonuria. Jocelyn is examined and the results are normal. Both blood and urine specimens are obtained. Jocelyn's parents are informed about PKU and, in particular, taught about the phenylalanine-free diet (Figure 1-49). Jocelyn will be fed with a special low-phenylalanine formula. Her parents are reassured that Jocelyn can be expected to achieve essentially normal cognitive development if this diet is adhered to. Nevertheless, they are in a state of shock, and are frightened. They spend considerable time talking with the clinic social worker and are introduced to other parents who have children with PKU. This somewhat reassures them. The next day, they get a call from the clinic nurse, telling them that Jocelyn's phenylalanine level was 25 mg/dL (the normal level being <2 mg/dL), confirming the diagnosis. Two weeks later, the urine pterin analysis is reported as normal, indicating that Jocelyn has classic PKU and not a cofactor deficiency in which the increased phenylalanine would be a secondary finding.

The frequency of PKU is approximately 1 in 10,000 births. Prior to the advent of newborn screening and treatment, PKU was one of the most common causes of mental retardation. The enzyme responsible for the disorder is phenylalanine hydroxylase, which catalyzes the hydroxylation of phenylalanine to tyrosine (Figure 1-50). In the absence of this reaction, phenylalanine builds up to high levels. Phenylalanine is believed to be toxic at high concentrations. A major phenylalanine metabolite, phenylpyruvic acid, may also be toxic. The major target for toxicity is the nervous system. Also, there is relative deficiency of tyrosine, the precursor for the neurotransmitters dopamine and norepinephrine. A deficiency of these neurotransmitters is likely to be deleterious for the developing brain. Children with untreated PKU exhibit profoundly delayed cognitive development. They also tend to have fair hair and skin owing to relative deficiency of melanin (normally derived, in part, from tyrosine).

Although the majority of those with PKU are affected owing to mutations in the phenylalanine hydroxylase gene, a few have mutations in a different

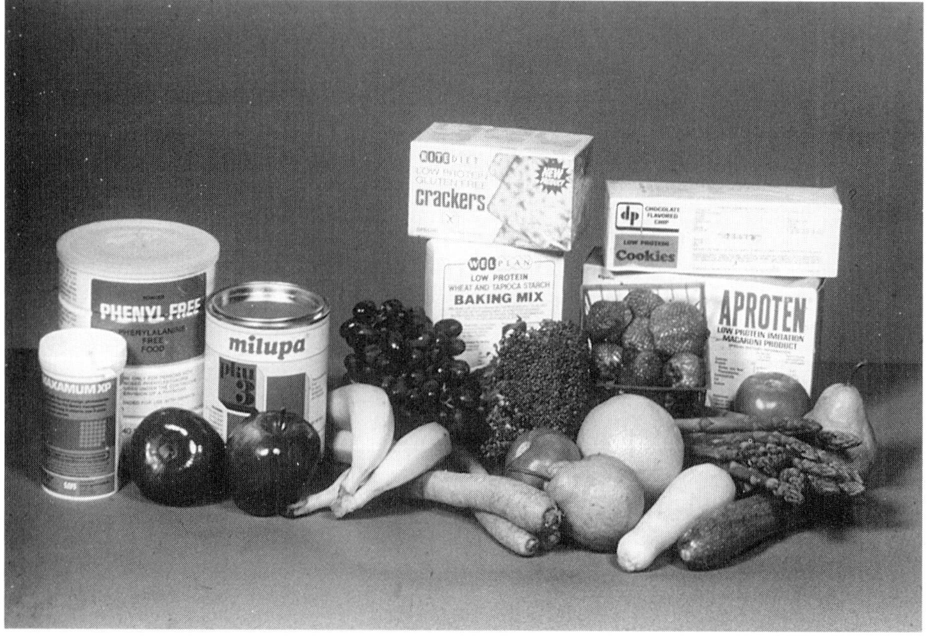

FIGURE 1-49 Example of foods permitted in phenylalanine-restricted diet.

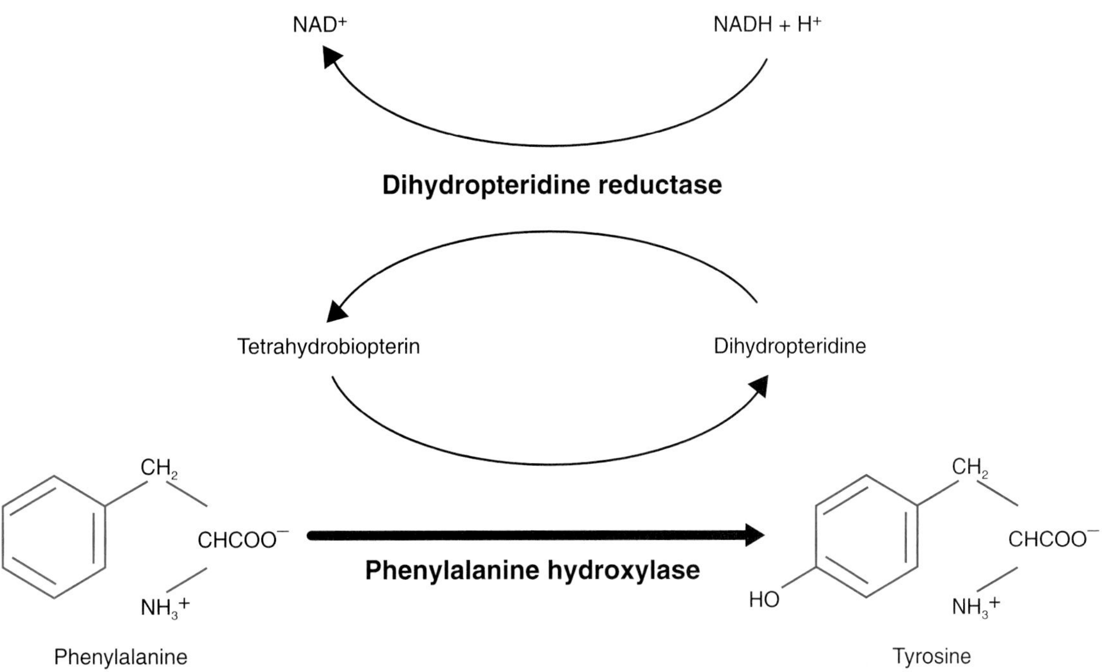

FIGURE 1-50 Pathway of phenylalanine metabolism.

enzyme, dihydropteridine reductase (DHPR), or in the synthesis of biopterin. DHPR is involved in the reduction of H2 biopterin into H4 biopterin; H4 biopterin is a cofactor required for the hydroxylation of phenylalanine to tyrosine. It also is required for hydroxylation of tyrosine and tryptophan, which are critical in the synthesis of vital brain neurotransmitters. Absence of DHPR activity results in deficient phenylalanine hydroxylation in the presence of normal phenylalanine hydroxylase enzyme and also in deficient neurotransmitter production. The diagnosis of DHPR deficiency is made by a blood filter paper assay. Defects in the synthesis of biopterin, which produce the same effects as DHPR deficiency, are identified by a urine assay of biopterin and neopterin.

Treatment of PKU is based on restriction of dietary intake of phenylalanine. The major protein source is a preparation of individual amino acids excluding phenylalanine and containing supplemental tyrosine as well as carbohydrate, fat, and minerals. Low-protein foods are allowed in measured quantities, but foods that contain significant amounts of protein (e.g., meat, fish, cheese, ice cream) are prohibited. Compliance with dietary treatment is monitored by testing blood phenylalanine levels. The goal is to maintain the level at less than 7 mg/dL, an amount considered safe for normal development. DHPR deficiency and biopterin synthesis are treated with H4 biopterin and supplements of neurotransmitter precursors (dopa, carbidopa, and 5-OH-tryptophan), perhaps in conjunction with a low-phenylalanine diet.

The care of a child with PKU imposes many demands on a family to maintain compliance with the diet. It is best to provide care in a multidisciplinary clinic, where a physician, a nutritionist, and a social worker, as well as (in some cases) other professionals might be involved. The family is likely to require considerable education and support.

Part III **April 1973**

Jocelyn is now 7 years old. She is in first grade and doing well in school. Increasingly, however, she is rebelling against her special diet, mainly because it is not pleasant-tasting, and she sees others her age eating tastier foods at school. After discussion with the metabolism clinic staff, her parents are told that strict adherence to the diet is less important by this age, because the brain is now fully developed. They take this as a sign that Jocelyn's diet can be liberalized. Over time, Jocelyn is less and less compliant, and gradually the family drifts away from the clinic.

The first few years of life are a critical time in terms of brain development. Compliance with the low-phenylalanine diet is especially important during this time. Formerly, it was considered safe to relax the diet later in childhood. The current view is that dietary relaxation during childhood or even in adolescence can lead to cognitive loss and emotional problems. Consequently, most centers now recommend at least some degree of compliance with a low-phenylalanine diet throughout life.

Part IV **March 1986**

Jocelyn is now 20 years old. She has completed high school, although her grades were quite poor. She has also had a history of psychological and behavioral problems. Jocelyn has just given birth to a baby boy. His birth weight is only 2000 g, although he was born at term. He is microcephalic and has a congenital heart defect (tetralogy of Fallot). His physicians are optimistic that they can repair the cardiac problem but are concerned that microcephaly predicts that his cognitive development will be delayed. His blood phenylalanine level is in the normal range at 2 weeks of life. He is being breast-fed. There is no history of PKU in his father's family. Jocelyn is afraid and confused. She is not married and relies on her parents for emotional support.

There are many possible causes of the birth of a child with microcephaly (small head size) and a congenital heart defect. In the child of a woman with PKU, however, by far the leading cause is a syndrome called *maternal PKU*. If the mother has not adhered to the low-phenylalanine diet, there will be high levels of phenylalanine in the fetal environment owing to the mother's defect, and the fetal phenylalanine hydroxylase activity will not be high enough to deal with this load. The consequence is an ironic outcome of treatment of a genetic disorder—toxicity to the next generation. In this case, the effects of high phenylalanine are particularly devastating. The fetus is exposed at a time when major organ systems, including the brain, are developing rapidly. It is common for these infants to have low birth weight, congenital heart disease, and inadequate brain development at the time of birth.

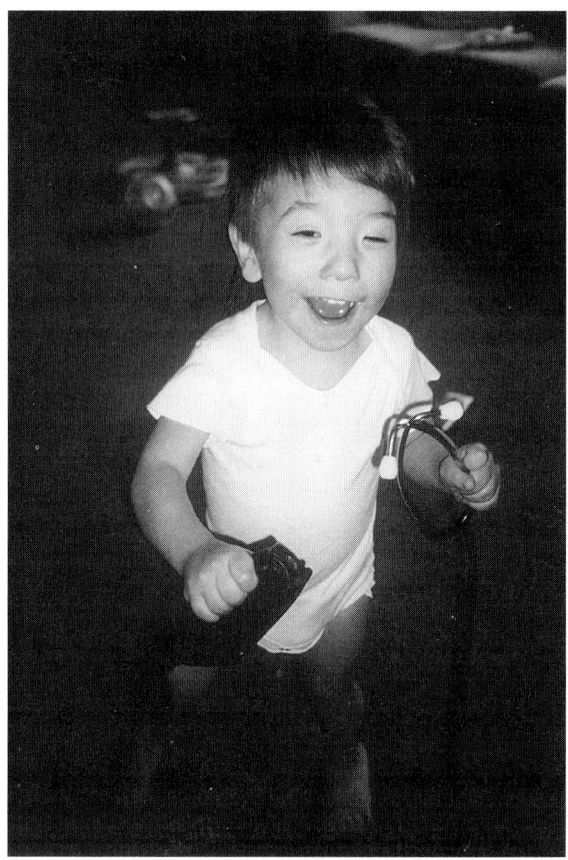

FIGURE 1-51 Child with dysmorphic facial features (broad nasal bridge, small nose) and microcephaly due to in utero effects of maternal phenylketonuria. (Courtesy of Dr. Harvey Levy, Children's Hospital, Boston.)

Epilogue **September 1988**

Jocelyn is pregnant again, by a different partner. Her son is now 2½ years old (Figure 1-51), and Jocelyn and the boy live with Jocelyn's parents. His heart defect has been surgically repaired, but he is severely developmentally impaired. He began walking only recently, and he is not yet talking. For this pregnancy, Jocelyn was counseled to start a phenylalanine-restricted diet prior to trying to conceive and to continue the diet throughout the pregnancy. Her compliance, which has been good, has been monitored by following blood phenylalanine levels. She is reassured that fetal damage from maternal PKU is very unlikely to occur in this pregnancy.

Maternal PKU is largely a preventable disease. The woman with PKU should be placed on a low-phenylalanine diet prior to the time of conception. The rationale for preconceptual treatment is that major events in organogenesis occur at a point very early in pregnancy when a woman may not realize that she is pregnant. Preconceptual treatment and continued close adherence to the diet throughout the pregnancy, on the other hand, have been shown to prevent the teratogenic effects of maternal PKU syndrome. Major efforts now are under way to identify women of childbearing age with PKU, to educate them regarding the need to maintain the diet, and to monitor them throughout pregnancy.

REVIEW QUESTIONS

1. A couple is seen for genetic counseling regarding the risk of having a child with Gaucher disease. Gaucher disease is recessively transmitted. The man has a brother who is affected, whereas the woman has no family history of the disorder. Both are of Ashkenazi Jewish background. Assuming a carrier frequency of 1 in 50 in this population, what is the risk to this couple of having an affected child? Draw a pedigree of the family.
2. A cDNA is cloned that corresponds to a protein believed to be responsible for a recessive genetic disorder. When restriction enzyme–digested DNA from a number of individuals with the disorder is hybridized with the probe on a Southern blot, the following pattern is seen:
   a. Why are four bands seen in most lanes?
   b. What might account for the difference in lane 3?
   c. Is this change a pathogenic mutation?

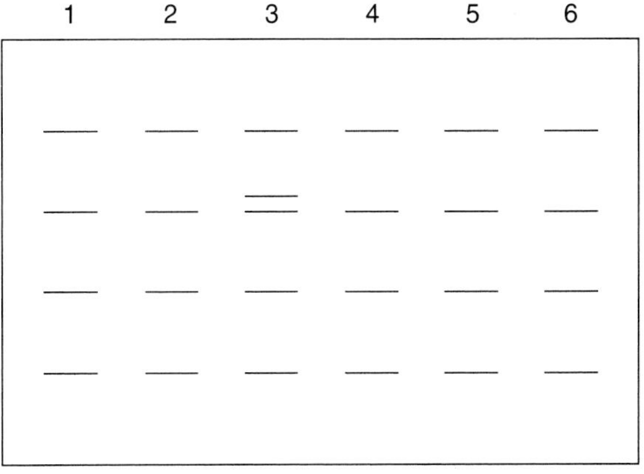

3. A PCR assay is set up for a recessive disorder, in which a region flanking a common mutation is amplified and the amplification products from each individual are applied to a nylon membrane. One membrane then is hybridized with DNA homologous to wild-type-labeled oligonucleotide and one to mutant oligonucleotide. Identify the genotypes of the individuals in the figure:

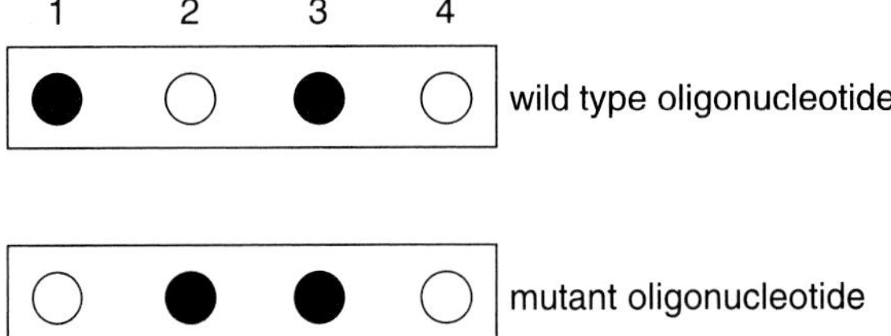

4. What would be the likely impact of the mutation shown on the expression of the protein?

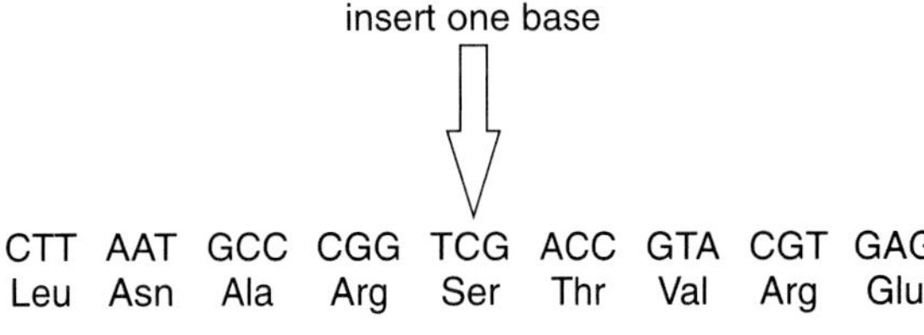

FURTHER READING

1.7 Tyrosinase Mutations

Spritz RA, Strunk KM, Giebel LB, King RA. Detection of mutations in the tyrosinase gene in a patient with type IA oculocutaneous albinism. N Engl J Med 1990;322:1724–1728.

1.8 Genetic Heterogeneity in Albinism

Oetting WS, Mentink MM, Summers CG, et al. Three different frameshift mutations of the tyrosinase gene in type IA oculocutaneous albinism. Am J Hum Genet 1991;49:199–206.

Spritz R, Strunk KM, Hsieh C-L, et al. Homozygous tyrosinase gene mutation in an American black with tyrosinase-negative (type IA) oculocutaneous albinism. Am J Hum Genet 1991;48:318–324.

1.9 Genotype-Phenotype Correlations in OCA

Giebel LB, Tripathi RK, Stunk KM, et al. Tyrosinase gene mutations associated with type IB ("yellow") oculocutaneous albinism. Am J Hum Genet 1991;48:1159–1167.

Giebel LB, Tripathi RK, King RA, Spritz RA. Tyrosinase gene missense mutation in temperature-sensitive type I oculocutaneous albinism. A human homologue to the Siamese cat and the Himalayan mouse. J Clin Invest 1991;87:1119–1122.

King RA, Townsend D, Oetting W, et al. Temperature-sensitive tyrosinase associated with peripheral pigmentation in oculocutaneous albinism. J Clin Invest 1991;87:1046–1053.

1.11 Treatment of Inborn Errors of Metabolism

Barton NW, Brady RO, Dambrosia JM, et al. Replacement therapy for inherited enzyme deficiency—macrophage targeted glucocerebrosidase for Gaucher's disease. N Engl J Med 1991;324:1464–1470.

Beutler E. Gaucher disease: new molecular approaches to diagnosis and treatment. Science 1992;256:794–799.

Figueroa ML, Rosenbloom BE, Kay AC, et al. A less costly regimen of alglucerase to treat Gaucher's disease. N Engl J Med 1992;327:1632–1636.

CASE STUDY

Koch R, Levy HL, Matalon R, et al. The North American Collaborative Study of Maternal Phenylketonuria: status. Am J Dis Child 1993;147:1224–1230.

Koch R, Levy HL, Matalon R, et al. The international collaborative study of maternal phenylketonuria: status report 1994. Acta Paediatr Suppl 1994;407:111–119.

Lenke RR, Levy HL. Maternal phenylketonuria and hyperphenylalaninemia. N Engl J Med 1980;303:1202–1208.

Matalon R, Michals K, Azen C, et al. Maternal PKU collaborative study: pregnancy outcome and postnatal head growth. J Inherit Metab Dis 1994;17:353–355.

Platt LD, Koch R, Azen C, et al. Maternal phenylketonuria collaborative study, obstetric aspects and outcome: the first 6 years. Am J Obstet Gynecol 1992;166:1150–1160.

Rohr FJ, Doherty LB, Waisbren SE, et al. New England Maternal PKU Project: prospective study of untreated and treated pregnancies and their outcomes. J Pediatr 1987;110:391–398.

2 Structural Gene Mutations

OVERVIEW

Genes encode the structure of proteins, but proteins can serve a wide variety of roles. One role—catalysis of biochemical reactions by enzymes—has already been considered. Other proteins play a structural role, making up parts of cells or being secreted into the extracellular matrix. Less is known about these proteins and the genes that encode them than about enzymes. This is changing rapidly, however, as the tools of molecular biology are being applied to the study of structural proteins. Molecular biologists are learning that mutations in genes for structural proteins often result in dominantly transmitted disorders—that is, disorders expressed in heterozygotes. In part, this may be because structural proteins tend to interact with one another. The structure may be critically weakened if 50% of a protein is defective; the result may be an abnormal phenotype.

This chapter will explore the genetic and biochemical mechanisms of structural gene mutations. First we will consider the history of a child with a disorder of bones, osteogenesis imperfecta. Osteogenesis imperfecta is transmitted as an autosomal dominant trait, so the principles of this mode of inheritance will be reviewed. The basis for osteogenesis imperfecta was initially worked out biochemically, revealing deficient production of the protein collagen.

Collagen is the major structural protein of the body and actually represents a complex family of subunits that aggregate in different combinations in different tissues. Understanding the genetic basis for osteogenesis imperfecta, however, required identification of the genes for the various collagen subunits. We will look at how these genes were cloned and then show how they were found to be abnormal in individuals with osteogenesis imperfecta. We will examine evidence gained from genetic linkage studies and then study the mutations themselves. Ultimate demonstration that a gene mutation is pathogenic requires that the disease be reproduced in an experimental system. In this instance, a collagen gene mutation was introduced into a mouse, resulting in the phenotype of osteogenesis imperfecta. We will see how transgenic mice are produced and how they can be valuable tools for creation of animal models of human disease.

Genetic heterogeneity occurs in osteogenesis imperfecta just as it does in albinism. The diversity of collagen mutations will be considered. Osteogenesis imperfecta also illustrates the rare but important genetic phenomenon of somatic mosaicism—the occurrence of a mixture of wild-type and mutant cells in the same individual. We will review evidence for this phenomenon and its consequences for disease transmission in a family.

Mutations in collagen genes also can cause phenotypes other than osteogenesis imperfecta. We will consider some of these other disorders that stem from collagen gene mutations. Finally, some of the clinical management issues raised by these complex disorders will be discussed. The Perspective describes living with osteogenesis imperfecta. We end the chapter by considering another structural gene disorder.

2.1 Osteogenesis Imperfecta

This is Carol's second pregnancy, so when fetal movements seem to decline during the sixth month, she becomes concerned. An ultrasound examination is performed and reveals that the fetus has short femurs and a poorly ossified skull. There are also fractures of several long bones. No other anomalies are noted, and fetal movements are seen. Carol and her husband, Neal, are referred to an expert in perinatal medicine, who discusses with them the implications of these findings. A problem with bone development seems most likely, although a specific diagnosis is difficult to make prenatally. Arrangements are made for Carol and Neal to meet with an orthopedist, and plans are made for the child to be delivered at a major medical center rather than a community hospital so that supportive care can be initiated promptly. Delivery is accomplished by cesarean section under spinal anesthesia. The baby is a boy, and he is named Lawrence. Holding Lawrence in the delivery room, Carol notices that he has a large head. One of his legs is fractured, and x-rays reveal diffuse undermineralization of the bones and several other fractures (Figure 2-1). The sclerae are bluish, and serum alkaline phosphatase activity is slightly elevated. It is determined that Lawrence has osteogenesis imperfecta.

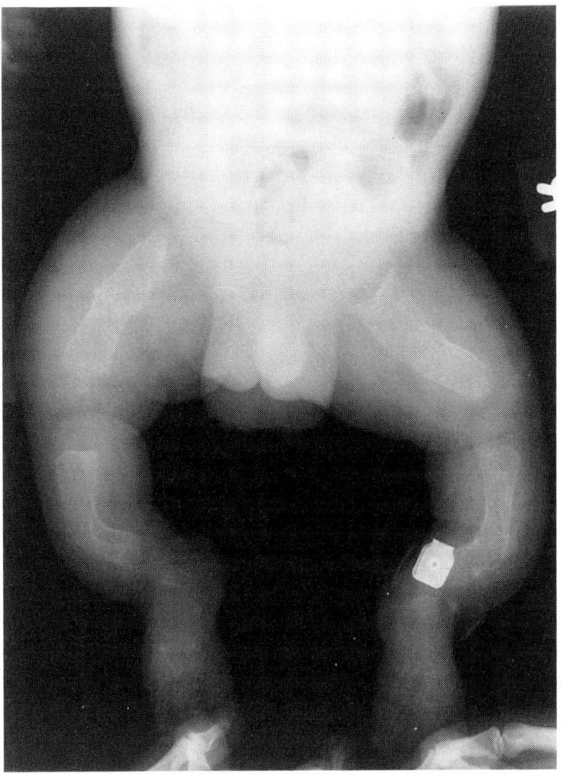

FIGURE 2-1 X-ray of infant with osteogenesis imperfecta, showing multiple fractures.

Osteogenesis imperfecta (OI) is a disorder marked by extreme fragility of bones. Bones consist of a protein matrix, composed largely of collagen, and a mineral component, mainly calcium phosphate. In OI, the bones are undermineralized and fracture easily. Immediate orthopedic management is important to treat congenital fractures, and the child must be protected as well as possible from trauma that can precipitate additional fractures.

Relatively few disorders other than OI are likely to present at birth in this way. One alternative is hypophosphatasia, an autosomal recessive disorder resulting from deficiency of the enzyme alkaline phosphatase. Alkaline phosphatase is necessary to mobilize phosphate stores and make the phosphate available to complex with calcium. Children with hypophosphatasia have defective mineralization of bones, with bowing of long bones and fractures. Those with OI have normal or slightly elevated alkaline phosphatase levels, ruling out hypophosphatasia. It is also common for children with OI to have blue or blue-gray sclerae. This reflects the fact OI is a disorder of connective tissue and therefore affects tissues other than bone. The sclerae are thinner than normal and consequently do not appear white.

At 1 year of age, Lawrence has made good developmental progress. He is alert and sociable and is beginning to say a few words. He has good coordination of his hands but is unable to sit without support. For this reason, a back brace is made to help him to sit. His health has been good, although he has had a number of upper respiratory infections. Lawrence has had numerous fractures of his bones, including both humeri, his right clavicle, and his left femur. He has spent most of his time in casts, which may be impeding his gross motor development. Lawrence feeds well, but his length and weight have grown slowly (both fall below the fifth percentile for age). His skull is fully ossified, but the anterior fontanelle has not yet closed. A few teeth have erupted, which appear gray in color.

Cognitive development is normal in children with OI. The bone defect leads to abnormal growth of the skull and late closure of the anterior fontanelle (the gap between the growing plates of skull bones in the front of the head that usually closes during the first year of life). The major problem faced by these children, however, is repeated bone fracture. Fractures can occur with minimal trauma, even in the course of ordinary events, such as changing diapers. Despite the best efforts to protect the child, fractures are inevitable in children with severe OI. This necessitates frequent casting, which can interfere with motor development.

OI may affect as many as 1 in 10,000 persons worldwide. The manifestations can be different in different individuals but, within a family, the expression is usually similar from person to person. A classification scheme has been devised that includes four clinically distinguishable forms of OI (Table 2-1). These forms differ in the degree of severity of bone fragility, whether the fragility improves with age or whether it leads to progressive bony deformity and short stature, whether blue sclerae are present, and whether teeth are involved (referred to as *dentinogenesis imperfecta*). Hearing loss may occur in any of the forms of OI. This results from defective sound transmission in the middle ear due to abnormality of the bony ossicles.

TABLE 2-1 Sillence classification of osteogenesis imperfecta

Classification	Clinical Features
I	Fractures during childhood, fewer after puberty; normal stature, minimal deformity; blue sclerae; normal teeth; some have hearing loss
II	Severe congenital bone deformities and fractures; usually lethal in perinatal period due to respiratory insufficiency
III	Congenital fractures with progressive growth failure and fractures; abnormal teeth, variable blue sclerae; possible hearing loss
IV	Mild or moderate bone fragility; abnormal teeth; normal or grayish sclerae; possible hearing loss

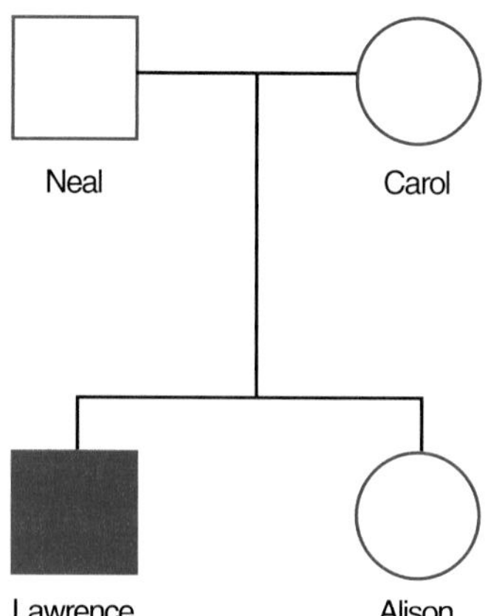

FIGURE 2-2 Family pedigree for Neal and Carol.

2.2 Genetics of Osteogenesis Imperfecta

Carol and Neal ask to speak with a genetic counselor regarding their risk of having another child with osteogenesis imperfecta. Carol is of French Canadian descent, and Neal's ancestry is English. There is no history of osteogenesis imperfecta on either side of the family. Both Carol and Neal are of average stature, and neither has ever had a bone fracture. They have one other child, a 4-year-old girl, Alison, who is in good health and has had no problems with her bones. Carol is employed as a psychologist, and Neal as an engineer (Figure 2-2).

Each of the forms of OI has a genetic basis, and the most common mode of transmission is autosomal dominant. An autosomal dominant trait is expressed in both heterozygotes and homozygotes (Figure 2-3), although the phenotype may be more pronounced in homozygotes. For rare traits, an affected individual is most likely heterozygous for the mutant gene and has a 50% chance of passing that gene to any child (Figure 2-4). Because the gene is not located on the X chromosome, the risk of transmission is the same regardless of which parent carries the mutation and regardless of the gender of the child (Figure 2-5).

In some families, OI is transmitted from generation to generation as a typical autosomal dominant trait. Approximately 50% of the offspring of an affected person inherit the gene and express the trait.

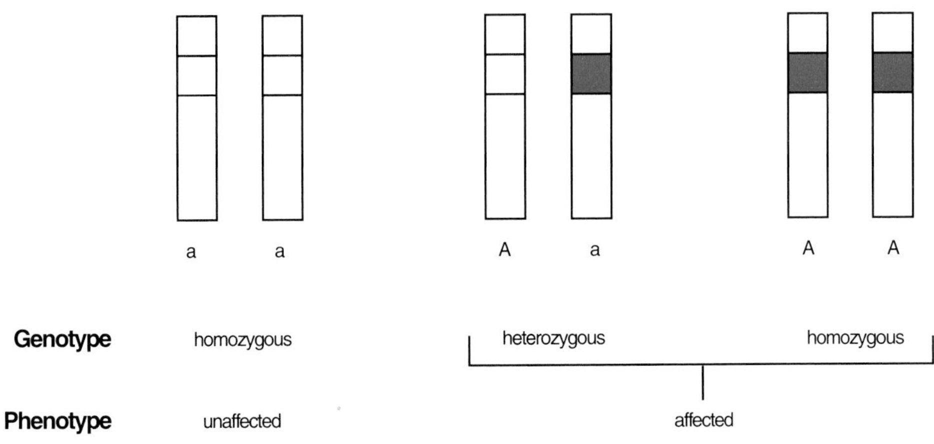

FIGURE 2-3 The phenotype of a dominant allele is expressed in individuals who are either homozygous of heterozygous for the allele.

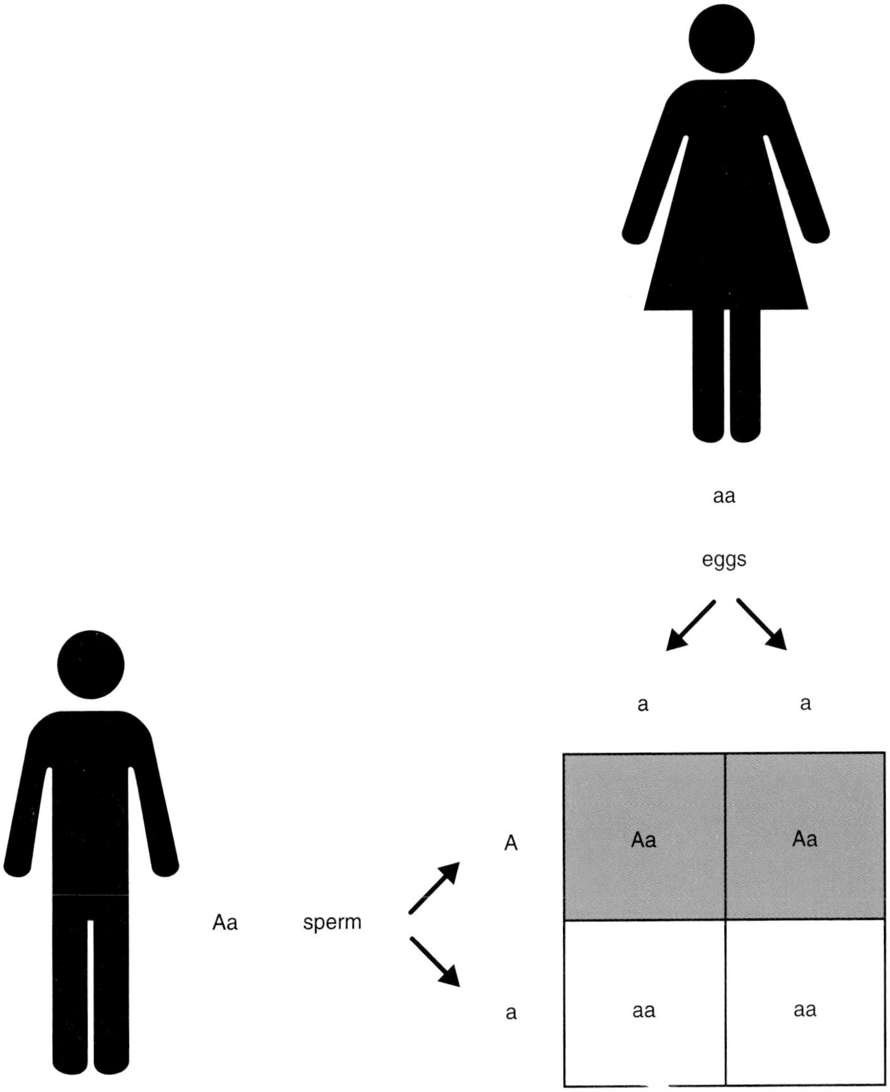

FIGURE 2-4 Individuals who are heterozygous for a dominant allele (the father in this figure) produce equal numbers of germ cells with either the dominant or recessive allele. Half the offspring will likewise be heterozygous.

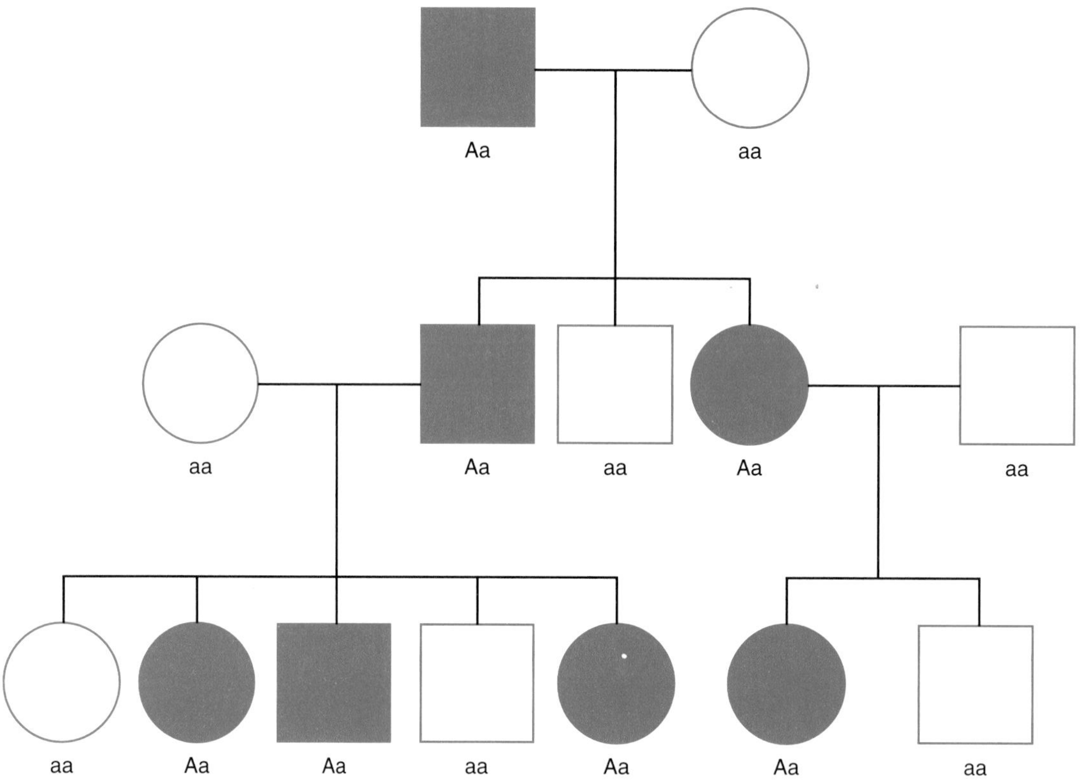

FIGURE 2-5 Pedigree illustrating autosomal dominant transmission. The dominant allele *A* is passed from generation to generation.

Often, however, the disorder appears for the first time in a family in a young child, and no prior history of OI is found, as was the case for Carol and Neal. This can create a dilemma in providing genetic counseling.

Assuming that all cases of OI have a genetic basis, a number of alternative explanations can be offered to account for sporadic occurrence of an affected child. For example, one of the parents of the affected child actually may be heterozygous for an OI mutation, yet for some reason may not display the phenotype of the disorder. This is referred to as **nonpenetrance**, defined as the absence of phenotype in a person known to carry a specific mutant gene. Nonpenetrance has been demonstrated to occur with many genetic traits and can be most easily inferred when a grandparent and child have a disorder that does not appear to be expressed in the parent (Figure 2-6).

In deciding whether a person is nonpenetrant for a genetic trait, it is important to define the phenotype carefully. With OI, for example, some individuals are mildly affected. They do not have blue sclerae or bone fractures and, from a clinical point of view, may be classified as nonpenetrant. If an x-ray is taken, however, these individuals have undermineralized bones characteristic of OI. From the radiologic point of view, therefore, they do express a phenotype. Indeed, the more carefully a suspected gene carrier is examined, the lower is the chance that he or she will be found to be nonpenetrant, although there are instances in which individuals with a mutant genotype are totally unaffected.

Sometimes a rate of penetrance will be specified for a disorder. This rate applies to a population, not to an individual. For example, if 60% of individuals who carry a mutant gene express the phenotype, the rate of penetrance is 60%. A particular individual, however, either expresses the phenotype or does not. The individual is either penetrant or nonpenetrant; it does not make sense to speak in terms of

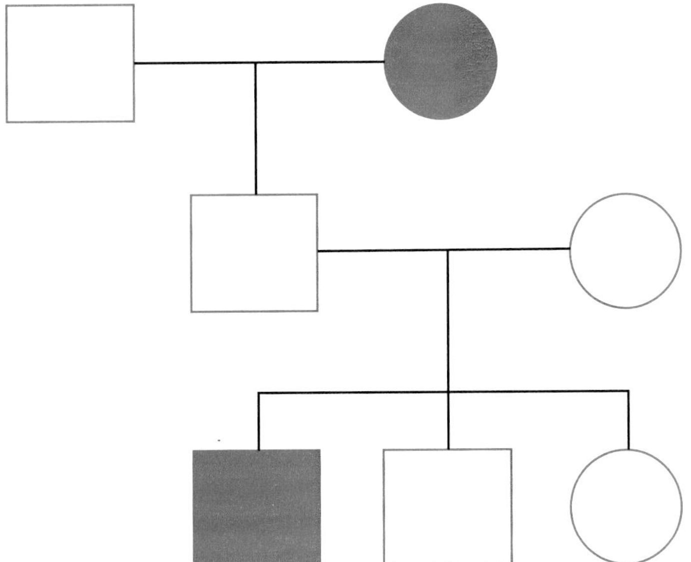

FIGURE 2-6 Example of nonpenetrance, in which an autosomal dominant trait is expressed in a grandparent and child but not in the parent. The unaffected parent must carry the mutant gene and is said to be *nonpenetrant.*

partial penetrance, although the person may have mild expression of the trait.

Penetrance sometimes is confused with another genetic term, **expressivity**. Expressivity refers to the degree of phenotypic expression of a genetic trait. Many genetic traits exhibit a wide range of expressivity, which means that the characteristics of affected individuals differ from person to person. This is illustrated by the autosomal dominant disorder neurofibromatosis type 1 (NF1). Affected individuals develop tumors along peripheral nerves, as well as patches of brown pigmentation on the skin (café-au-lait spots). Other features include bone deformities, learning disabilities, and brain tumors, but manifestations vary widely from person to person, even within the same family. Some have innumerable skin tumors and life-threatening malignant growths, whereas others have only a few skin spots (Figure 2-7). NF1 exhibits a wide range of expressivity, yet the penetrance is high: Virtually all persons who carry the NF1 gene mutation express at least some signs of the disorder.

Returning to the issue of the sporadically affected child with OI, it is possible, therefore, that one of the parents carries the mutant gene but is nonpenetrant. For a child with a severe, early-onset form of OI, however, this is unlikely, as the disorder has been found to have a high rate of penetrance.

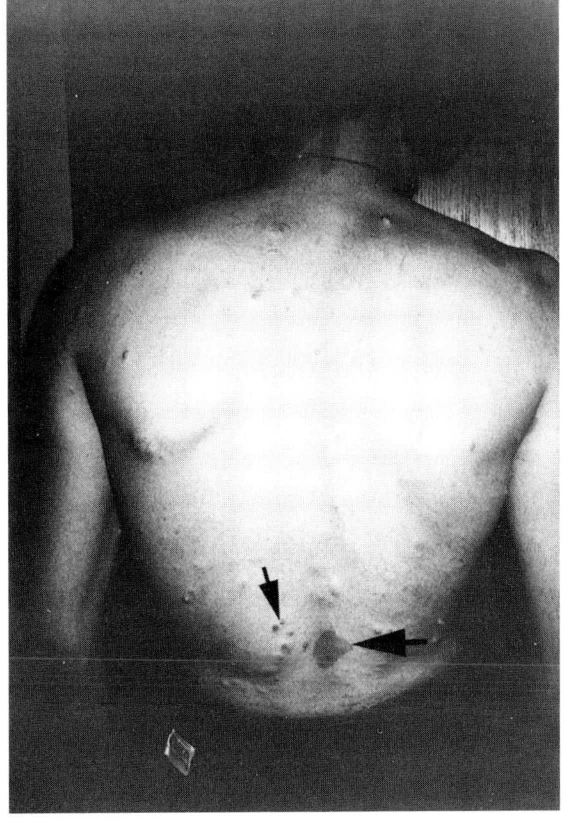

FIGURE 2-7 Individual with neurofibromatosis type 1, illustrating café-au-lait spots (*large arrow*) and skin neurofibromas (*small arrow*).

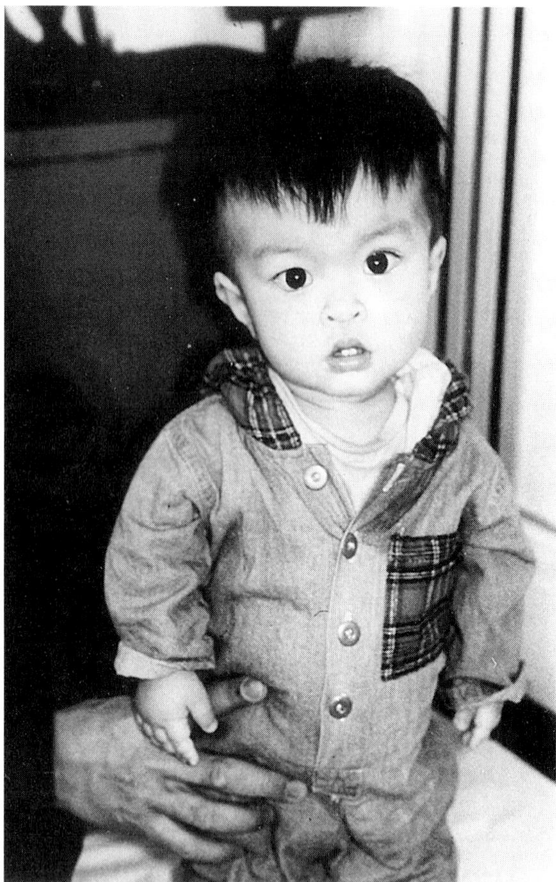

FIGURE 2-8 Young child with achondroplasia.

A second explanation for the sporadically affected child is that he or she represents a new mutation for the gene in question. This would imply that the gene mutation was not carried by either parent but instead arose in the sperm or egg cell that produced the child. It might be expected that a substantial proportion of children with the severe forms of OI (types II and III) might arise by new mutation. The severity of OI tends to be similar in different affected members of a family, suggesting that the different forms of the disorder are due to different mutations. The severe forms are not compatible with reproduction, so these cases would have to arise by new mutation, assuming that they are due to dominant mutations.

Another dominantly inherited disorder with a high rate of new mutation is achondroplasia. Persons with achondroplasia display abnormal bone growth and have extreme short stature (Figure 2-8). Affected individuals can reproduce and have a 50% chance of transmitting the condition to any offspring, but nearly 80% of cases of achondroplasia are sporadic (i.e., due to new mutation). It has been found that the fathers of sporadically affected children with achondroplasia tend to be older than the population mean for fathers at the time of conception of the child. No such effect is seen for maternal age. This has been interpreted in light of differences in the maturation of male and female germ cells. In the male, spermatogonia divide throughout adult life, beginning at puberty, whereas in the female, all oogonial mitoses are completed during fetal life (Figure 2-9). If each mitotic division is viewed as an opportunity for a mutation to occur, the probability of mutant germ cells will increase with paternal age but not with maternal age.

NF1 is another condition with a high rate of new mutation. NF1 affects approximately 1 in 4000 individuals, and approximately half the cases are sporadic. The penetrance of NF1 is nearly 100%, so sporadic cases are assumed to represent new mutations. This means that the mutation rate for this gene is approximately 1 in 16,000 germ cells (the calculation is detailed in Chapter 10), making it the highest rate known in humans. Interestingly, there is no definite paternal age effect on mutation rate in NF1. The mechanism underlying this high mutation rate is not known but is being investigated at the molecular level.

The case of a child with OI who is affected due to new mutation implies that the risk of recurrence for future offspring of the parents would be very low, probably no higher than for the general population. The affected child would have a 50% risk of passing the trait on to each of his or her future offspring. One special case of new mutation that would have different implications for the parents' risk of recurrence needs to be considered. This is **somatic** or **germ-line mosaicism.** It has been found that some individuals carry a mutation in some but not all cells of their bodies (Figure 2-10). If some germ cells carry the mutation, there is a risk of transmission of the trait to an offspring. Somatic mosaicism implies that the individual has both wild-type and mutant cells in his or her body. If mutation-bearing cells are sufficiently prevalent, clinical signs of the disorder may be present. The mutation must have arisen during embryonic development. Germ-line mosaicism

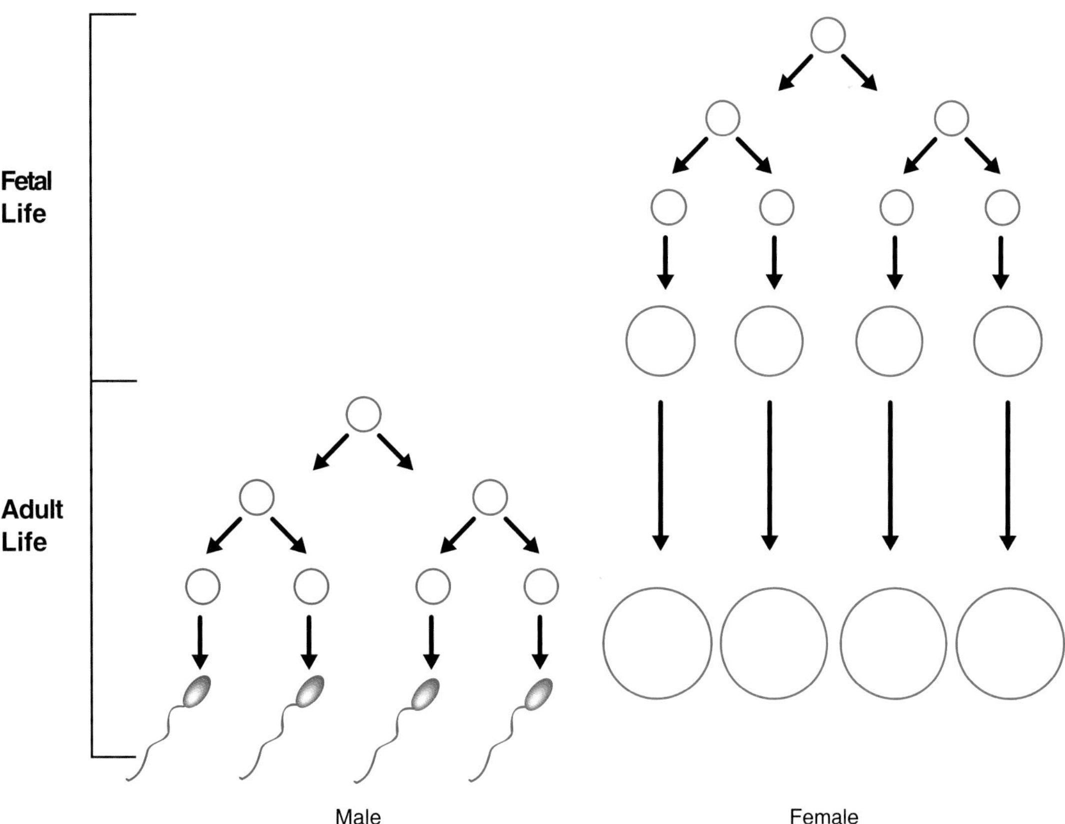

FIGURE 2-9 Differences in maturation of sperm (left) and eggs (right). Sperm undergo multiple rounds of mitosis, beginning at puberty and continuing through adult life. Eggs, in contrast, complete mitotic divisions during fetal life.

implies that the mutation is present only in cells of the germ-cell lineage, having arisen at some point during germ-cell development.

Recognition of the potential for mosaicism can have important implications for genetic counseling, because it is possible for more than one child to be affected even if neither parent has signs of the disorder. Instances of mosaicism have been identified in parents of children with OI, accounting for some families in which unaffected parents have more than one affected child. In the past, such families were explained by assuming that OI might sometimes be transmitted as an autosomal recessive trait. It is difficult to disprove this clinically, because often the children with the disorder are affected too severely to reproduce, and hence demonstrate dominant transmission. The lack of an increased rate of consanguinity in such families argues against autosomal recessive inheritance. The ultimate proof of somatic mosaicism, however, was obtained from molecular studies of mutations responsible for OI, which will be described later.

Autosomal Dominant Transmission

- Expressed in a heterozygous or homozygous individual
- Affects individuals of either gender
- Transmitted to 50% of offspring, regardless of gender

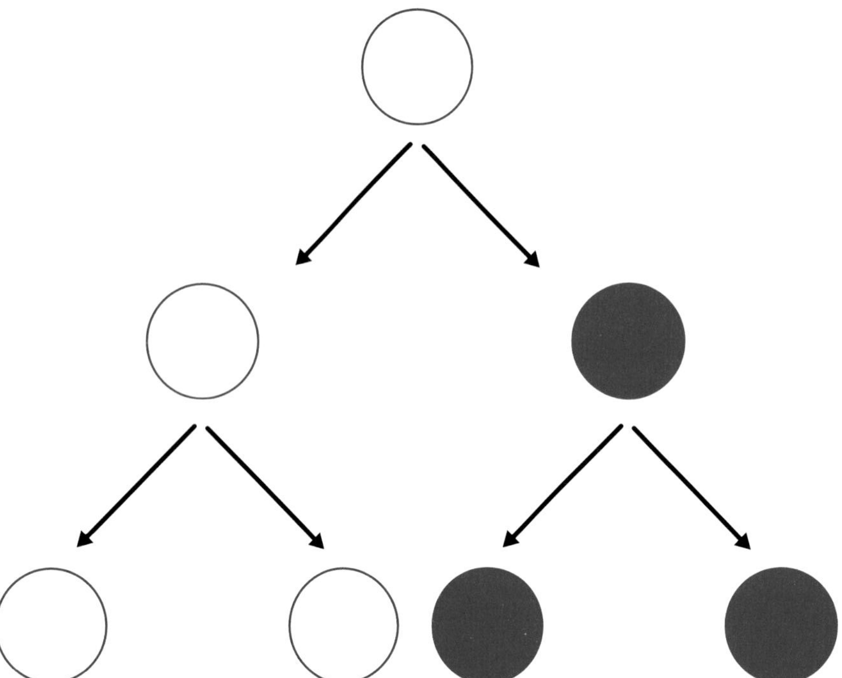

FIGURE 2-10 Somatic mosaicism arises by new mutation occurring during embryonic development. The individual contains a mixture of mutant and wild-type cells. If some germ cells carry the mutation, the mutation can be transmitted to an offspring.

2.3 Biochemical Basis of Osteogenesis Imperfecta

Lawrence is now 2 years old. He has continued to have multiple fractures, although his health has been good. He is not yet walking but is beginning to say a few words. Carol and Neal are told that in order better to classify Lawrence's osteogenesis imperfecta, and to provide more precise genetic counseling, a skin biopsy will be necessary. Lawrence has an inguinal hernia, and the biopsy is taken during surgical repair of this hernia..

Although bone is the most prominently affected tissue in persons with OI, there are many extraskeletal manifestations as well. Blue sclerae and abnormal teeth have already been mentioned. In addition, the skin tends to be thin and susceptible to injury, joints may be hyperextensible, and hernias occur commonly. These features indicate that a generalized defect in the structure of connective tissue might underlie the disorder. Beginning in the 1970s, biochemical and histologic studies strongly implicated one of the most abundant proteins in connective tissue, collagen, as being the basis for OI.

Collagen is actually a term used to describe a group of closely related fibrous proteins that are widely distributed in the body. The prototype is referred to as *type I collagen*. Type I collagen is composed of three protein chains that are intertwined in a triple helix (Figure 2-11). Two of the protein chains are identical and are referred to as $alpha_1$(I); the third chain is $alpha_2$(I). The triple-helical domain extends over 3000 Å as a compact, rod-shaped molecule 15 Å in diameter. Type I collagen molecules aggregate into fibrils which, in turn, bundle into larger fibers. These are located extracellularly and lend strength to tissues such as tendon, skin, and bone.

In the region that forms the triple helix, both the $alpha_1$(I)- and the $alpha_2$(I)-chains have the amino acid sequence glycine-X-Y repeated more than 300 times, where *X* and *Y* represent various other amino acids. X often is proline, and Y often is proline or lysine. As the procollagen chains are synthesized

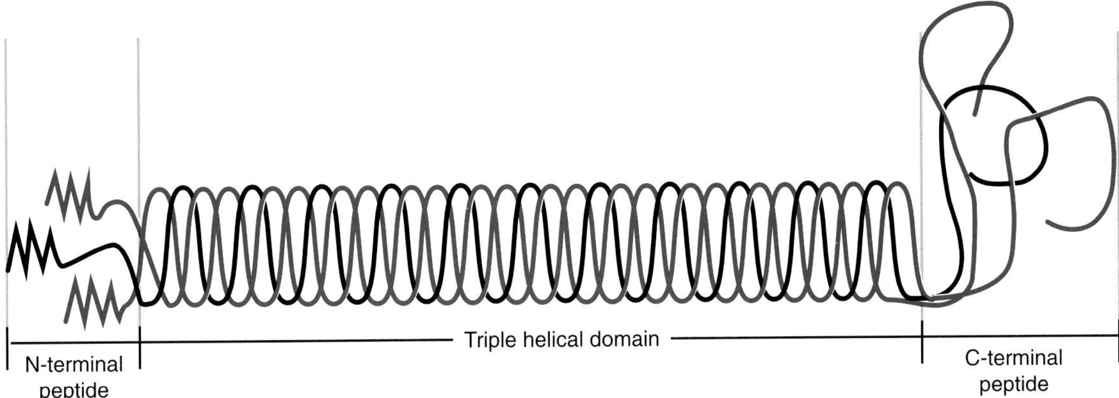

FIGURE 2-11 Structure of collagen molecule. The largest domain is the central triple helix, which consists of two alpha$_1$ I and one alpha$_2$ (I)—collagen molecules. The N-terminal and C-terminal peptides are cleaved during processing of the collagen.

on the ribosome (Figure 2-12), proline residues in the Y position are hydroxylated by the enzyme prolyl-4-hydroxylase, and some lysines in the Y position are hydroxylated by lysyl hydroxylase. Some of the hydroxylysines then are glycosylated with glucose or galactose. The newly synthesized chains are referred to at this point as pro-alpha-chains, because they have peptides at their N and C termini that are not represented in mature collagen. The pro-alpha-chains associate to form triple helices while still in the endoplasmic reticulum. The process begins with the formation of disulfide bonds forming in the C-terminal peptides, but the helices remain stable because of hydrogen bonds between glycine and hydroxyproline residues. The triple helices (procollagens) then are secreted from the cell, and the N- and C-terminal peptides are cleaved by specific proteases. The collagen molecules then aggregate in a staggered array into fibrils that are stabilized by cross-links between lysine or hydroxylysine residues in neighboring molecules that are catalyzed by the enzyme lysyl oxidase.

Many distinct types of collagen are found in different tissues (Table 2-2). The collagens are distinguished by the length and sequence of the triple-helical domains or the composition and size of the C-terminal or N-terminal peptides. They have different physical properties and are found in different quantities in different tissues. The tissue distribution of type I collagen mirrors the set of tissues affected by OI. This was one of the first clues that this type of collagen might be involved in OI. Histologic studies of tissues from persons affected with OI, moreover, show a disordered pattern of connective tissue (Figure 2-13).

Biochemical evidence implicating collagen as the defective protein in OI began to accumulate in the mid-1970s. Cultured skin fibroblasts from one infant with type II (perinatal lethal) OI were found to secrete abnormally low quantities of type I collagen. Electrophoretic analysis of pro-alpha$_1$(I)- and pro-alpha$_2$(I)-chains revealed two species of pro-alpha$_1$(I) of slightly different mobilities (Figure 2-14). The procollagen molecules were cleaved with cyanogen bromide (which cuts the proteins at methionine residues), and the fragments were studied by electrophoresis. An abnormally migrating fragment was found that corresponded with part of the pro-alpha$_1$(I)-chain. This mutant protein was found to differ from the normal by a deletion of 84 amino acids.

Similar findings occurred in a large number of samples from infants with OI type II. The difference in electrophoretic mobility of the normal and abnormal procollagen chains disappeared if the cells were treated with α-α'-dipyridyl, a chemical that inhibits posttranslational modification of collagen. Posttranslational modification normally occurs until formation of the triple helix. These data implied that the abnormal collagen molecules are unable to form

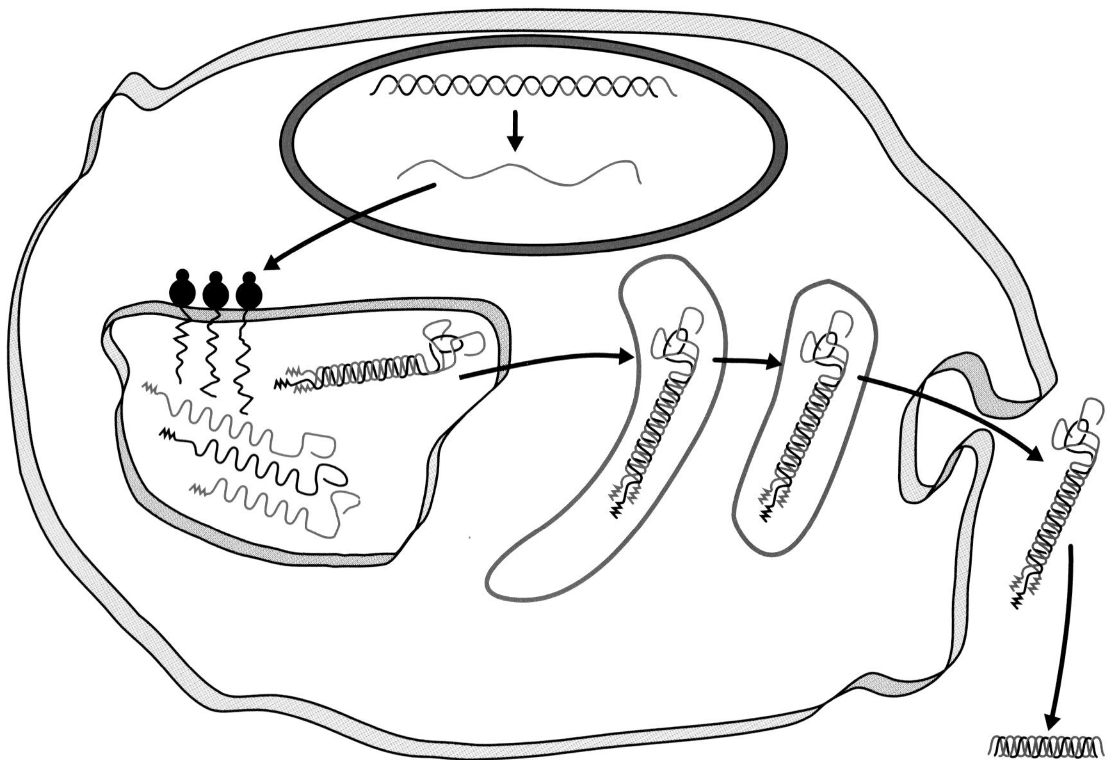

FIGURE 2-12 Processing of collagen. The procollagens are transcribed in the nucleus and translated on ribosomes in the endoplasmic reticulum. Here they are also chemically modified and glycosylated, and the three procollagens aggregate into a triple helix. The assembled molecules are secreted from the cell, and the N- and C-terminal peptides are cleaved.

TABLE 2-2 Types of collagens

Type	Chains	Tissue
I	$[\alpha1(I)]_2\alpha2(I)$	Skin, tendon, bone, arteries
	$[\alpha1(I)]_3$	
II	$[\alpha1(II)]_3$	Cartilage, vitreous humor
III	$[\alpha1(III)]_3$	Skin, arteries, uterus
IV	$[\alpha1(IV)]_2\alpha2(IV)$	Basal laminae
	$[\alpha1(IV)]_3$	
	$\alpha3(IV)\alpha4(IV)\alpha5(IV)$	
V	$[\alpha1(V)]_3$	Skin, placenta, vessels, chorion, uterus
	$[\alpha2(V)]_3$	
	$[\alpha1(V)]_2\alpha2(V)$	
	$\alpha1(V)\alpha2(V)\alpha3(V)$	
VI	$\alpha1(VI)\alpha2(VI)\alpha3(VI)$	Ubiquitous
VII	$[\alpha1(VII)]_3$	Epithelial-mesenchymal junctions
VIII	$\alpha1(VIII)\alpha2(VIII)$	Ubiquitous
IX	$\alpha1(IX)\alpha2(IX)\alpha3(IX)$	Cartilage
X	$[\alpha1(X)]_3$	Cartilage
XI	$\alpha1(XI)\alpha2(XI)\alpha1(II)$	Cartilage

Source: Modified from Byers P. Disorders of collagen biosynthesis and structure. In: Scriver CR, Beaudet AL, Sly WS, Valle D. The metabolic and molecular bases of inherited disease. New York: McGraw-Hill, 1995. Pp. 4029–4077.

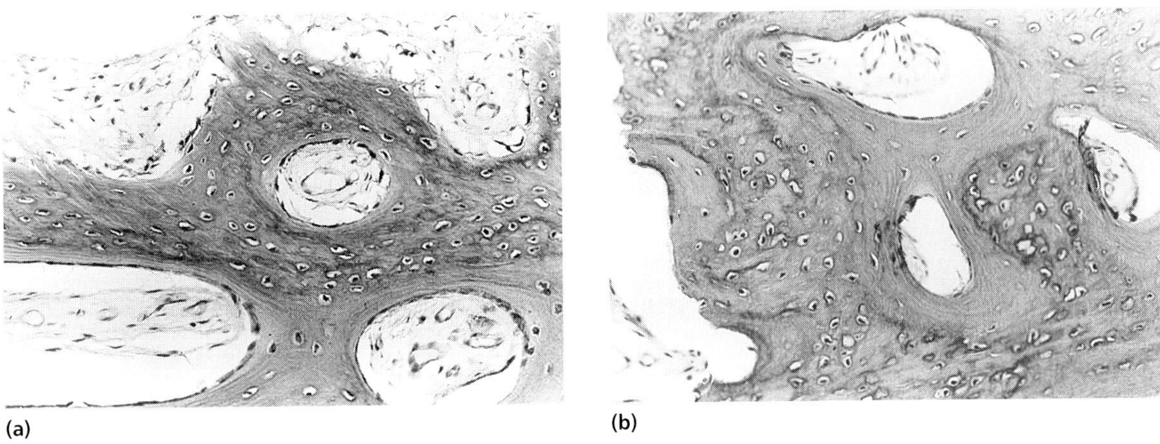

FIGURE 2-13 Photomicrograph of normal (**a**) or osteogenesis imperfecta (**b**) bone. Note the disordered bone matrix in the osteogenesis imperfecta tissue. (Courtesy of Dr. Frederic Shapiro, Children's Hospital, Boston.)

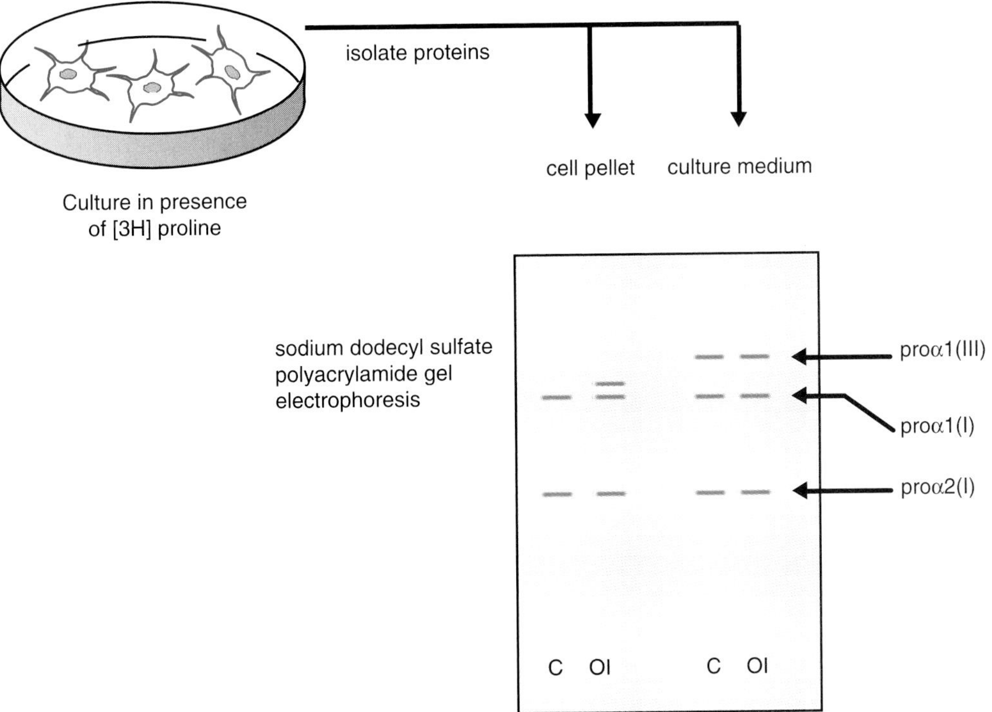

FIGURE 2-14 Cells were cultured in the presence of tritiated proline to label all proteins. Proteins were isolated from the cells and from the culture medium of both a control sample (*C*) and a sample from the child with osteogenesis imperfecta (*OI*) and subjected to electrophoresis in a polyacrylamide gel. The detergent sodium dodecyl sulfate denatures the proteins and allows separation on the basis of molecular weight. The gels then are stained to visualize the proteins. The sites of normal pro-alpha$_1$(I) and pro-alpha$_2$(I) are indicated. In the OI patient, an extra band is found in protein isolated from the cell pellet, which migrated more slowly than normal procollagens. A pro-alpha$_1$(III) band is normally seen only in culture medium with this assay. (Data from Barsh GS, Byers PH. Reduced secretion of structurally abnormal type I procollagen in a form of osteogenesis imperfecta. Proc Natl Acad Sci USA 1981;78:5142–5146; and Penttinen RP, Lichtenstein JR, Martin GR, McKusick VA. Abnormal collagen metabolism in cultured cells in osteogenesis imperfecta. Proc Natl Acad Sci USA 1975;72:585–589.)

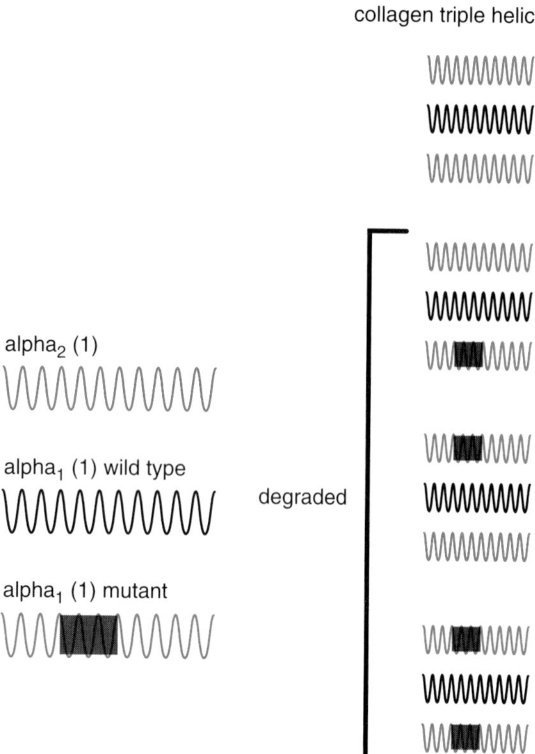

FIGURE 2-15 "Protein suicide model." Although only one allele for alpha$_1$(I)-procollagen is mutant, 75% of the triple-helical molecules incorporate at least one mutant protein and are degraded in the cell. This is referred to as a *dominant negative effect.*

normal triple helices, which leads to excessive modification of the protein chains.

Biochemical evidence for deficient type I collagen production was also found in samples from individuals with other types of OI. It appeared likely that mutations affecting pro-alpha$_1$(I) lead to deficient production of type I collagen. Because most forms of OI are transmitted as dominant traits, it might be asked why heterozygous mutation would lead to a phenotype. One answer lies in the fact that type I collagen is composed of two alpha$_1$(I)-chains and one alpha$_2$(I)-chain. A mutation in one of the alpha$_1$(I)-chain genes causes 50% of the pro-alpha$_1$(I) produced to be of the mutant form. When procollagen molecules form triple helices, three-fourths of the molecules will contain at least one mutant alpha$_1$(I)-chain (Figure 2-15). If the presence of even a single mutant chain renders the collagen nonfunctional, this means that a mutation affecting 50% of the gene product will affect 75% of the final protein, which is sufficient to result in a phenotype. This is in sharp contrast to the situation with mutations that affect enzymes, where 50% reduction of enzyme protein in a heterozygous carrier has no phenotypic effect.

In addition, other factors may explain the dominant phenotype. Because multiple collagen molecules aggregate into fibrils, if a proportion of molecules are defective, the strength of the entire tissue may be weakened. Thus, even mutations that impair alpha$_2$(I) production lead to an OI phenotype.

2.4 Collagen Genes

Each collagen chain is encoded by a separate gene. In 1979, Yamamoto and colleagues isolated type I collagen cDNA by reverse transcription of mRNA obtained from chick embryo bones (Figure 2-16). These tissues synthesize large quantities of type I collagen, and therefore a high proportion of mRNA from these cells encodes this protein.

Having the cDNA clones enabled the subsequent isolation of genomic clones from a genomic library. The pro-alpha$_1$(I)-chain is encoded by an 18-Kb gene with 52 exons. In the triple-helical coding domain, these exons contain 45, 54, 99, 108, or 162 base pairs, begin with a glycine codon, and end with the codon for a Y amino acid, and therefore encode a series of Gly-X-Y residues. Genes for many of the other procollagen chains have also been cloned.

Collagen types I, II, III, V, and XI have a generally similar structure and are referred to as *fibrillar collagens*. Their procollagen genes likewise have similar structures and thus constitute a *gene family*. It is believed that these genes were derived from a single ancestral gene more than 50 million years ago. The procollagens presumably evolved by successive duplications of primitive procollagen genes, with subsequent sequence divergence (Figure 2-17). Evolution by gene duplication, followed by accumulation of mutations in the resulting genes, has occurred in many genetic systems. It accounts for a number of multigene families—for example, the globin genes (encoding the various types of hemoglobin proteins)

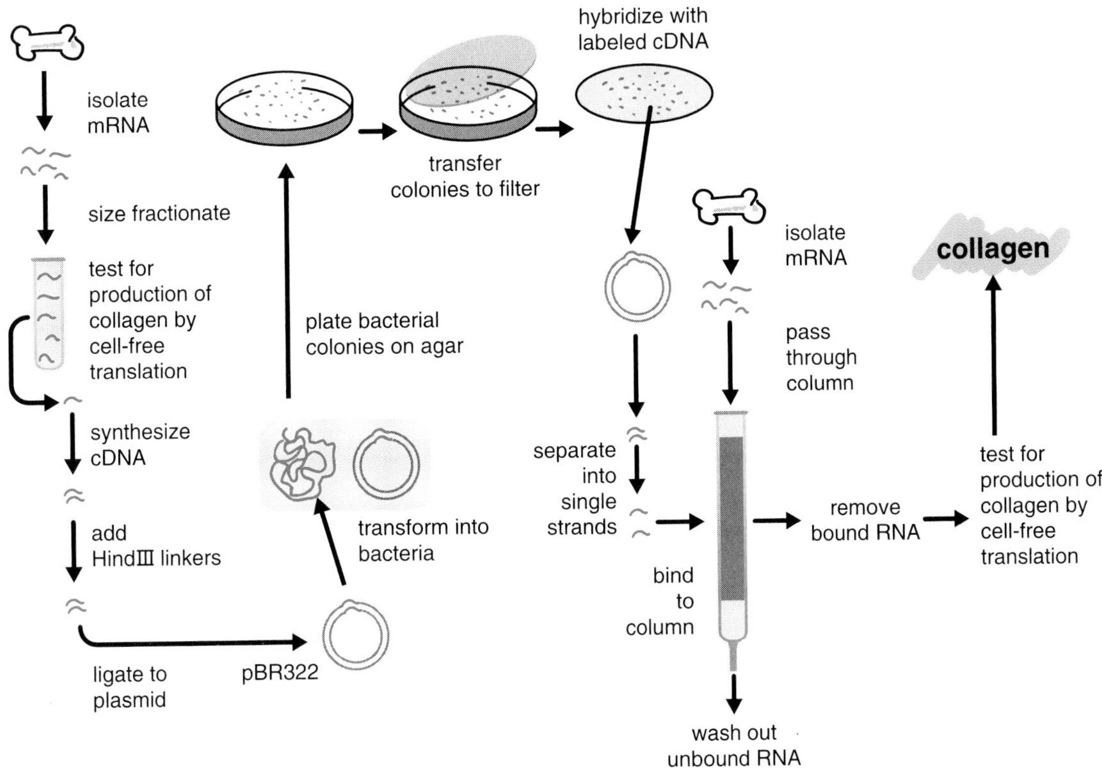

FIGURE 2-16 The mRNA was size-fractionated by centrifugation through a sucrose gradient, so that larger molecules sedimented to deeper regions of the gradient. Different fractions then were tested for ability to encode procollagen in a cell-free translation system. This system is produced from wheat germ and consists of ribosomes and enzymes necessary to translate protein if provided with a source of mRNA. Fractions that contained collagen-producing mRNA then were used to synthesize cDNA. cDNA molecules were ligated to oligonucleotides corresponding with the recognition site of the enzyme *HindIII* and then were cloned into the *HindIII* site of the plasmid pBR322. Recombinant plasmids were transformed into *Escherichia coli*. Bacterial colonies were screened by hybridization with radioactively labeled cDNA from the collagen-producing fraction, and a positive colony was selected. To demonstrate that the DNA inserted into the plasmid actually represented a collagen cDNA, the plasmid insert was separated into single strands and chemically bound to cellulose and placed in a column. Total RNA from chick embryo bones was passed through the column. RNA homologous to the cloned DNA formed a stable double helix with the cloned DNA and therefore was retained in the column. After extensively washing out any unbound RNA, the bound RNA was released from the column and used for protein synthesis in a cell-free translation system. The polypeptide pro-alpha$_1$(I) was synthesized, indicating that the cDNA for that protein had been cloned. (Data from Yamamoto T, Sobel ME, Adams SL, et al. Construction of a recombinant bacterial plasmid containing pro-alpha$_1$(I)-collagen DNA sequences. J Biol Chem 1980;255:2612–2615.)

and immunoglobulin genes. Other collagen types include nonfibrillar collagens, in which the triple-helical domains are interrupted by non-triple-helical regions. There are also short-chain collagens, in which the triple-helical region is shorter, and type VII collagen, which is a long-chain collagen that anchors epithelial cells to underlying mesenchymal cells. The genes for these collagens bear similarity to the fibrillar collagens but exhibit greater sequence divergence.

Although they appear to be derived from a single ancestral gene, the procollagen genes are dispersed in the human genome. The gene for alpha$_1$(I) was localized by the technique of somatic-cell hybridization. It

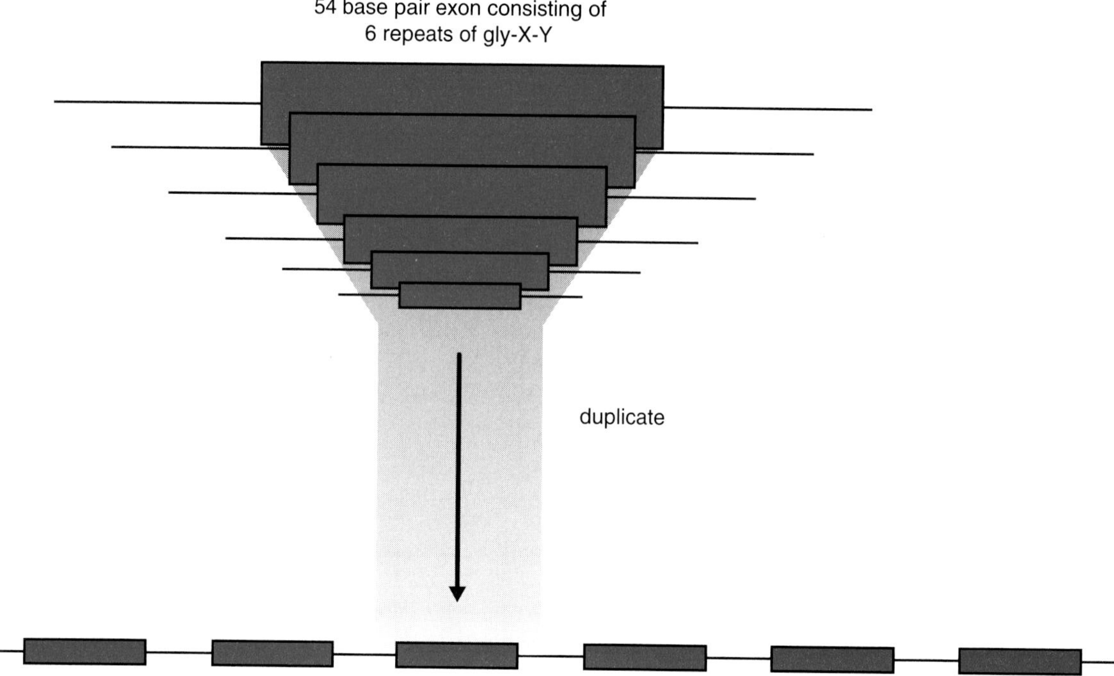

FIGURE 2-17 Evolution of collagen gene by duplication of primordial segment consisting of 54 base pairs with six gly-X-Y repeats. (Modified from Yamada Y, Avvedimento VE, Mudryj M, et al. The collagen gene: evidence for its evolutionary assembly by amplification of a DNA segment containing an exon of 54 bp. Cell 1980;22:887–892.)

has been found that cultured cells of different species can be fused together when treated with Sendai virus or polyethylene glycol. The two genomes occupy a single nucleus, but often the chromosomes of one species are lost during cell division. When human cells are fused with rodent cells, usually it is the human chromosomes that are lost. After many rounds of cell division, hybrid cells contain a full rodent complement along with only one or a few human chromosomes. In 1982, Huerre and colleagues used this approach to determine that the alpha$_1$(I)-gene is located on chromosome 17 (Figure 2-18).

The gene assignment was confirmed using a different approach, in situ hybridization. This is a more direct gene-mapping method, in which a cloned DNA sequence is tagged (in this case with a radioactive nucleotide) and hybridized with DNA on preparations of chromosomes fixed onto slides. The hybridized sequences are detected with autoradiography; silver grains superimposed on chromosomes are sought. Radioactive in situ hybridization has relatively poor sensitivity for detecting single-copy genes and requires examination of many chromosome preparations to identify sites where silver grains are found most consistently. More sensitive methods using fluorescent-labeled DNA probes are now available and will be described later. Nevertheless, the approach of radioactive in situ hybridization confirmed that the alpha$_1$(I)-gene is located on chromosome 17 (Figure 2-19).

The other procollagen genes have been localized using similar techniques. The genes are designated COL1A1 for alpha$_1$(I), COL1A2 for alpha$_2$(I), and so on. COL1A2 is on chromosome 7, COL2A1 on chromosome 12, COL3A1 on chromosome 2, and COL5A2 on chromosome 2. COL3A1 and COL5A2 are near one another on chromosome 2, but the other fibrillar collagen genes are widely dispersed.

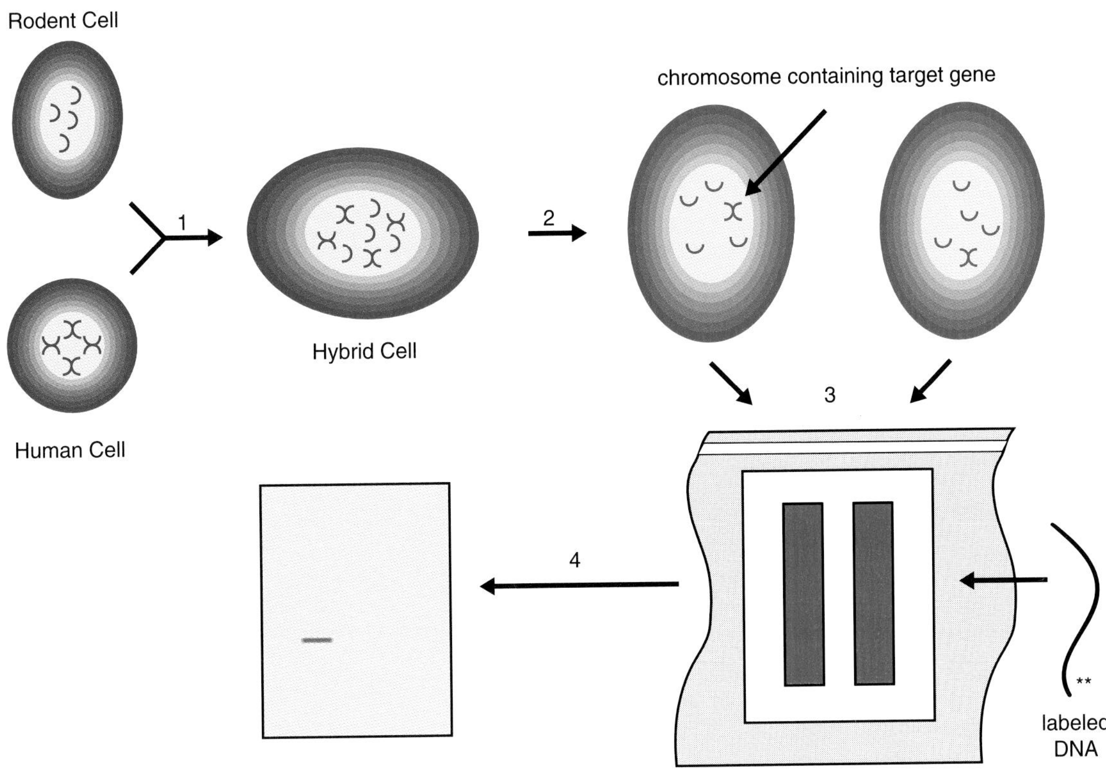

FIGURE 2-18 Human-mouse and human–Chinese hamster hybrid cells were prepared. After treatment with Sendai virus or polyethylene glycol (1) the cells were grown in culture medium containing hypoxanthine, aminopterin, and thymidine (so-called HAT medium). The rodent cells used had mutations in the genes for the enzyme hypoxanthine phosporibosyl transferase (HPRT), thymidine kinase (TK), or both. Aminopterin inhibits endogenous purine and pyrimidine synthesis. In the presence of this drug, cell growth depends on utilization of hypoxanthine and thymidine in the culture medium. This requires HPRT to make purines and TK to make pyrimidines. In the presence of HAT medium, the rodent cells that have not fused with human cells die for lack of one or both enzymes. Hybrid cells contain the enzymes from the human genomes. The unfused human cells are selected against by their sensitivity to the drug ouabaine, to which rodent cells are resistant. Hybrid clones were isolated that contained just a few human chromosomes (2), although the specific human chromosomes in each hybrid clone were different. DNA was obtained from these cell clones and was digested with *HindIII*, and a Southern blot was prepared (3). The probe in this case was radioactively labeled DNA corresponding with the human alpha$_1$(I)-cDNA. When the autoradiograms were developed, a signal was seen only in lanes corresponding with hybrid cells that contained a copy of human chromosome 17. This indicated that chromosome 17 was likely to be the site of the alpha$_1$(I) in humans. (Data from Huerre C, Junien C, Weil D, et al. Human type I procollagen genes are located on different chromosomes. Proc Natl Acad Sci USA 1982;79:6627–6630.)

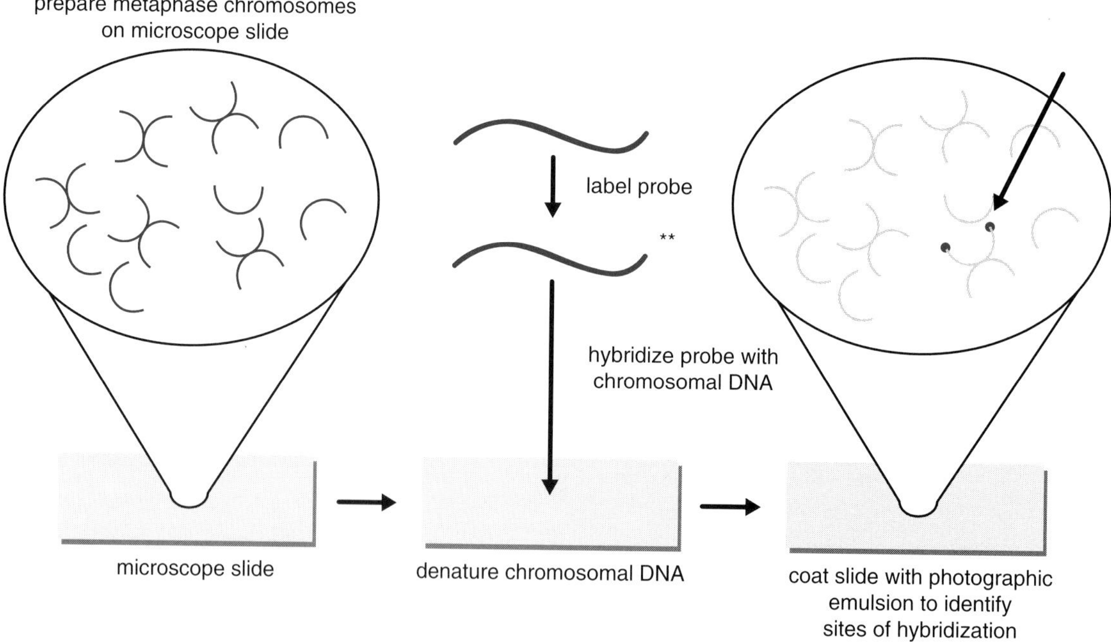

FIGURE 2-19 Human white blood cells were grown in culture and were treated with the drug colchicine, which disrupts the mitotic spindle and therefore arrests cell division at metaphase. At this point chromosomes are condensed, and easy to visualize. The cells were swollen with a hypotonic solution to spread the chromosomes and then were fixed onto glass slides. The chromosomal DNA next was separated into single strands by treatment with NaOH. A radioactively labeled alpha$_1$(I)-cDNA, also rendered single-stranded , was incubated with the chromosome preparations on the slides. The probe bound to the site of the normal alpha$_1$(I)-gene, and the site of binding was visualized by autoradiography. In this instance, autoradiography was done by coating the slide with a photographic emulsion and then "developing" this emulsion to reveal silver grains at the site of probe binding. Because there is only a single alpha$_1$(I)-gene on each haploid set of chromosomes, the hybridization signal with this approach tends to be very weak. It was necessary to examine many hundreds of cells to look for a site of consistent chromosome labeling and distinguish this from background labeling.

2.5 Linkage of Collagen Genes with Osteogenesis Imperfecta

Although biochemical evidence indicated that abnormalities of collagen occur in OI, it was possible that the primary genetic defects in some affected individuals might reside in genes for other proteins that interact with collagens. Establishing that the collagen genes themselves are the sites of mutations responsible for OI required that the gene mutations be identified. Finding mutations in large molecules such as the collagens is a laborious task, however. Prior to this undertaking, genetic linkage analysis was used to determine that a collagen gene must be the site of mutation in members of a family who have OI.

Genetic linkage analysis was accomplished with a restriction fragment–length polymorphism. Restriction enzymes recognize specific DNA base sequences, usually four to eight bases in length, cutting at that sequence and only at that sequence. Even a single base change at a cutting site destroys the ability of the enzyme to cleave the DNA at that location. Variation in base sequence from individual to individual is not uncommon, as has already been pointed out for the case of point mutations in the tyrosinase gene. Often these sites reside in introns or between genes. Such a base sequence variant that occurs within a restriction enzyme cleavage site leads to loss of the enzyme's ability to cut at that site. An individual can be homozygous for having the cutting site or for not having the cutting site or can be heterozygous,

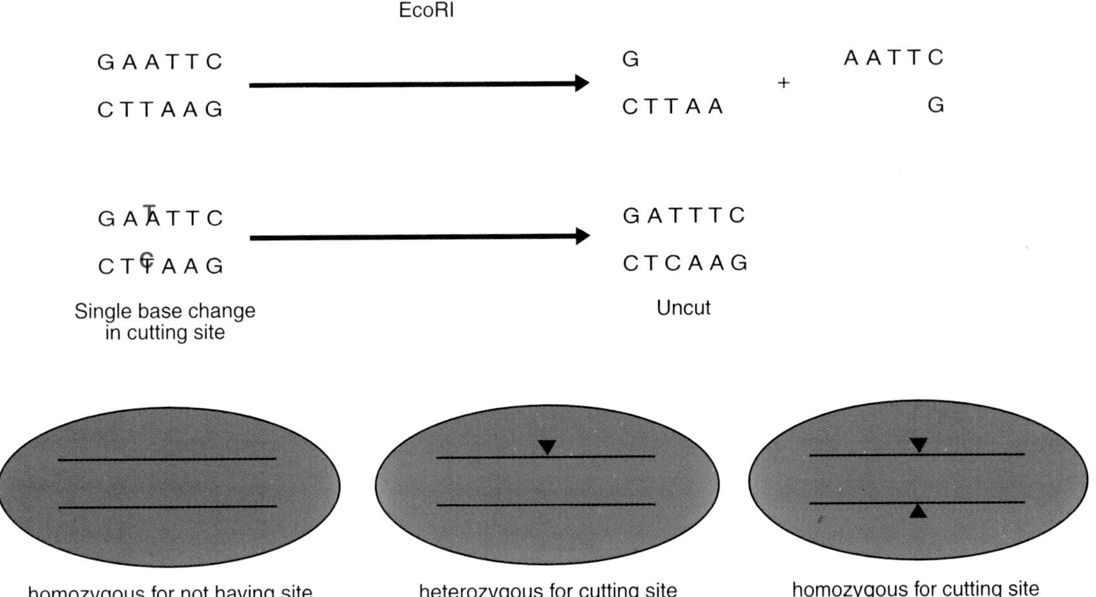

FIGURE 2-20 Restriction fragment–length polymorphism. The enzyme *EcoRI* recognizes the base sequence GAATTC and cuts both strands of DNA at this site (top). If a single base change occurs in the recognition site, the enzyme will not cut. An individual can have the cutting site intact on both chromosomes and be homozygous for having the site (bottom right), or can be heterozygous for the cutting site (bottom middle), or can be homozygous for not having the site. This constitutes a simple mendelian system for detecting a change of a single base of DNA.

having the intact site on one allele and absence of the site on the other (Figure 2-20).

The genotype of an individual with respect to such a polymorphism can be determined by Southern blotting. A cloned DNA sequence homologous to the region near the polymorphism will hybridize to a larger fragment if the enzyme does not cut at the polymorphic site and a smaller fragment if the enzyme does cleave there (Figure 2-21). Restriction fragment–length polymorphisms (RFLPs) have been instrumental in gene-mapping studies (see Chapter 3).

Application of this approach to OI required identification of an RFLP within one of the collagen genes. Tsipouras and colleagues were successful in finding such an RFLP in COL1A2. A fragment of the cDNA for this gene was hybridized with DNA from randomly selected individuals and digested with various restriction enzymes. When the enzyme *EcoRI* was used one of three patterns was seen. Sometimes a single 14-Kb band was noted on the Southern blot. Other times this band was missing, and two bands of 10.5 and 3.5 Kb were seen. Still other times all three bands were present. These findings were due to a polymorphic *EcoRI* site in the COL1A2 gene. The polymorphism was shown to be transmitted as a simple mendelian trait in families. A heterozygous parent would transmit to any child either the chromosome with the cutting site or the chromosome without the cutting site.

The segregation of this polymorphism was studied in a family with autosomal dominant OI (Figure 2-22). An individual in the first generation was found to be heterozygous, and her partner was homozygous for absence of the cutting site. Her two children with OI inherited the chromosome with the cutting site, and her two unaffected children inherited the other chromosome. The segregation of OI and the RFLP was traced through two additional generations, and in every case the disorder was passed along with the chromosome that had the intact *EcoRI* site.

The family of individuals III-3 and III-4 in the pedigree illustrate an important point about genetic linkage studies. Both III-3 and III-4 are heterozygous for the RFLP, but only III-3 is affected. This RFLP

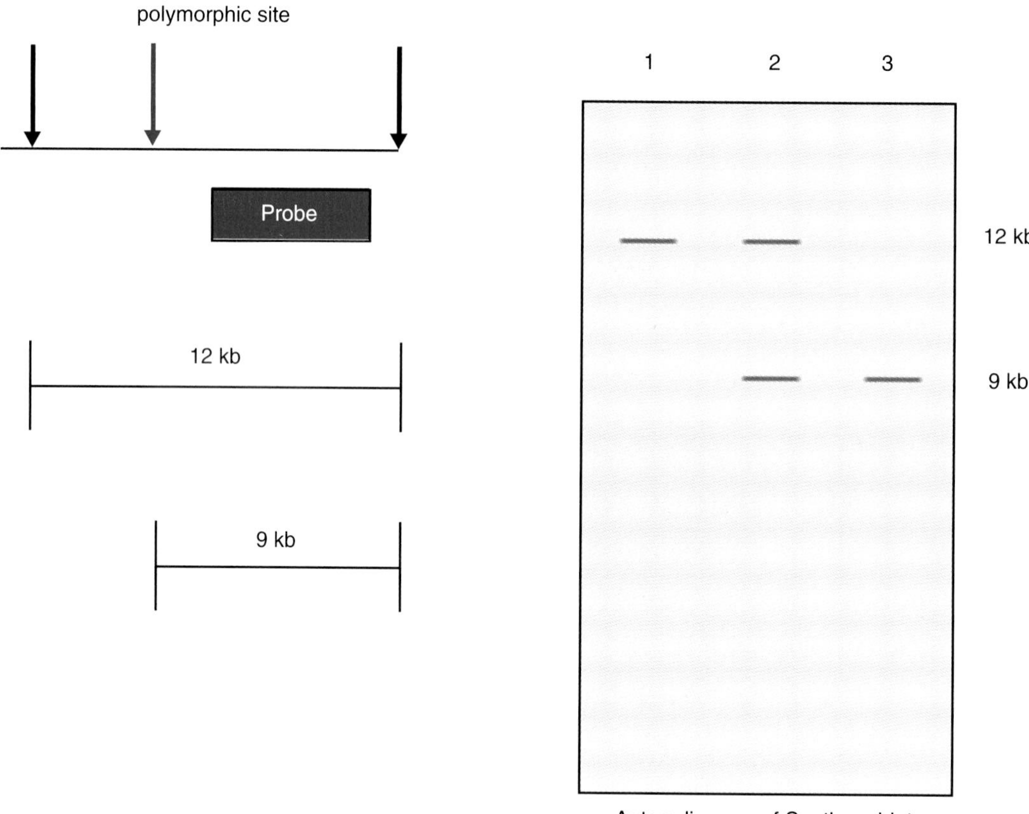

FIGURE 2-21 Southern blot analysis for restriction fragment–length polymorphism. The *solid arrows* denote constant cutting sites, and the *blue arrow* in the center a polymorphic site. The location of the cloned DNA probe is indicated. When the polymorphic site is present, the probe hybridizes with a 9-Kb fragment, whereas when the polymorphic site is absent, the probe hybridizes with a 12-Kb fragment. The Southern blot shown on the right is prepared by first cutting DNA with the appropriate restriction enzyme, then separating the fragments on an agarose gel, and finally by blotting these onto a membrane. The DNA in the membrane then is hybridized with radioactively labeled, cloned probe DNA. DNA in lane 1 is from an individual who is homozygous for not having the middle cutting site, and hence only a single 12-Kb band is seen. The DNA in lane 2 is from an individual who is heterozygous for the polymorphic site, and therefore both a 12-Kb and a 9-Kb band are seen. The DNA in lane 3 is from an individual who is homozygous for having the middle site, and a single 9-Kb band is seen.

must not be confused with the mutation that is responsible for OI in the family. Indeed, it is not a cause of OI, as III-4 has the change yet does not have the disorder. The nucleotide sequence change that leads to the RFLP has no phenotype; in this case it appears in intron 5 of the COL1A2 gene on chromosome 7. It is used in this study only as a marker to trace the inheritance of the two copies of the COL1A2 gene in the family. Individual II-2 has inherited the chromosome with the cutting site from I-2 along with the disease and hence, if the disease is due to mutation in COL1A2, the mutant copy of COL1A2 in II-2 must also have the intact *EcoRI* cutting site. She inherited a chromosome without the cutting site from her unaffected parent and passed that chromosome on to III-1, who does not have OI. Individual III-3 got the chromosome with the OI mutation and the *EcoRI* cutting site from his mother. His partner, III-4, is heterozygous also, owing to the fact that the polymorphism for the *EcoRI* site is common in the population. Their affected daughter is homozygous for the cutting site, but she must be

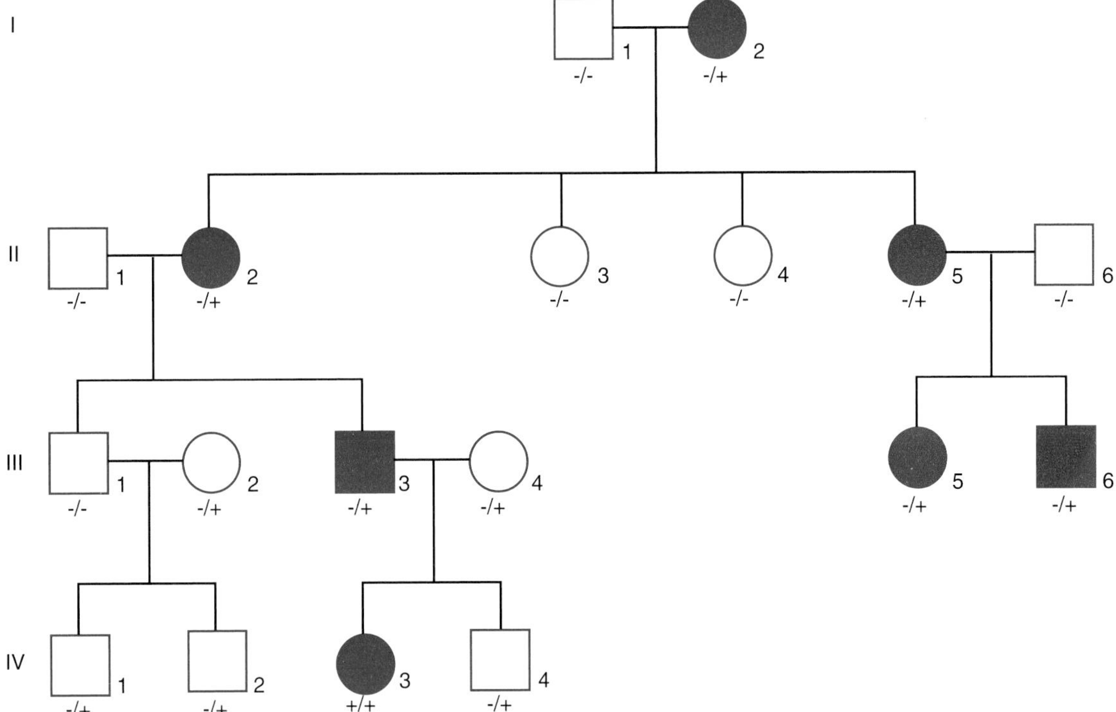

FIGURE 2-22 Linkage of restriction fragment–length polymorphism (RFLP) in COL1A2 gene with osteogenesis imperfecta. Affected individuals are indicated by darkened symbols. Genotypes for the RFLP are shown below the symbols. (+ denotes presence of the polymorphic cutting site; – denotes its absence.) Individual I-2 is heterozygous for the RFLP, and passed the + allele to both of her affected children. This same allele is transmitted together with osteogenesis imperfecta through two additional generations, suggesting that the copy of the COL1A2 gene with the polymorphic cutting site is associated with osteogenesis imperfecta in this family. (Data from Tsipouris P, Myers JC, Ramirez F, Prockop DJ. Restriction fragment length polymorphism associated with the pro-alpha$_2$(I) gene of human type I procollagen. J Clin Invest 1983;72:1262–1267.)

heterozygous for the OI mutation. Their other child is heterozygous for the cutting site and does not have OI, indicating that he must have inherited the *EcoRI* site from his mother, not his father.

This study indicated that, at least in this particular family, the polymorphism in the COL1A2 gene segregates with the OI phenotype. Methods of quantifying linkage data will be considered in the next chapter, but it should be clear that in this family the probability is low that the polymorphic cutting site would segregate with the OI trait in all members merely by chance. This provided evidence that an OI mutation must reside somewhere in the COL1A2 gene in this family. Indeed, the mutation subsequently was found to affect a splice junction at an exon-intron boundary. Similar studies done in families with OI type I have revealed that most are linked to COL1A1, but some are linked to COL1A2. Conversely, most families with OI type IV show linkage to COL1A2, but some are linked to COL1A1. Linkage studies have not been done for types II and III OI, because most cases are sporadic or occur only in small families. Precise definition of the various mutations responsible for OI required more detailed molecular genetic studies.

Restriction Fragment–Length Polymorphism (RFLP)

- Variation in base sequence that is common in the population and affects the recognition site of a restriction enzyme

2.6 Genetic Basis of Osteogenesis Imperfecta

Fibroblast cells are grown from the skin biopsy taken from Lawrence, and these are incubated with [^{3}H]proline. Ascorbic acid is present in the culture medium to enhance the production of collagen. The culture medium and cell layer then are harvested separately, and proteins are analyzed by electrophoresis. The quantity and quality of type III collagen is found to be normal. Two populations of type I procollagen are found, however. One is of normal electrophoretic mobility, and one migrates more slowly than normal. This slower-migrating collagen is selectively retained within the cells. Also, a new collagen chain approximately twice the size of normal alpha$_1$(I)-chains is found. The collagen molecules are cleaved at methionine residues with cyanogen bromide, and fragments are separated by electrophoresis. The new large chain is found to consist of a dimer of two collagen molecules connected by a disulfide bond between cysteines located between residues 123 and 401, where normally there are no cysteines. The slower migrating collagen is overmodified from residue 401 to the amino terminus. These findings suggest that Lawrence has a substitution of cysteine for glycine in the triple-helical domain of one COL1A1 gene.

An infant with OI type II and a mutant alpha$_1$(I)-chain having a deletion of 84 amino acids has already been mentioned. Chu and associates hypothesized that there might be a deletion in the DNA from that child. To test this hypothesis, they examined DNA from cultured fibroblasts from this child by Southern blotting, using fragments of COL1A1 cDNA as probes (Figure 2-23). Further mapping of the genomic COL1A1 sequence revealed a 0.5-Kb deletion that resulted in loss of 252 nucleotides in three complete exons, along with intronic sequences.

Large rearrangements of procollagen genes have turned out to be rare causes of OI. Another child was found to have a deletion of seven exons from the COL1A2 gene, leading to loss of 180 amino acids from the triple-helical domain. A third had an insertion of 600 base pairs, which caused duplication of 60 amino acids. All three of the children with multiexon

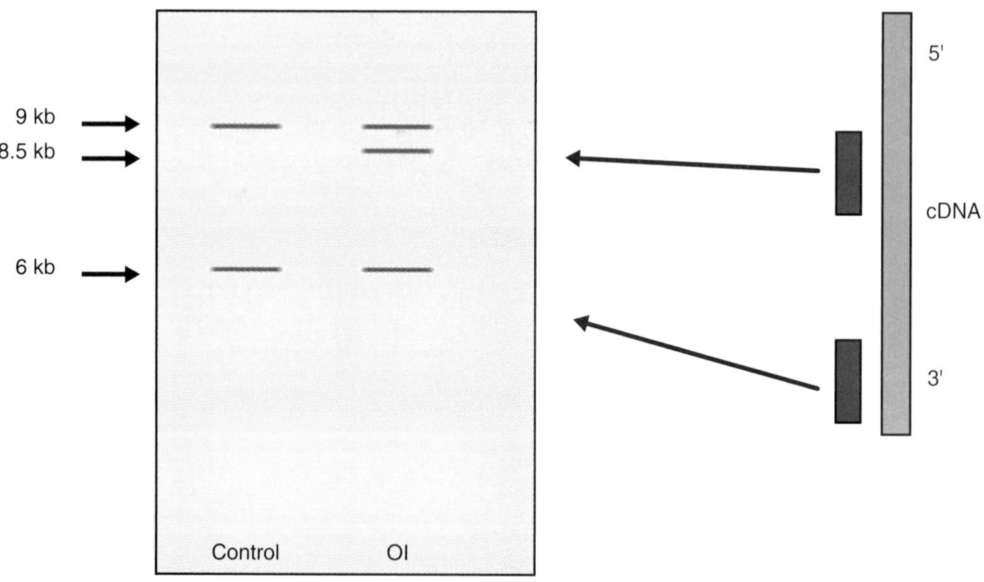

FIGURE 2-23 A 1.8-Kb *HindIII* fragment from the 3′ end of the cDNA hybridizes to a 6-Kb fragment in *HindIII*-cut DNA. When a *HindIII* cDNA from the middle of the alpha$_1$(I)-cDNA was used as a hybridization probe, a 9-Kb fragment was seen in controls but, in DNA from the child with osteogenesis imperfecta, two bands were seen: 9-Kb and 8.5-Kb. This is due to the presence of a 0.5-Kb deletion in the coding sequence from the child with osteogenesis imperfecta. (Permission from: Chu M-L, Williams CJ, Pepe G, et al. Internal deletion in a collagen gene in a perinatal lethal form of osteogenesis imperfecta. Nature 1983;304:78–80.)

rearrangements had the perinatal lethal form of OI—that is, OI type II. Apparently, these major changes in a collagen chain severely impair the stability of a majority of the collagen molecules. On the other hand, a 36–amino acid deletion of the alpha$_1$(I)-protein has been found in one family with OI type IV, indicating that not all deletions are necessarily lethal.

Point mutations in the collagen genes have been shown to be a far more common cause of OI than are large rearrangements. These mutations are diverse and, due to the large size of the collagen molecule, can be difficult to find. One of the first to be identified was localized by protein studies of an abnormal alpha$_1$(I)-chain from a child with OI type II. Analysis of cyanogen bromide–cleaved alpha$_1$(I) revealed evidence of an abnormal cysteine in the carboxyl terminal region of the triple-helical domain. The region of the gene that encodes this part of the protein was cloned from a lambda-phage genomic library prepared using DNA from the affected child. Lambda clones were selected that hybridized with a genomic clone corresponding to the region of interest. Three clones were obtained and sequenced. Two were found to contain the wild-type sequence, and one had a single base change of G to T, causing the glycine at position 988 to be replaced by a cysteine. It is expected that this would disrupt the triple helix in three-fourths of the collagen molecules, leading to the lethal phenotype.

A large number of point mutations that substitute various amino acids for glycine have been identified in the COL1A1 and COL1A2 genes. The phenotype of OI type I, II, III, or IV depends on the site of mutation and on the specific amino acid substitution. Phenotypes tend to be more severe when the substitution occurs nearer the carboxyl terminal end of the protein, and all substitutions of aspartic acid for glycine have been lethal. Substitutions outside the triple-helical domain have also been found but are less well characterized.

In addition to amino acid substitutions, some abnormal collagens appear to result from the loss of single exons. This might be due to deletion of the exon from the gene or, more commonly, to point mutations that lead to abnormal splicing. In general, mutations affecting COL1A1 tend to be associated with a more severe phenotype than those affecting COL1A2, and the phenotypes are more severe for mutations near the 3′ end of the triple helix than for those near the 5′ end.

2.7 Mouse Model for Osteogenesis Imperfecta

The collagen mutations found in persons with OI seemed likely to be the changes responsible for the phenotype. They tend to occur at sites of the collagen molecule that correlate with biochemical abnormalities of the triple helix and are not found in individuals who do not have OI. Proof that a mutation in COL1A1 can be pathogenic was provided by experiments in which mutant alpha$_1$(I)-chains were expressed in a mouse.

Stacey and coworkers inserted a mutant COL1A1 gene into fertilized mouse oocytes that were then brought to term (Figure 2-24). Two hundred eggs were injected with mutant COL1A1, and 220 were injected with a wild-type COL1A1. Forty-one mice were born from the mutant injections, and 47 from the control. Nine of the mutant mice died just prior to or shortly after delivery. Southern blot analysis revealed that these animals carried up to six copies of the mutant COL1A1 per cell. They had flabby limbs, soft skulls, and poorly mineralized, malformed bones. Thirty-two mice resulting from eggs injected with the mutant COL1A1 developed normally, and no mutant sequences could be detected. Likewise, none of the animals resulting from the wild-type COL1A1 injections displayed any physical abnormalities, although injected sequences could be detected. The mice expressing the mutant COL1A1 had reduced levels of type I collagen and alpha$_1$(I)-chains with abnormal electrophoretic mobility.

Mice containing foreign genes are referred to as **transgenic mice**. The formation of transgenic mice has been an important approach in the development of mouse models of disease and will be considered again later in this book.

2.8 Genotype-Phenotype Correlations in Osteogenesis Imperfecta

Lawrence's collagen changes are compared with those of other children with osteogenesis imperfecta. Cysteine-for-glycine changes at positions 988, 904, and 748 result in lethal OI type II. Substitution

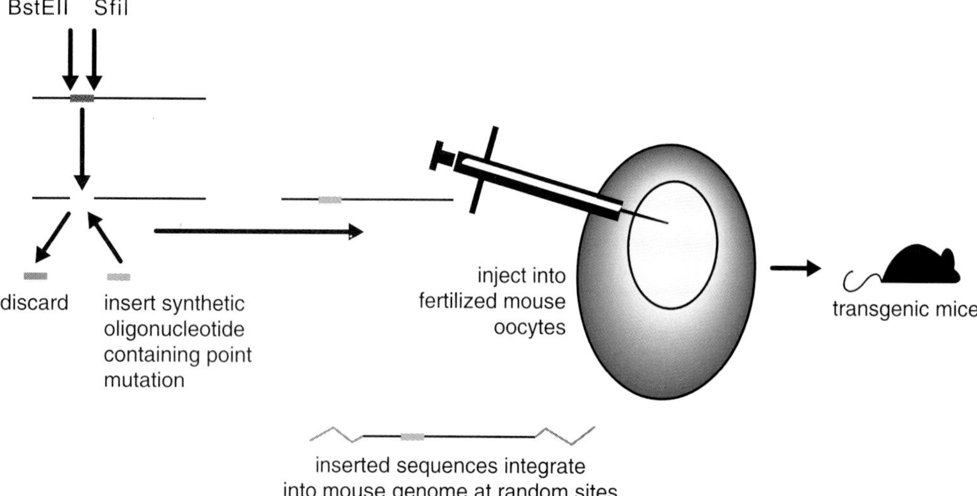

FIGURE 2-24 Transgenic mouse model for osteogenesis imperfecta. A COL1A1 cDNA was created containing a glycine-for-cysteine change at position 859. The base substitution was achieved by removal of a fragment of the clone by restriction enzyme digestion and replacement of the fragment with a synthetic double-stranded DNA containing the mutant sequence. The mutant full-length genomic clone was linked to the mouse COL1A1 promoter sequence and injected into fertilized mouse eggs. Many copies of the injected gene were inserted into the zygotes and were replicated with each cell division. The foreign DNA integrates at random into the mouse genome and is replicated along with mouse genes. The eggs were inserted into foster mothers and were allowed to develop to term in utero. The mice resulting from these eggs were examined phenotypically and at the DNA and RNA levels for presence of the mutant gene and transcript. (Data from Stacey A, Bateman J, Choi T, et al. Perinatal lethal osteogenesis imperfecta in transgenic mice bearing an engineered mutant pro-alpha$_1$(I) collagen gene. Nature 1988;332:131–136.)

at position 526 results in OI type III. Substitution at position 94 results in mild OI type I and, in one family with substitution at position 175, OI type IV was seen. Like Lawrence, several children have been found to have the change in the interval between positions 123 and 401, and generally the phenotype has been one of mild to moderate severity.

The ability to study collagen at the biochemical and molecular levels has made it possible to explore the molecular basis for the clinical types of OI. Osteogenesis imperfecta type I is the mildest form, associated with normal stature but increased bone fragility. The biochemical phenotype is decreased quantity of type I collagen. COL1A1 mutations responsible for OI type I tend to be chain-termination or splicing mutations. These result in a decreased quantity of type I collagen production, but that which is produced is of normal structure. This is in marked contrast with mutations responsible for OI type II, where abnormal alpha$_1$(I) aggregates with normal chains to affect three-fourths of the mature collagen fibrils (see section 2.3).

Gene rearrangements and point mutations responsible for perinatal lethal OI type II have already been described. All the point mutations in COL1A1 lie between amino acids 97 and 1014 in the triple-helical domain. More rarely, mutations can reside in the COL1A2 gene, producing defective alpha$_2$(I)-chains. One interesting instance has been reported of a child with compound heterozygosity for two different mutations in the COL1A2 genes, having inherited one mutant gene from each parent.

OI type III, the progressive deforming form is genetically heterogeneous. Both autosomal dominant and autosomal recessive transmission have been documented for this disorder. The autosomal recessive form generally is less common, except in certain populations, such as in South African blacks, where the autosomal recessive form is more common, presumably due to a founder effect. In this population, the mutation may be in a gene other than the COL1A1 or COL1A2 genes of type I collagen. Mutations of either COL1A1 or of COL1A2 have been identified in different individuals with OI type III. These mutations differ from those responsible for OI type II by their location in the molecule. In one study, mutations were found in a region of the alpha$_1$(I)- or alpha$_2$(I)-chains that are involved in bone mineralization and do not disrupt collagen stability at all.

Most individuals with autosomal recessive OI type III have normal type I collagen, indicating that mutations must reside in genes for other proteins. One exception is a child with OI type III who was homozygous for a four-base deletion in the COL1A2 genes. The deletion occurred near the 3′ end of the gene and led to a frameshift that truncated the protein by 33 amino acids. This ablated the ability of pro-alpha$_2$(I)-chains to participate in the formation of type I procollagen, resulting in the formation of pro-alpha$_1$(I)-homotrimers.

OI type IV is most often relatively mild, but there is a wide range of variation of severity, even within families. The cause of intrafamilial variability is unknown. Point mutations in either COL1A1 or COL1A2 have been identified in different families with OI type IV.

2.9 Somatic Mosaicism for Collagen Mutations

Biochemical studies performed on skin biopsies taken from both of Lawrence's parents are normal. Carol and Neal are counseled that there is still approximately a 6% risk of recurrence of osteogenesis imperfecta in future offspring. When Carol is pregnant again, cultured amniotic fluid cells are examined for collagen production. No evidence of abnormality is found, and a healthy baby girl, Lydia, is born.

OI type II usually affects one child in a family. Because it is lethal in the perinatal period, parent-to-child transmission generally is not observed. The evidence that it is due to a dominant genetic mutation is the observation of heterozygosity for collagen gene mutations in affected children. These changes usually are absent in parents' DNA and hence are acquired by new mutation.

In rare instances, however, OI type II occurs more than once in a sibship. Before it was possible to characterize collagen mutations, such families were thought to represent a rare recessive form of OI type II. Arguing against this hypothesis, however, was the fact that it is unusual to see families with one-fourth of the children affected. Molecular studies have provided a different explanation: One parent carries the mutation in some, but not all, cells.

One such family was studied by Cohn and colleagues. Two children who had the same father but different mothers were affected with OI type II (Figure 2-25). Both children were found to have the same mutation in the alpha$_1$(I)-chain: a substitution of aspartic acid for glycine at amino acid 883 in the triple-helical domain. At the DNA level, this represented a change from G to A, which disrupted a *BglI* enzyme recognition site. Polymerase chain reaction (PCR) primers were made that flanked the site of mutation and amplified a 225–base pair fragment that normally had two *BglI* sites, leading to fragments of 63, 9, and 153 base pairs after amplification and enzyme digestion. If the mutation was present, the fragment sizes were 72 and 153 base pairs. Using this assay, it was found that the father of the two affected children carried the mutation in lymphocytes, hair root cells, and sperm. The proportion of mutation-bearing cells was estimated to be one in eight based on the intensity of the 72–base pair band in the father compared with the children. The father did not manifest signs of OI, presumably because of the low proportion of affected cells. It also is possible that selection acted against mutant collagen-producing cells early in development, causing these to die and be replaced by normal cells.

Another family with somatic mosaicism for a COL1A2 mutation has been reported in which the mosaic individual had short stature and mild dentinogenesis imperfecta. The mutant allele was represented in various proportions in different cell types of this individual. Phenotypic effects probably are related to the proportion of mutation-bearing cells in critical tissues, such as bone, which is difficult to determine by direct testing.

The existence of somatic mosaicism raises an important issue in genetic counseling, indicating that a disorder such as OI type II, which usually is sporadic,

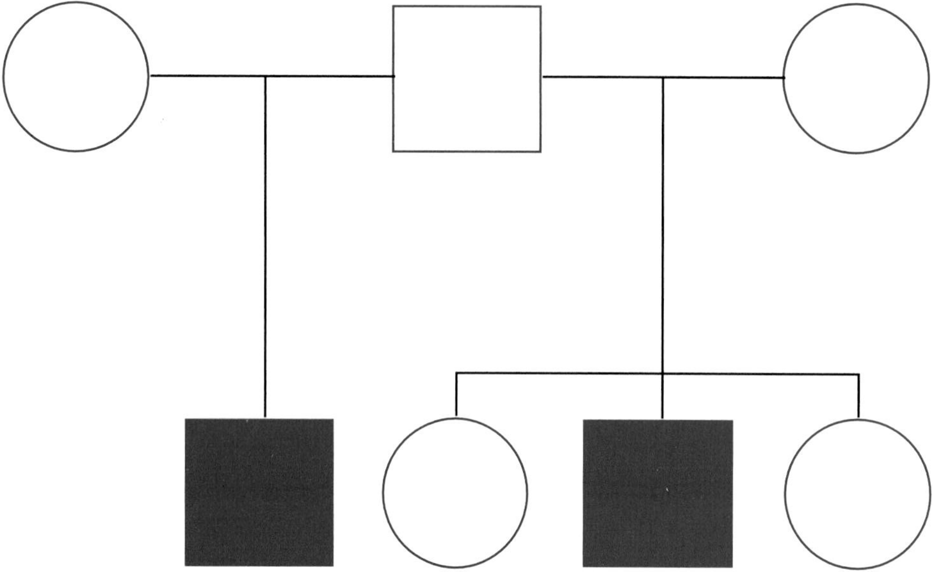

FIGURE 2-25 Family with two half-siblings having perinatal lethal osteogenesis imperfecta. No signs of the disorder were present in their father. (Data from Cohn DH, Starman BJ, Blumberg B, Byers PH. Recurrence of lethal osteogenesis imperfecta due to parental mosaicism for a dominant mutation in a human type I collagen gene (COL1A1). Am J Hum Genet 1990;46:591–601.)

can recur in a family. The recurrence risk of OI type II in a family with one affected child has been estimated to be approximately 6%. This risk figure factors in the possibility of mosaicism in some families and the fact that, in other families, a sporadically affected child may represent a new mutation. Mosaicism can occur in somatic cells or, in some cases, may be confined to the germ-cell line. A similar phenomenon has been seen in a number of other genetic disorders, such as Duchenne muscular dystrophy (see Chapter 4).

2.10 Other Disorders Associated with Defects in Collagen

Ehlers-Danlos syndrome is another disorder that has long been believed to represent a defect in connective tissue. The hallmark is hyperelasticity of skin and hypermobility of joints. Like OI, Ehlers-Danlos syndrome can be divided into a number of clinically distinct subtypes. The features of Ehlers-Danlos syndrome, the degree of severity, and mode of inheritance vary from type to type. Two forms, types IV and VII, have been associated with collagen defects. Type IV is referred to as the *vascular*, or *ecchymotic*, type. Skin is thin and translucent, with veins easily visible. Varicose veins may be present, and there is easy bruisability but minimal joint hypermobility. The major clinical problem associated with this disorder is spontaneous rupture of major arteries, the colon, and the uterus. These events can be life-threatening and can occur anytime, although they are most common after the third decade. Life expectancy rarely exceeds 50 years. There is no treatment other than prompt recognition of the symptoms of rupture and surgical repair.

Ehlers-Danlos syndrome type IV is due to mutations in the COL3A1 gene that encodes the chains of type III collagen. This collagen is prevalent in the walls of arteries, the bowel, and the uterus, coincident with the major tissues affected by the disorder. The biochemical phenotype is deficient production of type III collagen. As in OI, a wide variety of mutations occur in different affected individuals. Ehlers-Danlos type VII is manifested by congenital dislocation of the hips, dislocations of other joints, and extreme joint hypermobility. Most cases are sporadic, although parent-to-child transmission has been documented. In some affected individuals, this disorder is due to mutations in COL1A1 and COL1A2, the genes encoding the chains of type I collagen. The mutations occur in a heterozygous form, affecting one of the procollagen genes, and involve the cleavage sites for the N-terminal region of the molecule, thereby interfering with conversion of procollagen to collagen. This leads to defective collagen cross-linking and therefore reduces the tensile strength of tissues such as tendons.

Two other forms of Ehlers-Danlos syndrome are associated with defective collagen, although the mutations do not directly affect collagen genes. Ehlers-Danlos type VI leads to soft, hyperextensible skin, joint hypermobility, and ocular fragility. The ocular problem can lead to retinal detachment and rupture of the eye. This disorder is inherited in an autosomal recessive manner and is due to mutations in the gene encoding the enzyme lysyl hydroxylase, responsible for posttranslational modification of lysyl residues in collagen. This interferes with collagen cross-linking in the triple helix.

Ehlers-Danlos type IX is X-linked recessive and is characterized by lax skin, bladder diverticula and rupture, and skeletal deformities. It generally affects male individuals (as a result of X-linked recessive inheritance), and the biochemical defect appears to be an abnormality of copper metabolism. This leads to deficiencies of activity of enzymes that depend on copper binding, including lysyl oxidase. Lysyl oxidase is another enzyme involved in collagen cross-linking, and hence loss of activity is associated with generalized weakness of connective tissues.

The molecular basis of other types of Ehlers-Danlos syndrome is unknown. It is believed that defects in other proteins found in connective tissue may underlie these disorders or that other defects in collagen cross-linking may occur.

Three additional disorders have been found to be associated with collagen defects, at least in some individuals. Stickler syndrome is a dominantly inherited disorder characterized by myopia, hypoplasia of the lower jaw, and early onset of arthritis associated with dysplasia of the epiphyses. Type II collagen, a trimer of $alpha_1$(II)-chains, is abundant both in cartilage and vitreous (the fluid in the eye), and therefore was examined for mutations in persons with Stickler syndrome. Genetic linkage studies have indicated linkage of the disease with a polymorphism in COL2A1 in most, but not all, families with Stickler syndrome. Point mutations in COL2A1 have been identified in some families. The gene responsible for Stickler syndrome in other families is unknown. Type II collagen abnormalities have also been demonstrated in some individuals with Kniest dysplasia, a severe skeletal dysplasia characterized by short trunk and limbs, restricted joint mobility, myopia, cleft palate, and hearing loss. In one family, a child with Kniest dysplasia was found to have a deletion in one of her COL2A1 genes. Her mother had Stickler syndrome and carried the same deletion in some, but not all, of her cells. This is another example of somatic mosaicism, here causing a mild phenotype in the mosaic state and a more severe phenotype when all cells are affected.

Another disorder associated with mutant collagen is one of the dystrophic forms of epidermolysis bullosa, a heterogeneous condition in which there is formation of skin blisters. The dystrophic form is particularly disabling, as deep blisters occur from minor skin trauma and everyday movements and lead to permanent scars. A cycle of blistering, scarring, and healing leads to deformities, such as a fusion of the fingers into a mittenlike configuration. Both dominant and recessive forms of dystrophic epidermolysis bullosa have been observed. Linkage has been found between the gene COL7A1 and the dominant form of dystrophic epidermolysis bullosa in some families, and a point mutation in COL7A1 has been found in affected siblings in one family with recessive dystrophic epidermolysis bullosa. COL7A1 encodes the $alpha_1$(VII)-chain of type VII collagen, which anchors the epidermis to the underlying dermis. Mutations of COL7A1 presumably weaken these anchoring fibrils, leading to separation of the epidermis and consequent blistering.

A mutation in type X collagen occurs in some individuals with yet another skeletal dysplasia, Schmid dysplasia. Type X collagen is abundant in cartilage, and this disorder is characterized by defective maturation of collagen into bone. Given the long list of various types of collagen, it is likely that additional forms of connective tissue disorders will continue to be matched with mutations in the many genes that contribute to this complex system.

2.11 Management of Connective Tissue Disorders

OI and other connective tissue disorders cannot be cured. No medication can prevent or reverse bone fractures in OI or prevent arterial or bowel wall rupture in Ehlers-Danlos syndrome type IV. Dietary supplementation with calcium in OI or vitamin C (which is a cofactor for one of the enzymes involved in collagen cross-linking) is not known to be effective. Current management for these disorders is based on prevention of complications and symptomatic treatment. For OI, this amounts to avoidance of activities that might lead to trauma and dealing with fractures as they occur. Individuals with Ehlers-Danlos syndrome tend to get skin lacerations easily, and weakness of the underlying connective tissue makes it difficult for sutures to hold. This must be kept in mind when treating persons with the disorder.

Medical professionals can play an important role in educating families about these disorders and providing anticipatory guidance. Often a family has little knowledge about a condition at the time of birth of an affected child. Establishing a correct diagnosis can help to predict the likely course of a disorder and identify medical problems for which the person is at risk. It can also dispel myths and misconceptions about the disorder and place the medical problems in

Living with Osteogenesis Imperfecta

JUDITH KONG

When I was 6 months pregnant, I found out my son Travis had osteogenesis imperfecta, commonly called OI. Diagnosis was made by ultrasound. Travis's head was 10 weeks ahead of development of his femurs. He had little ossification of his skull and multiple fractures. I was told that he might not be born alive. The 2 months prior to his birth were the most difficult. I walked around in a fog. I desperately wanted my son to be born alive, yet I was petrified of having him.

When Travis was born, my family consisted of myself, my husband, and our 2-year-old daughter Jasmine. My mother and three of my sisters and their families lived in the same town. Fortunately, we had a wonderful family support system.

At birth, Travis weighed 6 pounds. His face and temples were the only hard surfaces on his skull. Old rib fractures were seen. He had active fractures of his humerus and clavicle, and both femurs were crumpled. At this time, he was diagnosed with type II OI and was given a very poor prognosis. Fortunately, a few months later he was diagnosed by genetic testing as having type III OI. Even with this, we remained in unknown territory because we were told that this mutation was the first one identified of its precise nature.

These first few months were very difficult. Travis cried constantly, seldom sleeping more than 30 minutes at a time. It was hard to know if he had bone pain or colic. Day and night, Travis was almost constantly held and always loved.

Life has felt like a roller-coaster ride. We might have a month without a crisis, and I would be on an emotional high. A new fracture would occur, and I would feel very low. Progress was so hard to see. At a year old, Travis weighed 11 pounds. At 20 months, he rolled over. At 3½, he could hold up his head when held over your shoulder. At 5, he could hold up his head from laying prone and learned sitting balance. At 7, he is still unable to get to the sitting position independently and weighs only 22 pounds. Crisis is a very common occurrence; between bone

perspective. In some cases, a program of regular medical surveillance may be appropriate to identify problems at a presymptomatic stage. Individuals with OI, for example, are at risk for developing hearing loss. Careful monitoring of hearing can lead to early identification and management of this problem.

Genetic counseling involves more than identification of recurrence risks and discussion of options for prenatal diagnosis. It begins with making a diagnosis and includes education of the family regarding the natural history of the disorder and options for management. It is not strictly the domain of the geneticist but should be a component of medical care provided by all health professionals.

Case Study

Part I

James and Thomas are brought by their mother to their primary care physician for a routine annual examination. James is 12 years old, Thomas 15. This has been a difficult year for the family, as their father, Henry, died suddenly and unexpectedly at age 45 of a ruptured aorta.

The physician is aware of this event and reviews the family history with the boys' mother. Henry's mother died in her early fifties of unknown causes, and Henry had two brothers and a sister. One brother is also dead

fractures and respiratory problems, we seldom have a 3-month period without worrying about Travis. Sometimes we can lose up to a year's worth of progress with one major crisis.

Our orthopedist has been a savior. He made many a referral for this distraught mother before I was hooked into the appropriate networks. He received calls for anything from hernias to respiratory distress. Within hours, I would have the appropriate resource. For the first 2 years, Travis never waited to see the orthopedist. We would go to the casting room and page him and, within minutes, he would see Travis. He was readily available for each broken bone, even on Mother's Day. Last year, Travis was involved in a school bus accident and broke both femurs. This wonderful man returned from his vacation to care for Travis personally.

Our doctor in the genetics clinic is also a blessing. She took over Travis's respiratory issues when I felt that his pulmonologist was useless. That doctor never saw Travis during any of his frequent hospitalizations for respiratory distress, nor could she answer my questions as to why Travis's breathing was so much more labored when he was sitting. The geneticist at one point needed to take some time off. During this time, I felt that I had solved part of the mystery—fluid overload. The doctor who had never dealt with me disregarded this as a possibility. I bought a scale and recorded Travis's daily weight along with an accurate intake and output. Fortunately, the geneticist had come back, and she agreed with my conclusion. Travis was placed on a diuretic and became much more stable and required fewer hospitalizations. Two years after Travis started on his regimen, another child with OI was started on the same treatment. If the geneticist had not been willing to listen to a parent with a totally undocumented theory, I believe there may have been two fewer children in the world today. This doctor also found a pulmonologist for Travis who is a pleasure to work with.

In conclusion, I summarize my 7 years with OI as follows: Travis is a remarkably intelligent, vivacious child. He is small in body but large in spirit. All those who know him love him. Life is totally unpredictable; just when things are going smoothly, a crisis will develop. Many things have had to be canceled at the last minute. It is my belief that being treasured by a large network of family and friends, in combination with having remarkable doctors, has allowed my son not only to live but to thrive.

of cardiovascular problems. The other brother and sister are alive, but the sister has severe visual problems due to dislocation of the lens in both eyes. Similar visual problems were present in Henry and his mother.

The family history suggests the possibility of autosomal dominant transmission of a cardiovascular disorder, perhaps associated with ocular problems.

The fact that Henry died of ruptured aorta is particularly suggestive of a disorder involving connective tissue. Ruptured aorta is characteristic of a number of disorders, including Ehlers-Danlos syndrome (type IV), *Marfan syndrome*, and can exist in isolation as a familial disorder. Apparently, the genetic pattern has eluded detection in this family so far. The primary care physician can provide an important service to the family by recognizing that there is a familial disorder and evaluating James and Thomas for possible signs of the disorder. This presents the possibility of altering the boys' medical management to prevent a disastrous outcome, such as occurred to their father, and allows for genetic counseling to be provided.

Part II

Although he has examined the boys many times, the physician devotes special attention this year to cardiovascular problems. He hears a faint extra heart sound—a ''click''—in Thomas. Both boys are quite tall, exceeding the ninety-fifth percentile for their age. (This had been true since early childhood, but was not viewed as being remarkable because Henry was tall.) Thomas is found to have mild scoliosis (lateral curvature of the spine). Both boys also are found to have long fingers (Figure 2-26) and loose joints.

Examination of James and Thomas provides further evidence for a connective tissue disorder. The boys are tall, which could be an incidental familial tendency or could be part of a disorder. That they have long limbs and fingers and toes suggests the latter. Flat feet, scoliosis, and hyperextensible joints further indicate a laxity of connective tissue. The click heard in Thomas's cardiac examination is indicative of a floppy mitral valve, referred to as *mitral valve prolapse*. All these features are suggestive of Marfan syndrome.

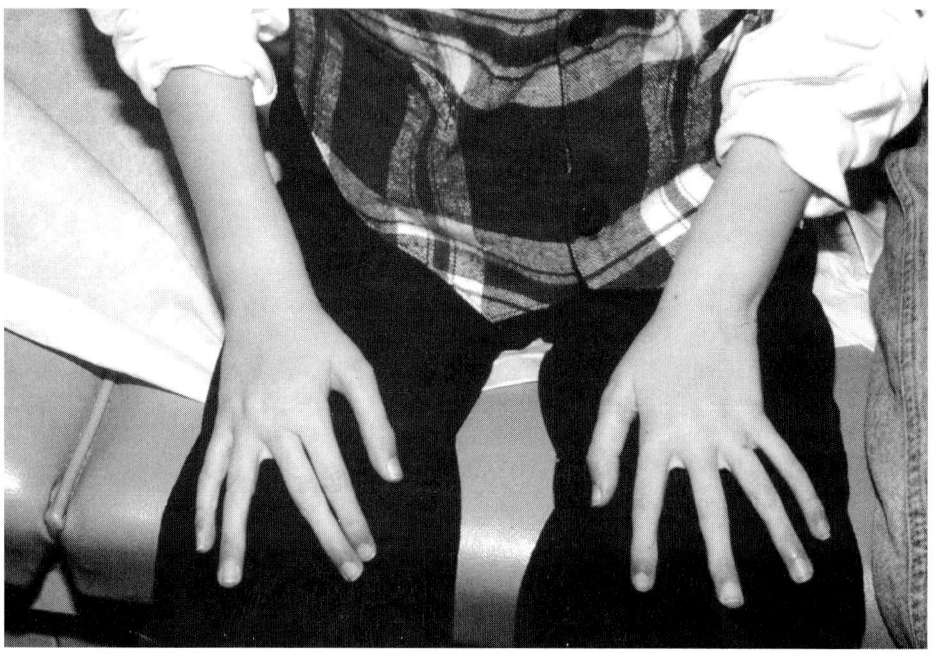

FIGURE 2-26 Arachnodactyly seen in a child with Marfan syndrome. (Courtesy of Dr. Ronald Laero, Department of Cardiology, Children's Hospital, Boston.)

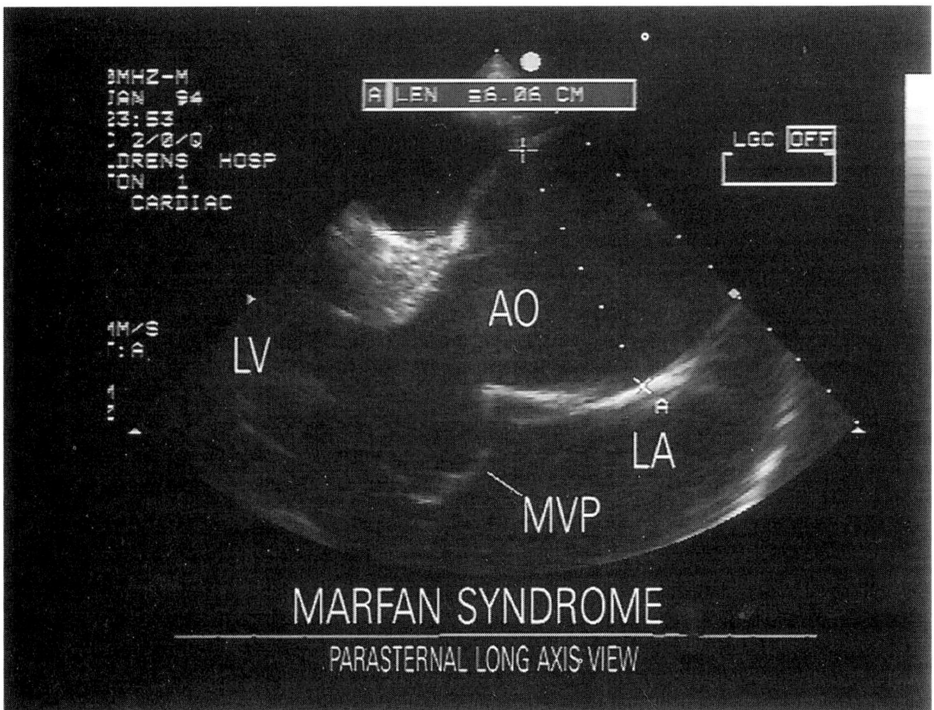

FIGURE 2-27 Echocardiogram showing aortic root dilation and mitral valve prolapse (*MVP*). (Courtesy of Dr. Ron Lacro, Department of Cardiology, Children's Hospital, Boston.) (*AO* = aorta; *LV* = left ventricle; *LA* = left artery.)

Marfan syndrome is an autosomal dominant disorder in which connective tissue is weakened. Skin may be soft but, unlike in Ehlers-Danlos syndrome, is not hyperelastic. Skin striae (stretch marks) tend to develop with age. Persons with Marfan syndrome tend to be tall and lanky, with especially long arms, legs, fingers, and toes. Elongation of digits is referred to as *arachnodactyly*. This growth pattern can be apparent even in the newborn period, although often it does not come to attention until later in childhood or adolescence. Joints are hyperextensible and may be prone to dislocation. Flat feet and scoliosis occur commonly. In the eye, the ligaments that support the lens may become weak and lead to lens dislocation. Myopia also is common. Cardiovascular manifestations include mitral valve prolapse and dilation of the root of the aorta, which may lead to aneurysm. Aortic dissection is the most life-threatening manifestation of Marfan syndrome.

Part III

In view of the family history and the finding of a click in Thomas, both boys are referred to a cardiologist. An echocardiogram reveals mitral valve prolapse in both and mild dilation of the root of the aorta in Thomas but not in James (Figure 2-27). The cardiologist suggests that the boys might have Marfan syndrome. He also prescribes a beta-blocker medication, atenolol, for Thomas. The cardiologist contacts the primary care physician to explain his findings and suggests that both boys be followed with annual echocardiograms. Also, both will require antibiotic treatment at the time of any dental procedures.

It is the cardiologist who puts the features together and suggests the diagnosis of Marfan syndrome for this family. Mitral valve prolapse is a relatively common problem, affecting nearly 5% of the general population. It consists of floppiness of the mitral valve, leading to regurgitation of blood into the left atrium during contraction of the left ventricle. It is characteristic of a wide variety of connective tissue dysplasias, including Marfan syndrome. It can also occur in isolation and can be a familial autosomal dominant trait. In isolation, it is most common in thin women. Mitral valve prolapse usually is asymptomatic, although it can be associated with fatigue, chest pain, or palpitations. Sudden death occurs rarely. The presence of a midsystolic click is the only physical sign on examination. Usually mitral valve prolapse does not require treatment. Those who have an audible click are at risk of bacterial endocarditis due to seeding of the floppy valve during times when bacteria are transiently present in the bloodstream. This occurs most commonly during dental procedures, so it is recommended that such individuals receive antibiotic treatment with such procedures.

Dilation of the aortic root, which is detected by echocardiography, is highly characteristic of Marfan syndrome. This has been attributed to stretching of the aorta due to the force of blood during systolic contraction of the heart. This can be a progressive problem, beginning in childhood or adolescence, and might be the harbinger of aortic dissection, which can lead to sudden death. It is important to recognize this problem in persons with Marfan syndrome, as careful surveillance can help prevent a catastrophic outcome. There is evidence that use of beta-blocker medications can reduce the likelihood of dissection. Such medications reduce the force of systolic contraction hence stress on the aorta. They are commonly used to reduce blood pressure in those with hypertension. Regular monitoring by echocardiography can lead to early detection of aneurysms, which, if present, can be treated surgically by replacement of the damaged vessel with a prosthesis.

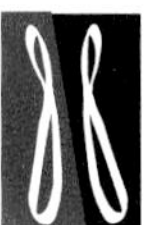

Part IV

The boys are referred to an ophthalmologist. No significant problems are noted. An orthopedist prescribes inserts for their shoes because of flat feet. Both boys have some degree of scoliosis, but neither requires treatment at present. The primary care physician speaks with James, Thomas, and their mother about Marfan syndrome. The family members have been seen by a number of specialists, and the family is getting a bit confused. Their physician refers them to a geneticist, who further explains the disorder and provides literature from the National Marfan Foundation.

The primary care physician plays a pivotal role in helping the family deal with a genetic disorder such as Marfan syndrome. Anticipatory guidance amounts to educating the family about possible problems and providing appropriate medical surveillance. Specialists may be called on to deal with particular problems. For Marfan syndrome, the major systems involved are cardiovascular, orthopedic, and ophthalmologic. Orthopedic problems include scoliosis, joint dislocations, and flat feet. Ophthalmologic problems include dislocation of the lens and myopia. Orthopedic and ophthalmologic complications of Marfan syndrome are not preventable by medical management, but prompt recognition of these problems is important to ensure the best outcome from symptomatic treatment.

Although Marfan syndrome can lead to lifelong medical problems, it is also compatible with a healthy, productive life. Aside from educating the family about the potential complications, the physician should place the disorder in perspective. There is a wide range of expression of Marfan syndrome; some individuals are obviously affected even in infancy, whereas others are so mildly affected as to be unaware of having the disorder. Medical management should be customized to the needs of the individual and normal activities encouraged as much as possible.

Part V

Medical review of the extended family reveals several others with Marfan syndrome (Figure 2-28).

Recognition of a condition such as Marfan syndrome can have implications for many members of a family. Other members may be affected and should be offered medical evaluation and genetic counseling. Usually this is communicated by family members themselves, who should encourage their at-risk relatives to seek medical evaluation. The process may be impeded, however, by emotional rifts within the family, or fear of medical problems or loss of insurance

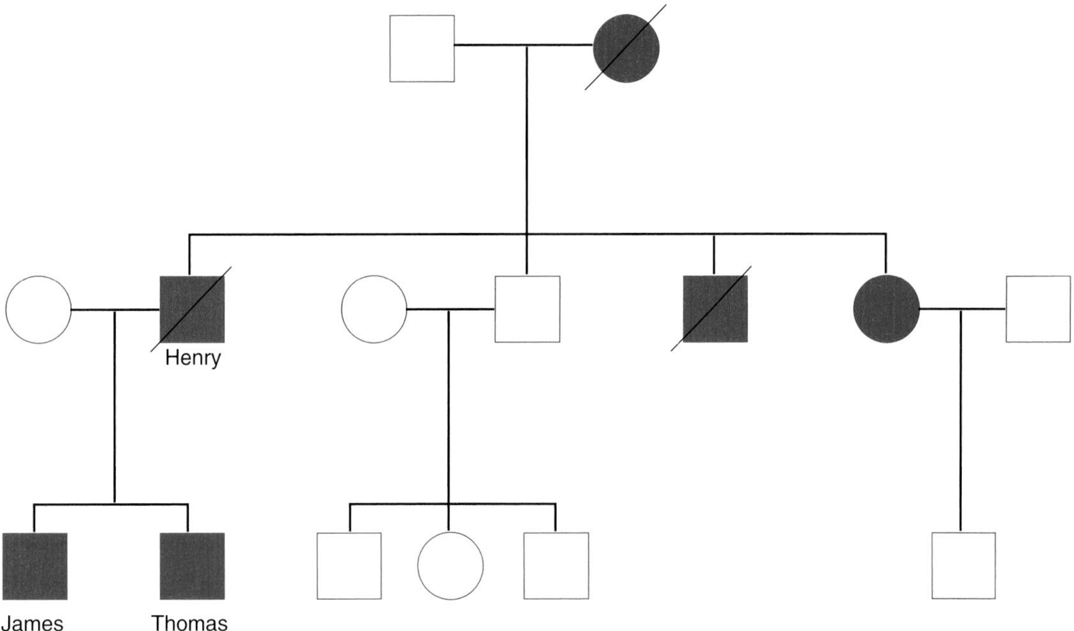

FIGURE 2-28 Pedigree for James and Thomas's family.

benefits on the part of some individuals. In some instances, this can create an ethical dilemma for the physician, who may have information that can be important for the health of a relative yet who is constrained by the confidentiality of the physician-patient relationship from divulging such information without permission of the patient.

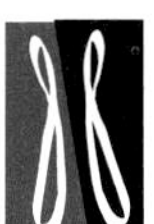

Part VI

The family is contacted by a research geneticist (through the family doctor) who is studying the molecular basis of Marfan syndrome. Skin and blood samples are obtained from James and Thomas. The skin samples are studied for fibrillin metabolism. Cultured cells are incubated in the presence of [^{35}S] cysteine for 30 minutes, after which they are grown in non–radioactively labeled cysteine an additional 0 to 20 hours. Extracts from cells, culture medium, and the extracellular matrix then are subjected to electrophoresis, and the radioactively labeled protein is visualized by autoradiography. Amounts of fibrillin in cellular, culture medium, and extracellular matrix fractions are determined by measuring the intensity of bands on the autoradiogram in James's and Thomas's samples as compared with controls. Overall fibrillin synthesis values for both boys' fibroblasts are approximately 50% of control values, and the amount of fibrillin deposited into the extracellular matrix is less than 30% of control. These results are shown in Figure 2-29. Immunofluorescence staining of fibrillin in skin fibroblasts is also performed (Figure 2-30).

The gene responsible for Marfan syndrome was identified through a convergence of genetic and histochemical analyses. The genetic approach amounted to linkage analysis using random genetic markers, which implicated a locus on chromosome 15 as being responsible for the disorder. Histochemical studies indicated a deficiency of the extracellular protein fibrillin in the tissues of individuals with Marfan syndrome. The cDNA for fibrillin was cloned, and the gene was found to map to chromosome 15 in the same region as the Marfan syndrome locus. Final proof that fibrillin represents the gene involved in Marfan syndrome was provided by identification of mutations in fibrillin in several individuals with Marfan syndrome.

Fibrillin is a 35-kDa glycoprotein found in connective tissue microfibrils. The protein contains 44 tandem domains homologous to a sequence found in epidermal growth factor. There are 65 exons encoded in a 9287–base pair transcript. There is a homologous gene on chromosome 5. This gene is referred to as *fibrillin 2*, and the gene responsible for Marfan syndrome on chromosome 15 as *fibrillin 1*.

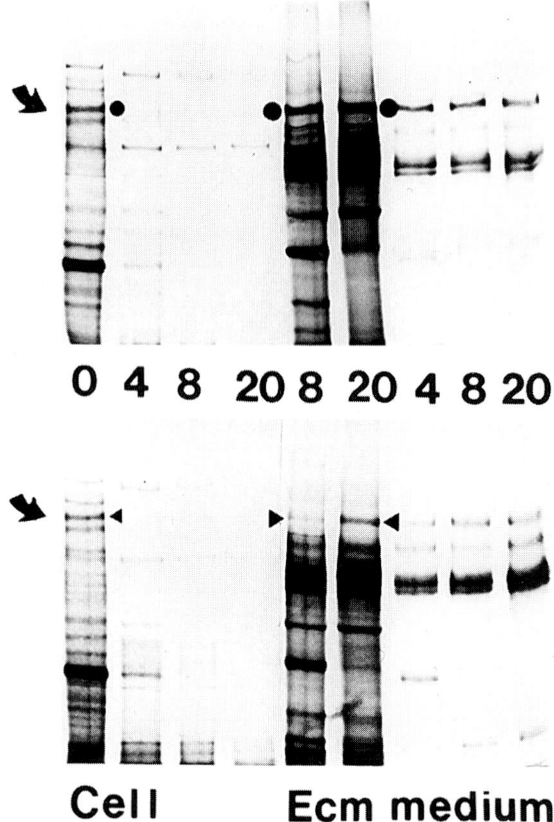

FIGURE 2-29 Autoradiogram of pulse-chase labeled control (top) and Marfan syndrome fibroblasts (bottom). Cells were labeled for 30 minutes with ^{35}S-cysteine and, after washing, the medium, cell extract, and insoluble material (*Ecm*) were dissolved in SDS at times 0, 4, 8, and 20 hours, and polypeptides were separated by gel electrophoresis. Fibrillin migrates as a single band of approximately 350 kDa (*curved arrow*). By comparison of fibrillin in the cell extracts and Ecm (*dots* versus *triangles*), the difference in synthesis and deposition between control and Marfan syndrome cells can be assessed. (Courtesy of Dr. Heinz Furthmayr, Department of Pathology, Stanford University.)

Fibrillin 2 has been implicated in a separate genetic disorder, congenital contractural arachnodactyly. In addition to having a role in Marfan syndrome, fibrillin 1 also appears to be responsible for a genetic form of ectopia lentis.

At the biochemical level, five types of fibrillin 1 abnormalities are seen. One type (type I) is characterized by decreased fibrillin synthesis, but normal proportions of fibrillin are secreted into the extracellular matrix. Total amounts of fibrillin in the extracellular matrix are approximately 50% of normal, proportional to the amount of fibrillin made. This biochemical phenotype is associated with a mild clinical phenotype. Type II involves reduced synthesis and secretion, associated with a more severe form of Marfan syndrome. Types III and IV involve normal rates of synthesis but moderate or severe reductions of deposition in the extracellular matrix. The final biochemical phenotype is entirely normal fibrillin production and deposition, in which case the cause of Marfan syndrome is unclear.

The mutation in this family results in a type II biochemical phenotype. As shown in Figure 2-29, there is marked reduction of total amounts of fibrillin both in the cellular fraction and in the extracellular matrix. The immunofluorescence photograph shows reduction in fibrillin staining.

FIGURE 2-30 Immunofluorescence photomicrograph of normal (**A**) and Marfan syndrome (**B**) fibroblast sample stained for fibrillin. (Courtesy of Dr. Heinz Furthmayr, Department of Pathology, Stanford University.)

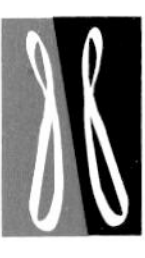

Part VII

DNA is obtained from peripheral blood lymphocytes from James and the fibrillin 1 gene studied by single-strand conformational polymorphism analysis (Figure 2-31A). An altered pattern, indicative of probable mutation, is found. The DNA is sequenced, and the result is shown in Figure 2-31B.

Genetic linkage studies have indicated that all cases of Marfan syndrome are likely to be due to fibrillin mutations. These mutations, however, are highly diverse and, therefore, locating an individual mutation can be challenging. Mutations responsible for reduced synthesis but normal deposition of fibrillin tend to be chain-termination mutations. The normal allele produces normal fibrillin, but the quantity is reduced to 50%. Mutations responsible for reduced deposition of collagen, with or without reduced synthesis, tend to be compatible with production of some abnormal fibrillin, which interacts with the product of the normal allele. Many of these are missense mutations.

In this family, the mutation was identified first by single-strand conformational polymorphism (SSCP) analysis and then was characterized by DNA sequencing. SSCP involves PCR amplification of segments of the gene (or cDNA), denaturation of the PCR products, and polyacrylamide gel electrophoresis under renaturing conditions (Figure 2-32). Rate of migration through the gel depends on conformation, which in turn reflects base sequence. Changes in sequence as subtle as a single base sometimes alter migration through the gel. In the example in the figure, the two strands of each PCR product are visible. Controls have two bands, whereas the Marfan syndrome patient has four, two corresponding with the wild-type allele and two with the mutant allele. SSCP provides a sensitive method for rapidly screening for mutations, although not all possible mutations will be detected.

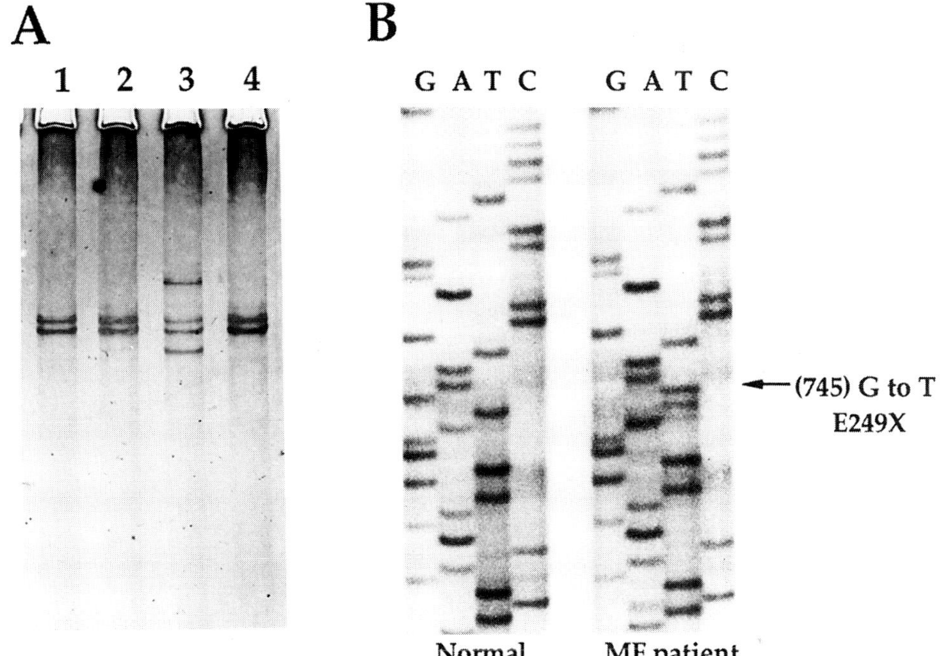

FIGURE 2-31 (**A**) Single-strand conformational polymorphism analysis of fibrillin 1 gene. The gene was amplified in small overlapping sections. Polymerase chain reaction (*PCR*) products were denatured and run on a polyacrylamide gel under renaturing conditions. Lanes 1, 2, and 4 are controls. Lane 3 is from Marfan syndrome DNA. (Courtesy of Dr. Uta Francke, Howard Hughes Medical Institute, Stanford University.) (**B**) Mutation analysis of fibrillin. The normal allele is sequenced on the left, the Marfan syndrome patient on the right. The mutation consists of a G-to-T transversion at position 745, which changes a glutamic acid codon (GAA) to a stop codon (TAA). (Courtesy of Dr. Uta Francke, Howard Hughes Medical Institute, Stanford University.)

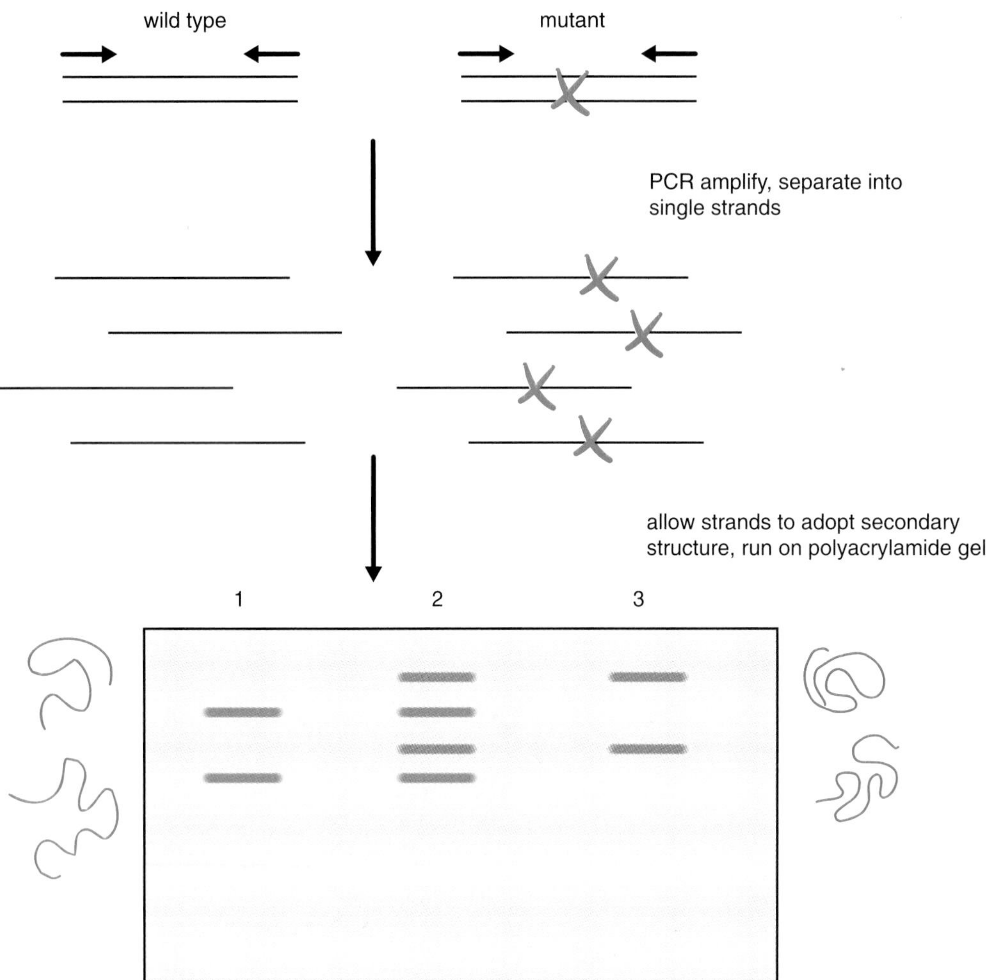

FIGURE 2-32 Diagram of single-strand conformational polymorphism analysis. DNA is amplified by polymerase chain reaction (*PCR*), and PCR products are separated into single strands. The strands then are allowed to adopt a secondary structure that depends in part on base sequence, and the strands are analyzed by polyacrylamide gel electrophoresis. The migration of single strands depends on conformation. A homozygote for a wild-type sequence (lane 1) has two bands, and a homozygote for a mutant sequence (lane 3) reveals two different bands. Both sets of bands are seen in a heterozygote (lane 2).

The sequence analysis shows that the patient has a G-to-T transversion, which changes a glutamic acid codon (GAA) to a stop codon (TAA). This causes truncation of the protein at amino acid 745. The truncated protein interferes with fibrillin deposition in a dominant negative fashion, explaining the less than 30% of normal deposition seen in Figure 2-29 and the low concentration of stained microfibrils in the immunofluorescence photograph (see Figure 2-30).

Other genotype-phenotype correlations have been made. The type I biochemical phenotype, in which there is decreased fibrillin synthesis, tends to be associated with mutations that either reduce levels of transcription or cause frameshifts that lead to low levels of product. Unlike the type II mutations, these type I mutations do not have a dominant negative effect, and the Marfan phenotype tends to be milder. Groups III and IV tend be associated with missense mutations. Type IV seems to be associated with mutations that inhibit microfibril assembly, also a dominant negative effect. Type III may be more heterogeneous pathogenetically. Both reduction in quantity of fibrillin produced and inhibition of assembly may occur in different cases.

REVIEW QUESTIONS

1. Would you expect a missense mutation in the fibrillin gene to result in a more or less severe phenotype than a frameshift occurring near the 5′ end of the gene?
2. A genomic DNA clone is found to hybridize to multiple bands on a Southern blot of restriction enzyme–cut DNA. How would you explain the presence of these multiple bands?
3. The penetrance of neurofibromatosis is essentially 100%. Would you give the same counseling to the parents of this affected child as you would to his sister, assuming that both parents and the sister have no signs of the disorder?

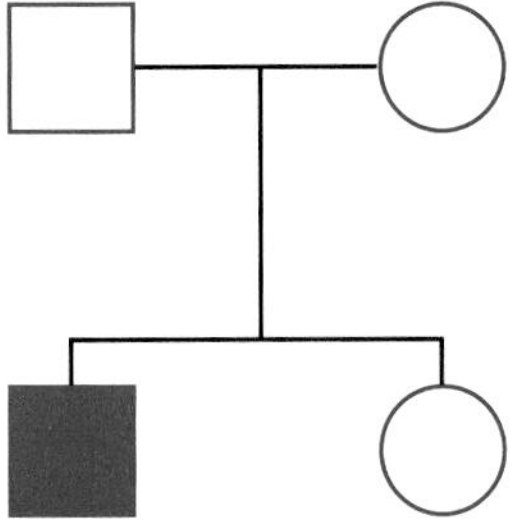

4. It sometimes is said that severe dominant phenotypes are more likely to be the result of new mutations than are severe recessive phenotypes. Why might this be so?

FURTHER READING

2.8 Genotype-Phenotype Correlations in Osteogenesis Imperfecta

Forlino A, Zolezzi F, Valli M, et al. Severe (type III) osteogenesis imperfecta due to glycine substitutions in the central domain of the collagen triple helix. Hum Mol Genet 1994;3:2201–2206.

Willing MC, et al. Osteogenesis imperfecta type I: molecular heterogeneity for COL1A1 null alleles of type 1 collagen. Am J Hum Genet 1994;55:638–647.

2.10 Other Disorders Associated with Defects in Collagen

Winterpacht A, Hilbert M, Schwarze U, et al. Kniest and Stickler dysplasia phenotypes caused by collagen type II gene (COL2A1) defect. Nature Genet 1993;3:323–326.

CASE STUDY

Aoyama T, Francke U, Dietz HC, Furthmayr H. Quantitative differences in biosynthesis and extracellular deposition of fibrillin in cultured fibroblasts distinguish five groups of Marfan syndrome patients and suggest distinct pathogenetic mechanisms. J Clin Invest 1994;94:130–137.

Aoyama T, Francke U, Gasner C, Furthmayr H. Fibrillin abnormalities and prognosis in Marfan syndrome and related disorders. Am J Med Genet 1995;58:169–176.

Dietz HC, Pyeritz RE. Mutations in the human gene for fibrillin-w (FBN1) in the Marfan syndrome and related disorders. Hum Mol Genet 1995;4:1799–1809.

Kainulainen K, Karttunen L, Puhakka L, et al. Mutations in the fibrillin gene responsible for dominant ectopia lentis and neonatal Marfan syndrome. Nature Genet 1994;6:64–69.

Kainulainen K, Sakai LY, Child A, et al. Two mutations in Marfan syndrome resulting in truncated fibrillin polypeptides. Proc Natl Acad Sci USA 1992;89:5917–5921.

Maslen CL, Glanville RW. The molecular basis of Marfan syndrome. DNA Cell Biol 1993;12:561–572.

Peltonen L, Kainulainen K. Elucidation of the gene defect in Marfan syndrome: success by two complementary research strategies. FEBS Lett 1992;307:116–121.

Rantamäki T, Lönnqvist L, Karttunen L, et al. DNA diagnostics of the Marfan syndrome: application of amplifiable polymorphic markers. Eur J Hum Genet 1994;2:66–75.

Tsipouras P, Del Mastro R, Sarfarazi M, et al. Genetic linkage of the Marfan syndrome, ectopia lentis, and congenital contractural arachnodactyly to the fibrillin genes on chromosomes 15 and 5. N Engl J Med 1992;326:905–909.

Tynan K, Comeau K, Pearson M, et al. Mutation screening of complete fibrillin-1 coding sequence: report of five new mutations, including two in 8-cysteine domains. Hum Mol Genet 1993;2:1813–1821.

3 The Human Genome

OVERVIEW

Identification of the gene responsible for a genetic disorder adds greatly to our understanding of the condition and our ability to deal with it. The cloned gene provides a basis for accurate diagnosis, including prenatal diagnosis. Genotype-phenotype correlations may be established, providing prognostic information along with diagnostic testing. The basic pathophysiology can be studied and animal models created to explore new avenues of treatment. Finally, genetic therapy becomes at least possible.

For the disorders considered so far, such as albinism and osteogenesis imperfecta, genes were cloned based on prior knowledge of the gene products. The cloning strategies made use of information about the protein products to identify the cloned genes. Although many other disorders have been approached in this way, the greatest challenge in molecular genetics is to clone a gene for a disorder in which the gene product is not known. This is true of most disorders and poses major problems in medical management. For the most part, we are ignorant about the underlying pathogenesis of these disorders, have only crude diagnostic tools, and lack any means of effective treatment. Recognition of genes for such disorders offers the opportunity to erase decades of ignorance and provide new tools for medical care.

The strategy for cloning genes whose products are unknown is referred to as **positional cloning.** First, the location of the gene on the chromosome is determined, and the entire region is cloned. This region then is searched for expressed sequences, and the gene of interest is identified on the basis of finding sequence differences between affected and unaffected individuals. The approach has also been referred to as *reverse genetics*, as it involves cloning a gene to identify the gene product rather than using the protein product to identify the gene. The term is a misnomer, however, since the genetic approach is in fact quite direct.

By any name, the positional cloning approach has led to major triumphs in human genetics since the early 1980s. This chapter will focus on one of those successes, the identification of the gene for cystic fibrosis. We will start with a case history and describe the clinical features of cystic fibrosis. Like albinism, cystic fibrosis is transmitted as an autosomal recessive disorder but, unlike albinism, it is relatively common, with a carrier frequency of 1 in 25 persons in the northern European white population.

The gene for cystic fibrosis was mapped to chromosome 7 by genetic linkage analysis. We will see how this mapping study was done and how linkage data is analyzed in humans, where studies must be done on relatively small families. Once genetic linkage was established and before the gene was identified, molecular diagnosis became possible using closely linked markers. We will review how that was accomplished for cystic fibrosis and how linkage-based diagnosis has been employed for other disorders. The high carrier frequency for cystic fibrosis is due, in large part, to a founder effect. There is still a weak association of the cystic fibrosis gene with alleles of nearby polymorphic loci, a phenomenon known as **linkage disequilibrium.** The basis for this effect, and its utility in genetic diagnosis, will be demonstrated. Then we will see how the cystic fibrosis gene was cloned, starting from a number of closely linked polymorphic loci, and how identification of the gene has led to improved diagnostic testing. Given the high carrier frequency of this disorder, such testing can be offered not only to affected individuals but also for carrier screening in the general population. This raises major logistical and ethical problems.

The ultimate hope for those affected with cystic fibrosis is the discovery of an effective treatment. Progress in molecular genetics has led to the possibility of gene therapy. We will look at the animal models that have been established for cystic fibrosis and review progress that is being made toward replacement of the defective gene.

An international effort now is under way to map the human genome and identify the 100,000 or more genes. This offers the promise of characterization of all genes responsible for genetic disorders. We will describe this effort and the strategies currently in use. We then will consider briefly a new means of prenatal diagnosis for cystic fibrosis, based on embryo biopsy. The Perspective explores life with cystic fibrosis, and the chapter closes with a clinical case based on another genetic disorder that has been the subject of a successful positional cloning effort.

3.1 Cystic Fibrosis

1978
Bill and Jane are concerned about the slow weight gain of their 6-month-old daughter Betsy. Betsy was healthy at birth, which was a relief because Jane had vomited persistently through the first 3 months of pregnancy. Several prenatal ultrasound examinations had been done and were normal. During the first 2 to 3 months of life, Betsy fed poorly and was irritable. Her stools were copious and loose. Although she was a healthy 8 pounds, 12 ounces (3.9 kg) at birth, she was slow to regain her birth weight and gradually fell off the growth curve. After several unsuccessful formula changes, the pediatrician begins an evaluation. Blood and urine tests are normal, but a sweat test reveals high levels of sodium chloride. Bill and Jane are told that Betsy has cystic fibrosis.

Cystic fibrosis is characterized mainly by the production of excessively thick secretions. The diagnosis usually is made in young children for one of two reasons: chronic malabsorption leading to failure to thrive or recurrent respiratory tract infections. Malabsorption results from deficient secretion of pancreatic enzymes, leading to inadequate digestion of proteins and fats, which are lost through the stool. Poor weight gain ensues, and stools tend to be copious and foul-smelling. Respiratory problems result from obstruction of small airways by thickened mucus and colonization by bacteria that eventually become resistant to commonly used antibiotics. There are frequent upper and lower respiratory infections, including colds and pneumonias. Aside from thickened secretions, there is excessive secretion of NaCl in sweat, which forms the basis for the most common diagnostic test for cystic fibrosis: Sweat production is stimulated by application of an electric current to a small area of skin, the sweat is collected, and NaCl content is determined. The test is inexpensive, sensitive, and specific.

Bill and Jane are in shock. They are worried that Betsy will die or be severely handicapped. Moreover, they cannot understand how this has happened. They have heard that cystic fibrosis is a hereditary disorder, but neither is aware of any history of the disease on either side of the family. Betsy is their first child.

Bill and Jane meet with the staff of the cystic fibrosis clinic. Betsy is started on an oral preparation of pancreatic enzymes and fat-soluble vitamins, and her weight gain improves. She is also treated with an oral antibiotic effective against Hemophilus influenzae *and* Staphylococcus, *which results in fewer respiratory infections. Her parents are reassured by her good progress over a period of several months but are still very upset.*

Management of cystic fibrosis is based on enzyme replacement for those with pancreatic insufficiency—not all with cystic fibrosis have pancreatic involvement—and antibiotic treatment for respiratory infections. The family is also taught chest physical therapy—how to pound on the chest to dislodge thickened secretions from small airways. Such efforts can reduce the burden of childhood illness considerably and improve nutrition.

3.2 Genetics of Cystic Fibrosis

Bill and Jane are told that cystic fibrosis is inherited as an autosomal recessive trait. They are both of Irish ancestry and are surprised to learn that 1 in 25 individuals of northern European background is a cystic fibrosis carrier. It is explained that they face a 25% risk of recurrence of cystic fibrosis in any future child.

Cystic fibrosis represents another example of a disorder subject to the founder effect, described in Chapter 1. Although the disorder is found in many races and in many parts of the world, it is most prevalent in whites of European background. In this population, the disease frequency is 1 per 2500.

Determination of the disease frequency is relatively straightforward, as most affected individuals sooner or later come to medical attention. It is carrier frequency, however, that is most useful for genetic counseling, and this is a more difficult value to measure. Cystic fibrosis carriers have no clinical or biochemical abnormalities. They do not have malabsorption, do not have increased respiratory infections, and do not secrete excessive NaCl in their sweat. Decades of effort to design a reliable biochemical carrier test for cystic fibrosis have been unproductive.

The frequency of cystic fibrosis carriers can be calculated, however, by inference from the disease frequency, using principles of population genetics. The basis for this calculation was established in 1908 independently by the English mathematician G. H. Hardy and the German physician W. Weinberg. Their formulation, which has come to be known as the *Hardy-Weinberg equilibrium*, states a simple relationship between the frequency of alleles at a genetic locus and the genotypes resulting from those alleles.

Consider a gene locus with alleles *A* and *a*. Let the frequency of *A* be designated by the variable p and the frequency of *a* by the variable q. If all alleles at this locus are either *A* or *a*, then $p + q = 1$. The frequency of sperm or egg cells in the population carrying *A* or *a* will thus be p or q, respectively (Figure 3-1). If we assume that the union of germ cells carrying either *A* or *a* is entirely random, we can easily calculate the frequency of zygotes having the genotype *AA*, *Aa*, or *aa*. The frequency of *AA* will be p^2 and of *aa* will be q^2. The frequency of heterozygotes will be $2pq$, reflecting that *Aa* individuals can arise in two ways: fusion of *A*-bearing sperm with *a*-bearing eggs, or vice versa.

The Hardy-Weinberg equilibrium depends on a number of assumptions. As already noted, mating must be random with respect to genotype. If there is preferential mating between *AA* and *AA* individuals, for example, there will be more homozygous individuals and fewer heterozygotes. Also, the population is assumed to be very large, so that statistical fluctuations will be negligible. Later we will explore the consequences of deviation from this assumption. There must be no mutation of *A* alleles into *a*, or *a* into *A*. Finally, individuals of all genotypes must be equally capable of reproduction (i.e., there must be no selection).

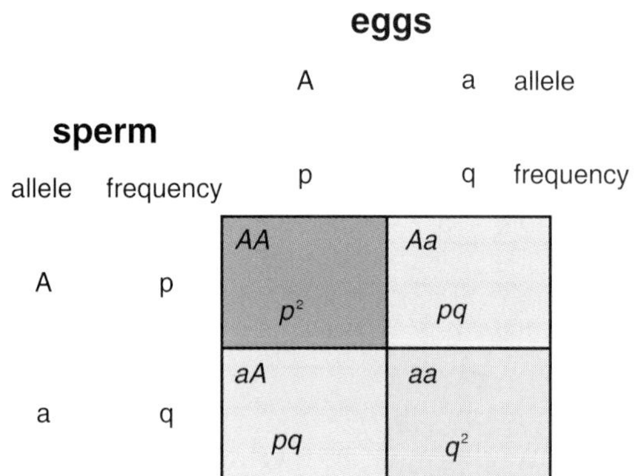

FIGURE 3-1 The frequency of *A* bearing sperm or eggs is p and of *a* bearing sperm or eggs is q. Assuming that mating is random with respect to genotype, that there is no mutation from *A* to *a* or vice versa, that there is no migration in or out of the population, and that mating efficiency is equal for all genotypes, the frequency of the genotype *AA* is p^2, of *Aa* is $2pq$, and of *aa* is q^2.

How is the Hardy-Weinberg equilibrium used to calculate carrier frequency of cystic fibrosis? In this case, the *A* allele is the wild-type and *a* is the cystic fibrosis mutation. The frequency of *aa*—that is, of individuals affected with cystic fibrosis—is 1 in 2500 in northern European whites. Thus $q^2 = 1/2500$, and hence $q = 1/50$. Because $p + q = 1$, p must be 49/50. The carrier frequency, then is $2pq = 2(49/50)(1/50) \cong 1/25$ in this population.

We are assuming, of course, that the cystic fibrosis gene obeys the assumptions of the Hardy-Weinberg equilibrium. The northern European population is very large, large enough to minimize random statistical fluctuation (such as the chance that only noncarriers happen to bear children in one generation). Mating may not be entirely random with respect to genotype. Some cystic fibrosis carriers may meet one another because of their affected siblings, for example, and choose either to mate or not to mate, having been brought together because of genotype. For the most part, however, cystic fibrosis carriers are not aware of their carrier status.

Two other assumptions clearly are not fulfilled. The ability of individuals with cystic fibrosis to have offspring is definitely impaired. Males with cystic fibrosis usually are infertile due to hypoplasia of the vas deferens. Females may be fertile, but reproduction is severely limited by the medical burden of the disorder. Many cystic fibrosis homozygotes therefore do not reproduce, which should cause loss of cystic fibrosis alleles in the population from one generation to the next. The rate of change of the frequency of the cystic fibrosis allele is very slow, however, as only one in 2500 individuals is subjected to this negative selection. Most cystic fibrosis alleles exist in heterozygous carriers who are not subject to selection.

Later in this book we will explore the forces that maintain cystic fibrosis alleles in the population. The high prevalence of the disorder in some populations and the severe clinical course have motivated an aggressive search for the gene. We will now turn to the means by which that search was conducted.

3.3 Mapping the Gene for Cystic Fibrosis

Betsy is now 3 years old and is doing well. She is growing and developing normally. In spite of antibiotic treatment, she has had many respiratory infections, but her parents have learned to deal with these. Betsy has had one hospitalization to allow intravenous antibiotic treatment for Pseudomonas *infection. Although they are encouraged by Betsy's progress, Bill and Jane have decided that they definitely do not want to have another affected child, but they are eager to have additional children.*

The risk of recurrence of cystic fibrosis for a couple with one affected child is 25%. Until the 1980s, there was little that could be offered to provide prenatal diagnosis. In the early 1980s, prenatal testing was developed based on measurement of intestinal microvillar enzymes in amniotic fluid. These enzymes normally are passed through the fetal intestines into amniotic fluid in small quantities. Some fetuses with cystic fibrosis excrete fewer of these enzymes, providing the basis for a prenatal test. The test is prone to false positive and negative results, however. A better prenatal test awaited the advent of genetic studies.

Finding the gene for cystic fibrosis was a formidable undertaking. Unlike the case for albinism or osteogenesis imperfecta, there were no clues to the nature of the gene product. The approach that ultimately was fruitful was based on molecular genetics and is referred to as *positional cloning*. The principle is that the location of a gene is determined, and DNA from that chromosomal region is cloned. The gene of interest is then identified within this cloned region, and the protein product is inferred from the base sequence of the DNA. The approach has been referred to as *reverse genetics*, to distinguish it from the more traditional approach of identification of a disease gene based on knowledge of the gene product. Although the term is somewhat colorful, it is a misnomer: The approach is really a direct genetic one rather than an indirect route to the gene through the gene product.

The positional cloning approach to cystic fibrosis began with mapping the gene. This was accomplished by genetic linkage analysis, using restriction fragment–length polymorphisms (RFLPs), such as were described in Chapter 2. The approach was complicated by the fact that there was no "candidate gene" to test for linkage, unlike the case with osteogenesis imperfecta, for which there was collagen. For cystic fibrosis, linkage had to be sought with markers throughout the genome.

The goal was to find a DNA sequence that segregates in a family along with the cystic fibrosis gene. A pair of alleles for two genes that are located adjacent to one another on the chromosome will segregate together to germ cells unless they are separated by recombination. Homologous chromosomes pair during meiosis and exchange segments. Although a meiotic exchange does not change the order of genes on the chromosome, it does reshuffle the specific set of alleles on a particular chromosome (Figure 3-2).

The frequency of recombination between a pair of genes is a function of distance; the farther apart they are, the more often a crossover event will occur between them. If two genes are very close together, recombination between them will be rare. A particular pair of alleles that are together on the same chromosome copy in an individual are said to be in

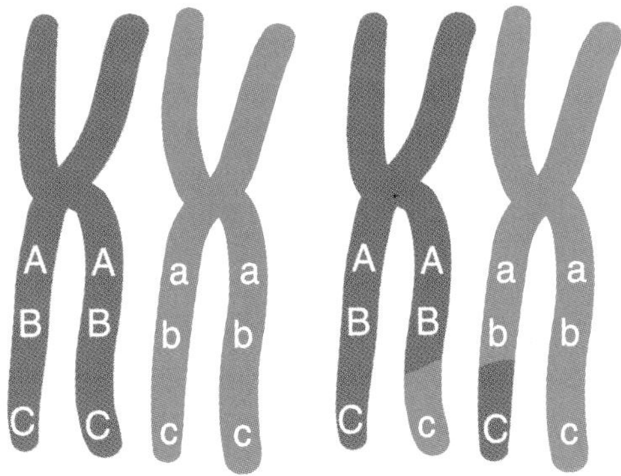

FIGURE 3-2 Crossing over during meiotic pairing of homologous chromosomes. Although the relative locations of genes A, B, and C are not changed, the particular sets of alleles on the two chromosomes change. Initially, alleles *A*, *B*, and *C* were on one member of the pair, and *a*, *b*, and *c* on the other. After the crossover, *A* and *B* are together with *c*, and *a* and *b* are together with *C* on one of the two chromatids of the recombined chromosomes.

FIGURE 3-3 A pair of homologous chromosomes with two gene loci, each having two alleles (*A* and *a*, or *B* and *b*). Alleles *A* and *B* are on the same chromosome and are said to be in coupling. The same can be said for alleles *a* and *b*. Alleles *A* and *b*, or *a* and *B* are said to be in repulsion.

coupling. In Figure 3-3, *A* and *B* are in coupling. The situation in which alleles are on opposite members of a pair of chromosomes is referred to as **repulsion**—*A* and *b* or *a* and *B* in Figure 3-3. For extremely closely linked genes, alleles that are in coupling tend to remain so from generation to generation. Only rarely will new combinations be created by crossing over (Figure 3-4).

The opposite situation is illustrated by genes that are located on different chromosomes. In this case, alleles segregate randomly to gametes. For an unlinked pair of loci, there is a 50% chance that the parental combination of alleles will be found in an offspring and a 50% chance that a nonparental combination will occur. Fifty-percent recombinant-nonrecombinant genotypes is the outcome of random segregation (Figure 3-5).

Suppose that two genes are separated by a distance such that recombination occurs between them 10% of the time. On average, then, 10% of germ cells will be recombinant and 90% nonrecombinant. This expectation would be realized if a very large number of offspring are sampled. The ideal mating experiment matches a doubly heterozygous individual with a homozygous partner and

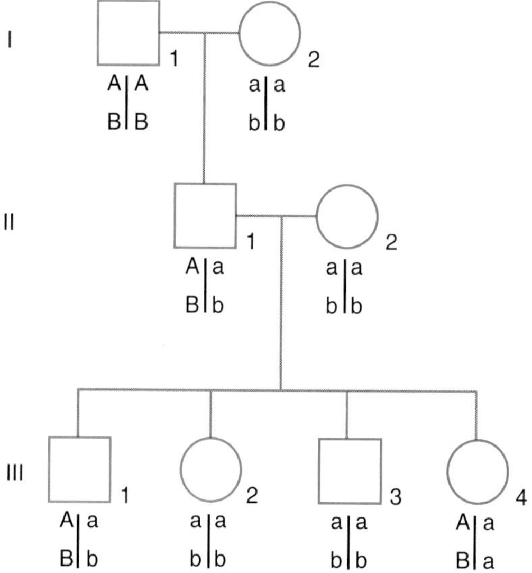

FIGURE 3-4 Complete linkage between a pair of loci, with no recombination between them. Individual II-1 is heterozygous at the two loci; his partner is doubly homozygous. Each offspring in generation III gets *a* and *b* from mother and either *AB* or *ab* from father. None gets the recombinant *Ab* or *aB* from father.

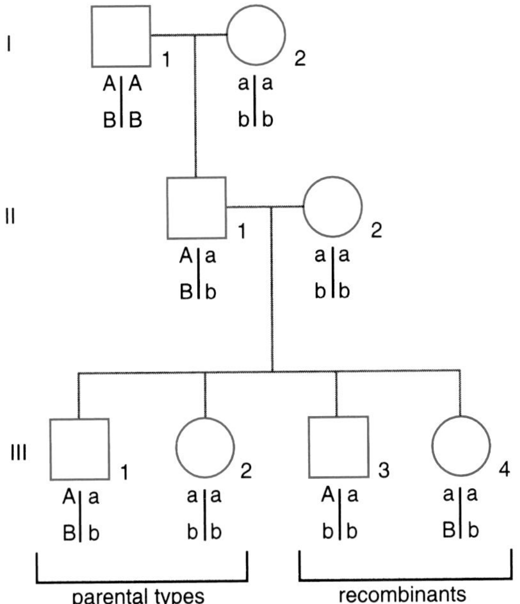

FIGURE 3-5 Random segregation of alleles at two unlinked loci. Individual II-1 is doubly heterozygous and produces four types of sperm: *AB*, *Ab*, *aB*, and *ab*. Each of these four combinations are represented in one of the offspring, resulting in equal numbers of parental (nonrecombinant) and recombinant offspring.

samples an extremely large number of offspring. Such experiments are commonly done with fruit flies or mice but are not useful for mapping the human genome.

To circumvent this difficulty, a statistical approach has been developed to extract linkage information from smaller families. What is calculated is the relative likelihood (the odds) that a particular set of family information would be obtained if a pair of genes is linked rather than if they segregate randomly. For a family with four children, the chance of any combination of recombinant and nonrecombinant genotypes given nonlinkage (i.e., the result of random segregation) is (1/2)(1/2)(1/2)(1/2) = 1/16. The chance of seeing any combination of recombinant and nonrecombinant genotypes if the genes are linked depends on how closely they are located. For example, if the rate of recombination between a set of genes is 10%, then the probability of a recombinant individual is 0.1 and of a nonrecombinant individual is 0.9. In a family of four (Figure 3-6), the probability of seeing one recombinant and three nonrecombinant offspring if the rate of recombination is 10% is (0.1)(0.9)(0.9)(0.9) = 0.073.

In this family of four with one recombinant and four nonrecombinant combinations of alleles, which is more likely: linkage with a recombination frequency of 10% or nonlinkage? We can compare the relative likelihood of these two possibilities by creating an odds ratio. In this case, 0.073/0.0625 = 1.168; linkage is slightly favored over nonlinkage. The same can be done for other values of the frequency of recombination between the two genes. The frequency of recombination usually is expressed as a variable θ, the **recombination fraction**. At $\theta = 0$, the probability of linkage will be 0 if any recombinants occur at all, because at this value of θ, no recombination is possible. At a value of $\theta = 0.5$, the odds ratio will be 1, because 50% recombination is equivalent to nonlinkage. Between $\theta = 0$ and $\theta = 0.5$, the probability of linkage depends on family size and the number of recombinant and nonrecombinant individuals. A family with some recombinant individuals, but fewer than expected by random segregation, will be better explained by an intermediate value of θ: If θ is too low, few or no recombinants are expected, and if θ is too high, more are expected.

Although this approach can help extract data from small families, there is a limit to what can be learned. In a family of four, chance segregation of alleles can lead to no recombinant offspring even if genes are unlinked. There is no substitute for the statistical power of large numbers. How can this be achieved?

One way is to pool data from many families—in effect, to consider many different sibships as if they were all one big sibship. In figuring the probability of linkage or nonlinkage for a family, each child is viewed as an independent statistical event, and the total probability of the recombinants and nonrecombinants in the sibship is the product of the individual probabilities for each offspring. If data from multiple sibships are obtained, the probability for each offspring can be multiplied together. This becomes unwieldy if many individuals are studied, but recall that adding logarithms is the equivalent of multiplying the numbers to which they correspond. Therefore, the odds ratio can be calculated for each sibship, the fraction converted to a logarithm, and then the log of the odds ratio added from family to family to obtain

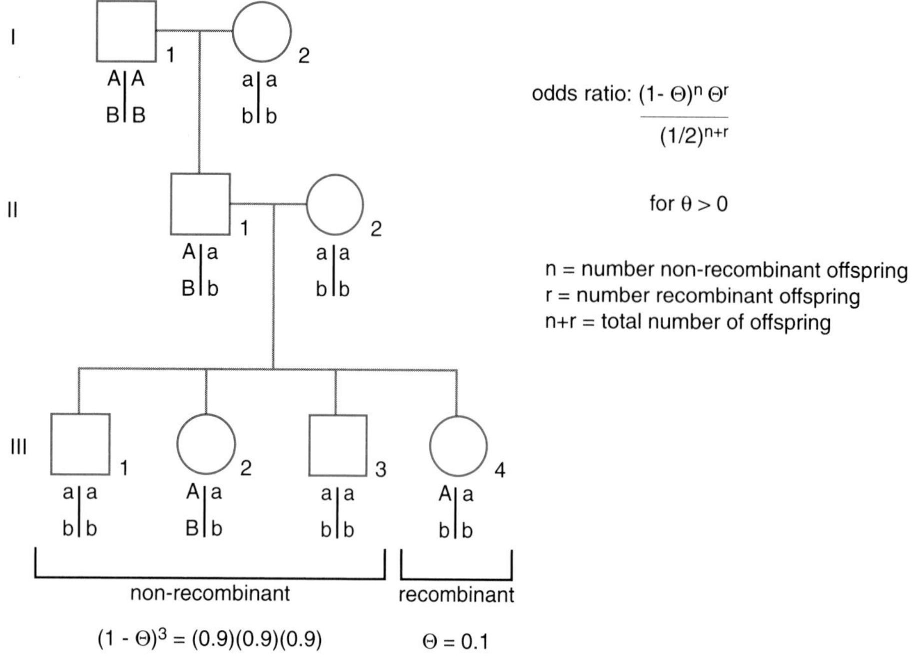

FIGURE 3-6 Calculation of odds ratio. In this sibship of four, three children are nonrecombinant and one is recombinant. The numerator of the odds ratio is the probability of seeing this number of recombinant and nonrecombinant offspring at recombination value of θ, and the denominator is the probability given random segregation (equivalent to $\theta = 0.5$). This odds ratio indicates the relative likelihood of this family's data given recombination at a set value of θ compared with random segregation. The odds ratio is computed for multiple values of θ.

a final log of the odds ratio for the entire data set. Log of the odds is abbreviated with the acronym *lod*.

Lod "scores" are calculated family by family for multiple values of θ and then are summed for each value of θ. How high does a lod score need to be to indicate probable linkage or nonlinkage of a pair of genes? Evidence for linkage generally is accepted when a lod score is 3 or greater, indicating at least a 1000 : 1 odds ratio favoring linkage. Lod scores of −2 or less are accepted as evidence against linkage, indicating at least a 100 : 1 ratio favoring nonlinkage. The value of θ at which the peak lod score is obtained is taken as the **maximum likelihood value of θ** (Figure 3-7).

There is one important caution that must be applied to the study of linkage by pooling data from many families. This is the possibility of genetic heterogeneity. If different genes at different locations are responsible for a disorder in different families, no one marker will be linked to them all. Data favoring linkage in one family will be canceled out by data favoring nonlinkage in another, with the net result that random segregation will be favored. This could have happened with osteogenesis imperfecta if a random search of markers had been done to look for linkage to osteogenesis imperfecta type I or II. Some families have mutations in COL1A1 and some in COL1A2, and no single marker would have been linked to the disease in all families. It is always safer to analyze linkage data from individual large families to avoid the pitfall of genetic heterogeneity.

The search for linkage requires the availability of a large number of polymorphic loci that are widely distributed in the genome. Prior to the discovery of DNA base sequence variants, only protein variants such as blood group antigens were available for mapping studies. RFLPs added substantially to the gene map, making it much more likely that a gene of interest could be mapped. Even more powerful linkage markers are available today, as will be seen soon, but it was with RFLPs that the cystic fibrosis gene was mapped.

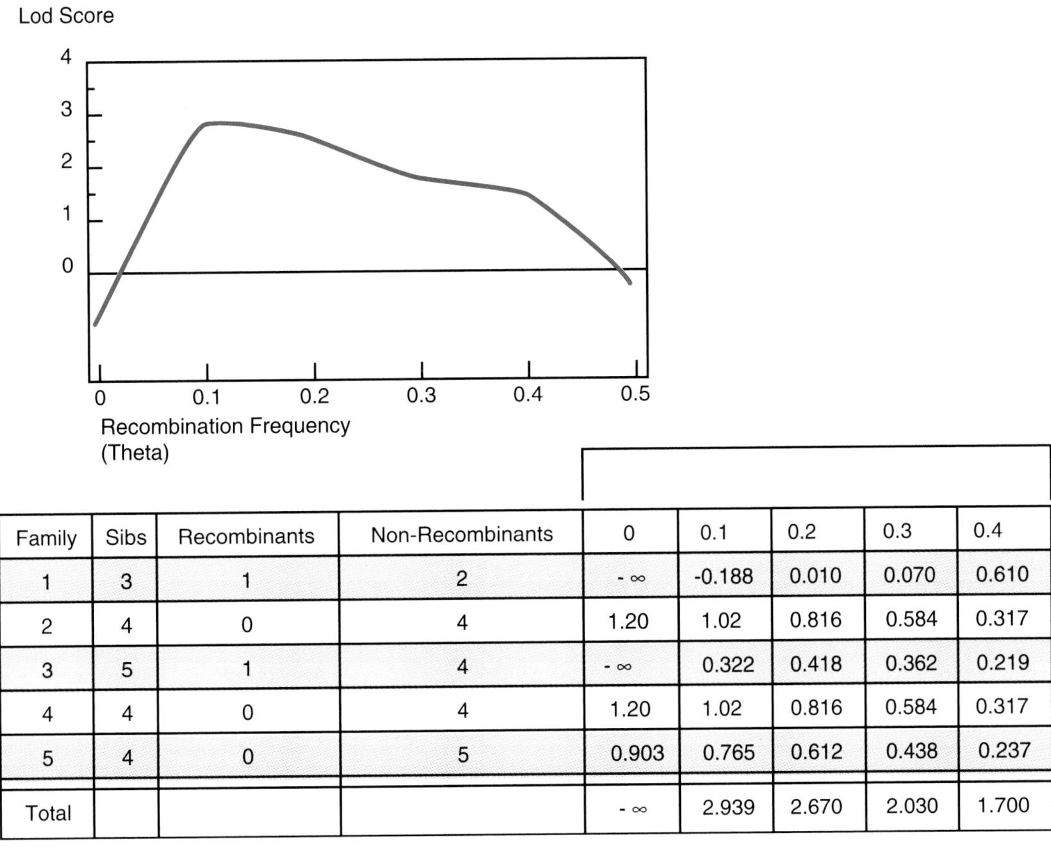

Family	Sibs	Recombinants	Non-Recombinants	0	0.1	0.2	0.3	0.4
1	3	1	2	- ∞	-0.188	0.010	0.070	0.610
2	4	0	4	1.20	1.02	0.816	0.584	0.317
3	5	1	4	- ∞	0.322	0.418	0.362	0.219
4	4	0	4	1.20	1.02	0.816	0.584	0.317
5	4	0	5	0.903	0.765	0.612	0.438	0.237
Total				- ∞	2.939	2.670	2.030	1.700

FIGURE 3-7 Analysis of linkage data from five families. The number of siblings in each family, as well as the number of recombinants and nonrecombinants in each family, is shown in the table. Lod scores were computed for the indicated values of θ and summed across all the families, and then a graph of lod score versus θ was made. The lod score peaked at just less than 3 at a value of θ of 0.1. This would be the maximum likelihood estimate for θ based on these data.

Even with a large number of closely spaced markers, it can be difficult to find linkage to a gene of interest if families are small. Because there were no plausible candidate genes for cystic fibrosis, a tedious marker-by-marker search for linkage was necessary. Many of the polymorphic markers tested for linkage did not represent genes but rather were bits of cloned DNA isolated from a genomic DNA library. Such "anonymous" markers are largely of unknown function. They most likely are derived from DNA between genes or from introns rather than from genes themselves. Their chromosomal locations can be identified using in situ hybridization or somatic-cell genetic studies (Chapter 2).

Linkage studies described thus far have involved following codominant markers through a family. Application to the study of a recessive disorder such as cystic fibrosis is based on similar principles, but there is the added challenge that coupling phase is not known (Figure 3-8). To deal with this, the linkage calculation is done twice, once with one assumption for coupling phase and again assuming the opposite phase. The lod score then is calculated as the average of the two.

The first conclusive sign of linkage to cystic fibrosis was found using a protein polymorphism for the enzyme paraoxonase (Figure 3-9). Soon after, a number of additional DNA markers were found that were linked to cystic fibrosis, and the genes were located on chromosome 7. Recombinants were found between any of these genes and cystic fibrosis, indicating that none was the cystic fibrosis gene itself.

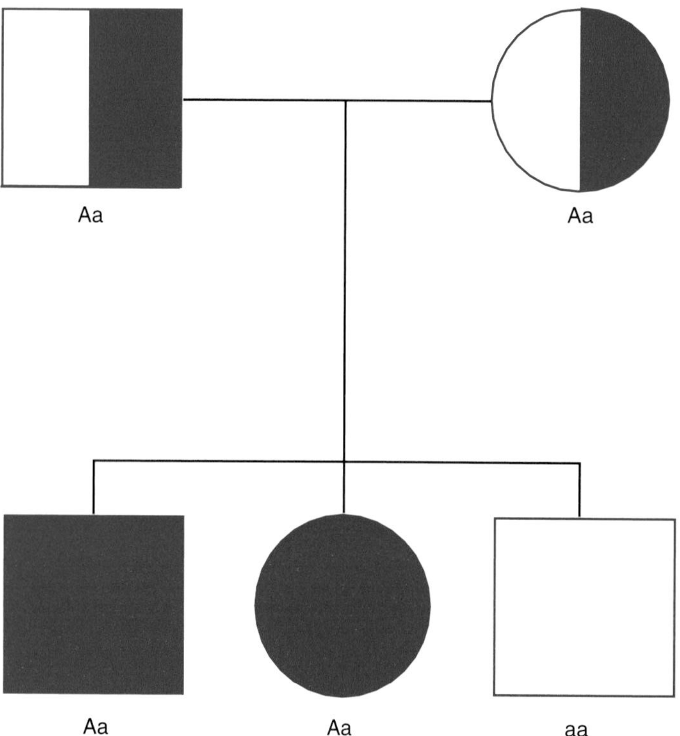

FIGURE 3-8 Linkage calculation for a marker with two alleles (*A* and *a*) in relation to a recessive disorder (darkened symbols indicate affected offspring; both parents are therefore carriers). In this family, the father is informative for the marker (i.e., he is heterozygous). He passes the *A* allele to each of the affected children and *a* to the unaffected child. We do not know, however, that *A* was in coupling with the disease allele in the father, so the linkage calculation is done twice—once assuming that *A* is in coupling with the disease and none of the children is recombinant, and once assuming that *a* is in coupling with the disease and all the children are recombinant. The final lod score is calculated as the average of these two.

Lod scores at recombination fractions (Θ) of:

Number of Families	0.01	0.05	0.10	0.15	0.20	0.25	0.30	0.35	0.40
36 Candian	-6.02	0.98	2.84	3.20	2.96	2.44	1.79	1.13	0.55
3 other families	0.14	0.69	0.79	0.75	0.66	0.53	0.39	0.25	0.12
Total	**-5.88**	**1.67**	**3.63**	**3.95**	**3.62**	**2.97**	**2.18**	**1.38**	**0.67**

maximum likelihood value of Θ = 0.14 at a maximum lod score of 4.13

FIGURE 3-9 Data demonstrating linkage of anonymous DNA marker DOCRI-917 and cystic fibrosis. Linkage data were obtained from 39 Canadian families and 3 other families and are shown separately. The totals are the sums of individual lod scores from the two data sets. The lod score exceeded 3 at recombination fractions of 0.10–0.20; the peak of 4.13 was obtained at θ of 0.14, making this the maximum likelihood value of θ. (Data from Tsui L-C, Buchwald M, Barker D, et al. Cystic fibrosis locus defined by a genetically linked polymorphic DNA marker. Science 1985;230:1054–1057.)

The closest marker was found to recombine with cystic fibrosis less than 1% of the time. One percent recombination is referred to as 1 centimorgan (1 cM) of genetic distance, after the geneticist Thomas Hunt Morgan, who did much of the early gene-mapping work with fruit flies. Therefore, the markers were less than 1 cM from cystic fibrosis (Figure 3-10). As will be seen, this map provided an important clinical tool, but some time would elapse before it enabled the cystic fibrosis gene itself to be found.

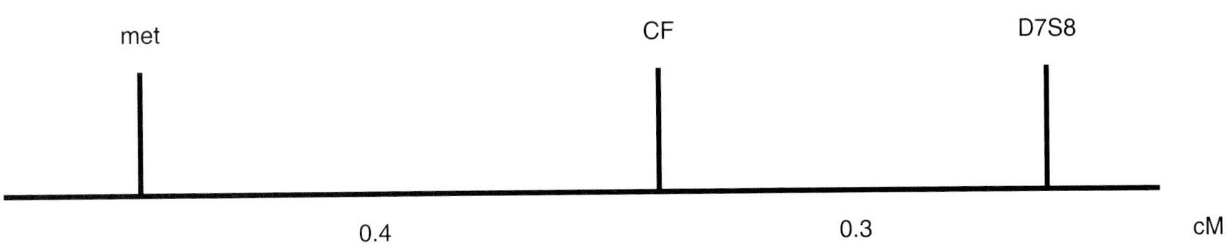

FIGURE 3-10 Genetic map of the region of chromosome 7 surrounding the cystic fibrosis gene. Cystic fibrosis gene (*CF*) is flanked by two markers—*met*, which is a gene involved in some kinds of cancers, and *D7S8*, an anonymous DNA marker. (The designation *D7S8* indicates DNA marker [*D*] on chromosome 7 [*7*], single copy [*S*], marker number 8 [*8*].) The genetic distances below the line are given in centimorgans and are maximum likelihood estimates. The gene order itself also is a maximum likelihood estimate. (Data from Beaudet A, Bowcock A, Buchwald M, et al. Linkage of cystic fibrosis to two tightly linked DNA markers: joint report from a collaborative study. Am J Hum Genet 1986;39:681–693.)

3.4 Diagnosis of Cystic Fibrosis by Linkage Analysis

1988

Bill and Jane are excited to learn that the cystic fibrosis gene has been located and that linkage-based prenatal diagnostic testing is possible. Blood is taken from both of them and from Betsy to determine if genetic linkage analysis will be informative for possible future prenatal testing.

Definitive molecular diagnostic testing requires that the gene for a disorder be cloned and that pathogenic mutations be easily identifiable. Knowing the location of a gene does not permit such direct testing but does allow linkage-based testing. Just as it could be used to find marker genes that segregate with a disease gene in a family, so linkage analysis can be used to track the disease gene through a family in relation to known linked markers.

The basic strategy is shown in Figure 3-11. Here an individual (II-1) has an autosomal dominant disorder for which no laboratory diagnostic test exists. He is heterozygous for a marker gene closely linked to the disease gene that has two alleles, *A* and *a*. His partner is homozygous *aa*. In this family, we know that *A* is in coupling with the disease because II-1 inherited both the disease and the *A* allele from his mother, I-2. There are five children. III-1 and III-5 both have inherited the *A* allele from father and, of course, *a* from mother. Because they received the marker allele in coupling with the disease, they would be predicted to be affected and, indeed, they are. III-2 and III-4 got *a* from both parents and are predicted to be unaffected, as they are. Child III-3 would seem to be problematic, however. She got the *a* allele from her father and is therefore predicted to be unaffected, yet she has the disorder. This illustrates an important pitfall of linkage-based diagnosis, which is genetic recombination.

We have already seen how genetic recombination can change the association of particular alleles on a chromosome without changing the order of gene loci. Recombination between a marker gene and a disease gene can lead to a diagnostic error in a family linkage study. There are a number of ways to deal with this possibility.

First, the probability of recombination between the marker and disease should be known from the studies that established linkage in the first place. This allows an estimate to be made of the accuracy of a genetic linkage study. In Figure 3-11, if the recombination frequency between the marker and the disease is a value θ, then the accuracy of a diagnosis in the third generation is $1 - \theta$. For example, if the marker is 3 cM from the disease gene, then the risk of disease in individuals III-1 and III-5 is $(1 - 0.03) = 0.97$; likewise, the risk of disease in individuals III-2, III-3, and III-4 would be 0.03 (or the likelihood of being unaffected is $[1 - 0.03] = 0.97$). Obviously, it is advantageous to use very tightly linked markers for diagnostic testing whenever possible.

Another strategy for dealing with recombination is to use flanking markers. This is illustrated in Figure 3-12. Here genetic recombination can be readily detected because the combination of flanking markers in an individual changes as a consequence of

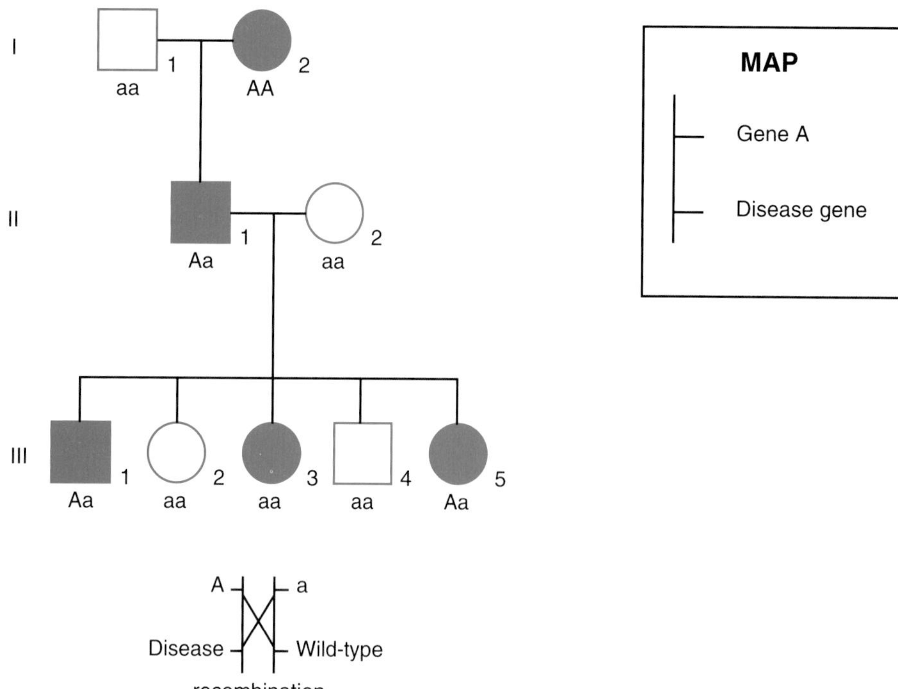

FIGURE 3-11 Linkage-based diagnosis in a family with autosomal dominant disorder. Father II-1 is heterozygous for a closely linked marker with alleles *A* and *a*. *A* is in coupling with the disease gene in him, as he inherited both *A* and the disease allele from his mother, I-2. Children III-1 and III-5 inherit both *A* and the disease, and children III-2 and III-4 inherit both *a* and the nondisease allele. Child III-3 is a recombinant, as shown.

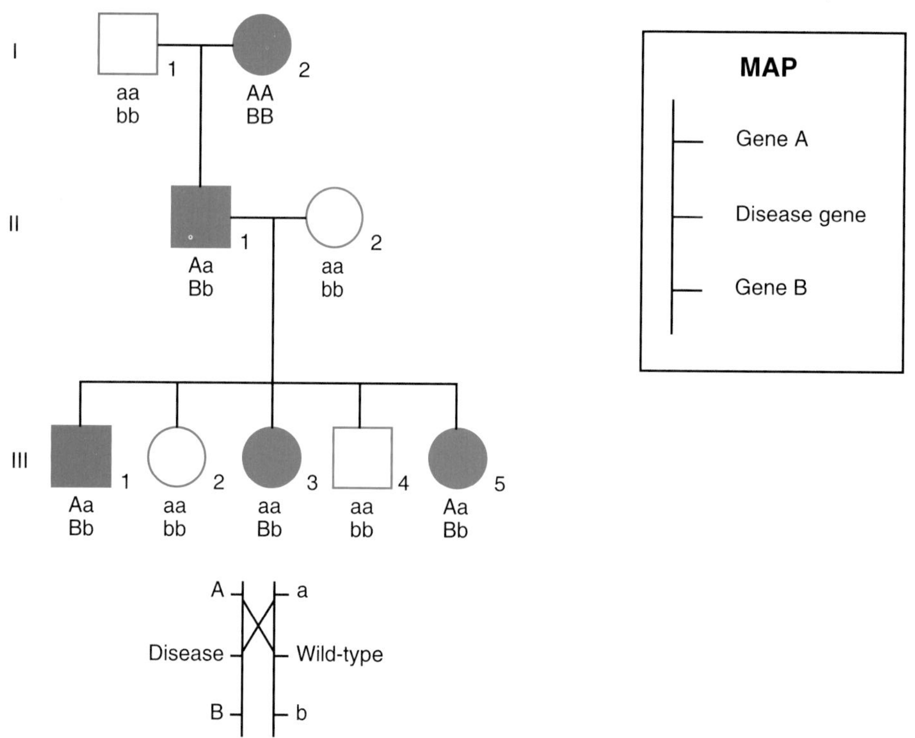

FIGURE 3-12 Use of flanking markers to detect recombination event. In this family, the recombination in III-3 is apparent from the fact that she inherited an *a* allele in coupling with a *B* allele from her father, which is different from the coupling of *A* with *B* in her father.

recombination. In this case, individual III-3 inherited *a* and *B* from father, whereas we know that in II-1 *A* and *B* were in coupling. Therefore, a crossover must have occurred somewhere between *A* and *B*. If *A* and *B* are approximately equidistant from the disease locus, we would have no way of knowing whether it is more likely that the crossover occurred between *A* and the disease or between *B* and the disease. The diagnosis in III-3 would therefore be indeterminate. If the markers are not equidistant from the disease gene, their relative distance from this gene can be used to estimate the likelihood that the crossover occurred on one side or the other, and hence whether or not the child is likely to be affected. In practice, though, markers are likely to be more or less equally separated from the disease gene. The use of flanking markers at least alerts one to the fact that recombination has occurred.

There are a number of other limitations of the use of genetic linkage data for diagnostic purposes. First, unlike most diagnostic tests, this approach requires study of a family, not just an individual. Relatives in two or more generations must be available and willing to participate in the study. Some may be deceased, or it may be difficult to motivate some to provide a blood sample, perhaps because of poor relations in the family or because of fear of what the test will show. Linkage studies have revealed, for example, a fairly high incidence of *nonpaternity*—that is, the stated father of a child is not the biologic father (inferred from the fact that the child has an allele that could not have been inherited from the father). Often in such cases, the stated father is not aware that the biologic father of a child is different, setting up a sensitive and awkward situation in genetic counseling.

Second, linkage testing is not informative for all families. The parent who carries a mutant gene must be heterozygous for the linked marker or markers, and the alleles in this parent must be distinguishable from those of the partner. These criteria are more likely to be met if there are many markers available that are linked to a disease gene of interest and that flank the disease gene. It also is helpful to have markers that are *highly polymorphic*, which means that there are multiple alleles at the locus and heterozygosity is common in the population. We will encounter examples of very highly polymorphic loci later in this chapter.

Third, and most important, is the issue of genetic heterogeneity. Because we are not determining directly the gene mutation, linkage-based testing relies on the assumption that the disease gene in the family is indeed linked to the marker gene. If diagnosis of the disease in the proband is incorrect, any inference of diagnosis based on linkage in another member of the family will also be incorrect. Even if the diagnosis in the proband is correct, however, genetic heterogeneity can provide misleading results. This would occur if there are many genes at different locations that can account for a given phenotype. If the wrong gene is tracked through the family, the results of linkage analysis will not indicate inheritance of the true disease-causing gene. Consider osteogenesis imperfecta, where mutations in COL1A1 or COL1A2 can be responsible for the disorder in different families. Tracking a polymorphism in COL1A1 will not be helpful if the mutation in a given family is in COL1A2. Families studied for diagnostic purposes rarely are large enough to independently confirm linkage to a particular locus. Therefore, if genetic heterogeneity is known to exist, linkage studies must be used with great caution.

Despite these limitations, linkage analysis can be a very powerful approach to providing genetic diagnosis in a number of settings. It can be used for prenatal diagnosis, as fetal DNA can be obtained from chorionic villus biopsy or amniocentesis. Also linkage analysis can provide presymptomatic diagnosis. In many disorders, signs and symptoms of disease may not be apparent early in life, yet linkage analysis can indicate who is likely to be affected in a family at any time. This can be useful in identifying those who need to be followed medically for potentially treatable complications and for providing genetic counseling. Such use, however, can also open Pandora's box, possibly creating major psychological and adjustment problems for a person in whom a diagnosis of being affected is made years or even decades in advance of the appearance of symptoms.

Both the benefits and dangers of linkage-based testing are illustrated by the example of Huntington's disease. Huntington's disease is a devastating neurologic degenerative disorder. Affected individuals are normal at birth and may remain symptom-free for the first few decades of life. Inevitably, however, they insidiously develop abnormal body movements, particularly sudden movements of the limbs called *chorea*. In addition, there is loss of cognitive skills and psychiatric disturbances appear, mainly depression. The disorder gradually progresses to the point of total dependence and eventual death.

Huntington's disease is a classic example of age-dependent penetrance, a phenomenon whereby the likelihood of exhibiting symptoms for those who inherit the gene increases with age (Figure 3-13). Very few patients manifest signs during childhood, and most do by age 70. Between these points, the frequency of symptomatic individuals gradually rises. One of the tragic features of the disorder is that most who carry the gene do not know that they are affected at childbearing age. Hence, they transmit the gene to offspring before they are aware of being affected themselves.

Huntington's disease was one of the first genetic disorders to be subjected to a linkage-based approach for diagnosis. Genetic linkage to markers on chromosome 4 was established in 1981. Prior to that time, there was no prenatal or presymptomatic diagnostic test. It was quickly established that all known cases of Huntington's disease were linked to the same markers, making genetic heterogeneity very unlikely. Gradually, a number of highly informative markers became available. Diagnostic testing in a family could then be offered to identify individuals who had inherited the mutant gene and to provide prenatal diagnosis for gene carriers.

Although such testing was technically feasible, its application in this case raised a number of social and ethical problems that have led to great caution in its use. There is no cure for Huntington's disease—treatment is aimed at amelioration of symptoms—but to this day there is no way to stem the inexorable progression. Therefore, presymptomatic diagnosis is of limited utility. It can reveal to a healthy person the fact that at some point—years or decades in the future—neurologic deterioration will occur, without providing any recourse to prevent the inevitable. Individuals with Huntington's disease vary with respect to their interest in having such information. Many have requested testing in the hope that they will learn that they have not inherited the disease. Others prefer to live with uncertainty rather than risk the possibility that they may receive the devastating news that they are affected. Even those found not to be affected face psychological hurdles, particularly guilt at having escaped a fate that their siblings may still face. There is also a risk of discrimination against those found to inherit the gene. They may be deemed unsuitable for holding certain jobs even though no neurologic symptoms are present and may be denied health or life insurance.

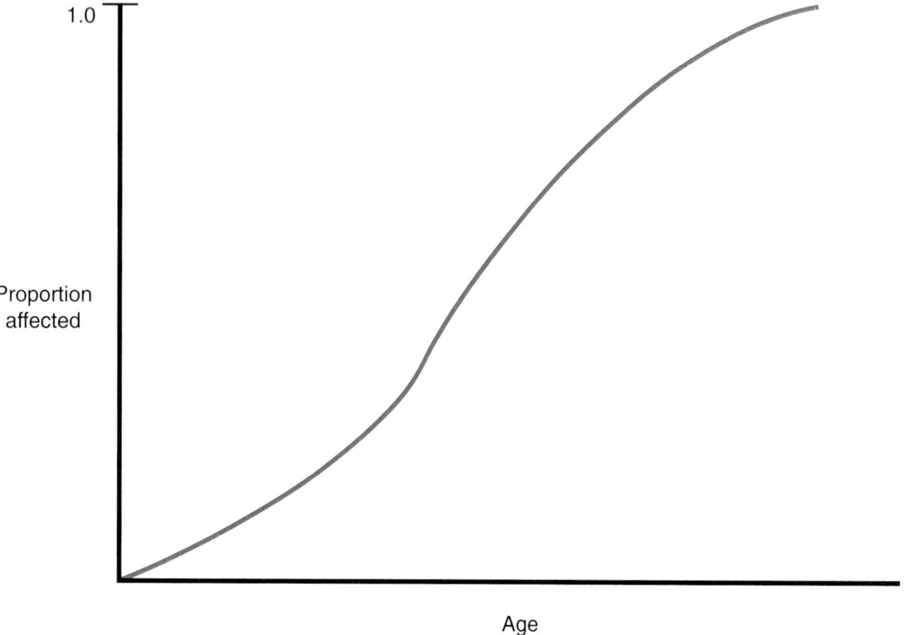

FIGURE 3-13 Age-dependent penetrance in Huntington's disease. The proportion affected increases with age. Very few patients show signs of Huntington's disease in the first decade of life, but by age 70, virtually all who have the mutant gene have signs of the disease.

These problems were recognized by investigators who developed the diagnostic test for Huntington's disease, and the test was implemented cautiously. Comprehensive psychological analysis and counseling are provided before any testing is initiated for an individual, and information is provided in a setting where counseling could be provided. Under these carefully controlled circumstances, the Huntington's disease testing program has proceeded successfully and has provided a model for implementation of genetic testing. As will be discussed later, the Huntington's disease gene has been cloned, and direct diagnostic testing now is possible. This increases the accuracy of testing and obviates the need for a family analysis, but the social and ethical issues of testing continue to be addressed.

Bill and Jane learn that they are informative for markers that flank cystic fibrosis. Jane is pregnant and undergoes chorionic villus sampling. The results of the prenatal test are depicted in Figure 3-14.

Thus far, linkage testing has been described for autosomal dominant disorders, but it is also applicable for recessively inherited disorders. In this case, the strategy is illustrated in Figure 3-15. Here, both parents, who are obligate carriers because they have an affected child, are heterozygous for a closely linked marker gene. The affected child has inherited the *A* allele from both parents, and therefore *A* is most likely in coupling with the disease allele in each parent. A sibling who inherits *AA* probably is affected, one who inherits *aa* probably is unaffected, and one who inherits *Aa* probably is a carrier (although which parent transmitted the mutant allele and which the wild-type allele is ambiguous). Risk assessment in this case is a bit less straightforward than for an autosomal dominant disorder. A child who inherits *AA* is most likely to have inherited the disease gene from each parent. The inference that *A* is in coupling with the disease, however, is based on the affected sibling having inherited *A* from each parent along with the disease allele. If that child represents a crossover from either parent, it is possible that *a* is really in coupling with the disease in either or both parents. Furthermore, even if *A* is in coupling with the disease in both parents, there is further chance of recombination occurring in either parent in forming additional sperm or egg cells. Thus, a child who inherits *AA* may be a carrier, or a child who inherits *Aa* may be affected. Exact calculation of risks is complex, but the use of closely linked flanking markers can go far toward dealing with this problem, resulting in highly accurate testing.

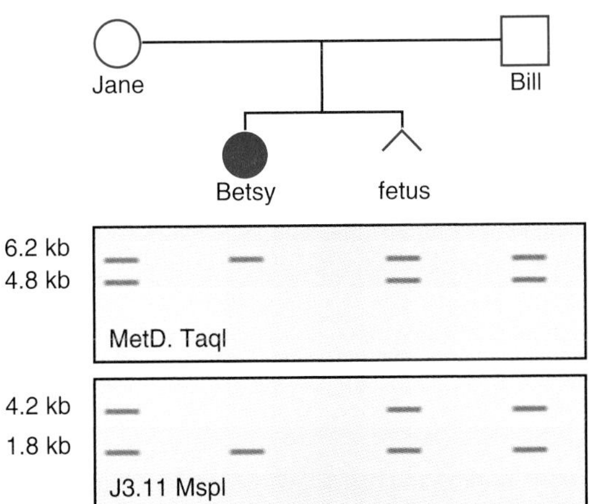

FIGURE 3-14 Result of Jane's prenatal test depicts autoradiograms for two flanking polymorphic markers, MetD and J3.11. Genomic DNA cut with TaqI and hybridized with MetD reveals either a 6.2-Kb or a 4.8-Kb allele; MspI-cut DNA hybridized with J3.11 reveals a 4.2-Kb and a 1.8-Kb allele.

Application of linkage diagnosis to a recessive disorder is subject to the same limitations as were noted for a dominant disorder, with one addition. It is vitally important that an affected child be available in order to offer linkage-based testing for future offspring. Families are most motivated to obtain prenatal testing for devastating, lethal genetic disorders. Sometimes the affected child is deceased before the family inquires about genetic testing, such that DNA probably cannot be obtained from that child to make testing possible, though in some cases autopsy material may be available to provide access to DNA from the affected child. Also, the use of polymerase chain reaction–based polymorphisms (to be described in detail later) allows testing to be done on very small quantities of DNA, such as may be obtained from blood spots on cards collected for newborn screening (see case at end of Chapter 1). Because of the possibility of linkage testing, though, it is wise to save DNA from a child with a genetic disorder, even if a genetic test is not available at the time of death. A number of commercial DNA banking services exist, and the DNA can be of great value if a diagnostic test eventually is developed.

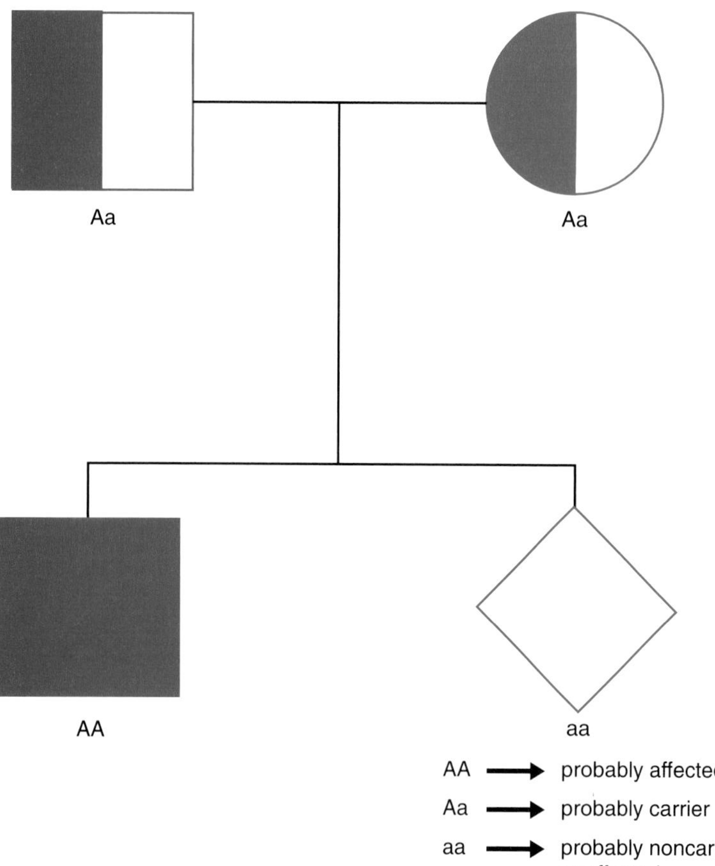

FIGURE 3-15 Linkage-based diagnosis for autosomal recessive disorder. Both parents are heterozygous for the disease gene as well as for the marker gene. The affected child has received the *A* allele from each parent, indicating that *A* probably is in coupling with the disease allele in both parents. Therefore, a child who has the genotype *AA* is likely to be affected, one with *Aa* likely to be a carrier, and one who has *aa* is likely to be an unaffected noncarrier.

The cystic fibrosis mutation is in coupling with the 6.2-Kb MetD allele and the 1.8-Kb J3.11 allele in both parents (as these are the alleles that Betsy inherited). The fetus is heterozygous for both alleles and therefore is probably a cystic fibrosis carrier but is unlikely to be affected. Bill and Jane are delighted, although they understand that the test is not 100% accurate.

3.5 Linkage Disequilibrium

Bill's brother Tim is about to be married. Tim and his partner, Holly, seek counseling about their risk of having a child with cystic fibrosis. Tim is told that his risk of being a carrier is 1:2 and that linkage testing can help refine that risk. The testing is done, and Tim is found to be a carrier. Holly has no family history of cystic fibrosis. The couple asks whether DNA testing is possible to determine Holly's risk of being a carrier.

In considering linkage data so far, we have assumed that the particular alleles for different loci on a chromosome are independent of one another. For example, in Figure 3-15, we inferred that both parents probably have an *A* allele in coupling with the cystic fibrosis mutation, because their child with cystic fibrosis is *AA*. The *A* allele, however, is not the cystic fibrosis gene itself but rather a closely linked marker. In another family, *a* could easily be in coupling with a cystic fibrosis mutation, and the vast majority of individuals in the general population who have an *A* allele are not cystic fibrosis carriers. It is necessary to determine empirically the association of a particular allele with a disease gene such as cystic fibrosis in each family being studied.

The assumption that alleles at linked loci will be associated in coupling at random is known as **linkage equilibrium**. The particular combination of alleles on one copy of a chromosome is referred to as **haplotype**. Linkage equilibrium implies that the frequency of any particular haplotype is the product of the frequencies of the individual alleles (Figure 3-16). Knowledge of the genotype at one locus does not provide information about the allele present at another linked locus.

Linkage equilibrium does not apply to all sets of linked genes; there are examples in which particular sets of alleles are more likely to be found in coupling than are expected due to chance. This is referred to as **linkage disequilibrium**. Linkage disequilibrium has been found for haplotypes involving the cystic fibrosis gene and two particular closely linked markers, designated XV-2C and KM-19. Each marker has two common alleles and, in one study of 127 unrelated individuals with cystic fibrosis, one particular haplotype was found to be associated with the cystic fibrosis mutation more than 85% of the time (Table 3-1).

The data in Table 3-1 indicate that haplotype *B* is far more likely to be associated with cystic fibrosis than any of the others, although it must be remembered that most individuals in the general population do not carry a cystic fibrosis mutation even if they have haplotype *B*. From these data, the probability that a randomly chosen chromosome of a given haplotype represents a cystic fibrosis chromosome was calculated. Also, it was possible to determine the odds that a person with a particular genotype is a cystic fibrosis carrier (Table 3-2).

How does linkage disequilibrium come about? Most likely it is a consequence of the founder effect. It has been pointed out that cystic fibrosis occurs most often in individuals of northern European ancestry, indicating that the most common cystic fibrosis mutation arose at some point in history in this population. Presumably, the mutation occurred in a cystic fibrosis gene that resided on a chromosome with the *B* haplotype for XV-2C and KM-19, by chance. For the next several generations, then, it would not be surprising to find that most of the cystic fibrosis mutations derived from this original mutation would also occur on *B* haplotype chromosomes. One might expect that, gradually, genetic recombination would occur between XV-2C and cystic fibrosis, or KM-19 and cystic fibrosis, placing the cystic fibrosis mutation on chromosomes with other haplotypes.

Given sufficient time, such recombination should lead to a random association of any particular haplotype and cystic fibrosis: That is, linkage equilibrium would be established. Therefore, linkage disequilibrium might occur for either of two reasons. One is simply that not enough time has passed for equilibrium to be reached. Both XV-2C and KM-19 are within a few tenths of a centimorgan from cystic fibrosis, so recombination between them and cystic fibrosis will be very rare. Indeed, it is likely that there has not been sufficient time since the cystic fibrosis mutation arose to reach equilibrium.

Another possible explanation for linkage disequilibrium is that there is some selective advantage to an individual who has one particular haplotype over another. This has been seen in the human leukocyte antigen (HLA) region, a set of closely linked genes involved in expression of cell surface proteins important for immunity. Here, combinations of alleles—that is, particular haplotypes—result in more effective function of the immune system and therefore

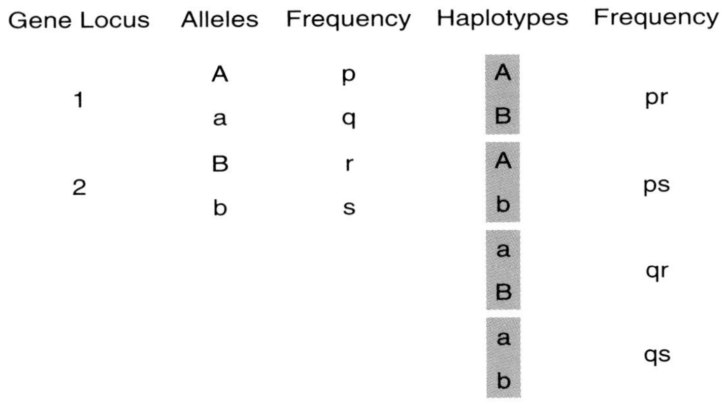

Gene Locus	Alleles	Frequency	Haplotypes	Frequency
1	A	p	A B	pr
	a	q	A b	ps
2	B	r	a B	qr
	b	s	a b	qs

FIGURE 3-16 Linkage equilibrium. For a pair of closely linked loci, each with multiple alleles (two alleles each in this example), the frequency of the various haplotypes is the product of the frequencies of the individual alleles in each haplotype. In other words, the association of any pair of alleles on the same copy of a chromosome is random.

TABLE 3-1 Haplotype analysis of 127 unrelated individuals with cystic fibrosis (CF) and 159 unaffected European individuals

			CF Chromosomes (n = 254 with CF mutations)		Normal Chromosomes (n = 318 wild-type)	
Haplotype	XV-2c Allele	KM-19 Allele	Number	Percent	Number	Percent
A	1	1	17	6.7	74	28.9
B	1	2	218	86.5	35	16.4
C	2	1	7	2.8	110	44
D	2	2	10	4	31	10.7
Uncertain			2		4	

Source: Redrawn from Beaudet AL, Feldman GL, Fernbach SD, et al. Linkage disequilibrium, cystic fibrosis, and genetic counseling. Am J Hum Genet 1989;44:319–326.

TABLE 3-2 Probability that a person unaffected with cystic fibrosis with a particular haplotype is a cystic fibrosis carrier (assumes 1-in-50 gene frequency for cystic fibrosis mutation)

Genotype	Probability	Odds
AA	0.0092	1:109
BB	0.199	1:5
CC	0.0026	1:384
DD	0.0124	1:81
AB	0.116	1:8.6
AC	0.0059	1:169
AD	0.0108	1:93
BC	0.113	1:8.8
BD	0.117	1:8.5
CD	0.00752	1:133
Population	0.0392	1:25.5

Source: Redrawn from Beaudet AL, Feldman GL, Fernbach SD, et al. Linkage disequilibrium, cystic fibrosis, and genetic counseling. Am J Hum Genet 1989;44:319–326.

are subject to favorable selection. No such selective advantage has been identified for haplotypes related to cystic fibrosis, however.

Whatever the biologic explanation for linkage disequilibrium, knowledge of the association of particular haplotypes with the cystic fibrosis mutation has been helpful for genetic counseling. Specifically, it can be used to refine the risk that an individual carries a cystic fibrosis mutation. A randomly chosen individual with no family history of cystic fibrosis who is of northern European ancestry has a 1 : 25 risk of being a carrier for the disease. If that person has a *BB* haplotype, however, for XV-2C and KM-19, the risk of being a carrier is approximately 1 : 5 (see Table 3-2). On the other hand, if the haplotype is *AA*, the risk of being a carrier diminishes to 1 : 109. Determination of these haplotypes is especially useful in counseling families in which one partner is known to be at high risk because of family history and the other has no such history but comes from a population known to be at risk.

Holly is tested for XV-2C and KM-19. She is found to have an AD *haplotype. Her risk of being a cystic fibrosis carrier is therefore 1:93. The risk to a child is calculated to be (1/2) (1/93) (1/4) = 1/744.*

3.6 Cloning the Gene for Cystic Fibrosis

Although the availability of closely linked markers enabled genetic diagnosis of cystic fibrosis in some families, this in no way substitutes for having identified the gene itself. With knowledge of the gene in hand, definitive genetic testing can be offered, even to establish the diagnosis of the disorder in a child without prior family history of cystic fibrosis. It offers the promise of establishing genotype-phenotype correlations and hence of predicting outcome. Most of all, given our ignorance about the pathophysiology of cystic fibrosis, identification of the gene means identification of the gene product, which might be the first step on the road to developing a cure. Even as linkage-based diagnostics were in hand, the cystic fibrosis community eagerly awaited discovery of the gene.

The strategy applied to this problem is referred to as *gene walking* (Figure 3-17). The idea is to start with a cloned marker that is closely linked to a gene of interest. This is used to identify other adjacent clones, eventually leading to the target gene. A DNA probe corresponding to the marker is used to screen a genomic DNA library. A cosmid library might be used, as it will have relatively large DNA inserts (upward of 50 Kb) (Figure 3-18). Cosmids that hybridize with the probe include substantial quantities of DNA to either side of the probe. DNA fragments from the ends of the cosmid then are used as probes to screen the genomic library once again, identifying cosmid clones. Cosmids that hybridize with these new probes, but not with the initial probe, contain DNA that overlaps with the first cosmid but extends in one or the other direction for some distance. Again, the ends of these new cosmids are used as probes for the library, and the process is repeated. After multiple iterations, hundreds of kilobases of

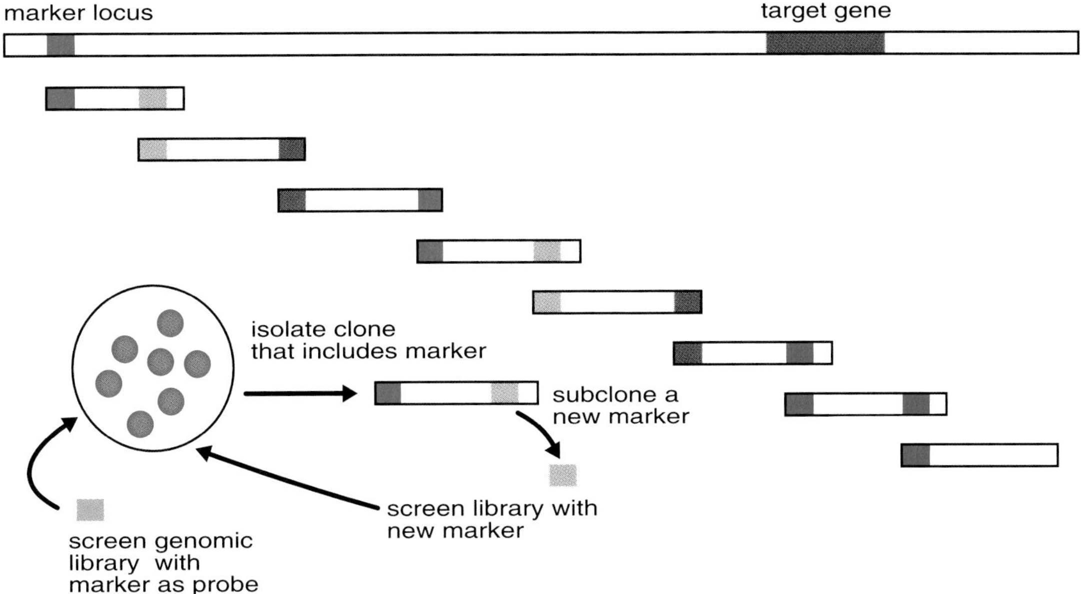

FIGURE 3-17 Schematic of gene walking, in which a linked marker is used to identify a clone that includes the marker, along with additional DNA. This process is iterated until a large region of overlapping clones has been obtained. Eventually, the gene of interest will have been cloned (although it remains to be identified). A unidirectional walk is illustrated here.

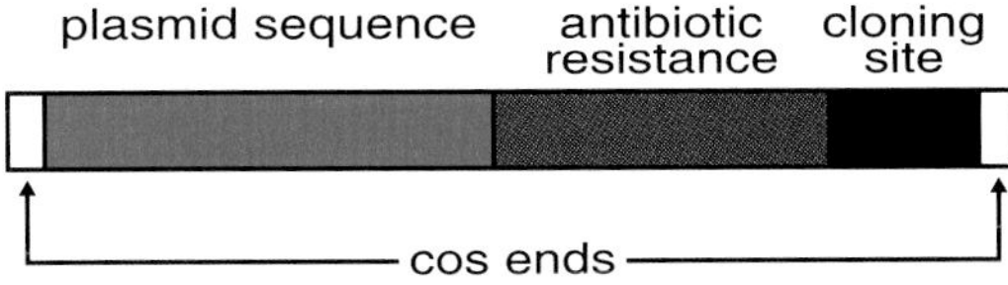

FIGURE 3-18 Cosmid cloning vector. The cosmid is an artificially created cloning vector; it does not exist in nature. It consists of a plasmid sequence but contains a DNA sequence at each end called a *cos* sequence. This is derived from a lambda phage and is the signal necessary to package the DNA into a phage head. This allows very large recombinant molecules to be placed in a phage coat and injected into a bacterial cell. Once inside, though, the recombinant molecule replicates like a plasmid. Cosmids have the advantage of ease of use, as do plasmids, but with the added advantage of large cloning capacity of phage. Because the phage DNA itself is eliminated, cosmids have very large cloning capacities—upwards of 50 Kb.

DNA can be cloned from a region surrounding a marker gene. If the marker is closely linked to the gene of interest, eventually that gene, too, will be cloned.

How close to the target gene does a marker need to be to ensure success? As a rough estimate, 1 cM of genetic distance corresponds with approximately 1 million base pairs of DNA. Actually, the correspondence of genetic distance with physical distance varies from one chromosome region to the next and may differ in male and female individuals even for the same region. Recombination tends to be suppressed near the centromere, for example. Even if the estimate of 1 cM = 1 Mb (megabase) is accepted, the enormous magnitude of a gene-walking project is apparent. A marker 1 cM from a disease gene requires at least 20 rounds of cosmid walking (and actually many more, considering that cosmids may contain fewer than 50 Kb DNA and the clones will overlap). Also, the walk may need to occur in both directions, unless there are multiple markers whose order relative to the target gene is known. Having such markers can help establish directionality of the walk toward the target gene.

The effort to clone the gene for cystic fibrosis began from two markers that were closely linked to cystic fibrosis and to one another. It was known from linkage data that the markers MET and D7S8 flank the cystic fibrosis gene. The walks were initiated from two other markers, D7S340 and D7S122, which were between MET and D7S8. Walking was done from several phage and cosmid libraries. A total distance of 249 Kb was cloned in 58 DNA fragments. A physical map of the region was generated using the technique of pulsed-field gel electrophoresis (Figure 3-19). This approach allows very large fragments of DNA, in the range of hundreds of kilobases, to be resolved on an agarose gel. A Southern blot prepared from this gel can be probed with various cloned sequences to determine whether these clones reside on the same large fragment of genomic DNA. The *long-range restriction map* obtained with pulsed-field gel electrophoresis helped to establish the directionality of the cloning process from the marker loci to the cystic fibrosis gene (Figure 3-20).

A gene-walking project can be complex and tedious. One obstacle is that not all regions of DNA will be represented in a genomic library. Some genomic regions tend to be unstable when cloned in bacteria and must be cloned instead in bacteria that are specially designed to reduce the likelihood of DNA rearrangement. Probably these DNA regions have structures that make them prone to rearrangement (and may, in fact, be "hot spots" for mutation). Another cloning strategy, referred to as *chromosome*

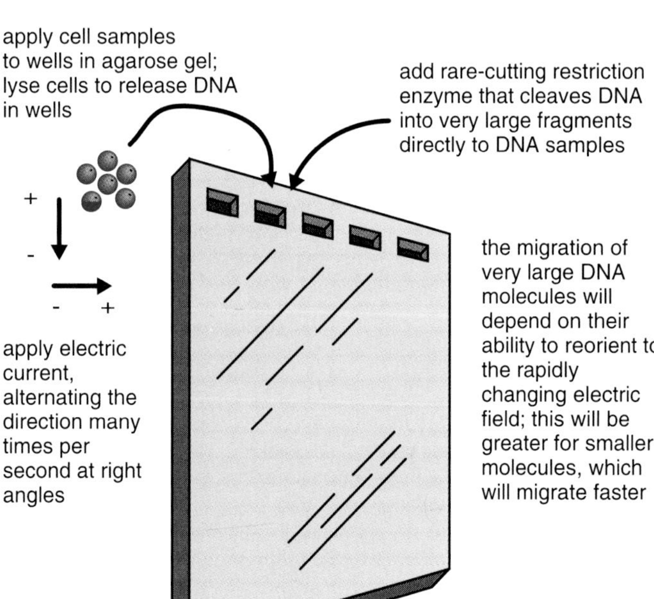

FIGURE 3-19 Pulsed-field gel electrophoresis. Large fragments of DNA are generated by cutting with enzymes that recognize rare sequences (usually of eight bases) and therefore cut only rarely. Digestion is carried out on DNA samples released from cells applied directly in the wells of the agarose gel. This minimizes the handling of DNA, which can lead to shearing into smaller fragments. The DNA fragments are likely to be in the size range of tens to hundreds of thousands of base pairs and cannot be resolved by ordinary electrophoresis. If, however, the direction of the electric field is briefly switched from top-to-bottom to left-to-right, the DNA molecules must reorient in the direction of migration if migration is to occur. This reorientation occurs faster for smaller molecules, which will therefore migrate faster and thus farther. This technique therefore provides a way to separate very large DNA molecules.

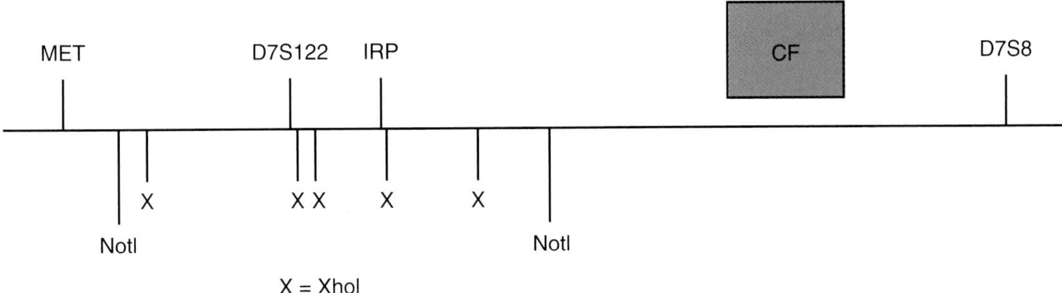

FIGURE 3-20 Long-range restriction map of cystic fibrosis region. Genetic linkage mapping indicated that the cystic fibrosis gene (*CF*) was between MET and D7S8, and finer mapping showed that it was between IRP and D7S8. The IRP gene was found to reside on the same NotI fragment as D7S122 but on different XhoI fragments. DNA between D7S122 and IRP was cloned, and this established the direction toward the cystic fibrosis gene. NotI and XhoI are two rare-cutting restriction enzymes used for long-range pulsed-field gel mapping. XhoI sites are shown only in the region of D7S122 and IRP. Other XhoI sites between IRP and D7S8 are not indicated. (Data from Rommens JM, et al. Identification of the cystid fibrosis gene: chromosome walking and jumping. Science 1989;245:1059–1065.)

jumping, was used to bypass areas that were difficult to clone. The principle is illustrated in Figure 3-21.

One of the greatest challenges in a positional cloning project is knowing when the gene of interest has been cloned, given that nothing is known about the structure of that gene. First, it is necessary to find genes in the cloned region, and then the specific gene of interest must be identified. To identify genes rapidly in the cystic fibrosis region, DNA clones were examined for sequence conservation across species. This approach is based on the assumption that segments of DNA that encode exons are likely to be conserved in evolution, whereas sequences between genes or in introns are less likely to be conserved. Areas of cross-species conservation are identified by using the cloned DNA as a probe on a Southern blot containing restriction enzyme–digested DNAs from various species (a "zoo blot," Figure 3-22).

Four regions in the area around the cystic fibrosis gene were found to display cross-species hybridization. The first was homologous with cloned cDNA from a human lung cDNA library but was known from linkage data to be distant from the cystic fibrosis locus (i.e., from a region known to recombine with the cystic fibrosis locus from family studies). The second region was found to correspond with a previously cloned gene also known to recombine with cystic fibrosis. The third cloned region did not correspond with cDNA clones and did not hybridize with mRNA on northern blots. This was not a good candidate for the cystic fibrosis gene.

The fourth region was found to contain a sequence that was highly GC-rich. Similar regions, called *CpG islands*, tend to occur at the 5′ ends of many genes, especially so-called housekeeping genes that encode proteins common to many cell types (as distinguished from tissue-specific genes). Although CpG dinucleotides often are modified by methylation of the cystosine residue, CpG islands tend to be hypomethylated. The restriction enzyme HpaII cuts at the site 5′-CCGG-3′ only if the DNA is unmethylated, giving tiny fragments in unmethylated CpG islands. Such tiny fragments were seen when the cloned DNA from region 4 was cut with HpaII, suggesting that a gene was located in the region. This was confirmed when the region was found to hybridize with cDNA obtained from a sweat gland cDNA library.

Region 4 was known from linkage data to reside in the area of the cystic fibrosis gene and corresponded with cDNA from a tissue involved in the disease. The initial cDNA clone comprised only 920 base pairs, whereas the full-length transcript consisted of approximately 6500 bases, according to northern blot experiments. Eighteen additional overlapping cDNA clones were isolated that enabled the full-length transcript to be sequenced. An open reading frame of 1480 amino acids was found. The predicted amino acid sequence included a domain that would span the membrane multiple times and several adenosine triphosphate–binding sites. Homology was found with other proteins involved in

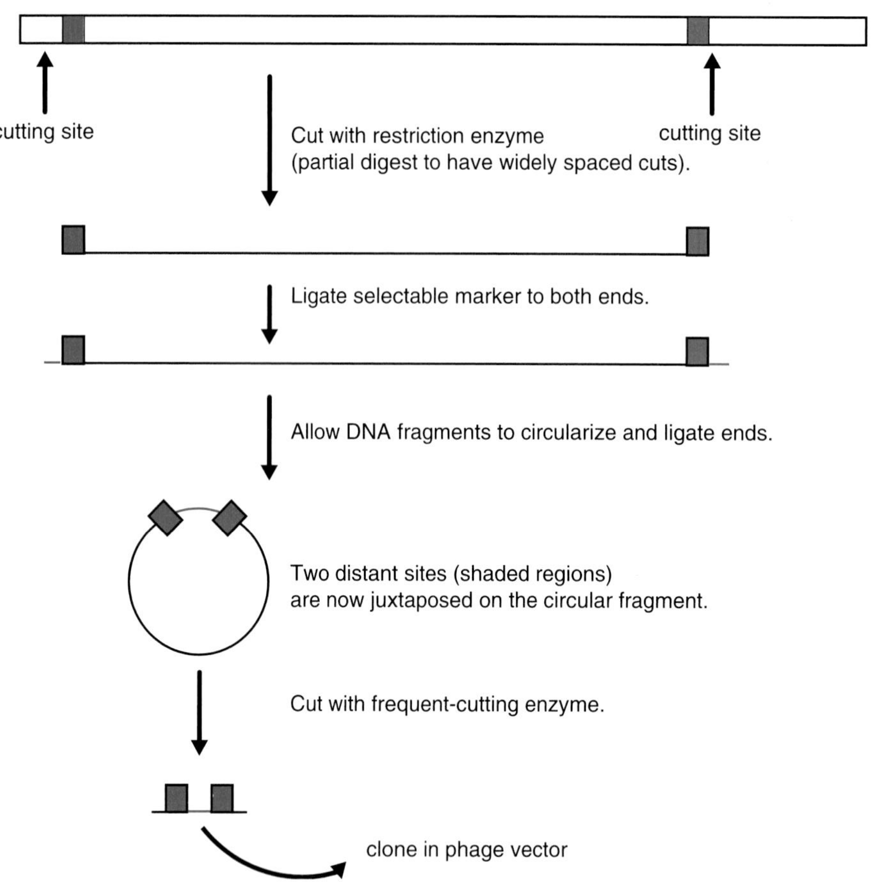

FIGURE 3-21 Principle of chromosome jumping. DNA is digested into very large fragments, which may be done with a frequent-cutting enzyme by digesting under conditions so that only a small proportion of enzyme sites are cut. This avoids a bias of cutting only at defined rare-cutting recognition sites. A marker then is ligated to each end of the fragments. The fragments have non–base paired complementary ends and are allowed to form circles, which then are ligated. In this way, two regions that previously were far apart in the genome are juxtaposed. Next the circles are cut with a frequent-cutting enzyme, and the resulting fragments are ligated into a phage vector. Phages are selected that contain the marker that was ligated to the ends of the DNA fragments (because these are the ones that incorporate two pieces of DNA that were previously far from one another). The result is a phage clone that juxtaposes parts of the genome that map tens or hundreds of kilobases apart. Screening the library with one DNA probe therefore identifies another probe some distance away. (Data from Collins FS, et al. Construction of a general human chromosome jumping library, with application to cystic fibrosis. Science 1987;235:1046–1049.)

transport of ions across the cell membrane (Figure 3-23). This seemed a plausible function, given the abnormal NaCl excretion from sweat in cystic fibrosis and the fact that the pathophysiology of the disease involves abnormal secretions.

If this gene were really the cystic fibrosis gene, it would be expected that affected individuals would have mutations. No major rearrangements of the gene, such as deletions or duplications, were found by Southern blot analysis. cDNA clones were isolated from affected and unaffected individuals and the sequences determined and compared. One mutation, a three-base deletion, was found in some affected individuals, which was predicted to cause loss of a phenylalanine residue from the protein (Figure 3-24). To determine the frequency of this mutation in a large number of individuals with cystic fibrosis, a polymerase chain reaction (PCR) assay was set up. Primers flanking the site were used to amplify the region, and the mutant or wild-type sequences were

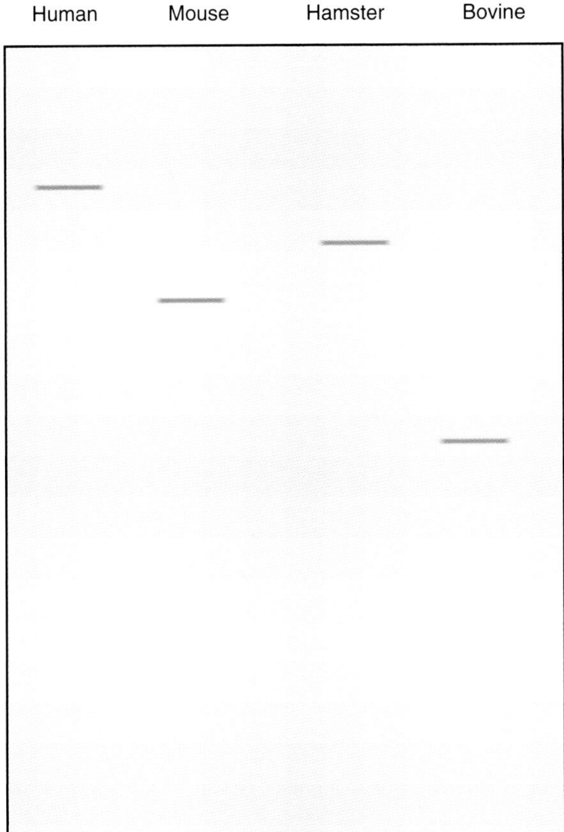

FIGURE 3-22 So-called zoo blot. DNA from different species is cut with a restriction enzyme, and a Southern blot is made and then hybridized with a cloned segment of DNA. If the sequence is highly conserved, hybridization will be seen with DNA from different species. The size of the fragments will differ because of sequence divergence between the species, leading to different locations of flanking restriction enzyme recognition sites. If the sequence is not conserved, hybridization will only be seen for DNA from the species from which the clone was isolated.

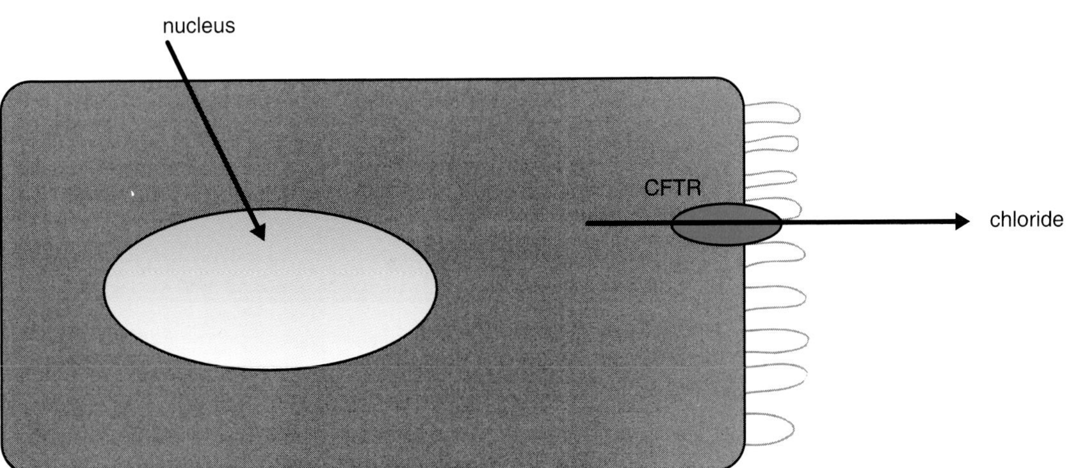

FIGURE 3-23 The cystic fibrosis transmembrane conductance regulator (*CFTR*) is a chloride channel that resides at the cell membrane and functions in the transport of chloride out of the cell.

ile phe
AAA GAA AAT ATC ATC TTT GGT GTT

AAA GAA AAT ATC AT- - -T GGT GTT

FIGURE 3-24 The ΔF508 mutation. The mutation is a deletion of three bases: CTT. This results in loss of a phenylalanine residue from position 508 in the CFTR protein. The new ATT codon still encodes an isoleucine at position 507.

detected with specific radiolabeled oligonucleotides (see Chapter 1). The mutation, referred to a *ΔF508* (for deletion of phenylalanine, abbreviated *F*, at codon number 508), was found on 145 of 214 cystic fibrosis chromosomes, or 68%. The ΔF508 mutation was found to reside on chromosomes with the haplotype of markers previously found to be in linkage disequilibrium with cystic fibrosis. It seemed highly likely that the cystic fibrosis gene had been found and that ΔF508 was the most common mutation. The gene product was named *cystic fibrosis transmembrane conductance regulator*, or CFTR.

3.7 Detection of Mutations in CFTR

The discovery of the gene for cystic fibrosis received widespread publicity and caused great excitement among the community of families affected with the disorder. It quickly became clear that, in some cases at least, accurate molecular testing could be offered for prenatal diagnosis. Bill and Jane inquired about this and arranged to have Betsy tested. Betsy is found to be a ΔF508 homozygote, and both Bill and Jane are heterozygous (Figure 3-25). Jane becomes pregnant, and fetal sampling is done by amniocentesis. The fetus is found to be a ΔF508 carrier, and the pregnancy is continued, resulting in the birth of a baby boy, James.

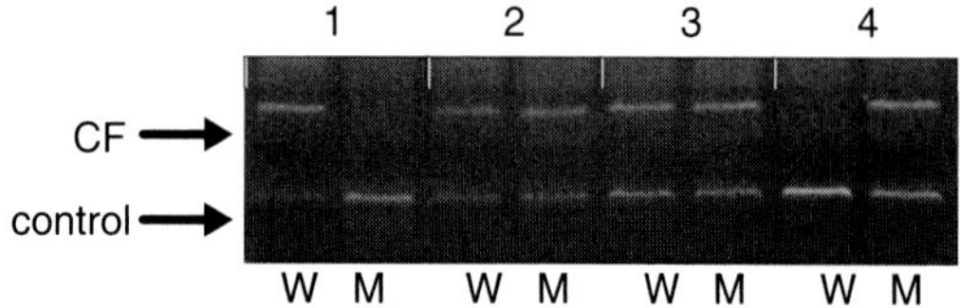

FIGURE 3-25 Identification of cystic fibrosis (*CF*) genotypes by amplification refractory mutation testing. DNA samples for a control (lane 1), Jane (lane 2), Bill (lane 3), and Betsy (lane 4) are subjected to PCR amplification using specific primers for the wild-type (*W*) or mutant (*M*) sequence for ΔF508.

The fact that ΔF508 accounts for almost 70% of cystic fibrosis mutations in the northern European population raised hopes that other common mutations would be found and that testing for just a few mutations would identify the majority of cystic fibrosis mutations. Such hopes were quickly dashed, however, as additional mutations were sought. In the northern European population, hundreds of diverse mutations comprise the additional 30% and, indeed, to this day not all have been discovered. Moreover, no single mutation other than ΔF508 accounts for more than a few percent of CFTR mutations. Whereas assay for ΔF508 detects 70% of cystic fibrosis chromosomes, it would require testing for more than a dozen mutations to raise the sensitivity to 85% and perhaps hundreds to detect more than 90% (Table 3-3). The prevalence of specific mutations also differs in different populations. In the Ashkenazi Jewish population, for example, five mutations account for 97% of

TABLE 3-3 Ten most common cystic fibrosis transmembrane conductance regulator mutations

Mutation	Relative Frequency
ΔF508	67.2
G542X	3.4
W128X	2.4
3905insT	2.1
N1303K	2.1
3849 + 10KbC T	1.8
R553X	1.4
621 + 1G T	1.3
1717-1G A	1.3
1078delT	1.1

Note: These frequencies represent an average over a large number of populations. Mutation nomenclature is based on the codon number and amino acids involved. An amino acid substitution is named by using the single-letter abbreviation for the wild-type amino acid, the codon number, and then the abbreviation for the substituted amino acid. X = a stop codon; ins = insertion. Mutations within introns that upset splicing are indicated by the last codon before the intron for a mutation near the 5′ end of the intron, or the first codon after the intron for a 3′ mutation, with a plus or minus sign indicating the number of bases into the intron and the specific base change.

Source: Based on data collected by the Cystic Fibrosis Genetic Analysis Consortium: Tsui LC. The spectrum of cystic fibrosis mutations. Trends Genet 1992;8:392–398.

cystic fibrosis mutations, whereas in American blacks, fewer than half the mutations are detected with a testing panel of more than 30.

CFTR mutations are widely distributed in the gene and are of various types. Large gene rearrangements such as deletions or insertions are rare. Rather, deletions or insertions of one or two bases or single base substitutions are the rule. The various mutations are not equivalent in their effects on the function of CFTR. Homozygotes for ΔF508 generally suffer pancreatic insufficiency, with consequent malabsorption. In contrast, compound heterozygotes with the genotype R117H/ΔF508 (R117H is a substitution of histidine for arginine at codon 117) usually have normal pancreatic function.

Bill's brother Tim is tested for ΔF508, and he is found to be a carrier, as had been predicted from linkage testing. Holly is not found to carry any of 32 mutations included in a testing panel offered by a commercial testing laboratory.

The large number of mutations to be tested has raised complex logistic and genetic counseling issues. From the point of view of the laboratory, it is necessary to devise an efficient and inexpensive testing scheme that detects the maximum number of mutations. From the counselor's point of view, families need to be informed that no currently available mutation detection scheme guarantees the identification of all possible CFTR mutations.

Several approaches for simultaneous testing of multiple CFTR mutations have been devised. One involves PCR amplification of several exons where mutations are known to reside, followed by hybridization of the PCR products with mutant or wild-type oligonucleotides (Figure 3-26). Another commonly used method relies on specific amplification of mutant or wild-type sequences with primers that bind to the site of mutation (Figure 3-27). It now is possible to obtain testing for more than 30 mutations at a cost of less than \$200. DNA is obtained from epithelial cells that line the inner cheek, obtained by painless brushing and thereby avoiding the sampling of blood.

The genetic counseling issues are in some ways more complex than the technical problems. Cystic fibrosis mutations are fairly common in the white population, and most families learn of their carrier status only after the birth of an affected child. The possibility of offering screening to couples from high-risk populations has therefore been raised. In principle, this offers the opportunity to identify heterozygous couples, who may be offered prenatal diagnosis. In practice, a number of social and ethical problems are raised.

A major issue concerns the sensitivity of the screening test. No mutation screen detects 100% of possible CFTR mutations, and therefore lack of detection of a mutation does not guarantee that an individual is not a carrier. Even if both partners are not found to carry a cystic fibrosis mutation, they are still at risk of having a child affected with a mutation not included in the testing. The impact of this can be seen if one compares the testing outcomes of a test that detects 70% of mutations (e.g., ΔF508 alone) and one that detects 90% of mutations (Table 3-4).

With either test, a finding that both partners are carriers leads to a 1 in 4 risk of cystic fibrosis in an

TABLE 3-4 Comparison of testing outcomes using a 70% or 90% sensitive mutation screen

	70% Detection Rate		90% Detection Rate	
Outcome	Frequency	Risk of CF in Offspring	Frequency	Risk of CF in Offspring
N/N	0.9447	1/30,000	0.929	1/250,000
N/C	0.054	1/300	0.069	1/1000
C/C	0.00078	1/4	0.0013	1/4

Note: N outcome is wild-type (i.e., no mutation found); *C* is mutant (i.e., carrier). Risks of cystic fibrosis (CF) in offspring are rounded figures. Figures assume a carrier frequency of all cystic fibrosis transmembrane conductance regulator mutations of 1 in 25 (or 0.04). The frequency of any outcome is calculated from the frequency of any detected allele. If the carrier frequency is 0.04 and 70% of mutations are detected, then the frequency of detected carriers is (0.04)(0.7) = 0.028. The frequency of C/C couples is (0.972)(0.972) = 0.9447; that of N/C couples is 2(0.028)(0.972) = 0.054, and so on. The risk of cystic fibrosis is calculated from the risk that a person is a carrier after DNA testing. See box for explanation of this calculation.

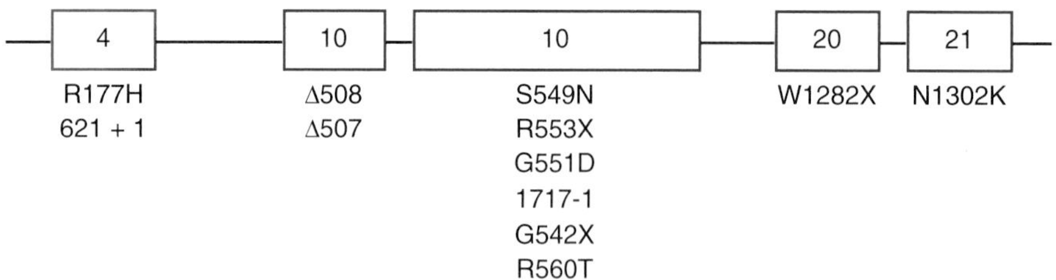

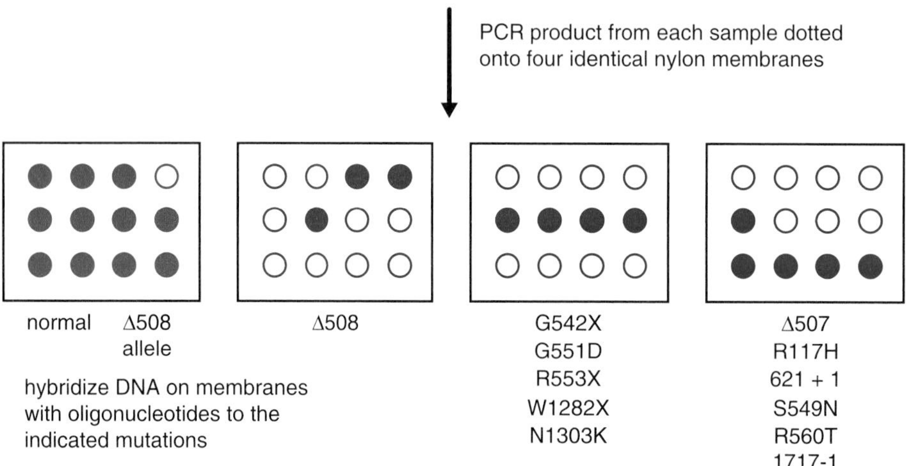

FIGURE 3-26 Detection of multiple CFTR mutations by oligonucleotide hybridization. In this scheme, 12 different mutations in five exons are tested. The five exons are first amplified by PCR, using primers that flank each exon. The PCR reactions all are done in the same tube, so that five different PCR products are obtained. Samples from each patient are dotted onto a nylon membrane, and then each membrane is hybridized with a pool of one or several radiolabeled oligonucleotides, each specific to a different mutation. The genotype of an individual is read by noting the pattern of hybridization obtained. For example, in the first row, the first two samples are homozygous wild-type for the ΔF508 mutation, the third is heterozygous for ΔF508 and wild-type, and the fourth is homozygous for ΔF508. In the second row, the first sample is a compound heterozygote for a mutation in the third and fourth oligonucleotide pools. The second sample is a ΔF508/third-pool heterozygote, and the third and fourth samples are homozygous either for a third-pool mutation or are compounds for a third-pool mutation and another unknown mutation. In the fourth row, all samples either are homozygous for a fourth-pool mutation or are heterozygous for a fourth-pool mutation and an unknown mutation. Samples positive for a third- or fourth-pool mutation then are tested individually for the different mutations in those pools. (Data from Shuber AP, et al. Efficient 12-mutation testing in the CFTR gene: a general model for complex mutation analysis. Hum Mol Genet 1993;2:153–158.)

offspring, and reliable prenatal testing is possible. Finding that neither partner is a carrier substantially reduces the risk of having an affected child, although that risk still is not zero. Couples need to be informed that this screening test does not detect all possible mutations and that some couples considered to be at very low risk may, rarely, still have an affected child. The most problematic scenario, however, is the middle one: One partner is found to be a carrier, and the other is not. This will be a common occurrence, happening in 5% or so of couples with a 70% sensitive screen and in more than 6% of couples with a 90% sensitive screen. In each case, the risk of having an affected child is increased over the population risk, as one partner is known to be a carrier and the other might still be a carrier for an undetected mutation. In such couples, mutation analysis will not identify cystic fibrosis in a child, as the second CFTR mutation is undetectable. Consequently, mutation screening has alerted the partners to their increased apparent risk yet has offered nothing further in the way of prenatal diagnosis.

Cystic fibrosis screening raises major challenges in both consumer and physician education. Those who become aware of the possibility of such screening may have unrealistic expectations. They will not realize, for example, that negative testing for a CFTR mutation does not rule out carrier status. Explaining the reasons that a child might still be born with cystic fibrosis despite negative screening requires mastery of modern molecular genetics, a skill not found among all current practitioners. On the other hand, if the couple is found to be at increased risk of having an affected child, the partners may not understand what cystic fibrosis is; indeed, they may have never heard of the disorder. They need to be made to understand the medical problems associated with cystic fibrosis and how these can occur to their child even though they themselves may be healthy and there is no prior family history of the disorder. It is not easy to explain this to a layperson who likely has little or no knowledge of genetics.

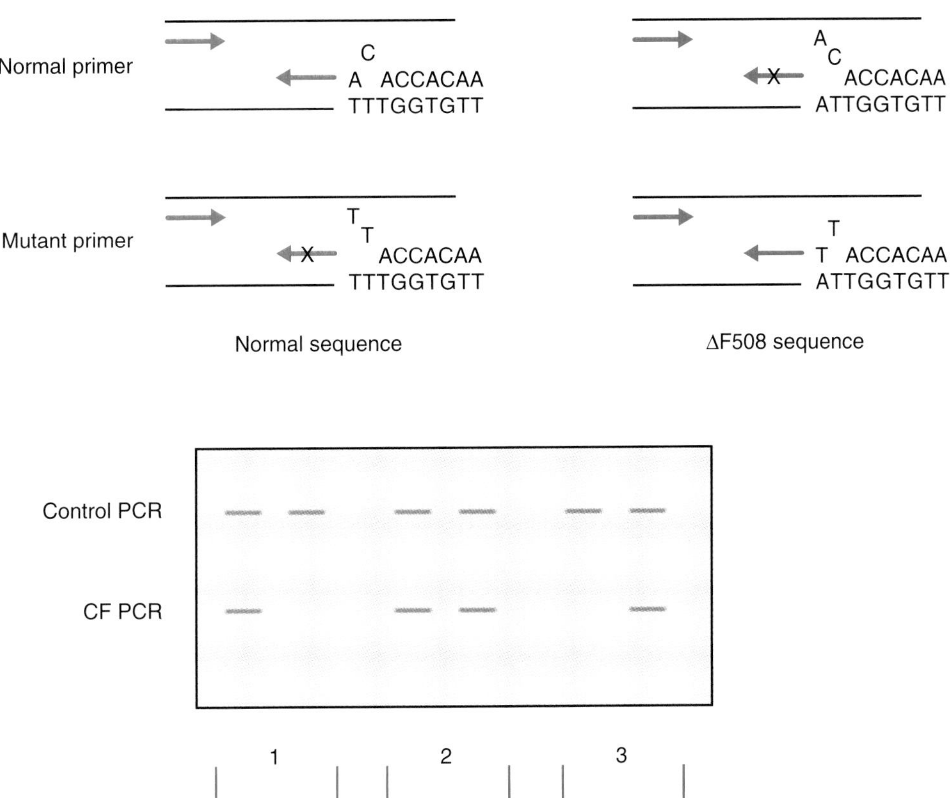

FIGURE 3-27 Mutation-specific PCR detection of ΔF508 genotype. Two separate PCR reactions are carried out for each patient sample. One PCR primer is common to both reactions, but the other PCR primer is different in the two reactions. In one reaction, this PCR primer is homologous to the wild-type sequence at position 508, whereas in the other, this primer is homologous to the mutant sequence. A PCR reaction will occur if the normal primer binds to the normal sequence or if the mutant primer binds to the mutant sequence, but not if the primers are mismatched. Note that the next-to-last base of the primers is intentionally mismatched to increase the discrimination of the reaction. The PCR products are run on an agarose gel. Each patient is represented in two lanes, one for the wild-type PCR (left) and one for the mutant (right). A control PCR reaction (outside the cystic fibrosis gene) is done for each patient. Patient 1 is homozygous wild-type at position 508, patient 2 heterozygous, and patient 3 homozygous for ΔF508. (Modified from Ferrie RM, et al. Development, multiplexing, and application of ARMS tests for common mutations in the CFTR gene. Am J Hum Genet 1992;51:251–262.)

Two approaches have been proposed for cystic fibrosis screening. One involves screening the entire population or at least that portion whose racial or ethnic background puts individuals at increased risk. This can be done by screening one member of a couple first and then offering testing to a partner if the first member is found to be a carrier (*stepwise testing*). Alternatively, the couple may be screened as a unit, offering counseling only if both partners are carriers (*couple testing*). Aside from the educational problems just noted, population screening raises financial questions: Who will pay for the tests? In the United States, individual insurance plans differ in their willingness to pay for such screening. Many people, therefore, including those with no insurance, would have to pay out of pocket for the testing. In essence, then, eligibility for testing becomes related to social class.

The other approach is sometimes referred to as *cascade testing*. Here, one begins with an individual affected with cystic fibrosis and then systematically works through the family, identifying carriers. First the parents are tested, then their parents and siblings, and so on. This has the advantage of being more targeted, and education is easier because there is a relative with the disorder. The testing can be less expensive, as the screened population is a subset of the general population and the most likely mutations are known in advance.

A final concern applies to any form of cystic fibrosis screening: that of stigmatization. It has been estimated that all persons carry several (perhaps 1–10) so-called lethal equivalents—that is, genes that would be lethal if present in a homozygous state. A CFTR mutation may be one of these, but a person who is not a cystic fibrosis carrier probably is a carrier for something else and does not yet know it. In spite of this, there is risk that a person found to be a cystic fibrosis carrier will suffer psychological damage as a result of this knowledge or will be viewed as a less suitable partner. There is also a risk of discrimination. Cystic fibrosis carrier status has no impact on a person's health or job performance, so it is doubtful that many will experience discrimination themselves. However, would partners found to be at high risk of having a child with cystic fibrosis be forced—by their health insurance carrier, a managed care system, or a government—to have prenatal diagnostic testing and to abort an affected fetus? The financial incentive to take the choice out of the partners' hands is great.

What is the risk of being a cystic fibrosis carrier if DNA testing is negative?

- Assume that 70% of all possible cystic fibrosis mutations are tested. Negative testing can occur in either of two ways: A person might be a carrier for a mutation not included in the testing panel, or the person is not a carrier at all. The risk that the person is a cystic fibrosis carrier can be calculated from the relative likelihood of these two possibilities.
- The likelihood that the person is a carrier for an untested mutation is the probability of being a carrier, which is 0.04, times the probability of having a negative mutation screen if the person is a carrier, which is 0.3 (i.e., $1-0.7$). Hence, the probability of this scenario is $(0.04)(0.3) = 0.012$.
- The likelihood that a person is not a carrier at all is the probability of not being a carrier, namely 0.96 (all such persons should have a negative carrier test).
- The relative likelihood of being a carrier, then, is $0.012/(0.012+0.96) = 0.012/0.972 = 0.0123$.

All these concerns notwithstanding, there is still a feeling that, sooner or later, cystic fibrosis testing may be inevitable, in some form at least. Many professional societies have issued statements urging caution, but at the same time commercial genetic testing laboratories are banking on the huge market that is represented by universal screening. Perhaps the major long-term question is whether a treatment will be developed for cystic fibrosis that will make moot the debate on screening.

3.8 Treatment of Cystic Fibrosis

Betsy is now 15 years old. She is in ninth grade and does well in school, in spite of having missed many days each year due to hospitalization. Most of these hospital stays have been required for "pulmonary cleanout," involving 10 to 12 days of intravenous antibiotic treatment and pounding on her chest to loosen secretions and open blocked airways. Betsy has a chronic cough but good exercise tolerance, and otherwise is a normal 15-year-old. She has many friends with cystic fibrosis, both from her visits to the clinic and to local support group meetings. She is aware that her health may deteriorate over the coming years. Both she and her parents are following very closely progress in the development of new treatment methods. Recently, Betsy has agreed to participate in a study of DNase treatment, which promises to loosen her lung secretions.

The life expectancy for persons with cystic fibrosis has improved dramatically since the 1940s. This is mostly due to the availability of antibiotic treatments and pancreatic enzyme replacement. DNase has been helpful in breaking up thick secretions in the lung. It is likely that further advances in medical therapy gradually will improve survival further, but major hope now is being directed toward using knowledge of molecular genetics to develop novel approaches, including the ability to correct the gene defect entirely.

Identification of the cystic fibrosis gene has led to major advances in understanding the pathophysiology of the disorder. The CFTR encodes a protein that spans the cell membrane and functions as a chloride channel. Transport of chloride ions across the cell membrane normally leads to passive diffusion of water into the extracellular space. Inadequate chloride transport draws less water and results in thickened secretions.

Four classes of mutations have been found that differ in their effects on the CFTR. One type leads to failure of production of the protein, generally due to chain termination or frameshift mutations. The second type, which includes the common ΔF508, disrupts processing of the CFTR protein. It is believed that such mutations lead to defective folding of the protein, interfering with normal processing in the Golgi apparatus and, ultimately, to failure of the protein to appear at the cell surface. The CFTR ion channel is regulated by binding of adenosine triphosphate and also by phosphorylation with a cAMP-dependent protein kinase. Rare point mutations have been found in regulatory domains of the CFTR. Finally, some point mutations have been found to disrupt the chloride channel itself.

Study of the pathophysiology of cystic fibrosis and possible therapeutic interventions has been facilitated by the development of mouse models of the disease. These have been created by transgenic approaches, as no natural mouse model of cystic fibrosis has been found.

The normal mouse gene that corresponds with the human CFTR was disrupted by a technique of targeted insertion in mouse embryonic stem cells (Figure 3-28). These are embryonic cells that can be grown in culture but, when injected into a mouse blastocyst, remain pluripotent and can contribute to the formation of the embryo, resulting in chimeric animals. The embryonic stem cells are derived from animals with a coat color different from the mice used to generate blastocysts, so the chimeras are readily identified by their patches of different coat colors. Some of these animals have mutant CFTR-bearing germ cells, and offspring are produced that are fully heterozygous for the CFTR mutation. These, in turn, are mated to produce homozygotes.

The insertional mutations lead to frameshifts and, therefore, to failure of CFTR protein production. In some instances, homozygous mice have displayed some features resembling cystic fibrosis in humans. The phenotypes differ in various gene-targeting experiments, using different gene disruption strategies and mice of different genetic backgrounds. The latter suggests that the expression of a mutant phenotype depends on the action of other genes, which may explain why natural animal models of cystic fibrosis do not exist and why even humans with the same mutation do not necessarily have exactly the same disease manifestations.

Because the most common CFTR mutations lead to absence of functional protein, promising new efforts at therapy have been directed toward replacement of the defective gene. In one experiment, cultured airway epithelial cells from a patient with cystic fibrosis were transfected with a recombinant vaccinia virus containing an intact CFTR gene. This resulted in correction of the defect of chloride transport. Similar results have been obtained using a retroviral vector to introduce the CFTR cDNA into cultured cells. These experiments have provided confirmation that the CFTR mutations are indeed pathogenic and that a foreign CFTR gene can lead to correction of the physiologic defect.

Unlike cells in culture, it will be difficult to target respiratory or gastrointestinal stem cells in vivo to introduce a new CFTR gene and expect it to be stably transmitted from cell generation to generation. An

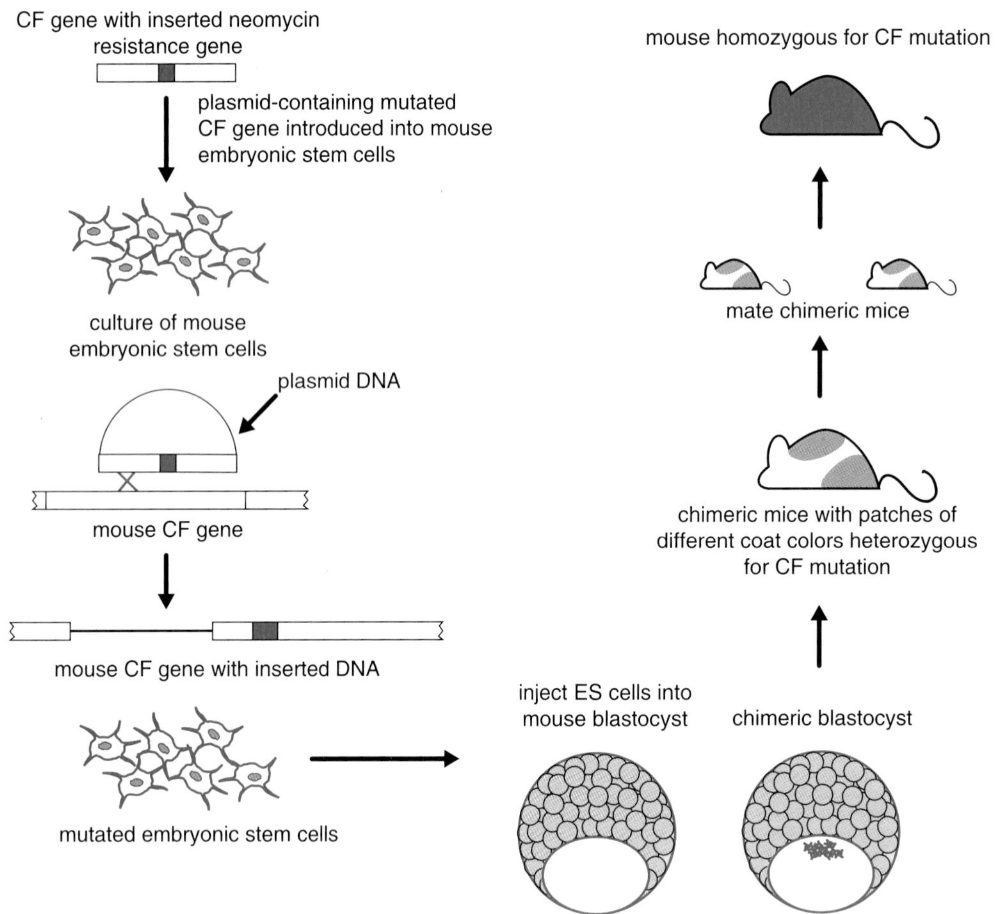

FIGURE 3-28 Creation of mouse model of cystic fibrosis (*CF*) by targeted gene insertion. A cystic fibrosis mutation is created by insertion of a neomycin resistance gene in CFTR cDNA. This is ligated to plasmid DNA and is inserted into mouse embryonic stem cells in culture. The gene is inserted by temporarily introducing pores into the cell membranes by an electric field (called *electroporation*). The vector DNA recombines with homologous mouse CFTR to cause an insertion into the gene. These embryonic stem cells (*ES cells*) are injected into mouse blastocysts and the chimeric blastocysts are brought to term by insertion into a mouse uterus. The chimeric mice have patches of different coat color, as the blastocysts are derived from albino mice, whereas the embryonic stem cells are from black mice. Some of the chimeric mice have germ cells derived from the embryonic stem cells. Mating these mice results in mice homozygous for the CFTR mutation. (Data from Dorin JR, et al. Cystic fibrosis in the mouse by targeted insertional mutagenesis. Nature 1992;359:211–215; and from Snouwaert JN, et al. An animal model for cystic fibrosis made by gene targeting. Science 1992;257:1083–1088.)

alternative that has been tried experimentally is to use a recombinant virus with an intact CFTR gene to infect differentiated cells. This may lead to correction of the gene defect in the infected cells but would last only as long as those cells survive. Because adenovirus is capable of infecting airway epithelial cells, a vector has been created in which some of the adenovirus genome is replaced with CFTR cDNA (Figure 3-29). Virus particles were produced by transfection of the recombinant genome into cells that contain the missing genetic information necessary for viral replication. Infection of the resulting virus into cultured cells from a cystic fibrosis patient resulted in physiologically significant chloride transport. The

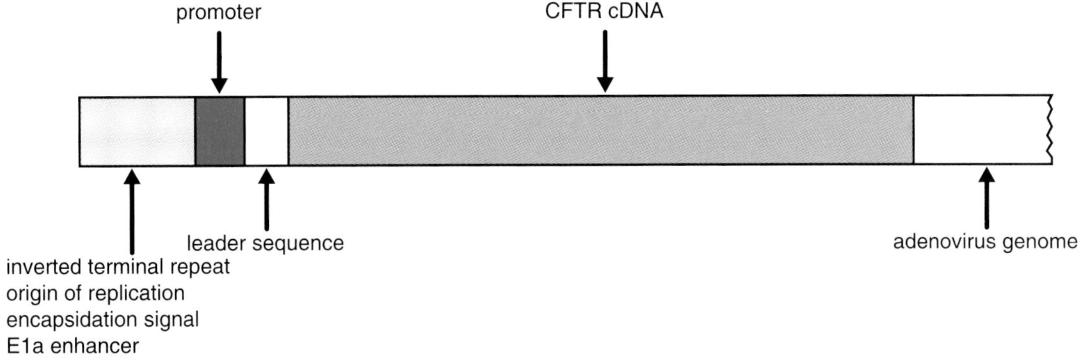

FIGURE 3-29 Adenovirus vector containing CFTR cDNA. A gene essential for adenovirus replication, E1a, has been removed and the CFTR has been inserted. Just 5′ to CFTR, there is a leader sequence, which encodes amino acids that enhance the stability of the translated protein; 5′ to that is a viral promoter sequence, and 5′ to the promoter are a set of sequences that enhance the efficiency of transcription (E1a enhancer) and provide a signal to ensure encapsidation of the viral DNA into viral coats, an origin of replication, and an inverted terminal repeat sequence necessary for viral replication. This recombinant virus was grown in a cell line that contains the E1a gene, permitting the DNA to be replicated and incorporated into viral capsids. These, in turn, were used to infect cystic fibrosis epithelial cells either in vitro or in vivo. Because the recombinant virus was missing E1a, however, the virus was unable to replicate in these cells. It did result in transcription of CFTR in the infected cells, however. (Data from Rosenfeld MA, et al. In vivo transfer of human cystic fibrosis transmembrane conductance regulator gene to the airway epithelium. Cell 1992;68:143–155; and Zabner J, et al. Adenovirus-mediated gene transfer transiently corrects the chloride transport defect in nasal epithelia of patients with cystic fibrosis. Cell 1993;75:207–216.)

virus also was used to infect the respiratory cells of rats, and production of human CFTR was noted to persist for several weeks.

These promising results have led to preliminary experiments in treating humans with cystic fibrosis. In one study, an adenoviral vector with a CFTR insert was administered to the nasal epithelium of 12 individuals with cystic fibrosis. There was a very low efficiency of transfer to epithelial cells, however, and no measurable change in chloride transport.

Another approach to introducing a gene, at least transiently, into differentiated cells has employed liposomes (Figure 3-30). These are artificially created vesicles consisting of a lipid bilayer resembling the cell membrane, which can fuse with the membrane and deliver the vesicles' contents into the cell. Transfection of respiratory epithelial cells of transgenic mice homozygous for a CFTR mutation resulted in restoration of normal levels of chloride channel activity.

Many technical issues remain to be worked out before gene therapy for cystic fibrosis will be practical. Most efforts to date have focused on respiratory cells, because these are most accessible for therapeutic intervention. Treatment of other affected tissues, such as gastrointestinal cells, will require other approaches. Nevertheless, there is optimism that a definitive therapy for some of the most devastating aspects of cystic fibrosis soon will become a reality.

3.9 The Human Genome Project

Cloning the cystic fibrosis gene was a tour de force in every sense: It required a large-scale international collaboration, first to establish the location of the gene and later to find the gene itself. Initial efforts at linkage were hampered by the relative paucity of informative markers and the need to work with small families. Unlike the case for some other genetic disorders, there were no examples of chromosomal rearrangements in rare affected individuals to point the way to the gene. That the gene was cloned in the space of only a few years is testimony to the skill and tenacity of the investigators involved, and their recognition of the importance of the task.

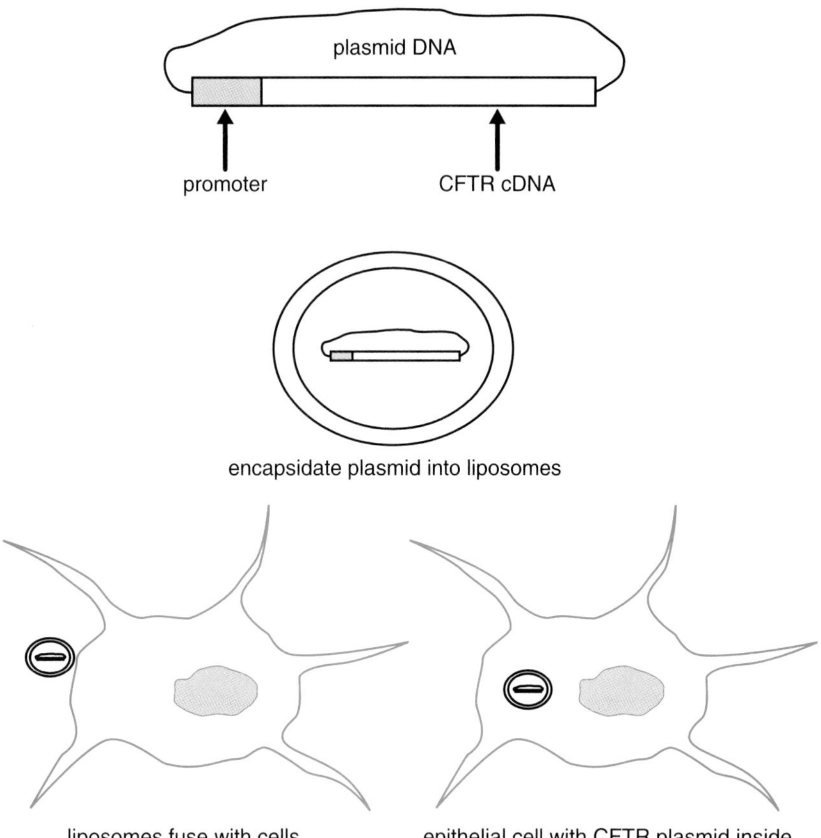

FIGURE 3-30 Liposome-mediated fusion of CFTR plasmid with epithelial cells. Liposomes are synthetic lipid bilayers. A recombinant plasmid was made that includes the CFTR cDNA, and this was placed inside liposomes. The liposomes were instilled into the airways of transgenic mice homozygous for a CFTR mutation and became fused with the plasma membranes of airway epithelial cells. The plasmids then were delivered inside the cells, where they led to expression of CFTR. Chloride channel activity was restored to normal levels in tracheae of transfected mice. (Data from Hyde SC, et al. Correction of the ion transport defect in cystic fibrosis transgenic mice by gene therapy. Nature 1993;362:250–255.)

Currently, there is an international effort under way to map the human genome and identify and sequence the genes. This effort is known as the *human genome project.* The motivation is enhanced understanding of human development and physiology through recognition of the location and structure of each of the 100,000 or more genes. It is expected that this will revolutionize diagnosis of genetic disorders, as has been the case for cystic fibrosis, and eventually will lead to insights that will favorably affect treatment. Moreover, it is likely that this effort will result in the identification of genes for previously unrecognized genetic disorders and help to unravel the genetic contribution to complex and common disorders such as heart disease and cancer. The goals of the U.S. effort in this project are outlined in Table 3-5.

The general approach consists of three stages. The first is to produce a high-resolution map of the human genome. This involves the generation of large numbers of polymorphic markers and their mapping by genetic linkage analysis. Having a gene map of 2- to 5-cM resolution will ensure that any new genetic trait can be readily positioned on the map. The second step is to produce a physical map of the genome and correlate it with the genetic map. This amounts to cloning large segments of DNA and locating the linked markers on the cloned segments. Finally,

TABLE 3-5 United States human genome project: 5-year goals

Mapping and sequencing human genome
- Genetic mapping: 2- to 5-cM-resolution gene map
- Physical mapping: 100-Kb-resolution physical map
- DNA sequencing

Gene identification
- Identify genes and place on physical map

Technology development
- Develop new approaches to DNA sequencing

Model organisms
- Mouse, *Escherichia coli*, yeast, *Caenorhabditis elegans*, *Drosophila*

Informatics
- Develop databases and software

Ethical, legal, and social implications
- Address public policy and social issues

Training
- Train scientists in disciplines related to genome research

Technology transfer
- Disseminate new technologies related to genome research

Outreach
- Share new developments and knowledge about genome

Source: Human Genome News, 1993;5(4):1–5.

expressed genes must be identified and sequenced. Most of the DNA in the genome does not encode genes but rather serves a structural role. Little is to be learned from sequencing this material, so the identification of expressed sequences is a prerequisite.

Work toward a high-resolution genetic map is well under way and has been facilitated by discovery of a class of highly polymorphic sequences that can be genotyped using PCR. These are simple sequence repeats, runs of two to four DNA bases repeated many times and dispersed widely in the genome. The exact number of repeats at a given site can vary from individual to individual and usually represents a stable mendelian trait (Figure 3-31). Repeat number can be determined by PCR amplification of the segment, using flanking primers and analysis of the size of the amplified product by polyacrylamide gel electrophoresis.

The gene map already is sufficiently dense that virtually any genetic trait can be mapped in a relatively short time. All that is required is access to a relatively large family, or a number of smaller families if there is no genetic heterogeneity. Thousands of genetic markers are available commercially. Soon it will be possible for multiple markers to be genotyped simultaneously on dozens of individuals, with computer methods used to determine genotypes and carry out linkage analysis.

The second stage involves cloning large segments to produce a physical map of the genome. Physical mapping also has been facilitated by the development of new technology, in this case cloning vectors that allow very large stretches of DNA to be isolated. The yeast artificial chromosome (YAC), for example, permits hundreds of thousands of base pairs of DNA to be cloned, representing entire genes or groups of genes (Figure 3-32). YACs consist of the essential elements for autonomous replication inside yeast cells, namely a centromere, a telomere (chromosome end), and an initiation point for DNA replication. All of the remainder of the chromosome is devoted to the sequence to be cloned. Another long-range cloning vector is the phage P1, which can accommodate tens of thousands of base pairs of DNA.

Genomic libraries have been made in both YACs and P1 and are available commercially. When a gene of interest has been mapped to a region with known genetic markers, these markers can be used to identify a particular YAC or P1 sequence. This greatly speeds the task of positional cloning, potentially obviating much of the labor that was required to walk from closely linked markers to the cystic fibrosis gene.

The third step, that of identifying and sequencing all genes, is the most ambitious. Despite major

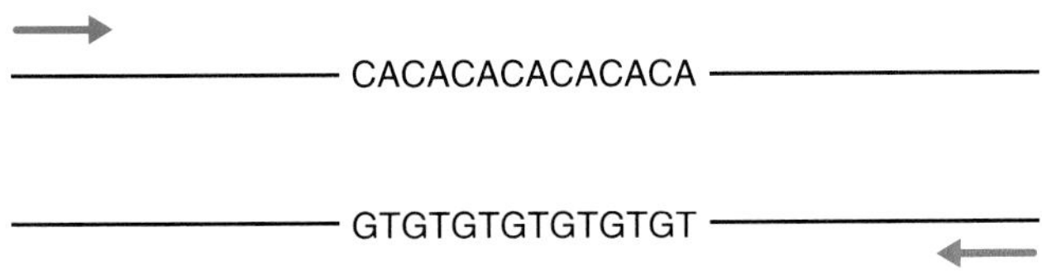

FIGURE 3-31 Simple sequence repeat. In this case, the dinucleotide CA is repeated several times. The allele shown has 7 repeats, but other alleles might have 5, 6, 7, 8, 9, or even more copies of the CA sequence. The copy number is determined by PCR amplification of the region (arrows denote PCR primers in flanking DNA) and determination of the size of the product by polyacrylamide gel electrophoresis.

Living with Cystic Fibrosis

JOAN FINNEGAN BROOKS

It is my best friend, yet it is my worst enemy. It makes me strong, yet it is my weakness. It is my constant companion, yet it makes me feel isolated and alone. It is hard to live with, yet I don't know how to live without it. *It* is cystic fibrosis.

I was born in 1960 and received a diagnosis of cystic fibrosis when I was just 1 month old. When I was adopted into my wonderful family, my parents had already suffered the heartache of losing their little girl, Peggy, to cystic fibrosis in 1952 when she was just 2½ years old. Those were the dark days in cystic fibrosis history; very little was known about the disease and even less about how to treat it. After Peggy died, my parents had another child, John, who also had cystic fibrosis. Only their oldest son, Pat, had been born without this disease.

I always knew I had cystic fibrosis: There was never a moment in time that I was told about my disease. It was always a part of my life and routine; living with cystic fibrosis seemed normal to me.

My brother John was 6 years older than me, and we went through the regimens of treating this disease together; morning and night aerosols; chest postural drainage sessions to help clear our lungs of thick mucus; daily doses of antibiotics, vitamins, and pancreatic enzymes; and a restricted low-fat, high-protein diet. Potato chips, hot dogs, and ice cream were just a few foods that were off-limits to us, since we couldn't digest the fat contained in those foods. Both John and I struggled to achieve healthy body weights, but accomplishing this was made more difficult by not being able to consume calorie-packed dietary fat.

Sadly, John lost his battle with cystic fibrosis when he was just 15 years old, in 1969. His death affected me profoundly, as I had to face the rest of my life without my brother, partner, and best friend. To this day, John's lost struggle against cystic fibrosis strengthens my determination to live life to the fullest, in honor of his memory.

Although I avoided any significant lung involvement growing up, I was severely underweight and constantly battled digestive problems. I was always the smallest student in the class, and I cringe when I see photos of myself with my normal-sized classmates from those days. I'm convinced that growing up too thin and small is just as traumatic as growing up overweight. As an adult, I am kindly labeled a "petite woman," and I no longer struggle with being underweight, thanks to the improvement in pancreatic enzymes that help me digest dietary fat.

A key development in my adolescence was my participation in field hockey during high school.

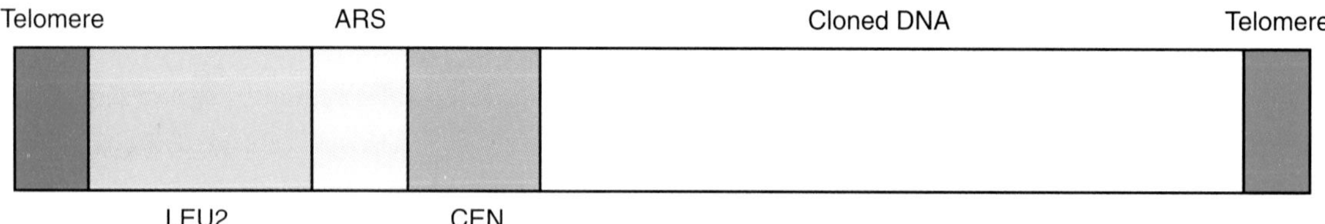

FIGURE 3-32 Yeast artificial chromosome, a linear structure capable of independent replication within a yeast cell. The two ends contain telomere sequences, which are necessary to maintain the structural integrity and replication of the chromosome. The leu2 gene (*LEU2*) is a selectable marker to identify yeast that have incorporated the recombinant chromosome. *ARS* stands for "autonomously replicating sequence" and contains an origin of replication. *CEN* is a centromere, which ensures stable division of newly replicated chromosomes to the two daughter cells. Much of the remaining material can be reserved for cloned DNA, permitting hundreds of thousands of base pairs to be cloned.

Surprisingly, the difficult running requirement was remarkably effective in helping me clear secretions from my lungs. From this experience, I developed a life-long discipline of vigorous, aerobic exercise. Exercise also gave me added independence and control over my cystic fibrosis care. I no longer needed to rely on someone else to help me with chest postural drainage to clear my lungs. Instead, I could exercise to accomplish the same thing. Whether I train for biathlon races (which feature running and cycling distances), play racquetball, or lift weights, exercise is a key component of what I do to keep healthy. I don't have the luxury of being able to skip my exercise regimen; if I don't exercise, my lungs fill up with thick, sticky mucus, which is very difficult to clear, and the likelihood of developing a serious lung infection increases.

My first hospitalization for cystic fibrosis occurred when I was 26 years old. I struggled with a bad lung infection for weeks and coughed up blood before agreeing to be admitted for a course of intravenous antibiotic therapy. I thought this was the beginning of the end of my battle with cystic fibrosis. After regaining my strength and excellent pulmonary status, I left the hospital and rededicated myself to making health my first priority. I went back to basics—eating well, adhering to my medication regimens, exercising regularly, and getting enough sleep. My attention to these details, combined with the unwavering support I received from my wonderful husband, Peter, and my family and friends, allowed me to make a complete recovery.

After this hospitalization, I also had a different outlook on how to manage my disease when my symptoms worsened. I learned that being hospitalized wasn't necessarily a harbinger of an inevitable pulmonary decline but simply another type of medical intervention that could help me stay well. Despite recognizing the value of hospitalizations in treating my disease, agreeing to be hospitalized is never an easy decision. Hospitalizations are disruptive to my career and home responsibilities and are always worrisome. Unlike most cystic fibrosis patients, I have only needed a few "tune-ups" with intravenous antibiotic therapy to maintain my pulmonary status.

Thankfully, I have survived beyond the mortality statistics that patients with cystic fibrosis face. Despite developing insulin-dependent diabetes in my mid-twenties, I'm lucky to be one of the pioneers who is living with cystic fibrosis into adulthood. In the coming years, I am confident that new treatments for cystic fibrosis will be developed. Since the discovery in 1989 of the gene that causes cystic fibrosis, new chapters in cystic fibrosis history are being written—chapters that chronicle improvements in the quality and longevity of life for patients and that describe cutting-edge research which offers unprecedented hope for a better future. I only wish that John and Peggy Finnegan had survived to share in the triumphs ahead.

advances in DNA sequencing technology, sequence determination is still slow and expensive. Major effort is therefore being devoted to developing new sequencing strategies and to developing computer databases capable of storing and retrieving the enormous volume of sequence data that will be obtained.

Two approaches are being used to identify expressed genes. One is to make a cDNA library from a particular tissue, such as brain, and then to examine each clone, one by one, for its position on the genetic and physical map. Eventually, the sequence of these genes can be determined, but for now only sufficient sequence to allow the gene to be identified by PCR is determined (that is, only enough to design primers to amplify a small part of the sequence). This approach has the advantage that it goes directly to expressed genes but the potential disadvantages of bias toward more abundantly expressed sequences and the fact that genes expressed in some tissues at a restricted time in development may be missed.

The alternative approach is first to develop a physical map and then to identify expressed genes in the cloned regions. Some means of identifying genes in cloned DNA have already been discussed, including searching for species conservation or looking for clusters of CpG dinucleotides. These, however, are laborious and, in the latter case, will miss many genes. Two other approaches are *exon trapping* and *cDNA selection*. Exon trapping is a cloning strategy using a vector that includes sequence signals needed for RNA splicing. The system is designed so that only DNA containing an exon, with proper splicing signals, can be cloned. This can be used to identify rapidly exons in a stretch of cloned DNA. The cDNA selection approach looks for homology between cDNA clones in a cDNA library and regions of cloned genomic DNA in a YAC or P1 vector. It can provide a rapid means of identifying expressed sequences in a cloned region that may correspond with a gene of interest.

The systematic effort to map and sequence the human genome already is having an impact on the identification of genes responsible for disease. Positional cloning now is vastly easier due to the availability of a dense gene map and libraries of YAC and P1 clones. In some cases, once an interval has been defined where a gene is known to reside, the task of cloning the gene amounts to determining which of many already cloned sequences in the region correspond to the gene of interest. This sometimes is referred to as the *candidate gene approach*. In the extreme, as more genes are cloned, this may be used even with rare disorders where gene mapping is difficult or impossible. Genes that seem plausible candidates based on physiologic data can be tested for mutation in affected individuals until the proper gene is found.

The human genome project raises many social and ethical issues. Among these is the question of distribution of resources toward the effort. The actual cost of the project is estimated to be in the hundreds of millions of dollars, although the final cost is difficult to know as changes in technology may lower the cost per gene. A more significant problem, though, is the risk that our knowledge of the human genome may accrue faster than society is able to deal with the implications of that knowledge.

The scope of DNA-based genetic testing will increase dramatically, making it possible to identify myriad genetic traits prenatally and to identify carriers. Currently, genetic testing approaches have been developed for a number of bona fide genetic diseases but, as more genes are identified, the distinction between a genetic trait and a disease is likely to become blurred. Is a gene that confers susceptibility to breast cancer a genetic disease? How do we classify a gene that leads to hypertension, or one that results in a reading disability? The opportunities for genetic selection in humans will likely outpace our realization of the full implications of determining the value of one trait over another.

We also are likely to be able to provide information about a person's genotype faster than we can help that person deal with the implications of this new knowledge. In some cases, knowing that one carries a disease gene helps to provide appropriate medical surveillance and genetic counseling. There are risks, though, of imposing a psychological burden on the individual that may outweigh these advantages. Also, as mentioned previously, there is a risk of what has been called *genetic discrimination*—for example, one's inability to obtain health or life insurance or employment because of one's genotype.

We are a long way from fully understanding the human genome and will remain so even after all the genes are cloned and sequenced. In defense of the human genome project, it can be noted that, unlike in other major technical projects such as the development of nuclear energy, social and ethical implications are being actively studied as the technical efforts are under way. We will see how much has been learned in recent decades about the responsible use of new scientific knowledge.

3.10 Preimplantation Diagnosis of Cystic Fibrosis

Bill and Jane learn that it is possible to establish whether an embryo will be affected with cystic fibrosis prior to implantation in the uterus. They travel to another city to participate in a protocol for preimplantation diagnosis.

The decision to terminate a pregnancy if a fetus is found to be affected with an inherited disorder can be difficult and traumatic. This is particularly true for second-trimester abortions but also applies to earlier terminations when diagnosis is made by chorionic villus biopsy. The advent of techniques of in vitro fertilization has led to the development of an alternative means of prenatal diagnosis of very early embryos. The idea is to fertilize several eggs in vitro and allow the first few cleavages to occur. One or two cells from each of the embryos then are biopsied and are subjected to genetic analysis. Only those found not to be affected with the disorder are implanted back into the uterus and brought to term (Figure 3-33). For some couples, this approach avoids some of the emotional traumas associated with pregnancy termination.

The application to cystic fibrosis is illustrated in Figure 3-34. It is necessary to know in advance the particular mutation for which the embryo is at risk and to have an assay that will work on individual cells. So far, relatively few disorders have been diagnosed by this method, but it remains a promising approach for future development. There are major technical challenges still to be addressed, and ethical issues have been raised regarding the expense involved in this complex procedure.

Jane gives birth to a healthy girl who is a cystic fibrosis carrier but who is not affected.

Case Study

Part I **August 1987**

Claire and Eric come for genetic counseling regarding myotonic dystrophy. Claire is affected with the disorder, as are her brother and her father. She has myotonia, facial weakness, and weakness of her hands, but otherwise has not experienced major problems due to myotonic dystrophy. Her brother has more severe weakness and cataracts, but he is able to work and otherwise is in good health. Their father is very mildly affected, having mild myotonia, frontal hair loss, and cataracts. Claire and Eric know that

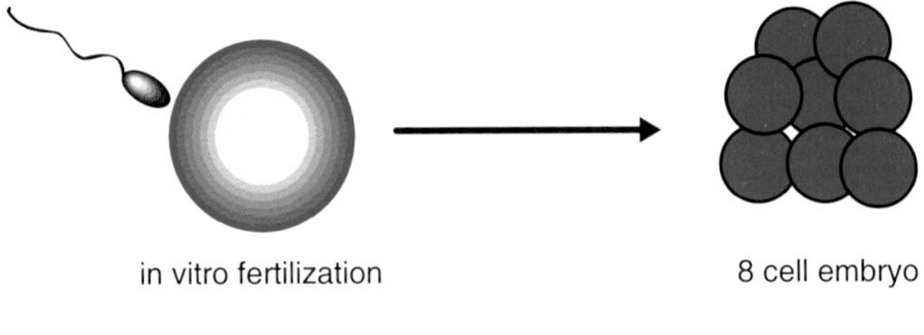

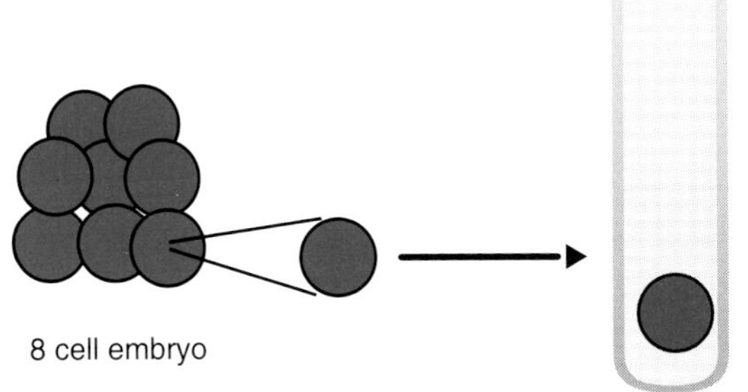

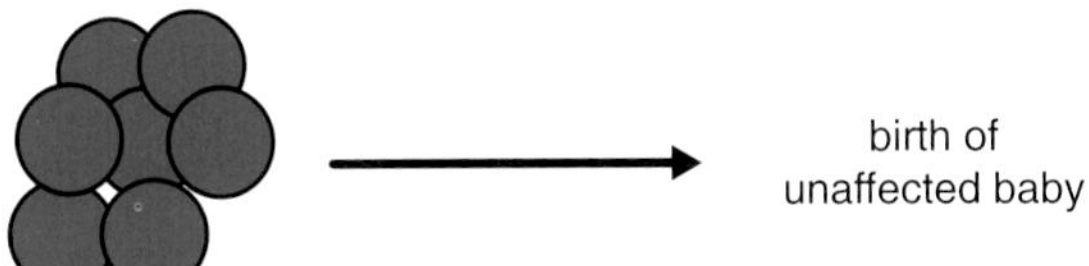

FIGURE 3-33 Scheme of preimplantation diagnosis. Egg cells are obtained from mother by insertion of a collecting catheter into the abdominal cavity, following hormone treatment to stimulate ovulation of multiple eggs. Fertilization of several egg cells is carried out in vitro. If both parents are carriers for a genetic disorder, some of the resulting embryos will be homozygous for the disorder, some will be heterozygous, and some will be homozygous unaffected. The embryos are brought to the eight-cell stage in vitro, and one or two cells from each embryo are biopsied using micromanipulation. DNA is extracted from the cells, and a sensitive PCR assay, capable of detecting a single cell's DNA, is used to determine the genotypes of the biopsied cells. Only embryos found to be unaffected (homozygous normal or heterozygous for an autosomal recessive trait) are implanted into mother's uterus and brought to term.

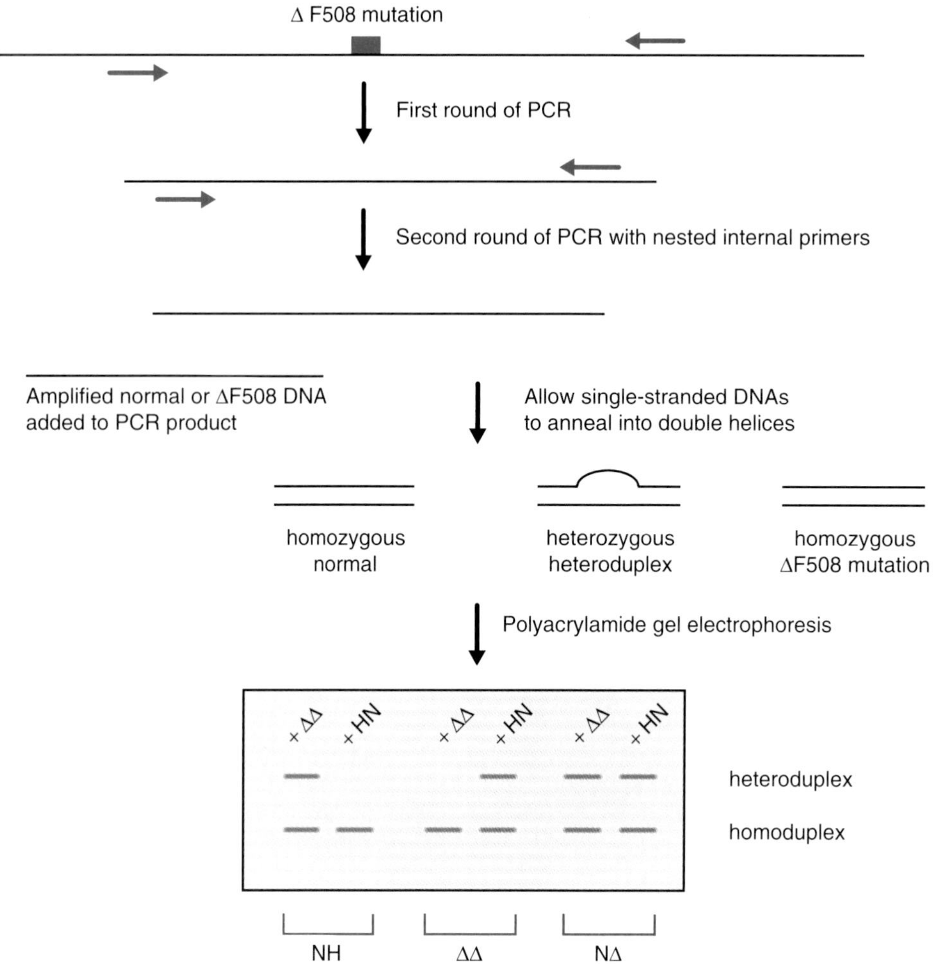

FIGURE 3-34 Preimplantation diagnosis of ΔF508 mutation. DNA from a single cell is amplified first, using a set of primers that flank the site of mutation in CFTR. The product then is subjected to a second round of amplification using primers that are nested internal to the original primers. This affords a high level of amplification of the single pair of CFTR alleles in the original DNA sample. Next the PCR product is mixed with PCR-amplified DNA from a known normal homozygote, or from a known ΔF508 homozygote, in separate reactions. The DNAs are separated into single strands and then allowed to re-anneal. Heteroduplexes between a wild-type DNA and a ΔF508 DNA will contain a base mismatch and form a small bubble, which leads to slower migration on a polyacrylamide gel. In the gel, the first sample is from a normal homozygote. Heteroduplexes occur only if ΔF508 DNA is added. The second sample is from a ΔF508 homozygote, and heteroduplexes form only with normal added DNA. The third sample is from a heterozygote, and heteroduplexes form with either added DNA. (Data from Handyside AH, et al. Birth of a normal girl after in vitro fertilization and preimplantation diagnostic testing for cystic fibrosis. N Engl J Med 1992;327:905–909.)

myotonic dystrophy is dominantly inherited and that they have a 50% chance of having an affected child. Their major concern is that a child might be born with more severe manifestations of the disorder than has been seen in the family so far.

Myotonic dystrophy is a disorder that affects many body systems but has its major impact on muscles. Myotonia is a phenomenon wherein muscles can contract but do not relax easily. A person with myotonic dystrophy experiences difficulty releasing a tight grip with the hands, for example. There is also muscular weakness, particularly involving facial and neck muscles and muscles in the limbs. In addition, there may be cataracts, diabetes mellitus, cardiac arrhythmia, gonadal atrophy leading to infertility or premature menopause, and frontal balding in men.

Myotonic dystrophy affects nearly 1 in 20,000 individuals. It is transmitted as an autosomal dominant disorder with variable expression but high penetrance. Expression can vary in a family, often displaying a tendency to become more severe from generation to generation. This phenomenon is referred to as **anticipation.** Anticipation used to be considered an artifact due to bias of ascertainment: A family in which a genetic disorder is manifest mildly in most individuals is more likely to come to medical attention if a more severely affected child is born. As will be seen later, in myotonic dystrophy, anticipation has a molecular genetic basis and is therefore a real biologic phenomenon.

Part II **September 1987**

The counselor explains that there is a risk of severe infantile myotonic dystrophy, which applies only if the myotonic dystrophy mutation is transmitted from the mother to a child. Approximately 20% of the children who inherit the mutation from their mothers develop this form of myotonic dystrophy, so Claire and Eric are quoted a risk overall of approximately 10%. Although they are not sure how they would use the information, Claire and Eric inquire about the possibility of prenatal diagnosis. They learn that linkage-based testing is possible, and so they arrange for blood to be drawn from both of them and from other members of Claire's family. A single marker that is 5 cM from the myotonic dystrophy locus is found to be informative. Alleles are indicated as 1 and 2 in the pedigree shown in Figure 3-35.

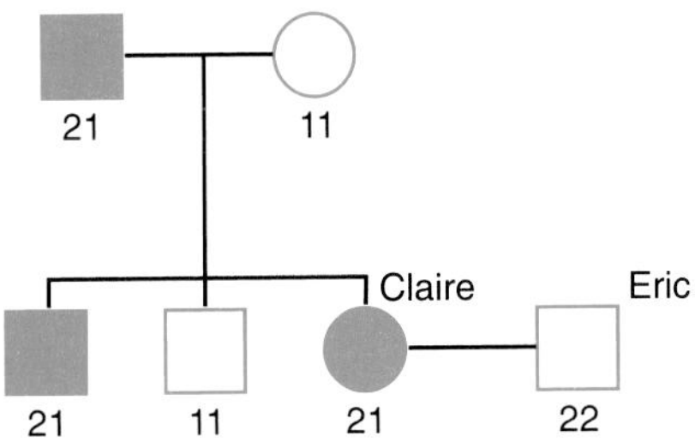

FIGURE 3-35 Pedigree for Claire and Eric.

Usually, the signs of myotonic dystrophy do not appear until after adolescence and progression is slow. Congenital myotonic dystrophy is a particularly severe form of the disorder that begins at birth. Infants with this condition are profoundly weak and floppy (hypotonic). Some have joint contractures, indicative of decreased movement during fetal life. They may have difficulty breathing independently and may require a ventilator. They also have difficulty sucking and swallowing, and poor gastrointestinal motility may result in constipation. Congenital myotonic dystrophy can be life-threatening, sometimes in spite of intensive neonatal support. Among survivors, though, some degree of clinical improvement may occur over a period of weeks. Eventually, muscle strength may improve so that walking can occur, although low muscle tone and weakness are likely to persist to some degree. It is common for these children to have delayed motor development, and many also have cognitive problems. It is not clear to what extent these are direct effects of the myotonic dystrophy versus secondary effects due to neonatal problems.

Congenital myotonic dystrophy occurs only in children who inherit the gene mutation from their mothers. Approximately 20% of the offspring of an affected mother who inherit the mutation will manifest this severe form. The likelihood of congenital myotonic dystrophy is greater if the mother's manifestations are more severe.

Linkage was established between myotonic dystrophy and DNA markers on chromosome 19 in the early 1980s. The informativeness study was done to determine whether Claire is heterozygous for a linked marker that is 5 cM from the myotonic dystrophy gene. Both Claire and her affected brother inherit the *2* allele from their father, and their unaffected brother inherits the *1* allele. The *2* allele must be in coupling with the disease in Claire, as she inherits a *1* allele from her mother.

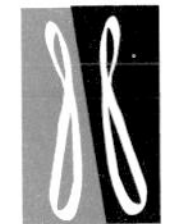

Part III **June 1988**

Claire is now pregnant and, at 16 weeks' gestation, an amniocentesis is performed. The fetal sample is grown in culture for 2 weeks and then is sent to the laboratory for DNA testing. The fetus is found to have the 2 2 *genotype for the marker.*

The fetus inherits the *2* allele from Claire, which is the allele in coupling with myotonic dystrophy.

The fetus is affected unless a recombination has occurred between the disease gene and the marker, which has a probability of 5%.

Part IV **July 1988**

It is determined, with about a 95% likelihood, that the fetus is affected,. There is approximately a 20% chance, then, that the baby will be severely affected at birth and an 80% chance of having milder and later-onset problems. Faced with a difficult decision, Claire and Eric choose to continue the pregnancy but switch their prenatal care to a university-affiliated high-risk obstetric service.

Decision making after a prenatal diagnosis has been established is a complex and personal process. Some couples choose termination of pregnancy, whereas others opt to continue despite the likelihood that the child will be affected with a genetic disorder. Often, the partners have not decided their course of action until after the diagnosis has been made, and some change their minds between initiation of the prenatal test and learning of the diagnosis. Factors that influence the ultimate decision include religious and philosophic beliefs, knowledge of the disorder, number of previous children, and degree of difficulty in conceiving a child.

In myotonic dystrophy, there is value to making a prenatal diagnosis even if the couple has decided not to terminate an affected pregnancy. Infants with congenital myotonic dystrophy require intensive medical support, which may be difficult in a community hospital. If the diagnosis is made prenatally, both the family and the medical team can be prepared for the birth of the child. A neonatologist can attend the delivery, which is best done at a center equipped to provide care in high-risk situations.

Part V **September 1988**

Claire has just delivered a baby boy. The couple was prepared for him to be severely affected because an ultrasound examination done at 33 weeks' gestation had demonstrated polyhydramnios, indicative of probable weakness of swallowing. The neonatal intensive care team was present at the delivery. The baby, named Joseph, made poor respiratory effort at birth and required immediate intubation. He is now in the intensive care unit, on a ventilator. He displays a paucity of movement and has difficulty swallowing.

Joseph displays the signs of congenital myotonic dystrophy. Late in the pregnancy, the fetus was noted to have an increased volume of amniotic fluid (polyhydramnios). Amniotic fluid consists mainly of fetal urine. Some is swallowed by the fetus, and fluid is reabsorbed and excreted through the placenta. If fetal swallowing is impaired, as is the case in congenital myotonic dystrophy, the volume of fluid tends to increase. Detection of polyhydramnios by ultrasonography therefore provides an indication of severe involvement of the fetus, although other disorders besides myotonic dystrophy can cause this finding.

Part VI **April 1992**

Joseph is now 3½ years old. He remained in intensive care for 3 weeks, gradually being weaned off the ventilator. Initially, he required nasogastric feedings, but eventually his suck and swallowing improved and he could be fed by bottle. His muscle strength also improved during the first few months of life. Still, his motor development has been delayed. He was not stable in a sitting position until past 1 year of age, and he is just now beginning to walk independently. Joseph is alert and sociable, although his language development has been delayed. He is just now beginning to put two or more words together. Claire and Eric have heard that the myotonic dystrophy gene has been identified and that new forms of testing are possible. They are interested in having additional children but would again like to have prenatal diagnostic testing done.

The gene for myotonic dystrophy was cloned essentially simultaneously in several laboratories and was reported in early 1992. By this time, very closely linked flanking markers were available, and a region of more than 300 Kb of DNA, including the flanking markers, had been cloned in YACs or cosmids. In different laboratories, the myotonic dystrophy gene was identified first either by scanning for a region of interspecies conservation or by hybridization of fragments with a cDNA library from brain. In this way, a DNA sequence was found that detects an EcoRI polymorphism in the general population, resulting in bands of approximately 9 and 10 Kb. Individuals with myotonic dystrophy, in contrast, were found to have larger bands, of 10 to 15 Kb.

Sequence analysis of the cloned region revealed the cause of the larger bands in persons with myotonic dystrophy. Near the 3′ end of the gene, in a region that is transcribed but not translated into protein, is a stretch of the triplet of bases CTG that is repeated 5 to 27 times (Figure 3-36). The exact number of CTG repeats varies in different individuals and represents a stable genetic trait. Thus, a person who has one allele with 7 repeats and another with 15 will transmit to a child either of these two alleles. Individuals with myotonic dystrophy, on the other hand, have 50 or more copies of the CTG repeat region, and this represents the mutation that upsets the function of their gene. The enlarged hybridization bands seen with Southern blotting are due to expansion of the CTG repeat, causing an increase in the size of the DNA fragment between EcoRI sites that flank the repeat region.

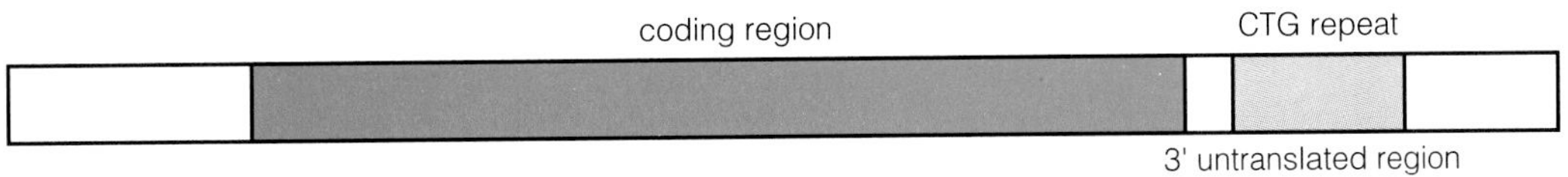

FIGURE 3-36 The CTG repeat is located in the 3′ untranslated region of the MD kinase gene.

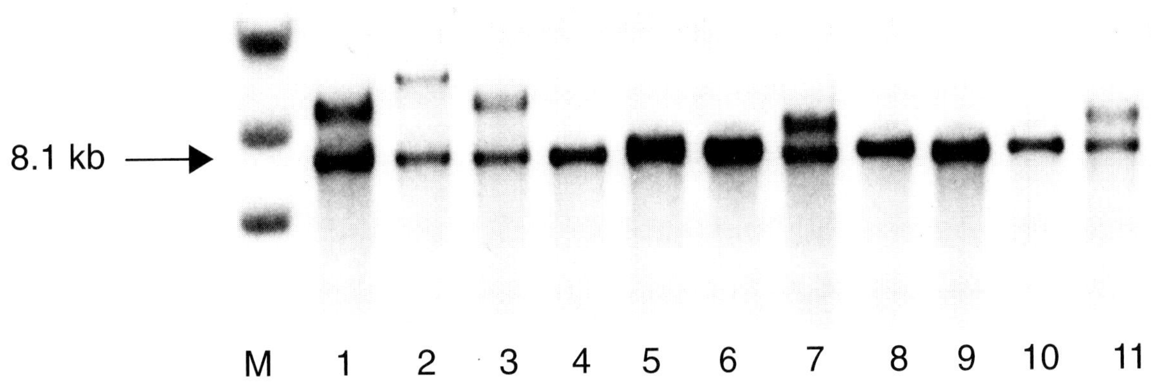

FIGURE 3-37 Southern blot of DNA samples digested with NcoI and hybridized with probe MDY-1 at the myotonic dystrophy locus. Eleven samples and size marker (M) are shown. The normal allele results in an 8.1 kb fragment. Individuals in lanes 1, 2, 3, 7, and 11 have, in addition to a normal allele, a higher molecular weight allele due to CTG repeat expansion. (Provided by Dr. Sue Richards, Baylor DNA Diagnostic Laboratory.)

Expansions of the CTG repeat not only disrupt the function of the gene but also render the region unstable and prone to further expansion. A parent with 60 repeats, then, may transmit the gene to a child with more than 80 repeats, for example. This accounts for genetic anticipation, as the onset of symptoms is earlier and progression is more rapid if the repeat region is larger. Interestingly, contraction of the CTG region has also been noted on rare occasions. Here, the CTG region may get smaller, so that a parent with myotonic dystrophy can transmit the affected copy of the gene yet the child will not be affected. Although rare, this indicates that genetic linkage analysis can give erroneous diagnostic results even in the absence of genetic recombination.

Alleles with 19 to 30 copies of the CTG repeat have been found to be in linkage disequilibrium with another polymorphism in the gene. All alleles with greater than 50 repeats arise from alleles with 19 to 30 repeats. The linkage disequilibrium suggests that the 19- to 30-repeat alleles arose by mutation from an allele with a smaller number of repeats (presumably 5 repeats, the most common allele) in a single individual long ago.

The gene for myotonic dystrophy encodes a protein that has been named *DM kinase*. It is expressed in brain, muscle, heart, and testes, all of which can be affected in myotonic dystrophy. Based on sequence analysis and comparison with other known proteins, the DM kinase protein has protein kinase domains, which can phosphorylate serine and threonine residues. Beyond this, the function of the protein is not known. The CTG repeat occurs near the 3′ end, in a region that is transcribed but not translated into protein. It is not clear how the CTG repeat expansion leads to the myotonic dystrophy phenotype. There appears to be reduction in the level of expression of

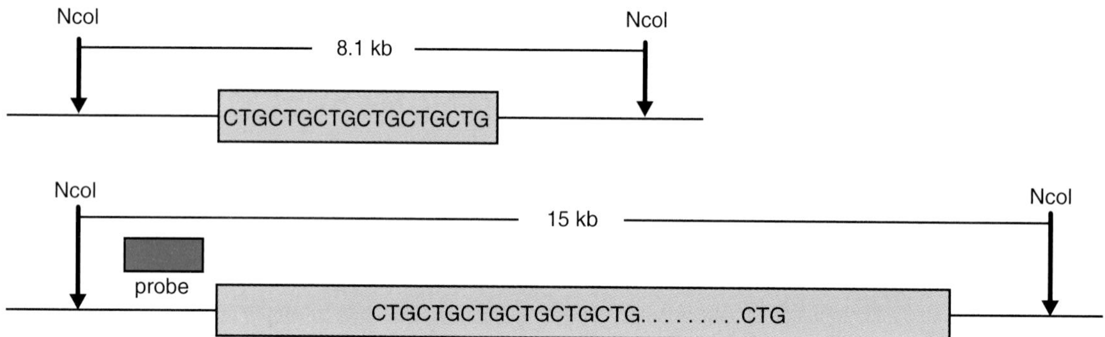

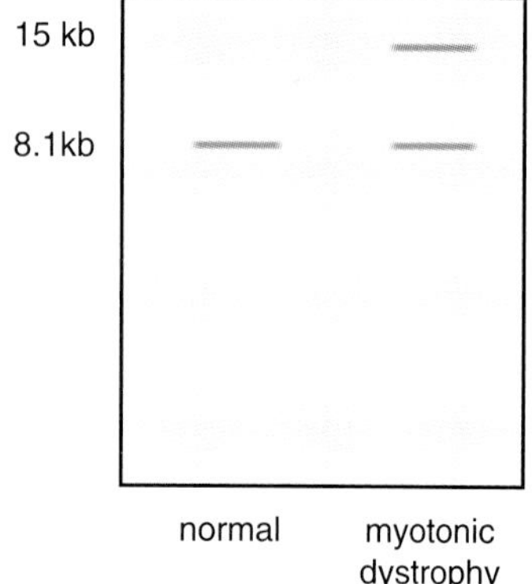

FIGURE 3-38 Southern analysis of CTG repeat. NcoI cuts DNA at sites flanking the repeat region. Normally this results in an 8.1 kb allele. An individual with myotonic dystrophy has, in addition to a normal allele, a second allele with an expansion of the CTG region, resulting in a larger band (15 kb in this case) on the Southern blot.

the DM kinase protein, and there is evidence that the presence of the repeat expansion somehow affects the levels of expression of other genes as well. This may explain the pleiotropic phenotype of myotonic dystrophy.

The expansion of a triplet of bases in a gene that results in mutation represents a motif that has been found in a number of other genes. Triplet-repeat expansion also accounts for Huntington's disease, fragile X syndrome, spinal and bulbar muscular atrophy, spinocerebellar ataxia, dentatopallidoluysian atrophy, Friedreich ataxia, and Machado-Joseph disease. It is likely that the list of triplet-expansion disorders will increase. Other examples of these disorders will be considered later in the book.

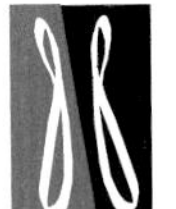

Part VII **December 1992**

Claire's myotonic dystrophy mutation is determined to be an expansion of the CTG repeat to approximately 700 copies (Figure 3-37). The same is true for her brother. Joseph is found to have more than 2000 copies of the repeat. Claire is again pregnant, and amniocentesis is performed. The fetus is found to carry a very large number of repeats, indicative of severe myotonic dystrophy. Claire and Eric choose to terminate this pregnancy.

Discovery of the triplet-repeat expansion in myotonic dystrophy has led to the development of accurate means of molecular diagnosis. The expanded region can be detected by protocols involving Southern hybridization or by PCR. In the former (Figure 3-38), a DNA probe adjacent to the repeat region is hybridized with DNA cut with *NcoI*. Normal alleles are smaller than 8.1 Kb, whereas expansions in the range that cause myotonic dystrophy tend to be larger. PCR analysis uses primers that flank the CTG repeat region (Figure 3-39). The amplified DNA is analyzed on a polyacrylamide gel to determine the size, which indicates the number of repeats. PCR provides more accurate size determination but may fail to amplify very large repeats. The Southern blot (Figure 3-37) shows a normal control, a moderate repeat

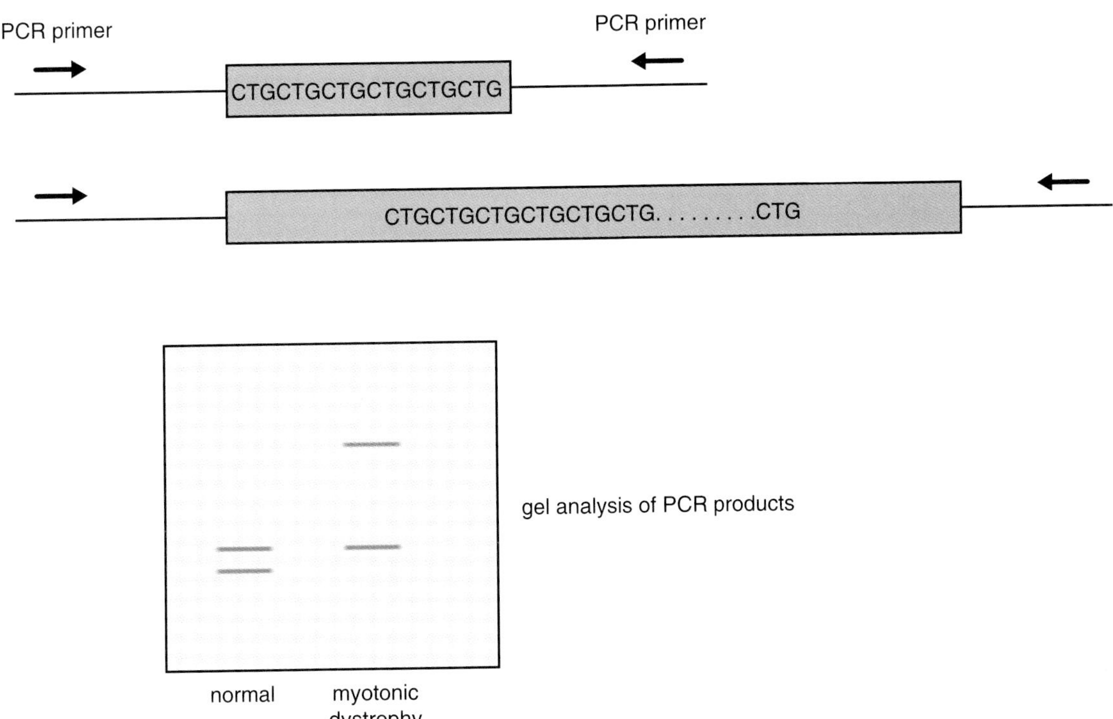

FIGURE 3-39 PCR analysis of CTG repeat. PCR primers flank the repeat region. PCR products are separated by size by electrophoresis. Two bands seen in an unaffected individual reflect polymorphism for the CTG repeat number. One of the bands will be substantially larger in size in an individual with myotonic dystrophy.

expansion of approximately 700 repeats, and a very large allele of more than 2000 repeats. In addition, there are other normal samples on this blot with alleles near the normal 8.1-Kb size.

REVIEW QUESTIONS

1. A polymorphic marker with two alleles, *A* and *B*, is closely linked to the gene for adult polycystic kidney disease. A family requests presymptomatic testing for their child, Tom, who, at age 5, does not manifest signs of the disorder. Based on the information shown below, is it likely that Tom is affected?

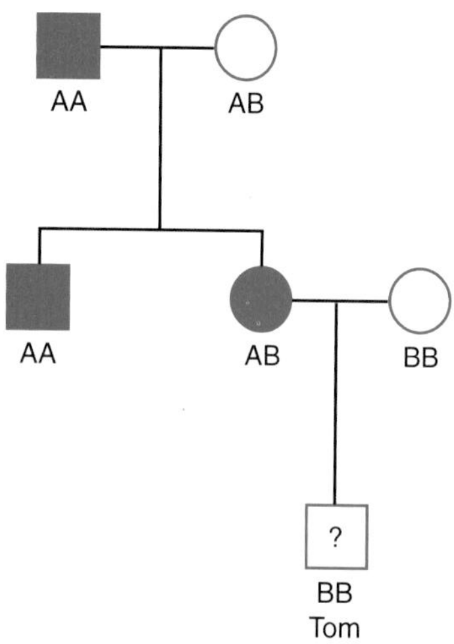

2. A linkage study of a candidate gene for a disorder reveals a lod score of $-\infty$ at a value of $\theta = 0$. What does this indicate about the potential candidate gene?

3. A polymorphic marker with alleles *1* and *2* is closely linked to the gene for spinal muscular atrophy, an autosomal recessive disorder.

   a. What is the most likely diagnosis in the fetus?

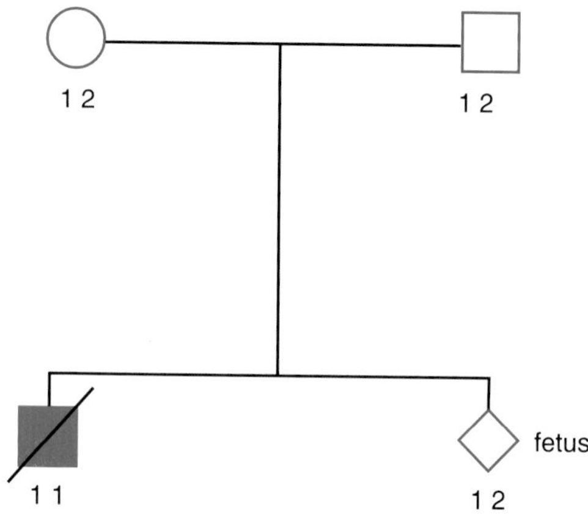

   b. How would the possibility of genetic heterogeneity influence the accuracy of diagnosis?

4. A family requests a linkage study for purposes of presymptomatic diagnosis of neurofibromatosis type 1 (NF1), which is an autosomal dominant disorder. A restriction fragment–length polymorphism within the NF1 gene is used, which detects either a 5-Kb or a 3-Kb allele in EcoRI-cut DNA, as shown in the map below. The depicted Southern blot is obtained.

   a. On careful inspection, you decide that this blot actually has revealed the likely NF1 mutation in the family. What is this mutation, and how is it demonstrated on this blot?

   b. Is it likely that II-2 is affected with NF1, based on the data shown?

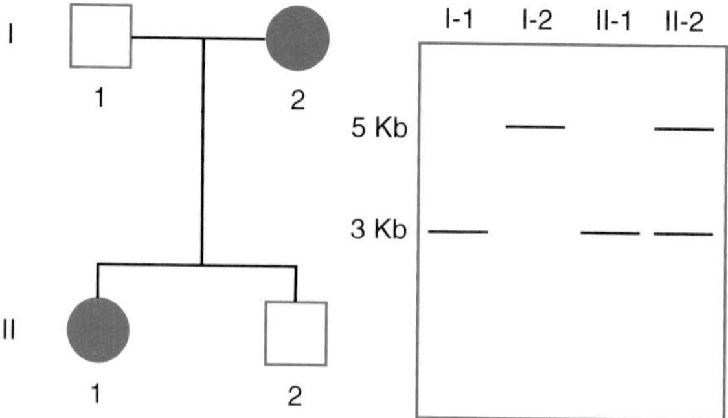

FURTHER READING

3.7 Detection of Mutations in CFTR

Miedzybrodzka ZH, Hall MH, Mollison J, et al. Antenatal screening for carriers of cystic fibrosis: randomised trial of stepwise v couple screening. Br Med J 1995;310:353–357.

Super M, Schwarz MJ, Malone G, et al. Active cascade testing for carriers of cystic fibrosis gene. BMJ 1994;308:1462–1467.

Williamson R. Universal community carrier screening for cystic fibrosis? Nat Genet 1993;3:195–201.

3.8 Treatment of Cystic Fibrosis

Drumm ML, et al. Correction of the cystic fibrosis defect in vitro by retrovirus-mediated gene transfer. Cell 1990;62:1227–1233.

Knowles MR, Hohneker KW, Zhou Z, et al. A controlled study of adenoviral-vector-mediated gene transfer in the nasal epithelium of patients with cystic fibrosis. N Engl J Med 1995;333:823–831.

Rich DP, et al. Expression of cystic fibrosis transmembrane conductance regulator corrects defective chloride channel regulation in cystic fibrosis airway epithelial cells. Nature 1990;347:358–363.

Welsh MJ, Smith AE. Molecular mechanisms of CFTR chloride channel dysfunction in cystic fibrosis. Cell 1993;73:1251–1254.

CASE STUDY

Aslanidis C, Jansen G, Amemiya C, et al. Cloning of the essential myotonic dystrophy region and mapping of the putative defect. Nature 1992;355:548–551.

Brook JD, McCurrach ME, Harley HG, et al. Molecular basis of myotonic dystrophy: expansion of a trinucleotide (GCT) repeat at the 3′ end of a transcript encoding a protein kinase family member. Cell 1992;68:799–808.

Buxton J, Shelbourne P, Davies J, et al. Detection of an unstable fragment of DNA specific to individuals with myotonic dystrophy. Nature 1992;355:547–548.

Caskey CT, Pizzuti A, Fu Y-H, et al. Triplet repeat mutations in human disease. Science 1992;256:784–789.

Harley HG, Brook JD, Rundle SA, et al. Expansion of an unstable DNA region and phenotypic variation in myotonic dystrophy. Nature 1992;355:545–546.

Harley HG, Rundle SA, MacMillan JC, et al. Size of the unstable GCT repeat sequence in relation to phenotype and parental transmission in myotonic dystrophy. Am J Hum Genet 1993;52:1164–1174.

Harper PS, Harley HG, Reardon W, Shaw DJ. Anticipation in myotonic dystrophy: new light on an old problem. Am J Hum Genet 1992;51:10–16.

Hofmann-Radvanyi H, Lavedan C, Rabès J-P, et al. Myotonic dystrophy: absence of GCT enlarged transcript in congenital forms, and low expression of the normal allele. Hum Mol Genet 1993;2:1263–1266.

Mahadevan M, Tsilfidis C, Sabourin L, et al. Myotonic dystrophy mutation: an unstable GCT repeat in the 3′ untranslated region of the gene. Science 1992;255:1253–1255.

Mahadevan MS, Amemiya C, Jansen G, et al. Structure and genomic sequence of the myotonic dystrophy (DM kinase) gene. Hum Mol Genet 1993;2:299–304.

Pizzuti A, Friedman DL, Caskey CT. The myotonic dystrophy gene. Arch Neurol 1993;50:1173–1179.

Reardon W, Floyd JL, Myring J, et al. Five years experience of predictive testing for myotonic dystrophy using linked DNA markers. Am J Med Genet 1992;43:1006–1011.

Reardon W, Newcombe R, Fenton I, et al. The natural history of congenital myotonic dystrophy: mortality and long term clinical aspects. Arch Dis Child 1993;68(2):177–181.

Redman JB, Fenwick RG Jr, Fu Y-H, et al. Relationship between parental trinucleotide CGT repeat length and severity of myotonic dystrophy in offspring. JAMA 1993;269:1960–1965.

Roig M, Balliu PR, Navarro C, et al. Barcelona presentation: clinical course and outcome of the congenital form of myotonic dystrophy. Pediatr Neurol 1994;11(3):208–213.

Sabouri LA, Mahadevan MS, Narang M, et al. Effect of the myotonic dystrophy (DM) mutation on mRNA levels of the DM gene. Nat Genet 1993;4:233–238.

Shelbourne P, Davies J, Buxton J, et al. Direct diagnosis of myotonic dystrophy with a disease-specific DNA marker. N Engl J Med 1993;328:471–475.

Wang J, Pegoraro E, Menegazzo E, et al. Myotonic dystrophy: evidence for a possible dominant-negative RNA mutation. Hum Mol Genet 1995;4:599–606.

4 X-Linked Genetic Transmission

OVERVIEW

4.1 Duchenne's Muscular Dystrophy What is Duchenne's muscular dystrophy?

4.2 X-Linked Recessive Genetic Transmission What are the unique properties of X-linked inheritance?

4.3 X Chromosome Inactivation How is dosage of X-linked genes controlled in females?

4.4 Mapping the Gene for DMD How was the Duchenne's muscular dystrophy gene localized on the X chromosome?

4.5 Cloning the Gene for DMD How was the Duchenne's muscular dystrophy gene identified?

4.6 Dystrophin: The Protein Product of the DMD Gene What is the structure of the dystrophin gene?

4.7 The Molecular Basis of Duchenne's and Becker's Muscular Dystrophy How do changes in the structure of dystrophin result in the phenotypes of Duchenne's or Becker's dystrophy?

4.8 Molecular Diagnosis of Duchenne's and Becker's Dystrophy How does knowledge of the dystrophin gene allow molecular diagnosis of muscular dystrophy?

4.9 Detection of Dystrophin Deletion Carriers How can heterozygous carriers for dystrophin deletions be identified?

4.10 Pathogenesis and Treatment of Muscular Dystrophy How has knowledge of the molecular basis of muscular dystrophy affected our knowledge of pathogenesis and changed clinical management?

Perspective Living with a child with Duchenne's muscular dystrophy

Case Study A problem that is more than skin deep

Thus far, we have considered two of the three major modes of single gene transmission: autosomal recessive and autosomal dominant. This chapter will describe the third major mode of transmission: X-linked. The special properties of X-linkage result from the fact that males are hemizygous for genes on the X chromosome: That is, they have only one copy of these genes, unlike the case for any autosomal gene. X-linked recessive traits are expressed in hemizygous males, and dominant traits often are more severe in the hemizygous state than in heterozygous females. X-linked traits are never passed from father to son. Another important feature of X-linkage results from the distinctive pattern of dosage compensation in females. One or the other of the two X chromosomes is largely inactivated in the cells of every female, beginning early in embryonic development. This equalizes the levels of expression of X-linked genes in males and females. Females who are heterozygous for X-linked mutations are literally mosaics for cells expressing the wild-type or the mutant allele.

We will begin this chapter with a case history of Duchenne's muscular dystrophy. This is a prototypic X-linked recessive trait that is lethal in males. The clinical course of Duchenne's dystrophy and an allelic disorder, Becker's dystrophy, will be described. The properties of X-linked inheritance will then be considered. Next, we will turn to the phenomenon of X chromosome inactivation and review recent data on the mechanisms that underlie this process. The Duchenne's muscular dystrophy gene was identified by positional cloning and, in fact, was one of the first genes to be found in this way. The gene was localized initially on the basis of finding chromosome abnormalities rarely in females affected with the disorder. We will see how this was done and then how the region surrounding the gene was mapped with restriction fragment–length polymorphisms (RFLPs). The gene itself was cloned by making use of another unusual chromosome abnormality—in this case, a large deletion. The means by which the gene was cloned and characterized will be reviewed.

The Duchenne's muscular dystrophy gene encodes an unusually large protein that has been named *dystrophin*. The properties of this protein will be explored, followed by a discussion of the means by which mutations of dystrophin account for the Duchenne's or Becker's dystrophy phenotypes. Cloning the dystrophin gene has provided important diagnostic tools, which will be described. We will see how these tools are used to identify both the disease in affected males and the carrier status in females. The discussion of muscular dystrophy will conclude with a consideration of future prospects for treatment, and we will consider a family's perspective on living with the disorder. The case will explore another X-linked disorder that displays a dominant, rather than a recessive, pattern of transmission.

4.1 Duchenne's Muscular Dystrophy

1982

James and Brenda, who are both in their mid-twenties, are referred to the genetics clinic. They are considering starting a family but are concerned because Brenda has a half-brother, Charles (same mother, different father), who has Duchenne's muscular dystrophy. This diagnosis was made at age 4, when Charles was evaluated for delayed language development. At that time, it was noted that he had achieved motor milestones normally in the first year of life but had difficulty climbing stairs and was considered to be clumsy in preschool. A serum creatine phosphokinase assay revealed a level of 11,000 mU/ml (normal <30), and a muscle biopsy showed a pattern typical for Duchenne's muscular dystrophy (Figure 4-1). Charles is now 18 years old. He has been wheelchair-bound since age 11 and has profound weakness of all proximal muscles (Figure 4-2). He has experienced several bouts of respiratory infections, the most recent of which required prolonged hospitalization. His cognitive function is impaired as well, his IQ having been measured as being in the eighties a few years ago, when his muscle strength was better.

Duchenne's muscular dystrophy (DMD) is characterized by progressive loss of muscle strength. Because it is inherited as an X-linked recessive trait, it occurs mostly in males (see section 4.2). The symptoms are noticed usually in the early years of life. A child with the disorder will learn to walk but develops clumsiness and may not get to running. The large muscles of the pelvic girdle usually are the first to display weakness. This is most obvious when the child is asked to stand quickly from a lying position. Weakness of the hip muscles causes the child to use his hands to help support the body as he rises (*Gower's sign*). The calf muscles are enlarged, although muscle mass is not increased (*pseudohypertrophy*), and he may walk on tiptoe. Gradually, strength diminishes in muscles throughout the body. Most boys with DMD become wheelchair-bound by late childhood or their early teen years. They develop weakness in the shoulder girdle and, eventually, in the arms and even the hands. Ultimately, they may be essentially paralyzed, except for finger and toe movements. Movements of the eyes, however, are not affected. The heart muscle also is involved, causing a dilated cardiomyopathy and leading to cardiac failure, which is exacerbated by severe scoliosis that limits breathing. There is an accompanying cognitive impairment found in many. Boys with DMD die in their late teens or early twenties. No medical treatment has been determined to be effective.

Increased levels of serum creatine phosphokinase (CPK) can be an indicator of muscle disease. CPK is an enzyme found in skeletal muscle, heart, and brain. When muscle cells are damaged, CPK leaks into the serum. Boys with muscular dystrophy have extremely high CPK levels. This is true even at birth, before muscle weakness is apparent. The definitive diagnostic test for muscular dystrophy is muscle biopsy. A small piece of muscle is removed and examined with the microscope. Dystrophic muscle is characterized by degenerating muscle cells and increased connective tissue and fat.

James and Brenda are concerned that they are at risk of having a child with DMD. Having read literature from the Muscular Dystrophy Association, they are also interested in knowing whether they could have a child with a milder disorder, Becker's dystrophy.

Becker's dystrophy is a related disorder, in which the same muscle groups are affected as in DMD but the age of onset is later and the rate of progression slower. It is also transmitted as an X-linked recessive trait. At the time when Charles's disease was diagnosed, there were no tests to distinguish Duchenne's from Becker's dystrophy, and it was unclear whether these disorders were due to mutations at the same or at different genes.

4.2 X-Linked Recessive Genetic Transmission

Further investigation of the family history reveals that Brenda's mother had a brother who also received a diagnosis of DMD. He had died at age 17 of pneumonia. Brenda's mother has a healthy brother as well. No other family members are known to have muscle disease (Figure 4-3).

Normal

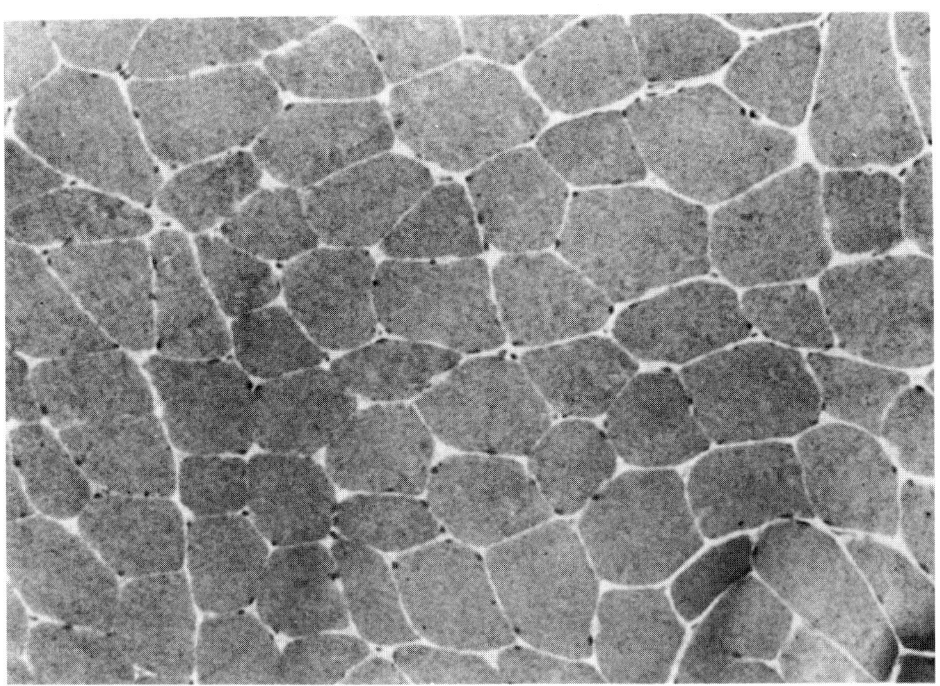

DMD

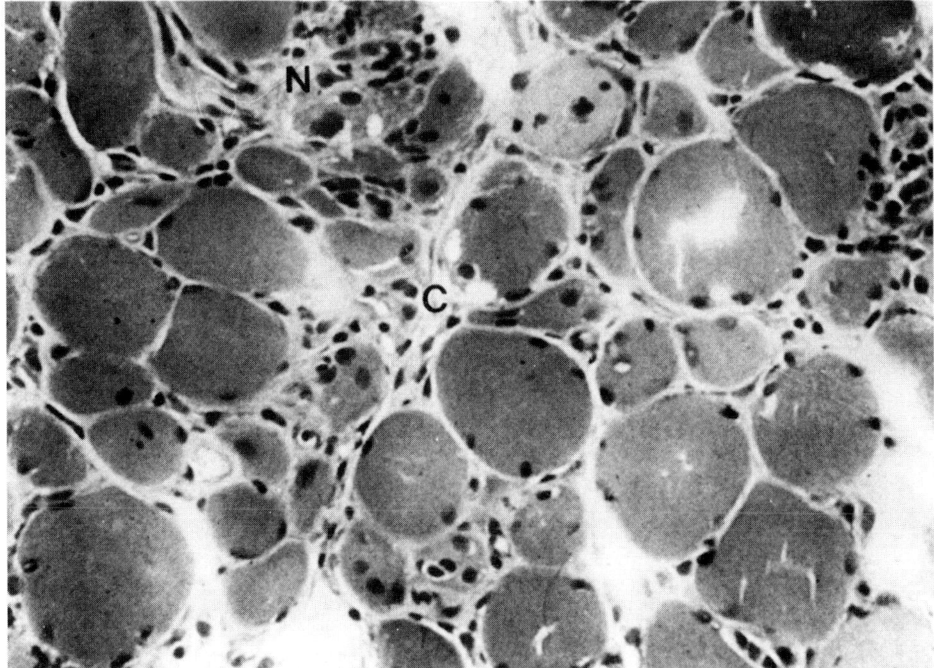

FIGURE 4-1 Normal muscle (top) and muscle from boy with Duchenne's muscular dystrophy (bottom). The dystrophic muscle contains muscle cells of various sizes with connective tissue (*C*) and a necrotic cell (*N*).

FIGURE 4-2 Boy with Duchenne's muscular dystrophy.

Genetic traits we have considered thus far have all been located on the non-sex chromosomes. Dominant traits are expressed in heterozygotes or homozygotes, whereas recessive traits are expressed only in homozygotes. The situation is different for X-linked traits. The X and Y chromosomes are responsible for gender determination: females have two X chromosomes, whereas males have an X and a Y chromosome. Most of the genes on the X are not found on the Y, and hence males have only a single copy of most X-linked genes (referred to as **hemizygosity**). A mutation in one of these genes, whether dominant or recessive, will be expressed in the male, who has only one copy of the gene. X-linked recessive traits are therefore more commonly found in males, because for rare mutations, homozygous females will be much less common than hemizygous males (Figure 4-4). The traits usually are passed from carrier females to, on average, half their sons. Inheritance of the mutant gene by a female offspring results in the girl being a carrier (Figure 4-5). If affected males are able to reproduce, they will pass the trait to all their daughters but to none of their sons. **The lack of male-to-male transmission is an important hallmark of X-linkage** (Figure 4-6).

It is not uncommon for X-linked disorders to occur in families with no known history of the condition. If the mother has few or no brothers or maternal uncles this may be due to chance: The mutant gene may not have been passed to any male relatives in recent generations, and therefore the phenotype may not have been encountered in the family. Usually, it is difficult to obtain reliable medical information for more than one or two generations in the past so, if males were affected in earlier generations, this

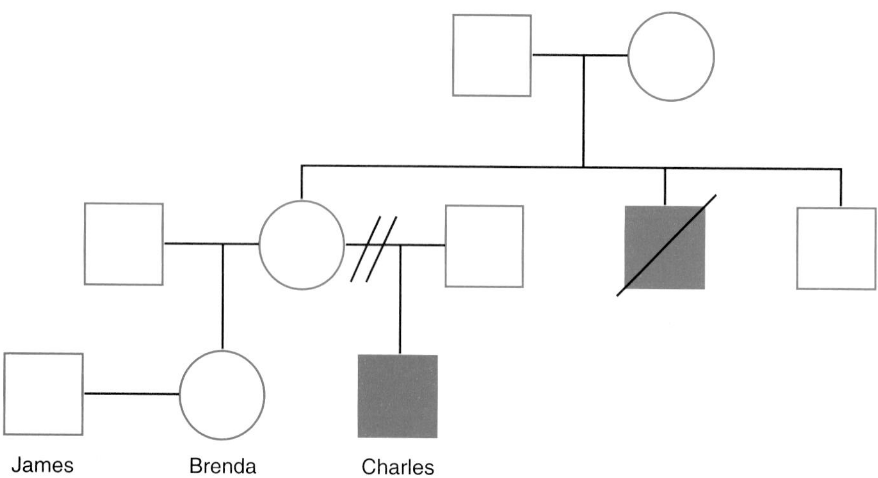

FIGURE 4-3 Pedigree of Brenda and James's family.

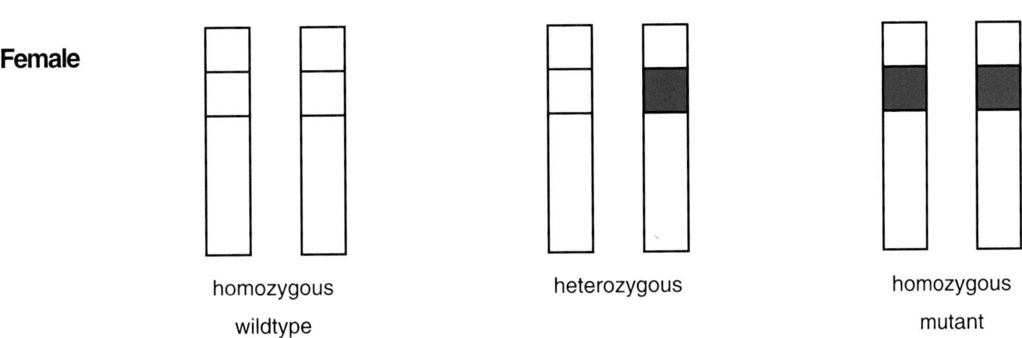

FIGURE 4-4 Males who carry a mutation on the X chromosome will express the phenotype associated with that mutation. Females can be homozygous for either the wild-type or the mutant allele or can be heterozygous. Whether they express the mutant allele depends on whether it behaves as a recessive or a dominant trait.

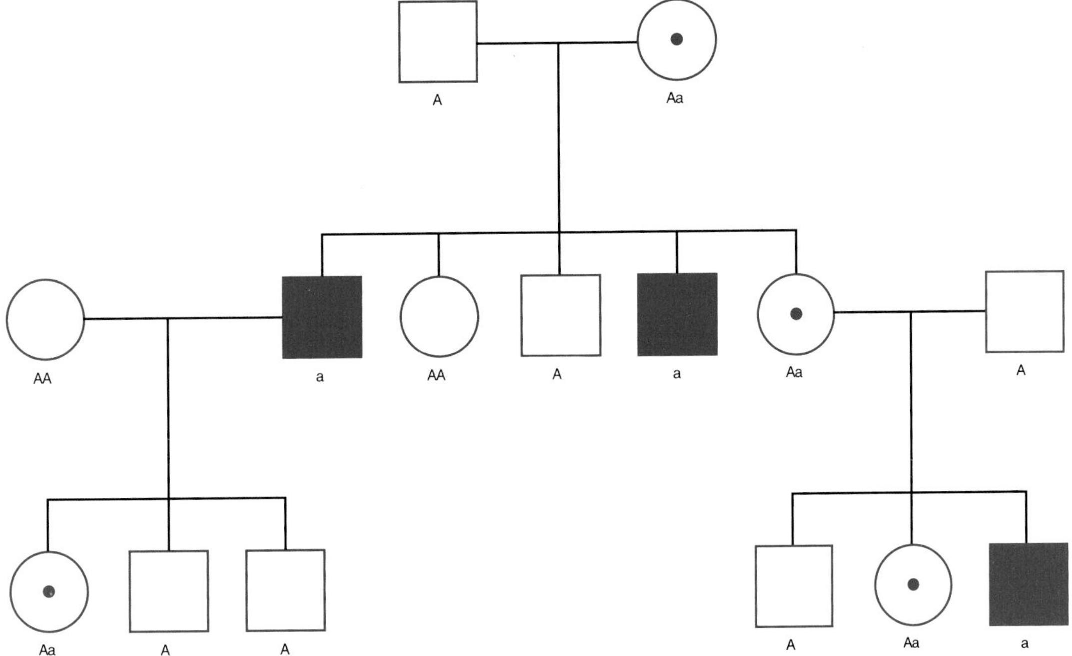

FIGURE 4-5 Pedigree illustrating X-linked recessive transmission. Note the absence of male-to-male transmission.

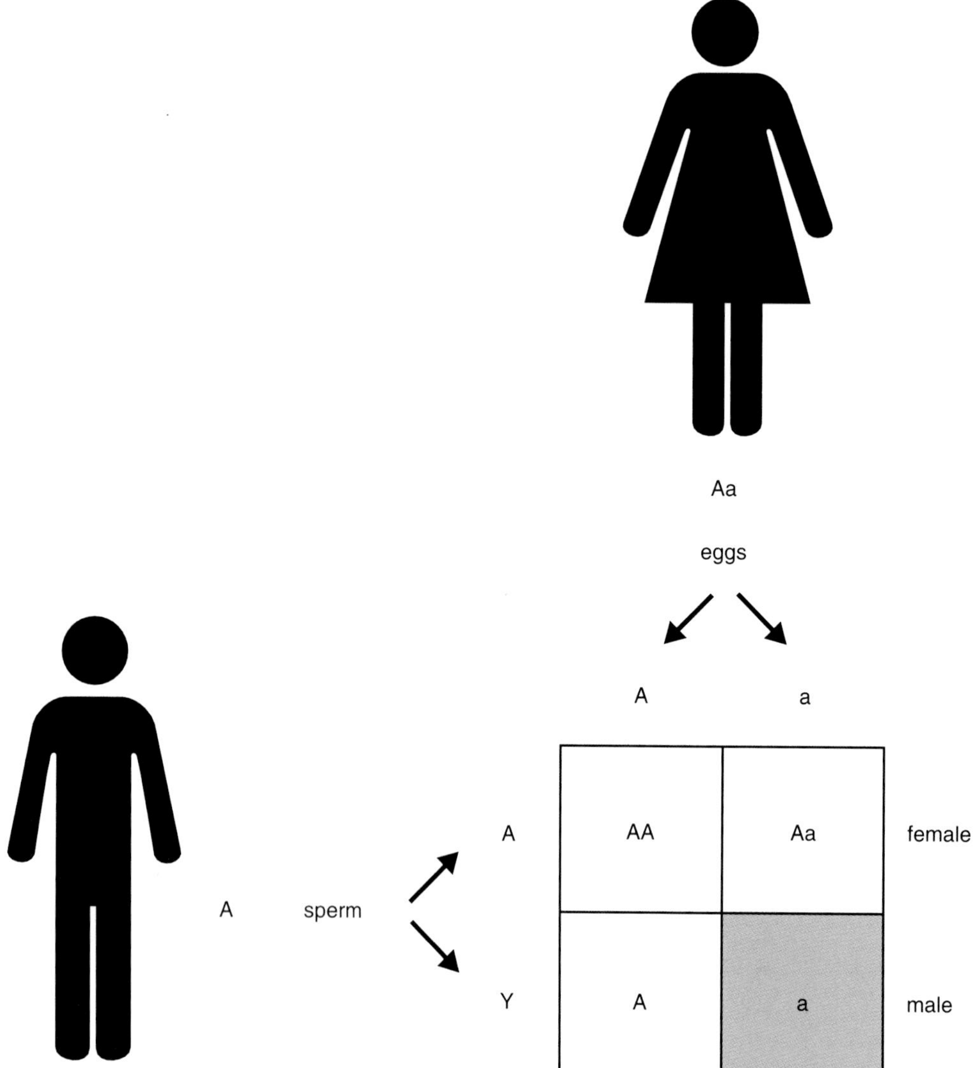

FIGURE 4-6 Heterozygous females produce eggs with either the dominant (*A*) or recessive (*a*) alleles. Half their male offspring will be hemizygous for *a* and have the trait, whereas half their female offspring will be heterozygous carriers.

information may not be known to current members of the family.

Alternatively, just as was the case for autosomal dominant traits, X-linked recessive traits may arise by new mutation. In this case, the mutation may have occurred in the egg cell that produced the affected boy or may have arisen in the sperm or egg that produced his mother. The distinction is important, because in the latter case the mother is at risk of having additional affected sons, whereas in the former case her risk is low. Germ-line mosaicism has also been seen for X-linked traits. We will explore options for carrier testing in the next section.

X-Linked Recessive Transmission

- Mostly affects males
- No male-to-male transmission
- Carrier females pass trait to half their sons and carrier status to half their daughters

4.3 X Chromosome Inactivation

1983

The genetic counselor explains to James and Brenda that Brenda's mother is an obligate carrier of DMD and therefore that Brenda is at 50% risk of being a carrier. The only means of carrier testing is blood CPK analysis. CPK testing is done on Brenda and on her mother and is repeated three times.

Although heterozygous carriers of a recessive disorder generally do not manifest signs of a mutant phenotype, carriers with partial expression are more common for X-linked than for autosomal disorders. The reason is related to the special manner by which X-linked genes are regulated in females.

Gene dosage is tightly controlled in humans and other animals. A person who has a single extra chromosome—for example, an extra copy of chromosome 21, as occurs in Down syndrome—displays physical problems. Having an extra copy of chromosome 21 somehow upsets a carefully tuned balance of gene expression. How, then, is gene dosage managed for X-linked genes, as females have two X chromosomes but males have only one?

The answer began to unfold in the 1960s, when it was first observed that one X chromosome remains condensed in female mouse cells but not in males. Then Mary Lyon noted that female mice heterozygous for genes on the X chromosome that determine coat color have a mottled phenotype (Figure 4-7).

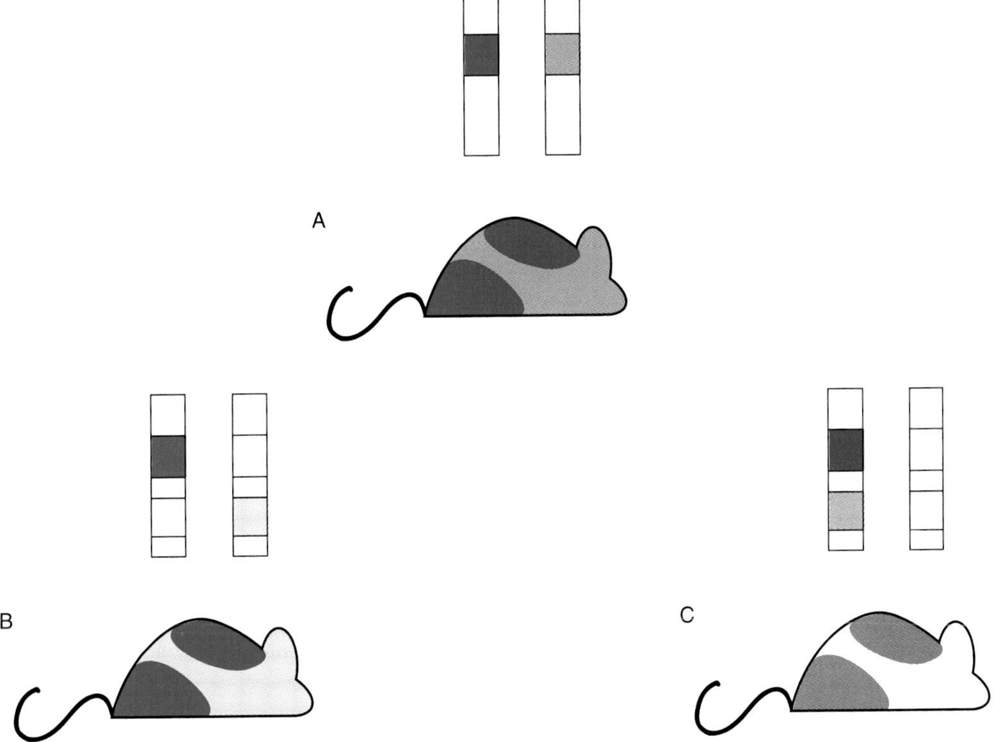

FIGURE 4-7 Experiments with mouse X-linked coat-color mutations, demonstrating X chromosome inactivation. (**A**) Heterozygous female mouse with patches of different coat colors. (**B**) Female mouse doubly heterozygous with different colors in repulsion, showing patches of the two colors. (**C**) Female mouse doubly heterozygous with alleles in coupling, showing patches containing both colors.

They have patches of fur of one or the other color rather than a blending of the two colors throughout their bodies. Lyon then bred mice that were heterozygous for two coat-color genes on the X chromosome. Again, a mottled phenotype resulted. If the two mutant alleles were on the same X, the two mutant phenotypes always were expressed in the same patch of skin. Conversely, if the two mutant alleles were on opposite X chromosomes, patches would display either one mutant phenotype or the other.

From these observations, Lyon formulated a hypothesis of dosage compensation for X-linked genes. She proposed that early in development one of the two X chromosomes is inactivated in each cell and that once this inactivation occurs, it is irreversible in all the descendants of that cell (Figure 4-8). A female is therefore a mosaic of cells expressing one or the other X chromosome. The Lyon hypothesis has been substantially borne out in all mammals, including humans.

In humans, as in mice, one of the two X chromosomes remains condensed through most of the cell cycle. In interphase cells, it forms a densely stained mass adjacent to the nuclear membrane, called the *Barr body* (after the neuroanatomist Murray Barr, who first observed this phenomenon in cat neurons). The Barr body is found only in female cells, never in males. Moreover, females heterozygous for X-linked genes also have mosaic expression. This is demonstrated by a polymorphism of the enzyme glucose-6-phosphate dehydrogenase, encoded on the X chromosome (Figure 4-9). The polymorphism involves a variant electrophoretic pattern: A female can be homozygous for a fast- or a slow-migrating band or can be heterozygous. When fibroblast cells from a heterozygous female are grown in culture and single cells are isolated and grown into colonies, the clones display either the fast- or the slow-migrating band but not both. X chromosome inactivation has been shown to apply to most X-linked genes but not all. In particular, there are a number of genes that are present on both the X and the Y chromosomes. As might be expected, these genes (sometimes referred to as **pseudoautosomal** because they can be transmitted from either the male or the female, like autosomal genes) escape X-inactivation.

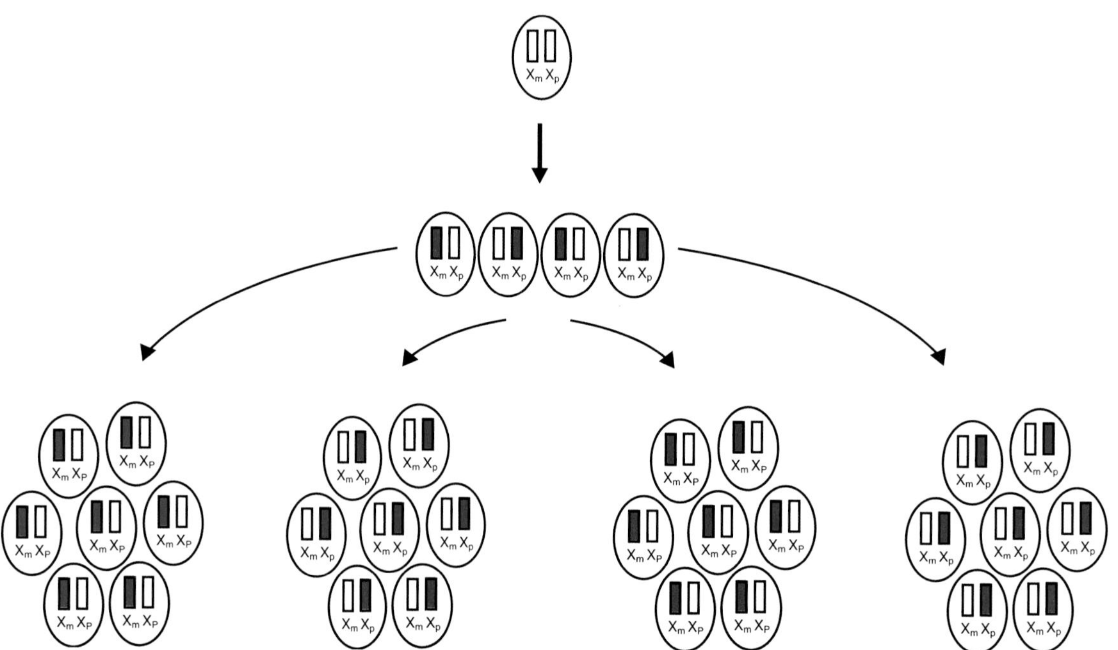

FIGURE 4-8 X chromosome inactivation. In the zygote, both the maternally and paternally derived X chromosomes (X_m and X_p) are active. Early in development, one of the two X chromosomes in each cell is inactivated (indicated as the dark chromosome). This X chromosome remains inactive in all the descendants of that cell.

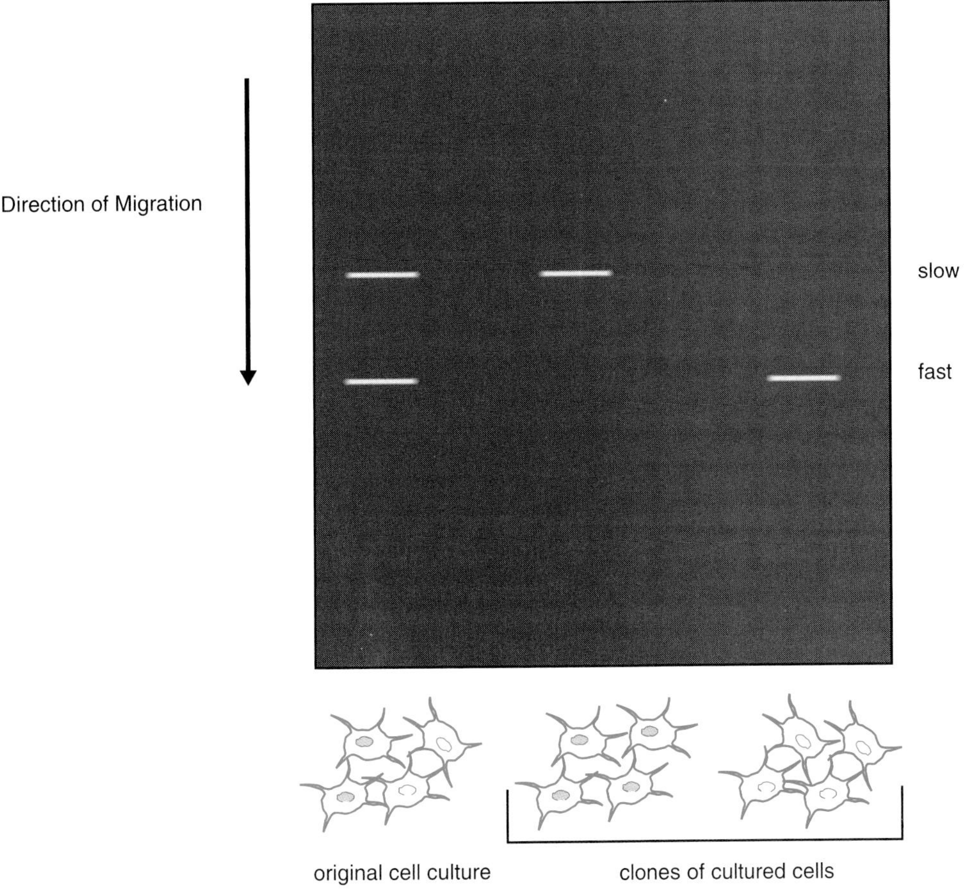

FIGURE 4-9 Electrophoretic separation of fast- and slow-migrating glucose-6-phosphate dehydrogenase (G6PD) alleles. Protein is extracted from cultured cells, applied to a gel, separated by charge by electrophoresis, and visualized by staining. Females heterozygous for a G6PD polymorphism show two bands. When individual cultured cells are isolated and grown in culture (cloned) from a heterozygous female, however, only one or the other G6PD allele is found. This is due to random inactivation of one of the alleles in each cell.

When and how does X-inactivation occur? It occurs at different times in the embryo and in extraembryonic tissues. In the mouse, the paternally derived X is preferentially inactivated in trophoblast tissue, whereas inactivation of the paternal and maternal X is random in the embryo. Random inactivation appears to apply to both embryonic and extraembryonic tissues in the human. In the embryo, the inactivation event occurs at different times in different cell layers. In the mouse, inactivation occurs in endodermal cells at the blastocyst stage and in ectodermal cells at about the time of gastrulation. Inactivation occurs at a point when multiple cells give rise to each tissue, so any tissue will be a mosaic consisting of cells that express one or the other X.

The precise mechanisms of X-inactivation are not known, but some understanding of the process at the molecular level has emerged. Many genes, particularly those expressed in a wide variety of tissues and referred to as *housekeeping genes*, have regions of DNA near their 5′ ends that contain many copies of the dinucleotide CpG. This was previously pointed out in discussion of the cloning of the cystic fibrosis gene, as these CpG islands can mark the sites of genes in a cloned region of DNA. The cytosine residues in these islands can be methylated, and extensive methylation of CpG dinucleotides is correlated with decreased levels of gene transcription. Genes on the inactivated X chromosome have been found to be extensively methylated, suggesting that methylation is involved in the maintenance of X chromosome inactivation.

The enzyme responsible for methylation of cytosine acts on double-stranded DNA. A CpG dinucleotide on one DNA strand corresponds with a CpG dinucleotide running in the opposite direction on the other strand. In newly replicated DNA, only one of the two strands will be methylated, but the enzyme recognizes the half-methylated double helix and methylates the newly incorporated cytosine (Figure 4-10). Completely unmethylated DNA is not recognized by the enzyme. This system provides a simple mechanism to preserve the methylation of one X chromosome in a clone of cells. Once the chromosome is methylated, methylation will occur on that same X in all descendants of the original cell.

The process of initiation of X-inactivation has also been examined at the molecular level. Studies of humans and mice with rearrangements of the X chromosome have revealed a region called the *X-inactivation center* (Xic) that is required for inactivation to occur (Figure 4-11). A gene has been found that maps to the Xic region, and this gene is expressed only on the *inactive* X (and hence is an exception to the usual rule of X-inactivation). The gene is referred to as *Xist*. Sequence analysis reveals no open reading frame, suggesting that the transcript is active as an RNA and does not encode protein. It has been proposed that Xist expression is responsible for initiation of the inactivated state on the X chromosome that expresses the gene. The effect of Xist, therefore, is postulated to be local. The means

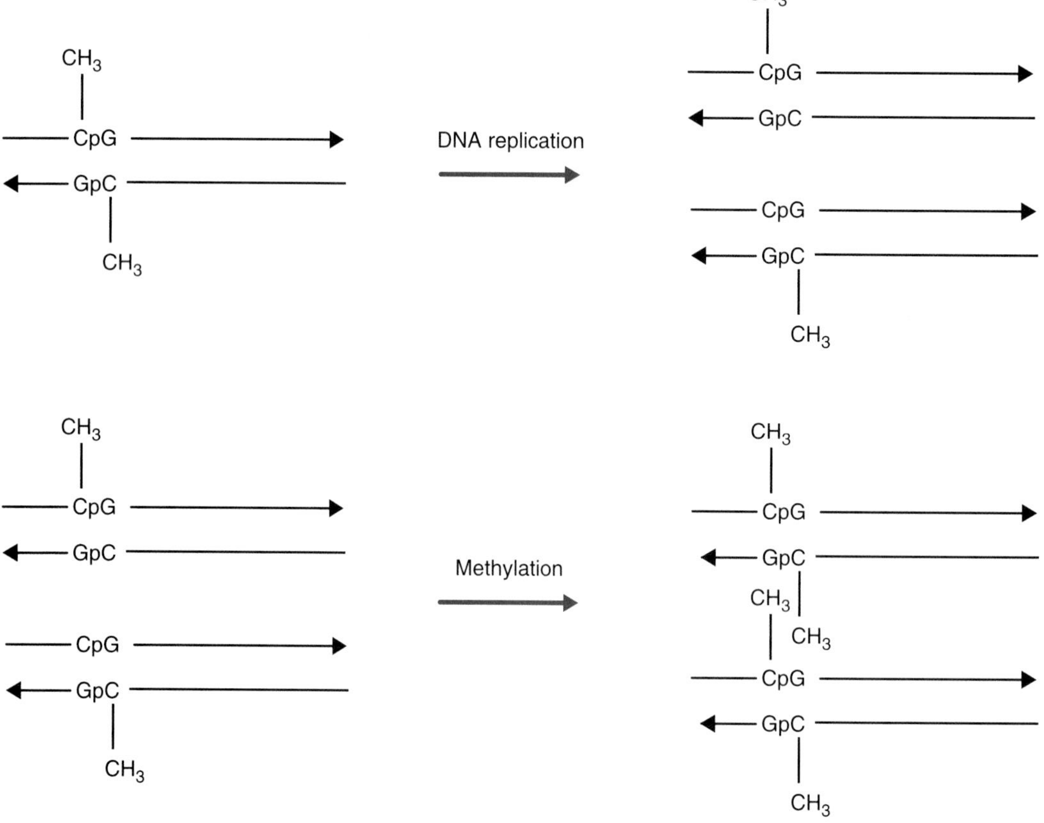

FIGURE 4-10 Cytosine residues adjacent to guanines may be methylated near the 5′ ends of some genes. When the DNA is replicated, only one strand will be methylated, but then an enzyme recognizes the single methylated strand and methylates the cytosines on the opposite strand.

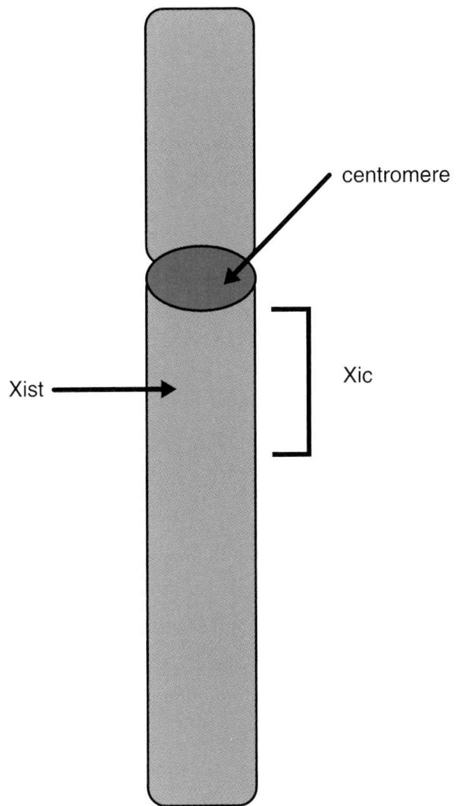

FIGURE 4-11 Diagram of human X chromosome showing approximate location of X inactivation center (*Xic*) and *Xist* gene.

whereby Xist expression occurs on one X but not the other remains to be elucidated.

Although selection of the X chromosome for inactivation is a random event, there are instances of nonrandom X-inactivation in females. This has come to attention because some females were found who manifest the phenotype of an X-linked recessive disorder. For example, women with DMD have been identified. These women are not homozygous for the muscular dystrophy mutation (which would require that their fathers have the mutant gene and transmit it, whereas males with DMD generally do not survive to reproduce). Some have had the phenotype of Turner's syndrome, in which females have congenital anomalies, ovarian dysgenesis, and short stature due to having only a single X chromosome and no other X or Y (hence, they have 45 chromosomes). These women are hemizygous for X-linked genes and therefore will express an X-linked recessive mutation such as DMD.

Other women with muscular dystrophy have a different form of chromosomal anomaly that has been particularly helpful in mapping the muscular dystrophy gene. These women have chromosome rearrangements, in which the X chromosome and a non-sex chromosome have exchanged segments (Figure 4-12). This is referred to as **translocation.** If the two chromosomes exchange segments, with no loss or gain of genetic material, the translocation is **balanced.** Balanced translocation might be expected to have no genetic consequences but, if the X is involved, the process of X-inactivation is affected. The part of the X that now is attached to an autosome will not be subject to inactivation if it is separated from the Xic. Meanwhile, the part of the autosome that is attached to the X with the inactivation center will be subject to inactivation along with the X material. Inactivation of the translocated X results in genetic imbalance wherein part of the X is inactivated and part of an autosome is inactivated. This alteration of gene dosage is lethal to the cell. Cells in which the normal X is inactivated, however, will have normal dosage of X-linked and autosomal genes and will survive. If the translocated X happens to carry a recessive mutant allele, that allele will be active in all surviving cells, and the female will manifest the mutant phenotype.

When females expressing the phenotype of DMD were examined for chromosomal rearrangements, it was noticed that many had translocations involving the X, and attachment of autosomal material occurred consistently at the same point in the short arm. These women had no prior family history of muscular dystrophy. It was suggested that the translocations in these individuals had occurred at the site of the DMD gene, disrupting the gene and resulting in mutation. This led to the hypothesis that the DMD gene resides at a point midway up the short arm of the X chromosome.

The CPK tests confirm that Brenda's mother is a DMD carrier, as was known from the family history. Brenda's results indicate that she is not a carrier, but she is told that these results do not definitively rule out carrier status.

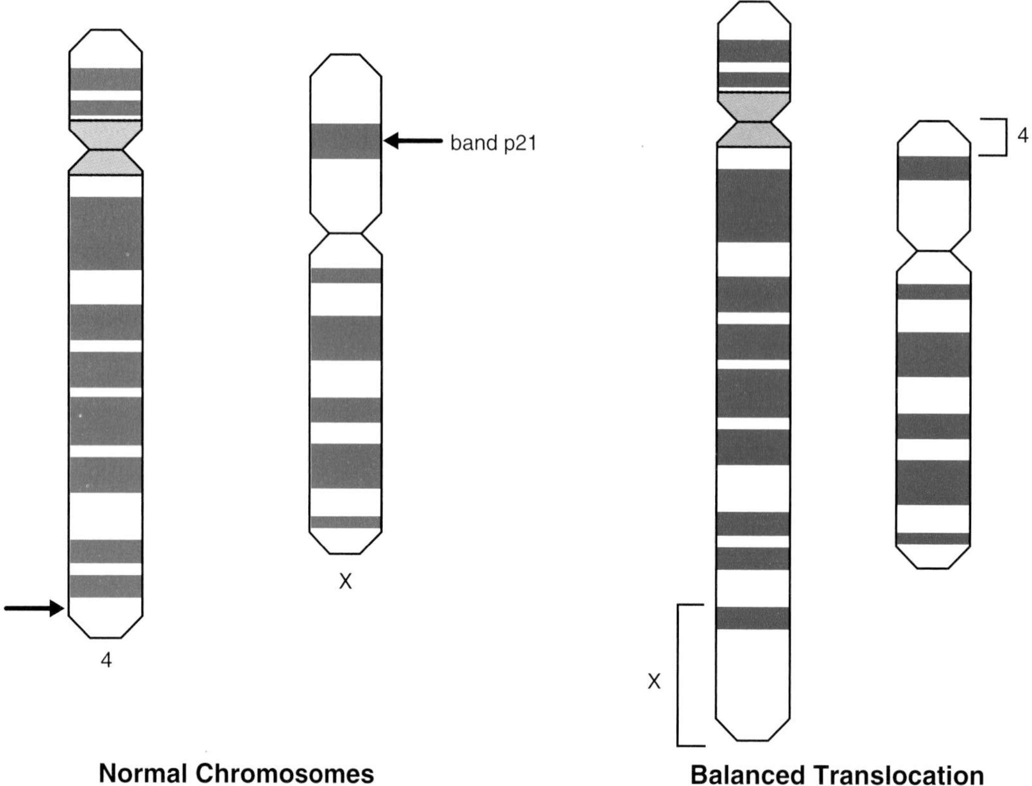

FIGURE 4-12 Translocation of chromosomes 4 and X, with breakpoint on X at band p21. The normal chromosomes are shown to the left, the translocated chromosomes to the right. Although no genetic material is lost or gained, the translocation disrupts the Duchenne's muscular dystrophy gene, which must therefore be located at band p21. For details of chromosome nomenclature, see Chapter 5.

Serum levels of CPK can be elevated in females who carry the DMD gene mutation. Muscle cells actually are a syncytium containing multiple nuclei derived from the fusion of immature myoblasts. Any individual myoblast in a carrier has a 50% likelihood of having an active mutant or normal X chromosome and therefore a 50% chance of expressing the muscular dystrophy mutation. Muscle cells with a chance preponderance of mutant-expressing nuclei will be prone to destruction and will release CPK. Muscle is capable of regeneration, though, so over time, damaged cells may be replaced by healthier muscle, causing the CPK values to decrease and thereby accounting for false negative tests. CPK can also be released from muscle cells damaged by simple trauma, including the trauma of vigorous exercise, thereby accounting for false positive tests. These testing problems are ameliorated somewhat by performing tests on several separate occasions, which allows the clinician to discount a single aberrantly elevated value. Clearly, CPK testing is an imperfect carrier test. Approximately two-thirds of DMD carriers have elevated CPK values, and approximately 5% of women in the general population have elevated values.

4.4 Mapping the Gene for DMD

1983

Based on the information they received in the genetics clinic, James and Brenda decide to try to become pregnant. In 1983, they telephone a genetic counselor at the clinic, who informs them that prenatal diagnosis of DMD has been done in a few families using RFLPs for a number of nearby loci. Blood is drawn from

Brenda, her mother, her father, Charles (who has DMD), Brenda's healthy maternal uncle, and both her mother's parents. Brenda turns out to be heterozygous at two loci that flank DMD, namely D2 and L128 (Figure 4-13). D2 recognizes two alleles in PvuII-cut DNA (6.0-Kb and 6.6-Kb); L128 recognizes two alleles in TaqI-cut DNA (12-Kb and 16-Kb).

The power of genetic linkage to map the DMD gene was recognized soon after the development of mapping strategies involving RFLPs. The task was substantially easier than for cystic fibrosis. The DMD gene was known to reside on the X chromosome, and experience with the rare individuals with chromosome translocations indicated that the gene might reside at a specific site on the short arm. Attention could therefore be focused on finding genetic markers in this region and testing them for linkage. Another advantage was that relatively large families were available, making easier statistical analysis of linkage data.

The first step was to find clones of DNA that mapped to the region of the X chromosome where the muscular dystrophy gene was believed to reside. One approach was to screen a human genomic library in lambda phage for clones that were linked to muscular dystrophy. Two strategies were employed to make the process more efficient. First, a library was made that was enriched for clones on the X chromosome by isolating X chromosomes from dividing cells and using these as the source of DNA to make the library. In one experiment, X chromosomes were purified by the technique of fluorescence-activating flow sorting (Figure 4-14). If the X represents approximately 5% of the DNA in a female cell, this provides a 20-fold enrichment for

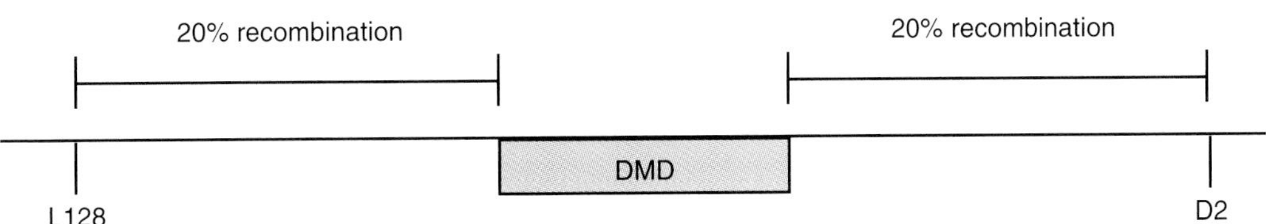

FIGURE 4-13 Map of region surrounding DMD gene, indicating flanking markers L128 and D2.

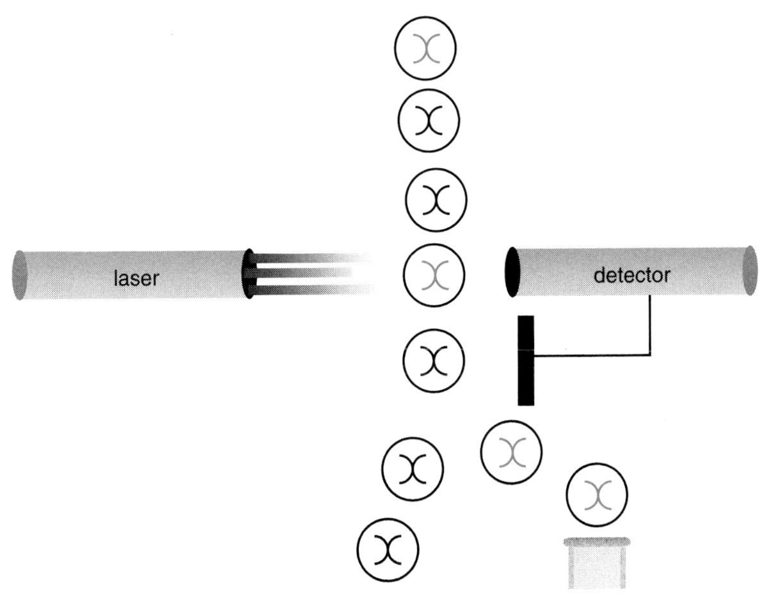

FIGURE 4-14 Fluorescence-activated chromosome sorting. Chromosomes are isolated from dividing cells and stained with a fluorescent dye. Individual chromosomes then are incorporated into tiny fluid droplets to which an electric charge is imparted. The droplets are passed in a stream before a laser. The laser excites the fluorescent dye, and the fluorescence is seen by a detector. Different chromosomes can be distinguished by size and degree of fluorescence. The detector is programmed to identify the characteristics of the X chromosome and, when an X is detected, to impart a charge briefly to a plate that draws the droplet into a collecting tube. Droplets that contain other chromosomes are discarded. After many millions of X-containing droplets are collected, the chromosomes in the tube are used to isolate DNA, which then is cloned in phage.

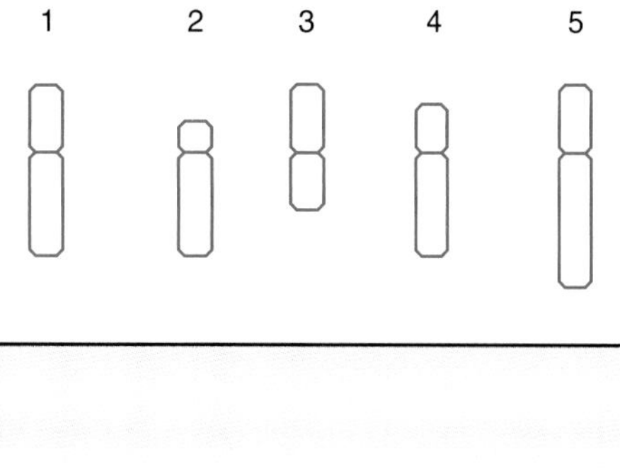

FIGURE 4-15 Southern blot analysis of a cloned piece of DNA hybridized with DNA from somatic-cell hybrids with X chromosome rearrangements. In this case, the cloned DNA comes from a region of the middle of the short arm of the X chromosome. Hybrids 1, 3, and 5 include this region, whereas 2 and 4 do not, due to loss of material from the X chromosome. The somatic-cell hybrids were made from human cells that carried various X chromosome rearrangements.

X-linked sequences over using a total human genomic DNA to make the library.

A second strategy was to use a set of somatic-cell hybrids to screen the library rapidly. The use of somatic-cell hybrids in gene mapping has already been described. Human-rodent hybrid cells tend to lose most of their human chromosomes, except for a random few that are retained. A Southern blot of restriction enzyme–digested DNA from a hybrid cell with the human X would recognize labeled DNA from the library only if the clone were located on the X. Regional mapping can be done if hybrid cells are made from human cells that contain X chromosome rearrangements (Figure 4-15). For example, in a person who carries an X-autosome translocation, the X is split into two pieces. If a hybrid is made using cells from such a person, hybrid colonies might include only a part of the short arm of the X. Cloned DNA will hybridize with DNA from the cells that contain only the region of the X from which the clone originated.

Using these strategies, many DNA clones were identified in the region of the muscular dystrophy gene. Next, it was necessary to determine which clones were polymorphic (Figure 4-16). Genomic DNA from many individuals was cut with a large number of restriction enzymes, each digestion done in a separate reaction. Digests then were hybridized with the cloned segments on Southern blots. A nonpolymorphic marker would yield identical hybridization patterns for all DNA samples cut with a given

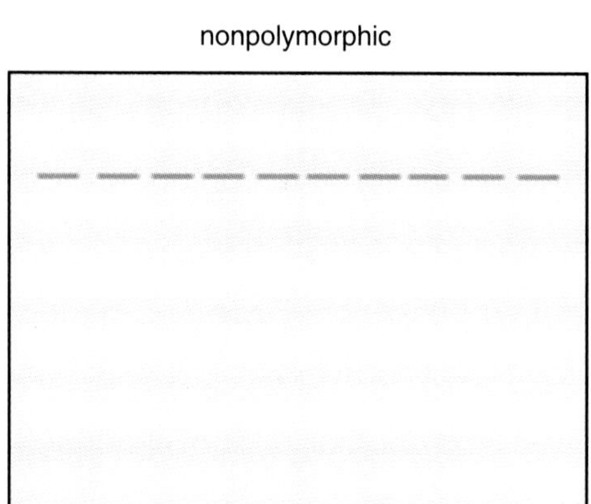

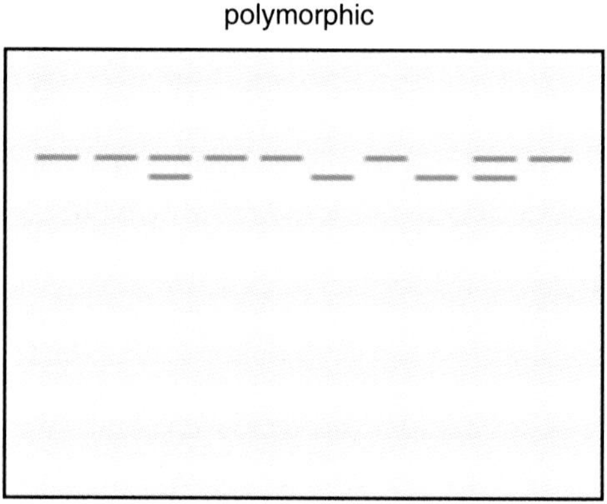

FIGURE 4-16 Southern blot of two different DNA probes, one that is not polymorphic (left) and the other that is polymorphic (right). The nonpolymorphic probe gives the same band size in each of many individuals, whereas the polymorphic probe shows two different band sizes in different individuals (some of whom will be heterozygous and thus have both bands).

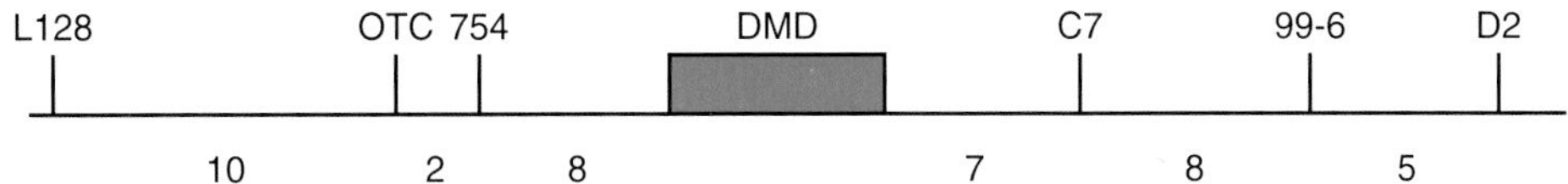

FIGURE 4-17 Approximate map of Duchenne's muscular dystrophy gene and flanking polymorphic markers. The marker designations are given at the top, and the distances between them in centimorgans at the bottom.

enzyme. A polymorphism would result in differences in the sizes of bands obtained with a given restriction enzyme.

The final step in mapping was to test the polymorphic markers for linkage to DMD. The segregation of marker alleles was tracked through large families, and lod scores were calculated. By 1983, the DMD locus was mapped to a precise region on the short arm of the X, at the site of the translocations in affected females, and a linkage map of the region was available (Figure 4-17).

This knowledge permitted carrier testing and prenatal diagnosis in some families, based on tracking the affected gene through the family. Flanking markers were available, increasing the accuracy of testing. Unfortunately, though, none of the markers was closer than approximately 5 cM, and recombination between flanking markers was relatively common. Therefore, many families received uncertain results after considerable effort had been expended to perform a linkage study. More powerful diagnostic tools based on identification of the actual gene mutations were needed.

Brenda is found to be "informative" for these flanking markers, so prenatal diagnosis is offered. The fetus from this pregnancy is determined, by karyotype analysis from an amniocentesis specimen, to be male, and DNA studies are performed on cells from the fetus (Figure 4-18).

The probable haplotypes with respect to DMD, D2, and L128 are shown in the pedigree. The DMD mutation is in coupling with the 6.0-Kb D2 allele and the 12-Kb L128 allele in II-2. This can be determined because II-2 had to receive the 6.0- and 16-Kb haplotype with the normal DMD allele from her father. A recombination must have occurred in I-2, leading to either II-2 or II-3, because in II-3 the maternal 6.0-Kb D2 allele is in coupling with the wild-type allele of the DMD locus. Because we do not know whether the crossover occurred in the meiosis leading to II-2 or II-3, the coupling phase of I-2 is uncertain, as indicated.

The consultand, III-1, received the 6.6-Kb D2 allele and the 12-Kb L128 allele from her father. She therefore must have received the 6.0- and 16-Kb alleles from her mother. Barring recombination between the DMD mutation and L128 in II-2, III-1 is likely not to be a carrier (actual risk of being a carrier is approximately 20% based on DNA analysis, less if negative CPK testing is taken into consideration).

The male fetus received the grandpaternal 6.6-Kb and 12-Kb alleles from III-1. He is therefore unaffected,

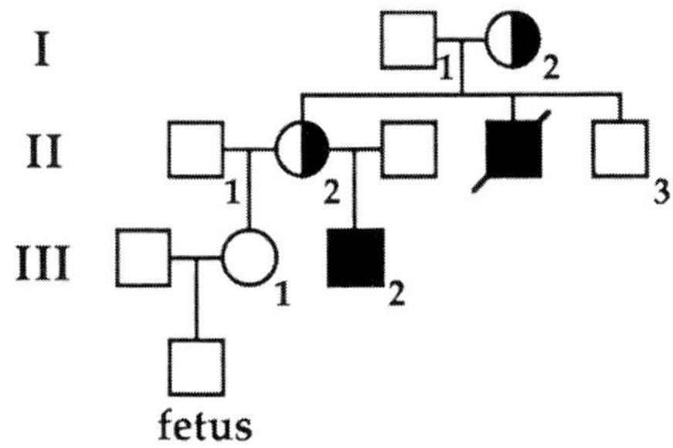

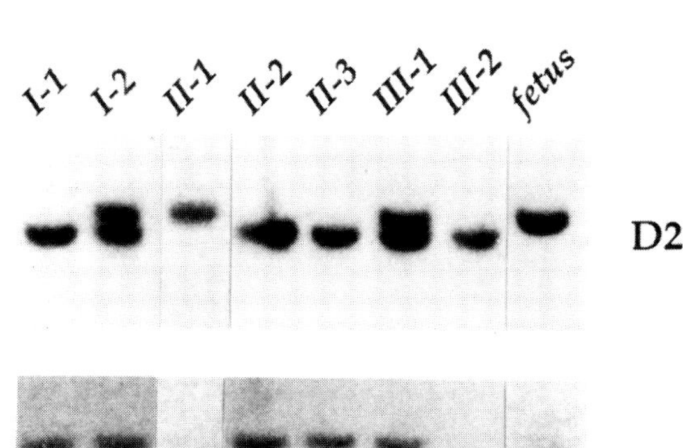

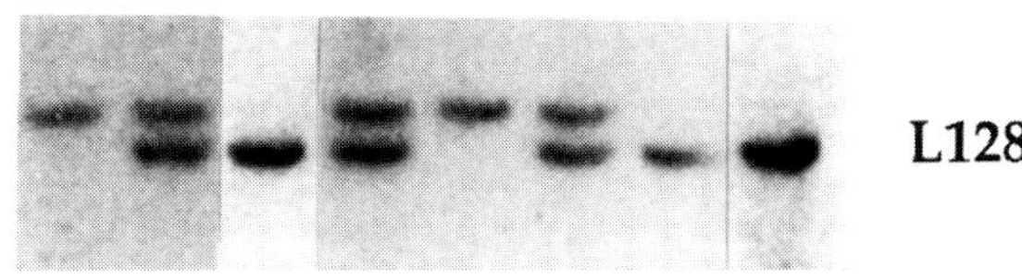

FIGURE 4-18 Results of Southern analysis with D2 and L128 probes. The pedigree is shown at the top, and the Southern blot results at the bottom.

unless his mother was a carrier ***and*** *a double crossover occurred between L128 and DMD on one side and D2 and DMD on the other [probability of (0.20)(0.20)(0.20) = 0.008]. He therefore has a greater than 99% chance of being unaffected. See Figure 4-19.*

4.5 Cloning the Gene for DMD

We have seen how finding rare individuals with a chromosome abnormality and muscular dystrophy provided a shortcut to mapping the gene. Still, the effort to clone the gene would have taken years had not another rare individual been found with a different kind of chromosome rearrangement. In this case, it was a male who had DMD in combination with three other genetic conditions: retinitis pigmentosum (a degenerative disease of the retina), chronic granulomatous disease (deficiency of granulocyte function leading to chronic pyogenic infections), and the McLeod phenotype (absence of a red blood cell membrane protein, resulting in hemolysis). It was hypothesized that the genes responsible for these conditions are contiguous on the X chromosome and that this person had deletion of the contiguous genes. High-resolution cytogenetic studies confirmed this (Figure 4-20).

The approach used to clone the DMD gene made ingenious use of this deletion (Figure 4-21). It was reasoned that the major difference between the DNA from this individual and from another person without the deletion would be the existence of DNA from within the deleted region. A plasmid library was made by subtracting the DNA from the deleted individual from normal DNA, cloning only the difference. The procedure resulted in a plasmid library that was greatly enriched for DNA in the deleted region.

As we have seen with cystic fibrosis, knowing when you have found the gene is one of the challenges of a positional cloning project. The first clue that the cloned sequences included the DMD gene was provided by a zoo blot. Some clones were found to hybridize with DNA from other species, including mouse, hamster, and chicken. Clones that displayed species conservation were likely to encode genes and represented candidates for the muscular dystrophy gene or other genes in the region.

The first clue that the gene for DMD was in hand came from Southern analysis of DNA from males with DMD. If one person could be affected owing to a very large deletion, it was reasoned that others'

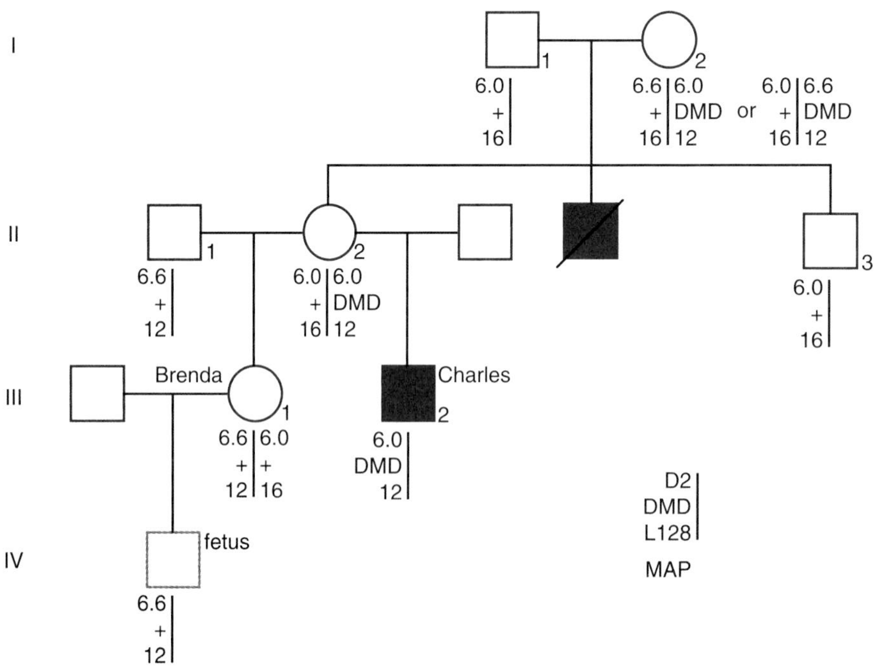

FIGURE 4-19 Pedigree indicating alleles for L128 and D2.

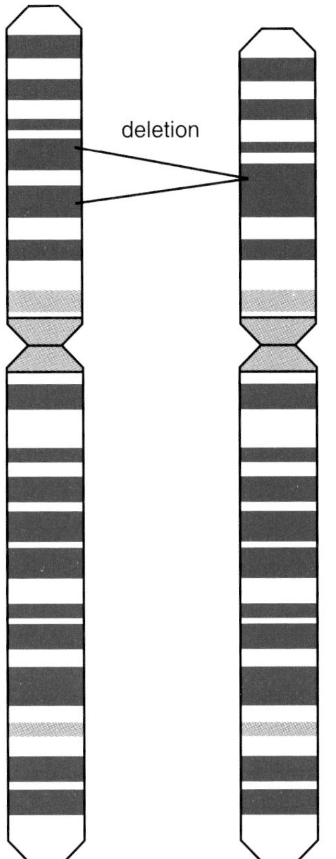

FIGURE 4-20 Diagram of deletion in individual with retinitis pigmentosum, chronic granulomatous disease, McLeod phenotype, and Duchenne's muscular dystrophy. (Francke U, Ochs HD, de Martinville B, et al. Xp21 chromosome deletion in a male associated with expression of Duchenne muscular dystrophy, chronic granulomatous disease, retinitis pigmentosum, and McLeod phenotype. Am J Hum Genet 1985;37:250–267.)

disorders might be due to smaller deletions, perhaps involving only part of the gene. A search of many individuals indeed revealed several with deletions of cloned DNA. An international collaboration was assembled through which 1345 samples were tested. Approximately 6.5% of these patients had deletions, providing strong evidence that the sequences resided in the region of the DMD gene. Moreover, deletions were found in males with either Duchenne's or Becker's dystrophy, indicating that the two conditions were allelic, or at least were very closely linked.

Having established that DNA had been cloned from the muscular dystrophy gene, it remained to clone a larger extent of the region to characterize the gene fully. This was accomplished by gene walking. Plasmid clones were used to screen a genomic library, resulting in isolation of larger phage clones. Cloned sequences were used to probe northern blots containing muscle RNA, which revealed hybridization to a very high-molecular-weight RNA—greater than 13 Kb. A muscle cDNA library was made, and clones homologous with sequences in the deleted region were isolated. Full-length transcripts rarely are isolated from a cDNA library, especially when the mRNA is so long. Multiple overlapping cDNA fragments were cloned and sequenced. The sequence was then examined for an open reading frame and an initiation codon (AUG for methionine).

Evidence was accumulating that the gene in hand was indeed the one responsible for DMD. It was expressed in muscle and, as more and more of the cDNA was cloned, an increasing number of deletions were found in muscular dystrophy patients. Most of these were intragenic deletions, making it very likely that the correct gene had been found. The search for the full-length genomic sequence produced some surprises. A 13-Kb message would be expected to occupy a larger genomic region due to the presence of introns. A "contig" of overlapping phage clones was constructed, but it was a long time before the contig included both the 5′ and 3′ ends of the gene. By this time, more than 2 million base pairs of DNA had been cloned, making this the largest gene yet discovered.

4.6 Dystrophin: The Protein Product of the DMD Gene

A 13-Kb mRNA would encode an unusually large protein. The predicted amino acid sequence was deduced from the triplet codons and was compared with other known proteins. A major region of homology was found with a protein called *spectrin*, which is an actin-binding protein located in the cytoskeleton of the red blood cell. The actin-binding domain is the N-terminal region. The bulk of the molecule (Figure 4-22) comprises a large rod domain consisting of 24 repeats of similar sequences of nearly 109 amino acids. A 150–amino acid cysteine-rich region is found after the rod domain. At the C terminus, there is a 420–amino acid region that is believed to interact with other membrane proteins.

The protein itself was identified by immunologic staining. Portions of the cDNA of the mouse

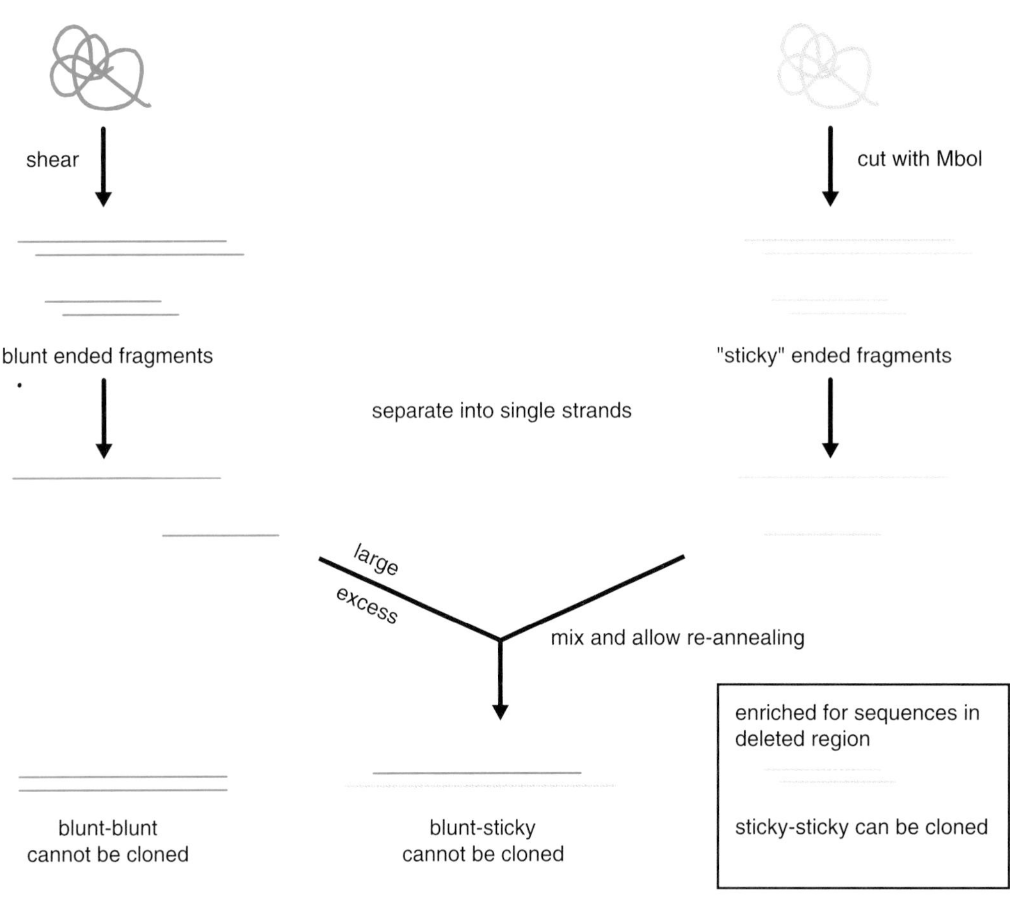

FIGURE 4-21 Cloning DNA from within X chromosome deletion. DNA was isolated from the male with an X chromosome deletion and sheared into fragments of random size. These fragments did not have unpaired bases at the ends (blunt ends). A small amount of DNA was obtained from an individual with a 49,XXXXY karyotype and was cut with the restriction enzyme MboI, resulting in sticky ends. The two sets of DNA fragments each were separated into single strands and mixed together under conditions that allowed re-annealing of DNA strands. Because most of the DNA present was blunt-ended (from the individual with the deletion), most double-stranded molecules contained either one or two blunt-ended strands. These could not be cloned in plasmid. Only double-stranded molecules with sticky ends on both strands could be inserted into a plasmid molecule. These would preferentially consist of DNA fragments from within the deleted region, as no blunt-ended DNA fragments would have been derived from this region. (Kunkel LM, Monaco AP, Middlesworth W, Ochs HD, Latt SA. Specific cloning of DNA fragments absent from the DNA of a male patient with an X chromosome deletion. Proc Natl Acad Sci USA 1985;82:4778–4782.)

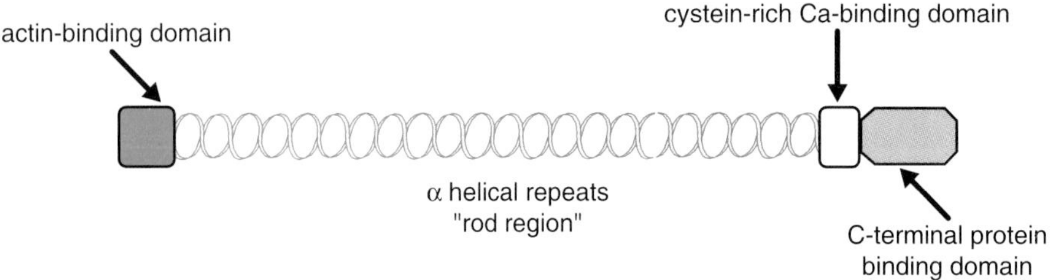

FIGURE 4-22 Diagram of dystrophin gene. (Redrawn from Ahn AH, Kunkel LM. The structural and functional diversity of dystrophin. Nat Genet 1993;3:283–291.)

muscular dystrophy gene were subcloned into plasmids that juxtaposed the cDNA with an inducible gene (trypE) (Figure 4-23). Stimulation of transcription in the bacteria resulted in production of a fusion protein, which then was purified and used to immunize rabbits. The resulting antisera were used to stain muscle protein samples separated by molecular weight in polyacrylamide gels. The antibody recognized a protein of molecular weight in excess of 400 kDa. This protein was expressed in muscle cells, including skeletal muscle, smooth muscle, and cardiac muscle. It also was found in brain tissue but nowhere else in appreciable abundance. Even in muscle, the protein was a minor species, comprising less than 0.002% of the total muscle protein.

Homology with spectrin suggested that the new muscle protein might also be associated with cell membranes. Antibodies were used to effect immunofluorescent staining of tissue sections for the protein. Fluorescent staining was found to outline each muscle cell in cross-section, confirming a cell membrane localization (Figure 4-24).

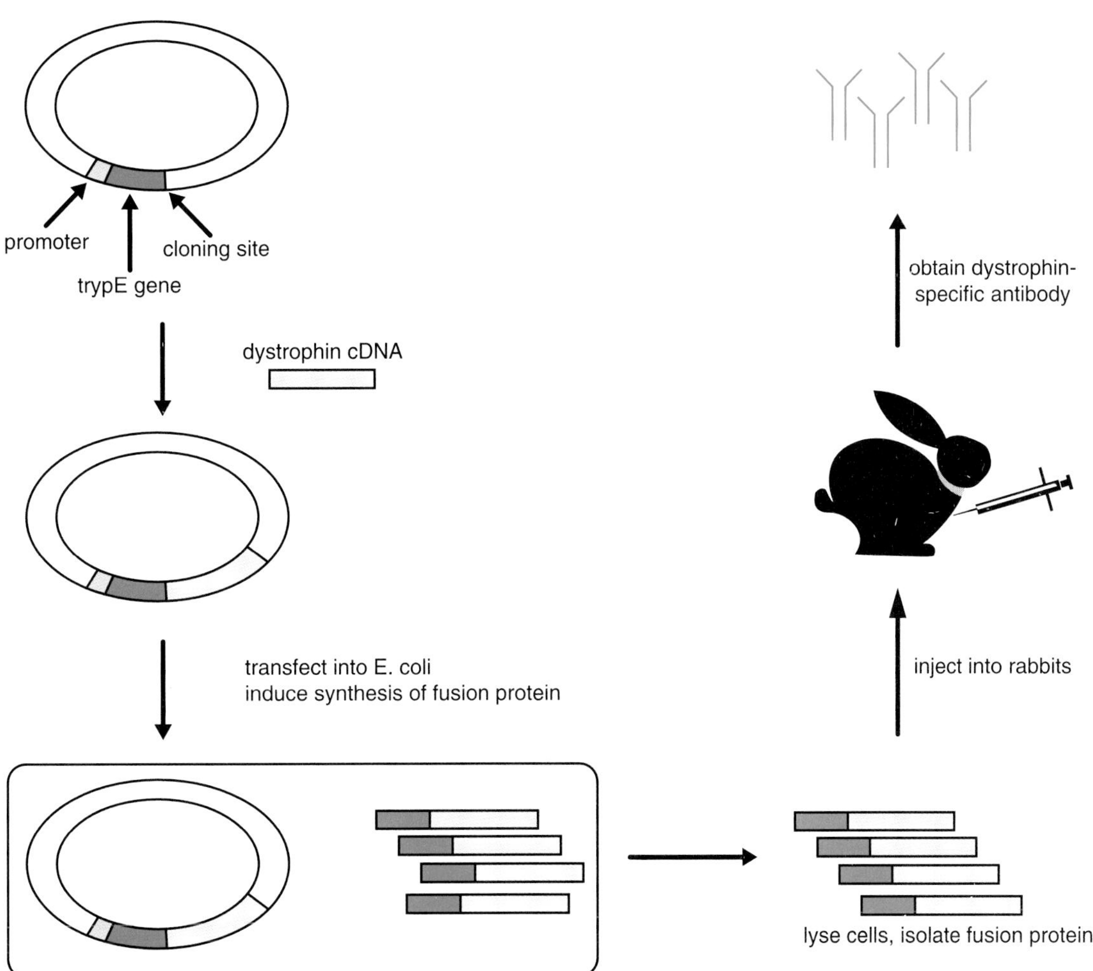

FIGURE 4-23 Preparation of antibodies to portions of dystrophin. Segments of the dystrophin cDNA were inserted into a plasmid containing part of the trypE gene adjacent to a promoter. The recombinant plasmids were transfected into *Escherichia coli*, and the promoter was induced with tryptophan. Fusion proteins were collected after lysis of bacterial cells and were injected into mice. Polyclonal antibodies were obtained that reacted with parts of the dystrophin molecule. (Data from Hoffman EP, Brown RH Jr, Kunkel LM. Dystrophin: the protein product of the Duchenne muscular dystrophy locus. Cell 1987;51:919–928.)

Normal

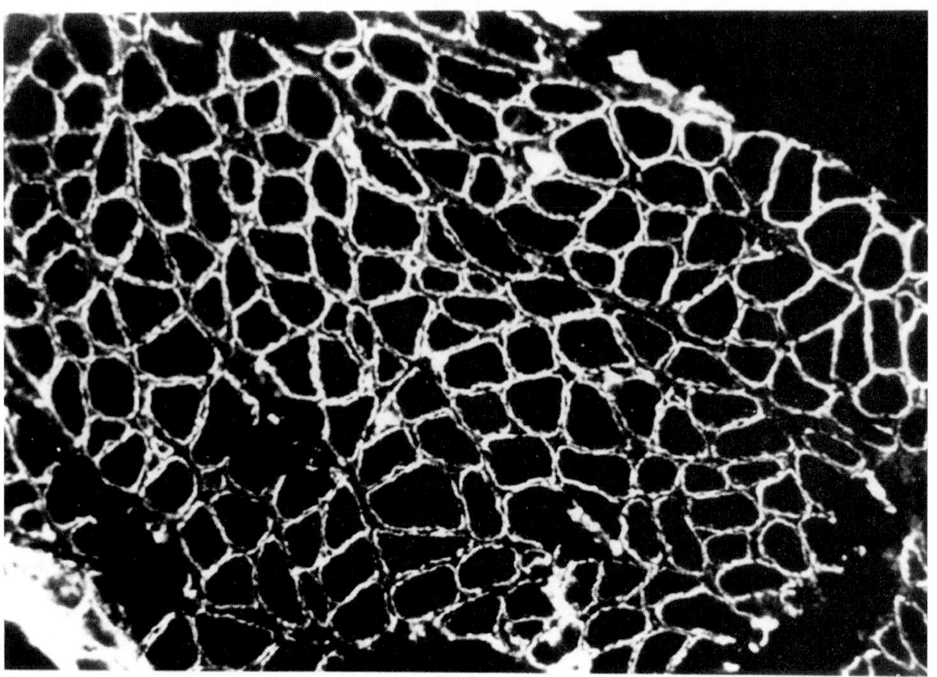

DMD

FIGURE 4-24 Immunofluorescent staining of dystrophin in normal muscle (top) and in muscle from a boy with Duchenne's muscular dystrophy (bottom). In the normal muscle, bright staining outlines each muscle cell in cross-section, indicating that dystrophin is located near the cell membrane. No dystrophin staining is found in the dystrophic muscle. (Courtesy of Dr. Louis Kunkel, Children's Hospital, Boston.)

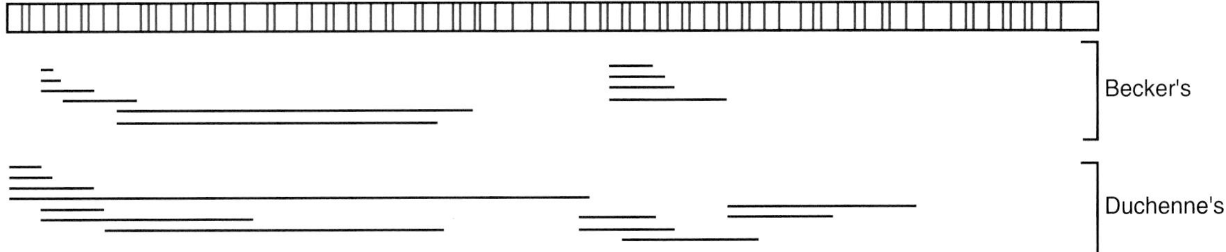

FIGURE 4-25 Map of dystrophin gene, with vertical lines indicating exons (not drawn to scale). Below the map, the extent of some deletions responsible for Duchenne's or Becker's dystrophy are indicated. The sizes of deletions are indicated approximately. (Data based on the work of Koenig M, Beggs AH, Moyer M, et al. The molecular basis for Duchenne versus Becker muscular dystrophy: correlation of severity with type of deletion. Am J Hum Genet 1989;45:498–506.)

This new protein had not been known to exist prior to its discovery through positional cloning of the muscular dystrophy locus. It was therefore dubbed *dystrophin*, though its function was still unknown. This precedent of naming a newly discovered protein after the disease that led to its discovery created some dissent (engendered by the argument that it made little sense to name a protein according to what it does wrong rather than what it does right), but the practice has continued for many other proteins discovered in the years since.

4.7 The Molecular Basis of Duchenne's and Becker's Muscular Dystrophy

1986

As part of a research study correlating dystrophin expression in muscle with clinical outcome in muscular dystrophy, a piece of frozen muscle from Charles's biopsy (obtained 15 years earlier) was found and tested for expression of dystrophin. The absence of dystrophin in Charles is interpreted as consistent with Duchenne's, as opposed to Becker's, muscular dystrophy, which in turn is consistent with the clinical history.

Studies of DNA from males with both Duchenne's and Becker's dystrophy revealed a high frequency of deletions in each disorder (Figure 4-25). This revealed that the disorders are allelic but, as deletions were mapped, no simple pattern emerged to distinguish Duchenne's patients from Becker's patients. Deletions were widely distributed over the length of the gene, with no clustering of deletions in Duchenne's or Becker's dystrophy. Size of deletion also did not correlate with phenotype: Large deletions were found in some with the milder Becker's dystrophy, whereas some with severe Duchenne's dystrophy had small deletions.

Phenotypic correlations were more successful at the protein level. Western blot studies revealed total absence of dystrophin in muscle biopsies from males with DMD (Figure 4-26). Immunofluorescence

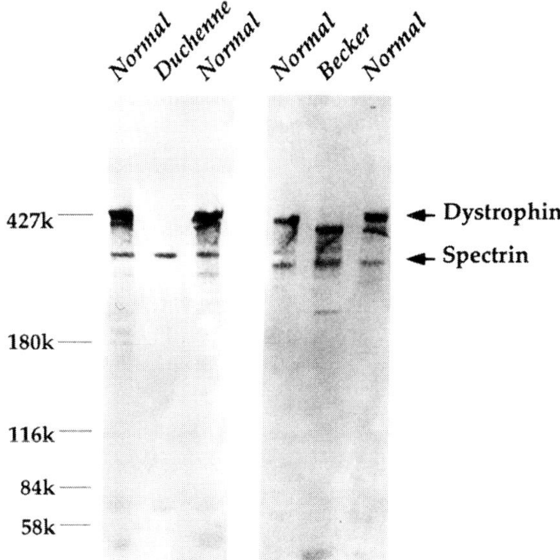

FIGURE 4-26 Western blot of Duchenne's and Becker's dystrophy muscle biopsies. The muscle protein spectrin also is stained, to serve as a control. (Courtesy of Dr. Louis Kunkel, Children's Hospital Boston.)

studies of DMD biopsies revealed muscle cells lacking the characteristic membrane-associated staining. Males with Becker's dystrophy, in contrast, were found to have dystrophin in their muscle, but this dystrophin was aberrant in quantity or quality (or both). Some were found to have deficient amounts of a more or less normal-sized dystrophin. Others had dystrophin of lower or, rarely, higher molecular weight than normal. Correlation of the presence or absence of dystrophin with the Duchenne's or Becker's phenotype was found to be so good that the protein test came to be incorporated into the diagnostic evaluation of boys with muscular dystrophy. Absence of dystrophin predicts DMD, the presence of abnormal dystrophin predicts Becker's dystrophy, and normal dystrophin means that some other diagnosis should be considered. This greatly improved the precision of pathologic diagnosis and the ability to predict disease progression.

It took some time to discern the molecular basis for the differences in dystrophin expression. The basis for genotype-phenotype correlations was found by analysis of the codons at the beginnings and ends of exons (Figure 4-27). Exons do not necessarily end at the third position of a triplet, completing the codon for an amino acid. Often the last base of an exon is the first or second base of a codon. As long as the first base of the next exon picks up where the previous exon left off to complete the codon, the reading frame of the protein is preserved. The protein will be shorter by whatever was lost in the deleted region but otherwise will be translated from end to end. If, however, two "noncompatible" exons are juxtaposed in the spliced message, a frameshift

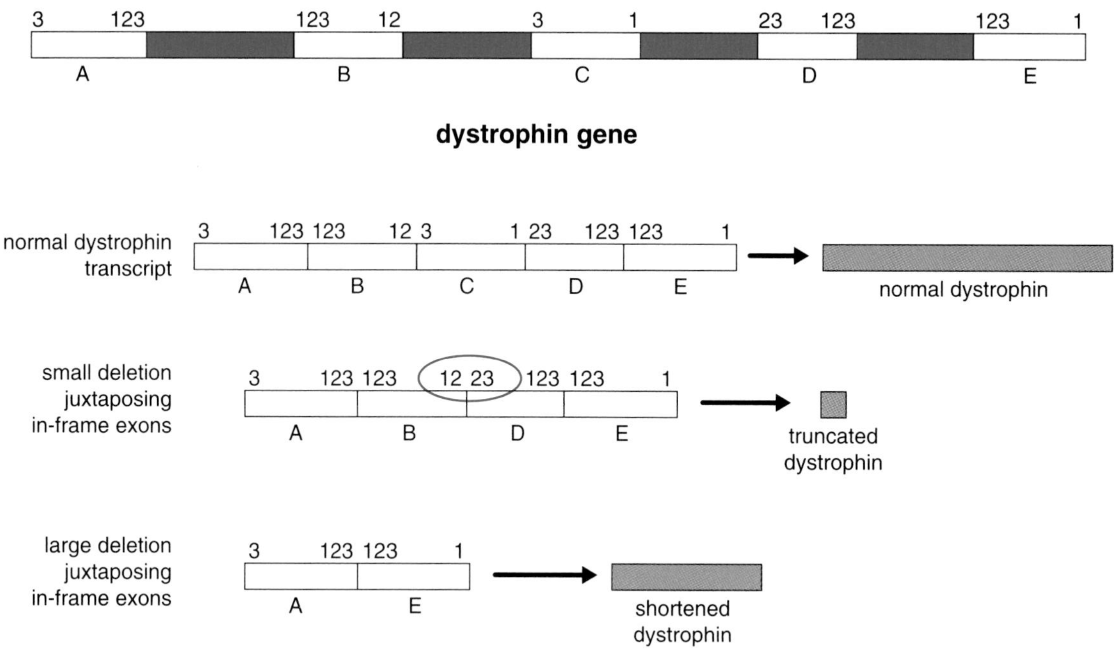

FIGURE 4-27 Reading-frame hypothesis. The reading frame at the beginnings and ends of exons (exons indicated by *white boxes* in gene) are shown at the top. The numbers *1*, *2*, and *3* correspond with the first, second, and third bases of a triplet codon, respectively. Hence, exon A, which ends with *123* includes the complete codon for an amino acid at its end; exon C, in contrast, ends with *1*, with exon D supplying the second and third bases to complete a codon. The normal dystrophin transcript is spliced together to provide a complete code for the dystrophin protein. A small deletion that removes exon C results in production of an out-of-frame transcript (see *circle*), producing a truncated protein. A larger deletion of exons B through D results in an in-frame transcript, which produces an intact, although shorter-than-normal, transcript. (Modified with permission from Monaco AP, Bertelson CJ, Liechti-Gallati S, Moser H, Kunkel LM. An explanation for the phenotypic differences between patients bearing partial deletions of the DMD locus. Genomics 1988;2:90–95.)

will occur. This inevitably leads to a stop codon after a short distance. The protein product will be severely truncated and probably is degraded in the cell.

Most cases of DMD are thus caused by deletions that create frameshifts resulting in complete absence of dystrophin. Becker's dystrophy usually is associated with in-frame deletions that allow production of an internally deleted but partially functional protein product.

The reading-frame hypothesis explains most, but not all, cases of Duchenne's or Becker's dystrophy due to deletion. Some very large in-frame deletions can produce DMD. Also, some out-of-frame deletions can produce a Becker's phenotype, probably due to exon-skipping events (Figure 4-28). When the dystrophin message is spliced, sometimes one or more exons are spliced out of the final message. Such **alternative splicing** events generate slightly different proteins and can occur as normal events in some tissues. Alternative splicing can result in juxtaposition of compatible exons despite the presence of a deletion, preserving the reading frame. In such instances, the protein will be of reduced size but will otherwise be intact, resulting in Becker's dystrophy.

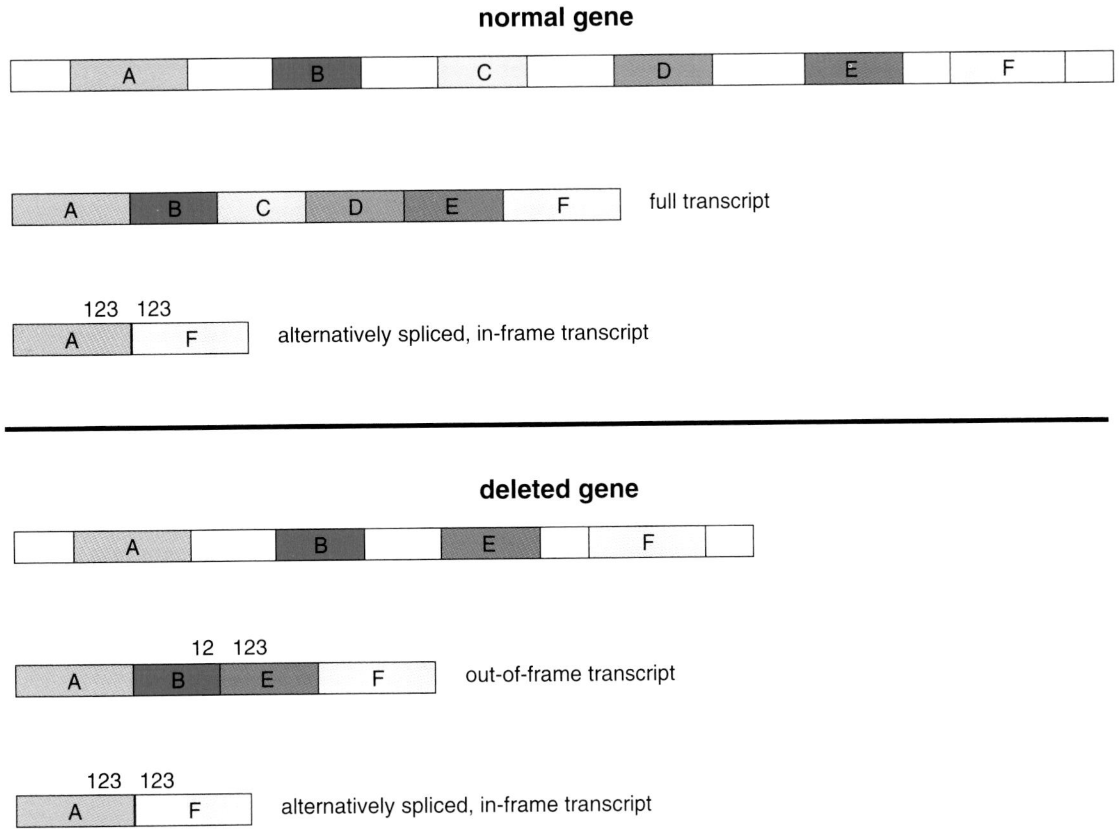

FIGURE 4-28 Exon skipping restoring reading frame from out-of-frame deletion. Normally (top), either a full-length transcript is made or an alternatively spliced transcript is made in which exons B through E are skipped. This alternatively spliced transcript is in-frame. A deletion of exons C and D results in an out-of-frame transcript, but this deletion does not affect the alternatively spliced transcript. (Modified with permission from Chelly J, Gilgenkrantz H, Lambert M, et al. Effect of dystrophin gene deletions on mRNA levels and processing in Duchenne and Becker muscular dystrophies. Cell 1990;63:1239–1248.)

4.8 Molecular Diagnosis of Duchenne's and Becker's Dystrophy

1986

James and Brenda read in the newspaper that the DMD gene has been cloned. They call the genetic counselor, who informs them that for some families direct detection of deletions, representing the gene mutation, is possible. This offers a chance for more precise prenatal diagnosis. Shortly afterward, Brenda becomes pregnant for the second time. DNA from Charles is still available in the laboratory and, after digestion with BglII, this is hybridized with cDNA probes from the DMD locus. On the basis of the findings from this study, prenatal diagnosis is undertaken, this time using material obtained at 9 weeks' gestation from a chorionic villus biopsy (Figure 4-29).

Although dystrophin analysis of muscle provides an accurate diagnostic test, it is invasive, with attendant high costs and risks. Also, dystrophin is not expressed in amniotic fluid cells or in chorionic villus tissue. Protein testing therefore does not provide a useful basis for prenatal diagnosis. We have already explored the power of linkage analysis for prenatal and carrier testing. The finding of a high frequency of gene deletion in affected individuals opened the way to direct mutation testing.

At first, deletions were identified by Southern analysis. A set of subclones of the cDNA were used as hybridization probes, producing a complex set of bands when genomic DNA was cut with a restriction enzyme, because each clone recognizes many (ten or more) exons. A male with a deletion lacks the bands that correspond with deleted DNA. A complete search of the dystrophin gene requires hybridization with 10 cDNA probes and scanning of dozens of bands for deletion. This is an expensive and laborious process, but it allows molecular diagnosis or prenatal diagnosis in two-thirds of cases.

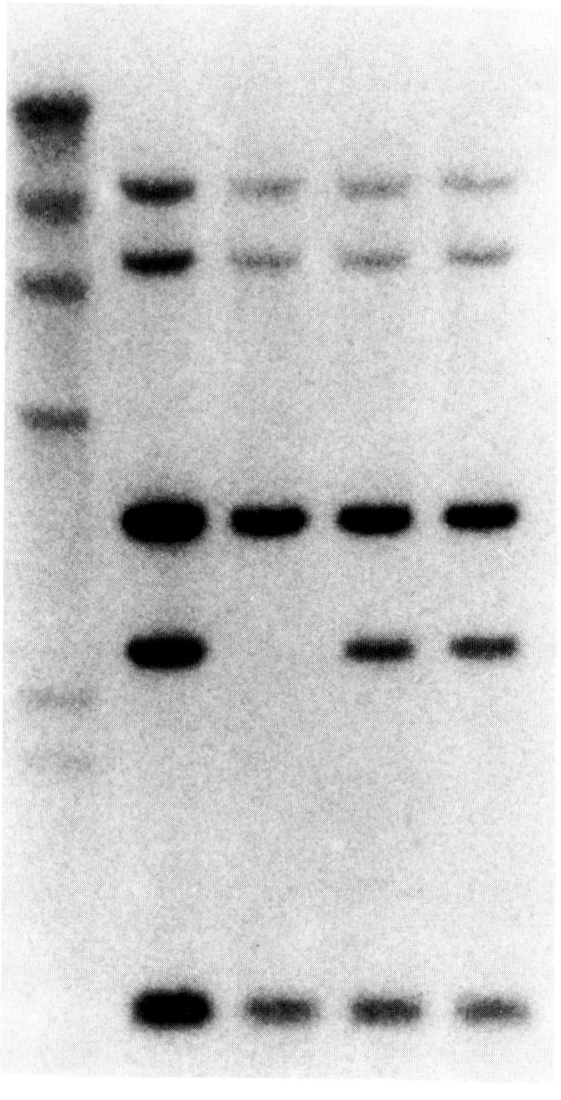

FIGURE 4-29 Identification of dystrophin deletion by Southern analysis. Lane 2 contains DNA from Charles, in which one band is missing. This corresponds with exon 51 of the dystrophin gene. His mother's DNA is in lane 1. Lanes 3 and 4 contain DNA samples from the previous pregnancy and the current prenatal sample, respectively.

1989

Brenda's pregnancy in 1986 resulted in the birth of a healthy son. A third prenatal diagnosis is undertaken in 1989. This time the polymerase chain reaction is used to detect the mutation in Charles and to study the fetal DNA obtained by chorionic villus biopsy (Figure 4-30).

The invention of the polymerase chain reaction (PCR) led to the development of an efficient, accurate, and inexpensive test for deletions. Oligonucleotide primers were made that were homologous either with intron sequences that immediately flank specific exons or with sequences at the beginning and

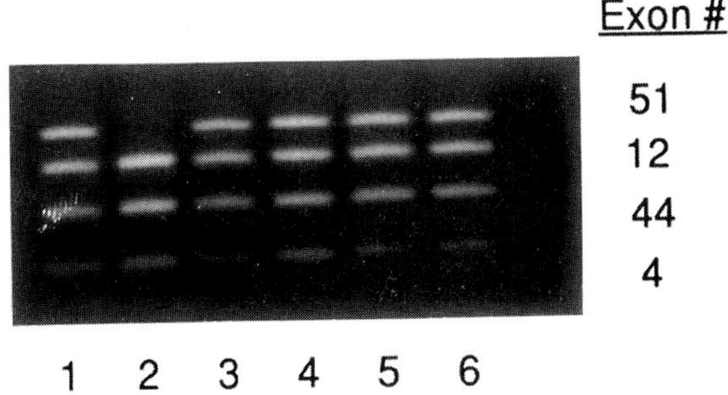

FIGURE 4-30 Multiplex PCR analysis of exons 51, 12, 44, and 4 of the dystrophin gene. Bands corresponding with each exon are indicated at the right. Charles's sample is in lane 2, showing an exon 51 deletion. DNA from the current pregnancy is in lane 3. Lane 1 is a control, showing all four bands. Other lanes contain samples for other males.

end of certain exons. This allowed PCR amplification of individual exons. The products could be visualized by separation in an agarose gel and staining with ethidium bromide (Figure 4-31). It was found that several exons could be simultaneously amplified in the same reaction, producing one band per exon. This is referred to as *multiplex PCR*. The order of bands in the gel is related to the exon size of the PCR product and not to the location of the exon in the gene. A male with a deletion would be missing bands corresponding with deleted exons.

A combination of 18 sets of PCR primers can identify 98% of all known dystrophin deletions. Diagnostic errors are very rare. The PCR test can be performed very rapidly—even within 1 day—and at

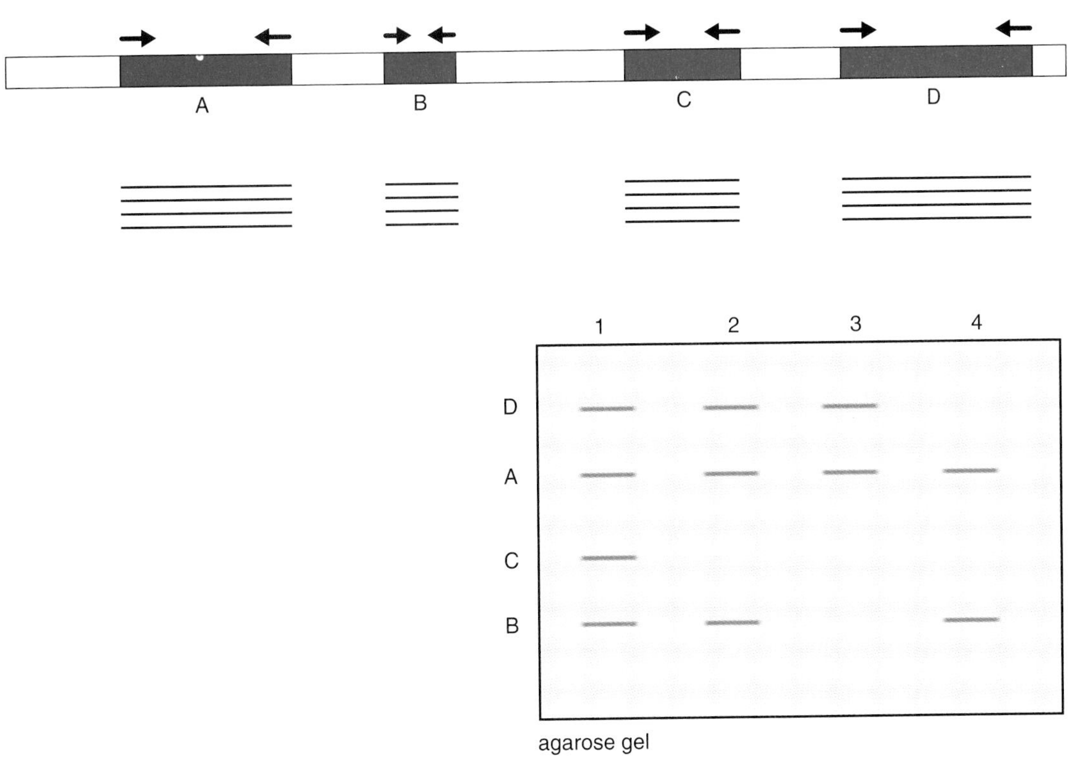

FIGURE 4-31 Multiplex PCR analysis of dystrophin gene deletions. Exons A, B, C, and D are amplified in a single PCR reaction (*arrows* indicate PCR primers). The products (shown below each exon) are separated by size on an agarose gel and are visualized by DNA staining. The order of exons in the gel is related to size, not position, in the gene. Lane 1 shows all four exons. Exon C is deleted in the sample tested in lane 2, exons B and C are deleted in lane 3, and exons C and D in lane 4. Note that exons that are adjacent in the gene are not necessarily adjacent in the gel. (Data from Chamberlain JS, Gibbs RA, Ranier JE, Nguyen PN, Caskey CT. Deletion screening of the Duchenne muscular dystrophy locus via multiplex DNA amplification. Nucleic Acids Res 1988;16:11141–11156; and from Beggs AH, Koenig M, Boyce FM, Kunkel LM. Detection of 98% of DMD/BMD deletions by PCR. Hum Genet 1990;86:45–48.)

a cost of less than $200. It can be done with very small quantities of DNA, including prenatal samples. Often both the 5′ and 3′ borders of a deletion can be inferred from the PCR analysis, allowing a distinction of Duchenne's from Becker's dystrophy by the reading-frame hypothesis. Many clinicians now rely on DNA testing before proceeding to muscle biopsy to diagnose Duchenne's or Becker's dystrophy, resorting to biopsy only if the deletion test is negative.

Approximately two-thirds of affected males have a dystrophin gene deletion. What about the others? Among the remaining one-third, nearly 15% have been found to have duplications of the dystrophin gene (Figure 4-32). Duplications have been identified by quantitative analysis of band intensity on Southern blots hybridized with cDNA probes. The analysis is difficult and time-consuming, as relatively small differences in hybridization intensity must be discerned among dozens of bands. The pathogenesis of dystrophin abnormalities due to duplication is similar to that of deletion: production of abnormal dystrophin in Becker's dystrophy or juxtaposition of out-of-frame exons leading to lack of dystrophin expression in Duchenne's dystrophy.

The remaining dystrophin mutations are slowly coming to light. Generally, they consist of mutations of one or a small number of base pairs, including small deletions, insertions, and single base changes. No simple technique has yet been devised to identify these mutations efficiently. Because of this difficulty, molecular diagnosis in the families where deletion, duplication, or point mutation cannot be found still relies on linkage analysis. In families where there is no prior history of DMD, and therefore where the mother may be a carrier or the affected son may be a new mutation, reliable molecular diagnosis may be impossible.

Because of the difficulty in finding nondeletion mutations, linkage analysis still is an important tool for genetic counseling in families with Duchenne's or Becker's dystrophy. Instead of using RFLPs, however, this is now commonly done using dinucleotide repeat polymorphisms (see Chapter 3). The advantage of using dinucleotide repeat polymorphisms is that differences among individuals are common, so it is easier to find informative markers. Dinucleotide repeat polymorphisms are available for many sites within the dystrophin gene.

4.9 Detection of Dystrophin Deletion Carriers

1992

It has long seemed likely that Brenda was not a DMD carrier, although prenatal testing was offered because it was difficult to be certain. Recently, an assay has been developed that provides more definitive carrier testing for dystrophin deletions.

The advent of a simple and reliable test for dystrophin deletions was a great help for diagnosis in affected individuals and for prenatal diagnosis. One of the difficult challenges in genetic counseling in muscular dystrophy is identification of carriers. The daughters of a carrier are at 50% risk of being carriers and often are interested in having their carrier status determined (Figure 4-33). Also, Duchenne's or Becker's dystrophy often arises as a sporadic event in a family. Approximately one-third of the time, this is due to new mutations, but two-thirds of the time, the mother of a sporadically affected male is found to be a carrier.

We have already seen that CPK testing provides a crude carrier test but is subject to both false positive

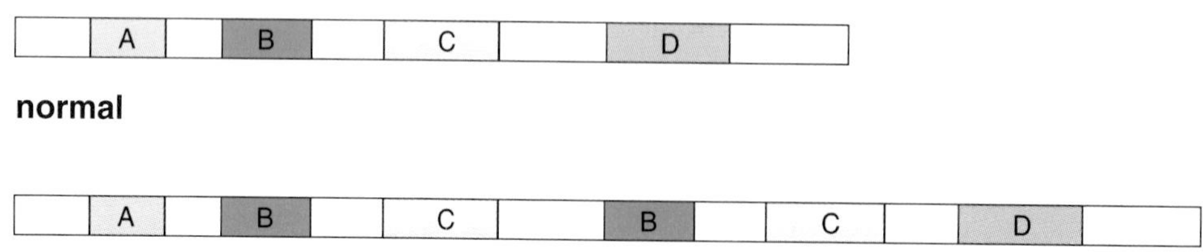

FIGURE 4-32 Diagram of duplication in dystrophin gene. The duplication juxtaposes exons C and B, which can result in an in-frame or an out-of-frame transcript.

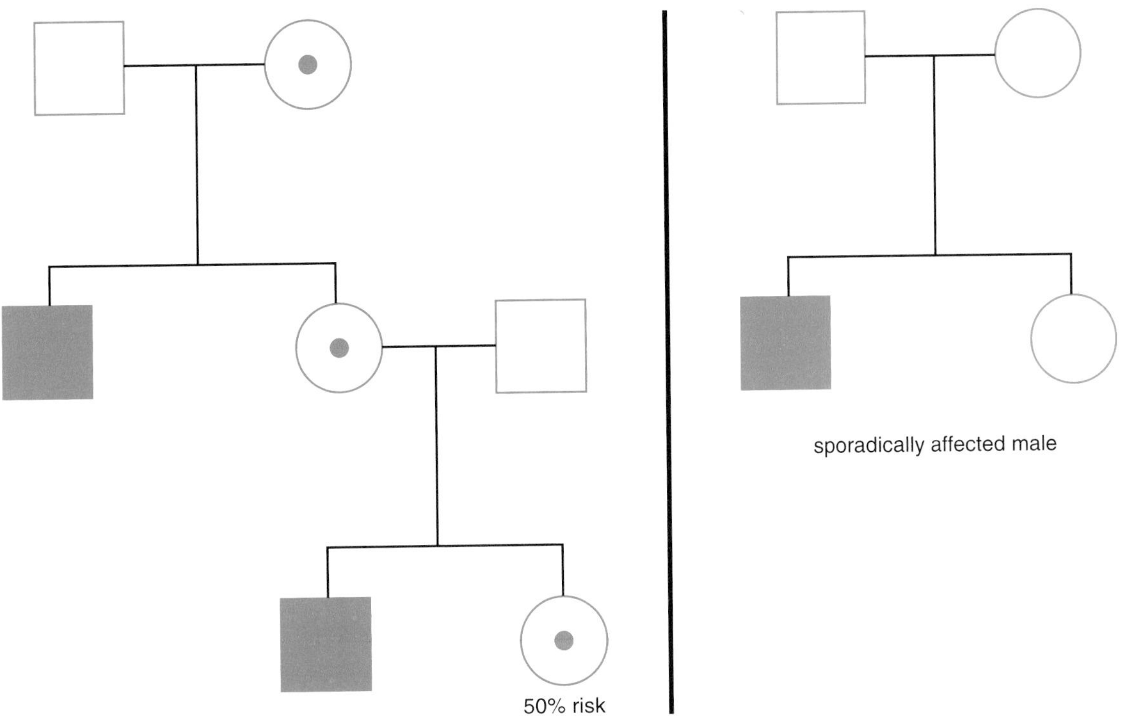

FIGURE 4-33 Two types of pedigrees encountered in families with Duchenne's or Becker's dystrophy. The family on the left includes two obligate carrier females and a woman at 50% risk based on family history. In the family on the right, there is one affected male. His mother has a 66% risk of being a carrier, and his sister, therefore, a 33% risk.

and false negative results. Linkage analysis sometimes helps to resolve carrier status but not if the proband is the only affected member of the family. For those families in which a deletion is found in the proband, direct analysis for carrier status should be possible.

Detection of heterozygous deletions is far more challenging than detection of hemizygous deletions. Quantitative analysis of band intensity on a Southern blot is technically difficult. Band intensity may vary from band to band on a blot, reflecting differences either in amount of DNA loaded on the electrophoresis gel or in efficiency of transfer of DNA fragments from gel to membrane. To carry out quantitative analysis, it is necessary to compare corresponding bands from a test subject and a normal control, using bands outside the deleted region to correct for differences in the amount of DNA present in the blot.

Quantitative analysis of PCR products also is difficult and prone to error. The quantity of PCR product doubles with each amplification cycle until enzyme or substrate is depleted. At this point, quantity of PCR products reaches a plateau (Figure 4-34). Plateau levels occur at product concentrations that are very near the point where product can be visualized with ethidium bromide. PCR products from a female who is heterozygous for a deletion may reach plateau one cycle later than products from one who is homozygous for the nondeleted allele, but if the reaction is sampled after plateau is reached, the deletion will be missed. Techniques have been devised to visualize PCR products at low numbers of amplification cycles, but generally these are difficult to perform. No approach has yet been devised that matches the simplicity and reliability of multiplex PCR detection of deletions in males.

Quantitative PCR is performed on a sample of DNA from Brenda. It is concluded that she probably is not a DMD carrier.

If carrier testing is difficult at the DNA level, what about protein testing? Immunofluorescent staining

Perspective

Living with a Child with Duchenne's Muscular Dystrophy

KATHY PICHEL

On March 9, 1981, we gave birth to a healthy baby boy. We were thrilled with our first addition to our family. Mark was a healthy toddler, except for numerous ear infections. When he was approximately 3½ years old, he became very ill, however, with vomiting and lethargy. He was brought to a hospital, where it was found that his liver enzymes and creatine phosphokinase levels (CPK) were slightly elevated. After he had recovered, the blood tests were repeated by his regular pediatrician and still were found to be elevated. Mark then was seen by a gastroenterologist, who asked about Mark's activity at home and whether he fell frequently or had trouble getting up once he fell. We considered him to be a normal 3-year-old. We next were sent to see a neurologist, who told us that Mark had some type of muscle disease, but he would need a biopsy to determine what kind.

The muscle biopsy showed that Mark had Duchenne's muscular dystrophy. We were told that Mark would be in a wheelchair by 8 to 9 years old and wouldn't live to be a teenager. Needless to say, we were devastated. I was pregnant at the time, and we were also told that the baby, if a boy, had a 50/50 chance of having the same disease.

When you are given such a grim diagnosis for your child, many things go through your mind. The most important one for us was that we were not going to let this destroy our family. Over the past 12 years, we have all been through quite a lot, good and bad, but through it all we've grown stronger and have tried to find the positive in any situation. The support of family and friends was very important to us, as was the support of our pediatrician. He called us the day we came home from the hospital, asked how we were doing, and told us to call him anytime, even if we just needed to talk. Even to this day—12 years later—when Mark sees him for his yearly checkup or if he's not feeling well, the pediatrician asks about Mark's physical and psychological well-being and also about the rest of the family.

Shortly after Matthew, my next son, was born, I returned to work as a nurse and had a long talk with a patient who had lost two sons with Duchenne's muscular dystrophy. She told me to remember that my child has a

of muscle can provide a means of carrier testing and dramatically demonstrates the principle of X chromosome inactivation. In a heterozygous female, approximately equal numbers of muscle nuclei should express either the normal or the mutated dystrophin allele in any muscle cell. When muscle cells are cut in cross-section and stained with dystrophin antibody, a mosaic pattern is seen of dystrophin-positive and -negative cells. This is diagnostic of carrier status. There are two problems with this form of testing, however. One is that it requires a muscle biopsy, which is an expensive, invasive procedure. The second is that, with time, the proportion of dystrophin-negative regions may decrease as defective muscle cells die and are replaced by those that, by chance, have a preponderance of normal-expressing nuclei. Therefore, false negative results can occur.

For females with sporadically affected sons, carrier testing, however it is done, can give deceptive results. Approximately 10% of such females have gone on to have a second affected son despite negative carrier testing (Figure 4-35). These women are presumed to represent mosaics, in whom either the mutation is confined to cells in the germ line or somatic cells with the mutation cannot be detected against the background of normal cells. The practitioner must bear this possibility of germ-line mosaicism in mind when counseling a family with one affected son.

muscle problem and not a mind problem. Many people had treated her boys as if they were mentally impaired, which was not the case at all. Speaking with this woman gave me a sense of strength that no one could feel unless they had been through a similar situation.

Over the years since Mark's disease was diagnosed, we have had three more children: a son, Matthew Charles, and two daughters, Kristine Marie and Courtney Rose. I had amniocentesis with each pregnancy, which showed that they did not have the gene change associated with muscular dystrophy. During our genetic counseling, we learned that Mark was probably a new mutation and that I probably was not a carrier.

When Mark was in first grade, he had surgery on both legs for muscle lengthening. He never complained through all of this; he just did what the physical therapists told him. It was difficult watching Mark fall and have trouble getting up, but we never let him give up trying until we knew he could no longer do something. I have to laugh now when I think of what people must have thought when he would fall and we would make jokes about it. Inside my heart was aching, but I knew that Mark had a long haul ahead of him and we couldn't overreact to things or he wouldn't have the strength to go through it all.

There was an incident at Mark's school once in which the school nurse insisted that I use the words *muscular dystrophy* with Mark. I tried to explain that a child in first grade has no concept of such terms but that he knew what he had because I had explained it in terms he could understand. At our next meeting, the school counselor was present because they felt that I was not accepting the diagnosis. Sometimes, it's so frustrating dealing with people's ignorance! My husband and I have always had Mark's best interest in mind and have never kept anything from him.

Mark's weakness progressed over the next few years, but his attitude always remained good. He has always been a happy child, but also a worrier at times. By fourth grade, he was in a wheelchair all the time. In the fifth grade, we ordered Mark's electric wheelchair but, because of insurance red tape and problems with the vendor, the chair did not come before he had outgrown the previous chair. One day, Mark and his brother were fooling around and Mark fell out of the chair and fractured both legs. He was in a cast for 4 weeks. I wondered how much a child could take, but Mark's attitude was remarkable.

The reason I feel it's important to write about these things is that, because of Mark's and our positive attitude, he has done very well. His upper body weakness has progressed, but that does not stop him. Regardless of what we were told about Mark's diagnosis, we truly believe that no one can predict what's in store for Mark or for us. Research has made great progress. We will never give up.

4.10 Pathogenesis and Treatment of Muscular Dystrophy

1993

James and Brenda have learned that Charles is now part of a study of the effectiveness of myoblast transfer as a treatment for DMD. They inquire about the status of this treatment and about other developments in the therapy for muscular dystrophy.

Prior to the discovery of dystrophin, there were essentially no clues to the pathogenetic mechanisms of muscle degeneration in Duchenne's or Becker's dystrophy. The major approach to management has consisted of supportive care: motorized wheelchairs, surgery for scoliosis, and cardiac and respiratory support. Affected muscle is infiltrated with inflammatory cells. This has led to the suggestion that inflammation may contribute to muscle degeneration, and to the trial use of corticosteroids, which inhibit the inflammatory response, as a means of treatment. Although the effects are subtle, some encouraging responses have been claimed, mainly the slowing of disease progression in some studies. Chronic corticosteroid use also presents long-term side effects, though, which must be weighed against any possible benefits.

The discovery of dystrophin has directed attention to the cell membrane as a site of pathology.

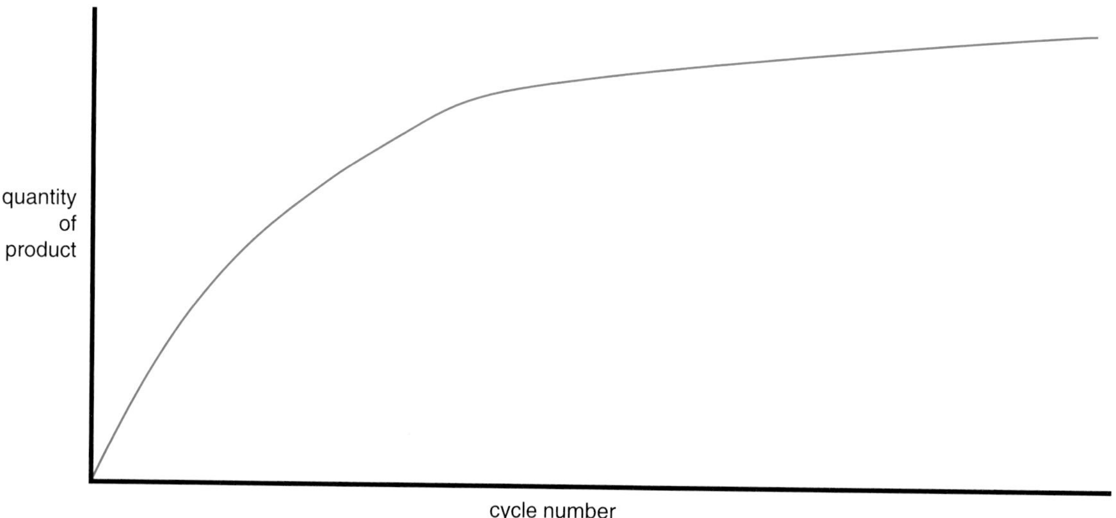

FIGURE 4-34 Graph of quantity of PCR product as a function of cycle number. Up to a point, there is an exponential increase in product quality. Eventually, however, a plateau is reached, when enzyme or substrate become depleted.

Muscle cells apparently can function in the absence of dystrophin, as males with absent dystrophin have normal strength at birth. Presumably, it is the survival of muscle cells after a period of activity that is impaired in the absence of dystrophin. The precise mechanism of muscle cell damage is unknown, however.

Since the discovery of dystrophin, efforts have been directed to developing a means of replacing the protein in defective muscle. One approach has been to inject normal myoblasts into dystrophic muscle, hoping that fusion of the injected myoblasts with existing muscle will at least partially restore dystrophin expression (Figure 4-36). There is a naturally occurring mouse model of DMD, and pilot experiments with these animals have produced encouraging results. Clinical trials have been done in humans with end-stage muscular dystrophy. It has been difficult to demonstrate any clinical benefit from this approach, and immune rejection of injected myoblasts seems to be a problem.

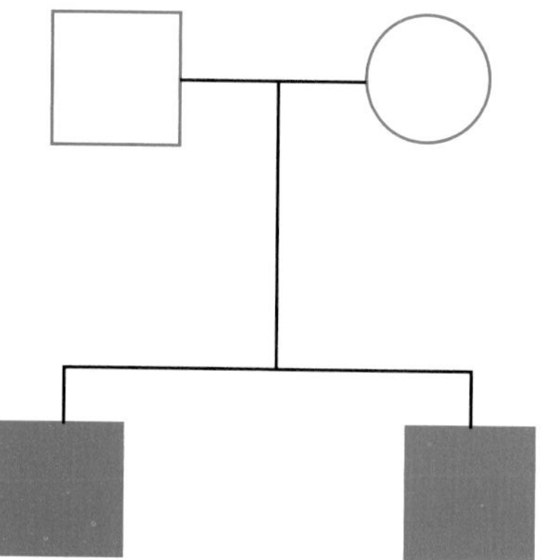

FIGURE 4-35 Family with two sons affected with muscular dystrophy due to deletion, in which the mother is not found to carry the deletion. This is attributed to germ-line mosaicism in the mother.

A more general treatment might be achieved if dystrophin expression could be restored by gene replacement. The large size of the dystrophin transcript makes this a challenging task, but some success has been achieved in pilot experiments. Eventually, there is hope that a vector may be produced that can infect muscle cells and restore dystrophin expression.

Dystrophin is one of a family of genes encoding actin-binding cytoskeletal proteins. Some of these are tissue-specific, but their mechanisms of regulation are unknown. It is possible that means will eventually be found to induce the expression of a dystrophin-related protein in dystrophic muscle, perhaps ameliorating the course of muscle cell degeneration.

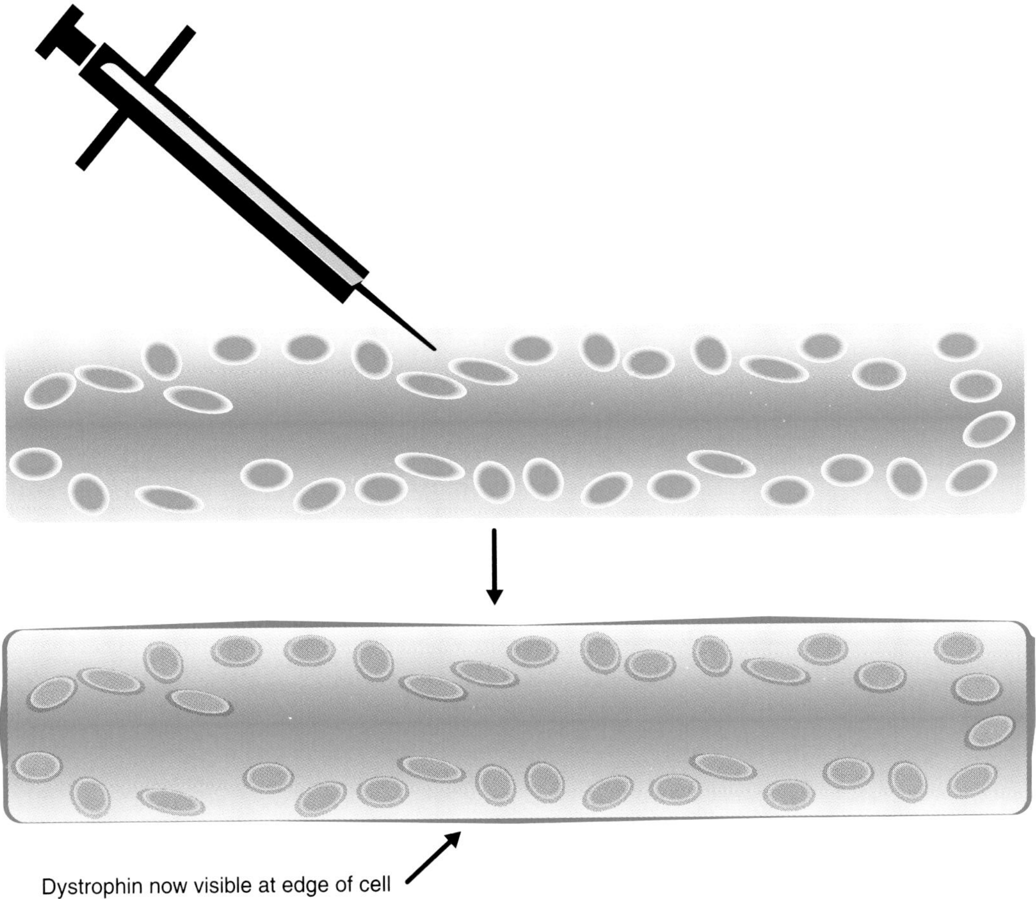

FIGURE 4-36 Myoblast transfer as approach to correction of Duchenne's muscular dystrophy. Myoblasts containing a normal dystrophin gene are injected into dystrophic muscle. These myoblasts fuse with the muscle fibers and express dystrophin. (Modified with permission from Partridge TA, Morgan JE, Coulton GR, Hoffman EP, Kunkel LM. Conversion of mdx myofibers from dystrophin-negative to -positive by injection of normal myoblasts. Nature 1989;337:176–179; and from Gussoni E, Pavlath GK, Lanctot AM, et al. Normal dystrophin transcripts detected in Duchenne muscular dystrophy patients after myoblast transplantation. Nature 1992;356:435–438.)

Although the phenotypes of Duchenne's and Becker's dystrophy are most prominently expressed in skeletal muscle, other organs may be affected as well. Cardiomyopathy is a common problem in affected individuals, though physical inactivity may delay the onset of symptoms. In rare cases, probably due to specific unique mutations, cardiomyopathy may be the sole phenotype. Another important affected tissue is the brain. Mild cognitive impairment is seen in approximately one-third of males with Duchenne's or Becker's dystrophy; there is a leftward shift of IQ values in affected individuals. Dystrophin is expressed in some neurons, although transcription in brain starts from a different promoter than that in muscle, and different patterns of alternative splicing occur, producing a slightly different form of protein (Figure 4-37). The function of this protein in brain cells and the pathogenesis of cognitive dysfunction is unknown. These will be important areas to understand as treatment is developed for the muscle disease. Formidable as it may be to correct gene expression in muscle, it is even more difficult to access brain cells.

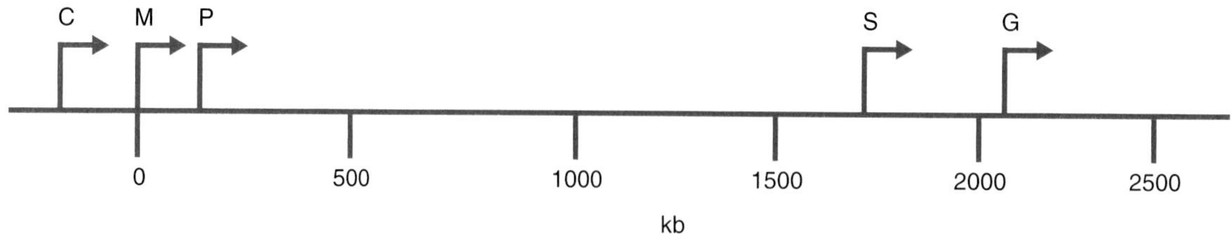

FIGURE 4-37 Sites of initiation of transcription of dystrophin. Near the 5′ end of the gene, different promoters initiate transcription in cortical neurons (*C*), muscle cells (*M*), and Purkinje cells (*P*). Nearer the 3′ end, there are two promoters for Schwann cells (*S*) and glial cells (*G*), which produce a shorter transcript. (Redrawn with permission from Ahn AH, Kunkel LM. The structural and functional diversity of dystrophin. Nat Genet 1993;3:283–291.)

Case Study

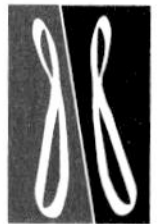

Part I

Carl and Sally are delighted when Sally gives birth to a baby girl, Julia. Julia weighs 7 pounds, 14 ounces, and is active and healthy. They take her home from the hospital after 24 hours. That night, however, they begin to be concerned when she displays some unusual twitching around the mouth and one foot. These episodes occur again the next day, and she also has some episodes of noisy breathing. The following day she will no longer breastfeed and has an episode during which she briefly turns blue. Julia is examined by her pediatrician, is found to be normal, and is sent home. That night, however, repeated episodes of cyanosis and twitching lead her parents to bring her to the hospital emergency room.

Julia is examined and noted to be afebrile. She has a faint red rash on her trunk, face, and extremities, which blanches with pressure. Her physical examination is otherwise unremarkable. Blood is drawn for studies, including electrolytes, glucose, calcium, magnesium, and culture. Urine also is taken for urinalysis and culture, and a lumbar puncture is done. Julia is started on antibiotics and is admitted to the hospital. Because there is concern that she might be having seizures, Julia also is treated with phenobarbital.

The clinical history suggests that Julia is having seizures. A seizure is an abnormal electrical discharge in the brain associated with symptoms related to the function of the part of the brain that is involved. In newborns, it can be difficult to relate seizure manifestations to one region of the brain. Often, seizure movements will be noticed in just one part of the body, even though abnormal electrical activity may be widespread. Moreover, seizures in the newborn can be subtle, so it is not unusual for them to go undiagnosed for a time.

Neonatal seizures can occur for many reasons. Complications of delivery, including central nervous system bleeds or asphyxia, usually cause seizures in the early hours of life and are associated with other problems such as abnormal muscle tone or poor feeding. Neonatal infection with bacteria or viruses must always be considered. Since the newborn has low resistance to overwhelming infection, it is common to treat babies who have seizures with antibiotics until blood, urine, and cerebrospinal fluid (CSF) cultures have been shown to be sterile. Metabolic disorders, including hypocalcemia, hypomagnesemia, electrolyte imbalance, hypoglycemia, or metabolic acidosis must also be ruled out. Finally, congenital malformations of the brain must be considered.

Part II

The cerebrospinal fluid examination is normal, as are electrolytes, calcium, magnesium, and the complete blood cell count. During the next 24 hours, Julia has numerous seizures. These have become more and more violent and dramatic. An electroencephalogram is done and reveals multiple seizure discharges coming from various areas of the brain. A computed tomography scan of the brain shows evidence of swelling but no other definite abnormalities (Figure 4-38). Julia continues on phenobarbital treatment, and

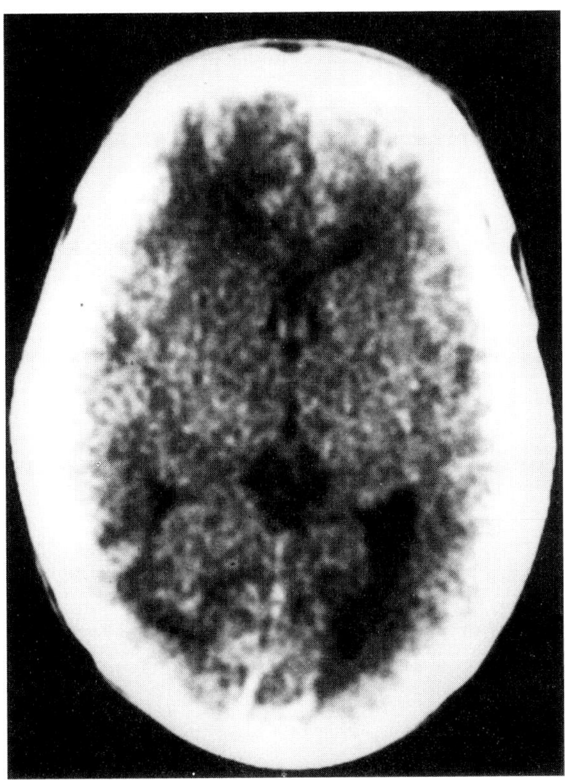

FIGURE 4-38 Computed tomography scan of brain. There is little demarcation of gray and white matter, and there is lucency (dark intensity on the scan) in the left frontal region and the right occipital region.

phenytoin has been added as well. All the cultures remain negative thus far, however. Julia is being treated with antibiotics as well as acyclovir, an antiviral medication, pending final culture results.

The electroencephalogram (EEG) shows that seizure discharges are occurring, but no one area of the brain seems to be responsible. The computed tomography (CT) scan is a method of imaging the brain in cross-section by computer analysis of x-ray absorption. The scan reveals evidence for generalized brain swelling but no other abnormalities. This might be seen in the setting of metabolic encephalopathy, but no metabolic disorder is demonstrated. It might also be seen in a child with widespread vascular occlusion, but it is not clear why this should be happening. At this point, then, the diagnosis is obscure, and treatment is limited to use of antiepileptic medication to treat the symptom of seizures.

Part III

Julia's condition remains tenuous for about a week. All her cultures have been negative, and antibiotics have been discontinued. Although her seizures continued for several days in spite of anticonvulsant medication, by the week's end they have stopped. She remains on phenobarbital. Julia is now somewhat more alert and is feeding by mouth again.

Her rash has also changed somewhat: In addition to the diffuse redness (erythema), now there are tiny blisterlike lesions on her face, trunk, and extremities (Figure 4-39). One of these is biopsied and sent for pathologic examination.

The seizures appear to settle down, and the acute phase of the illness seems to be passing. Still no diagnosis has been made. The only clue at this point is the continued presence of an unusual skin rash. The rash started as diffuse redness (erythema), which might occur in an allergic reaction or viral illness. Now it has evolved so that instead of being flat and red, there are blisterlike lesions. Similar lesions are seen commonly on the face of some children at a few days of age and are benign, referred to as *erythema toxicum*. This rash is different, however, in that it covers much of the body. A small piece of skin at the site of the rash is biopsied and sent for pathologic examination.

Part IV

The pathologist identifies the characteristic pattern of incontinentia pigmenti (Figure 4-40). This diagnosis confirms the clinical impression of the dermatologist who had been asked to see Julia. It also fits with Julia's clinical course. The diagnosis is explained to Carl and Sally.

The diagnosis of incontinentia pigmenti is arrived at by pathologic examination of the skin lesions. Incontinentia pigmenti is characterized by a skin rash that begins as diffuse erythema and then progresses first to vesicular lesions, next to wartlike (verrucous) lesions, and finally to regions of streaky hyperpigmentation. The name *incontinentia pigmenti* derives from the fact that, through the microscope, the characteristic finding is spillage of melanin from melanocytes in the extracellular space. Normally, melanosomes are transferred from melanocytes to keratinocytes, but here the melanocytes appear to be "incontinent" of melanin.

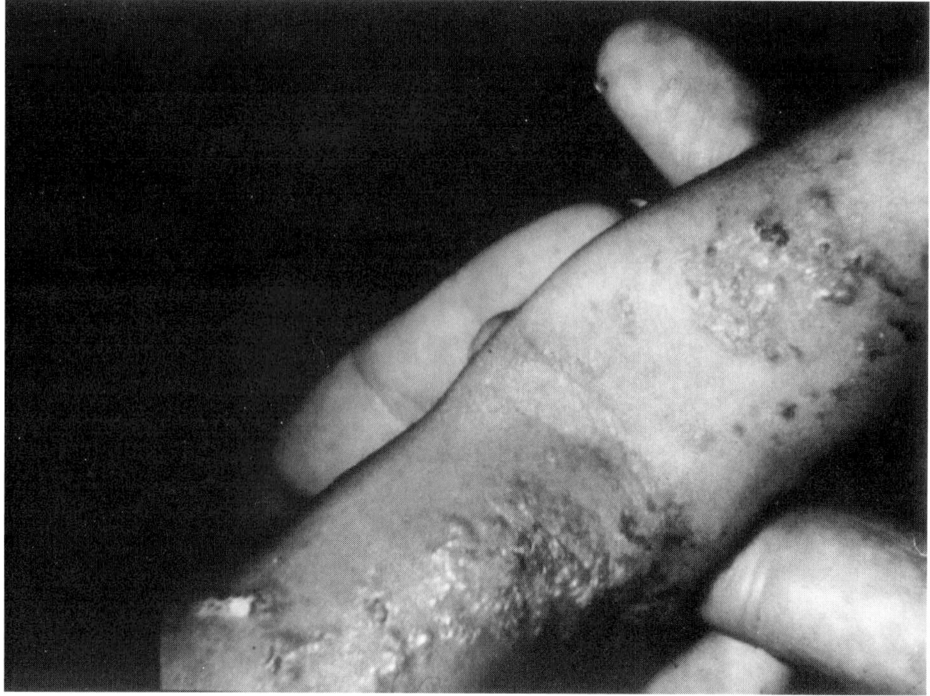

FIGURE 4-39 Vesicular rash on arm of infant. (Courtesy of Dr. Stephen Gellis, Children's Hospital, Boston.)

The skin involvement is not clinically dangerous, but neurologic complications can be more serious. When these occur—which is not always—they consist of seizures in the early days of life and may be accompanied by long-standing neurologic deficits, including abnormal muscle tone and motor and developmental delay. It is not clear what causes the neurologic problems, and it is not unusual for the seizures to be severe for a few days and then to settle down. Usually, the severe neurologic problems occur in the first days of life, if they occur at all. In many instances there are no associated neurologic problems.

Part V

The resident who is taking care of Julia asks for more information about family history. Sally is in good health and has had three other pregnancies. The first and third resulted in miscarriage in the first 4 months. The second resulted in the birth of a girl, Carol, who is now 3 years old and is healthy. Sally's mother is in good health and had three children, a boy and two girls, all of whom are well. Carl is in good health and has one brother. His father died of cancer, and his mother is alive and well. The dermatologist asks to examine Sally. He finds skin lesions indicative of incontinentia pigmenti on her legs and trunk. Sally has had these as long as she can remember but never thought much of them.

Incontinentia pigmenti follows a pattern of X-linked transmission but, unlike Duchenne's or Becker's dystrophy, is inherited as a dominant trait. It displays the usual rules of X-linkage, especially the lack of male-to-male transmission, but the phenotype is expressed in the heterozygous state. A true X-linked dominant trait would be expressed in both males and females. Affected males would transmit the trait to all of their daughters but to none of their sons, and affected females would pass the trait to half their offspring, male or female. Incontinentia pigmenti adds another twist: The disorder is lethal in males during in utero life. Males who inherit the gene die early in pregnancy and are spontaneously aborted. Carrier females therefore pass the trait to half their daughters, have fewer live-born sons

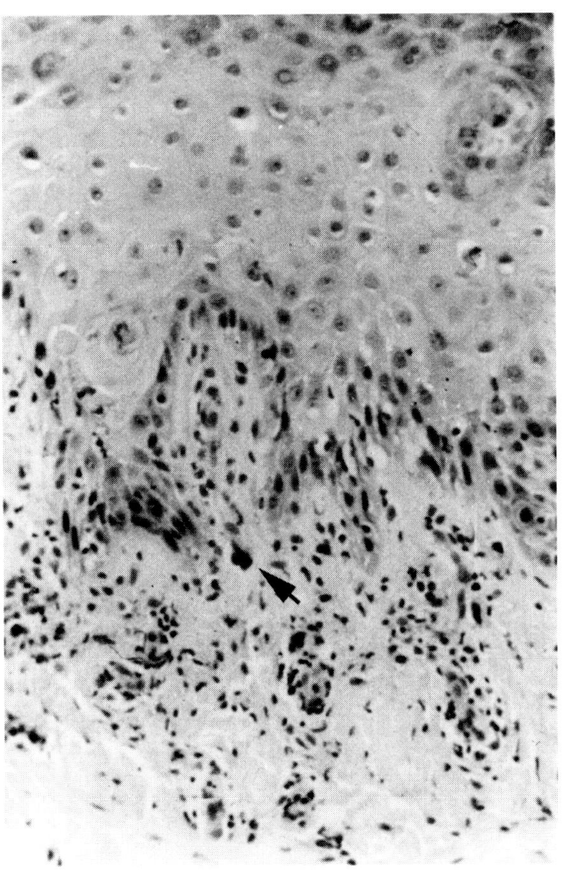

FIGURE 4-40 Photomicrograph of skin biopsy from child with incontinentia pigmenti, showing extracellular deposition of melanin (arrow) and abnormal keratinocyte morphology. (Courtesy of Dr. Stephen Gellis, Children's Hospital, Boston.)

than daughters, and have an increased frequency of miscarriage, the latter representing affected males.

Incontinentia pigmenti displays a wide range of expression, and new mutation is not uncommon. Often, there are no signs of the disorder in the mother of an affected female, and there is no family history of incontinentia pigmenti or recurrent miscarriage. Other times, subtle signs of the disorder may appear on the mother of a severely affected child and may also occur in other female relatives. Such women are at risk of having affected daughters, some of whom might be severely affected.

In Julia's family, Sally apparently carries the incontinentia pigmenti gene mutation (Figure 4-41). The two miscarriages may have been affected males. It is not known whether Sally represents a new mutation or whether her mother might have been affected.

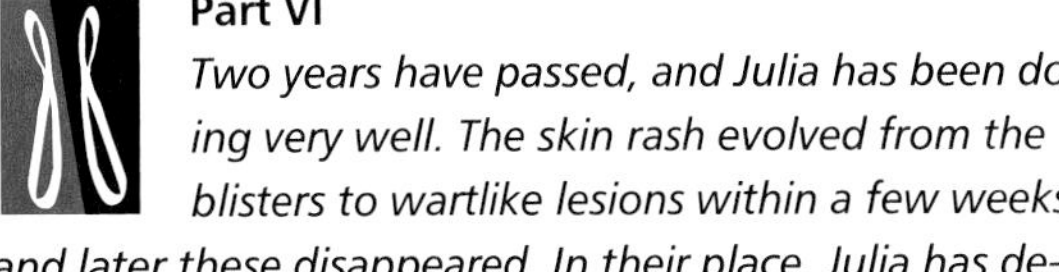

Part VI

Two years have passed, and Julia has been doing very well. The skin rash evolved from the blisters to wartlike lesions within a few weeks, and later these disappeared. In their place, Julia has developed streaky areas of increased pigmentation over parts of her body (Figure 4-42).

Julia has not had any additional seizures, and she is no longer taking medication. She is making developmental progress, although a bit more slowly than would be normal for her age. She began to walk at 19 months and is not yet saying words at 2 years. She has good understanding of language, though, and is a bright and alert child. She has been involved in physical therapy and has been noted to have increased muscle tone in her arms and legs. Julia's tooth eruption has been delayed, and some of her teeth are irregularly shaped. An eye checkup revealed that her vision is normal.

The skin lesions of incontinentia pigmenti continue to evolve from the wartlike lesions to streaky areas of hyperpigmentation. Eventually, these tend to fade, leaving streaks of atrophic, hairless skin. There may be patches of hair loss on the head, and fingernails and toenails can be ridged or pitted. In addition, the eruption of teeth may be delayed, and teeth may be misshapen. Some affected individuals have abnormal blood vessel proliferation in the retina, leading to retinal bleeding and blindness. Neurologic complications include seizures, motor dysfunction (including abnormal muscle tone or hemiparesis), and developmental delay. Neurologic symptoms generally are seen in those individuals who had neonatal seizures.

As has been noted already, there is a wide range of variability of expression of incontinentia pigmenti: Some individuals have neonatal seizures and other severe complications, whereas others have only a few skin lesions. It is not known why this variability occurs, but it does not seem to be explained by different mutations, as severely and mildly affected individuals can occur in the same family.

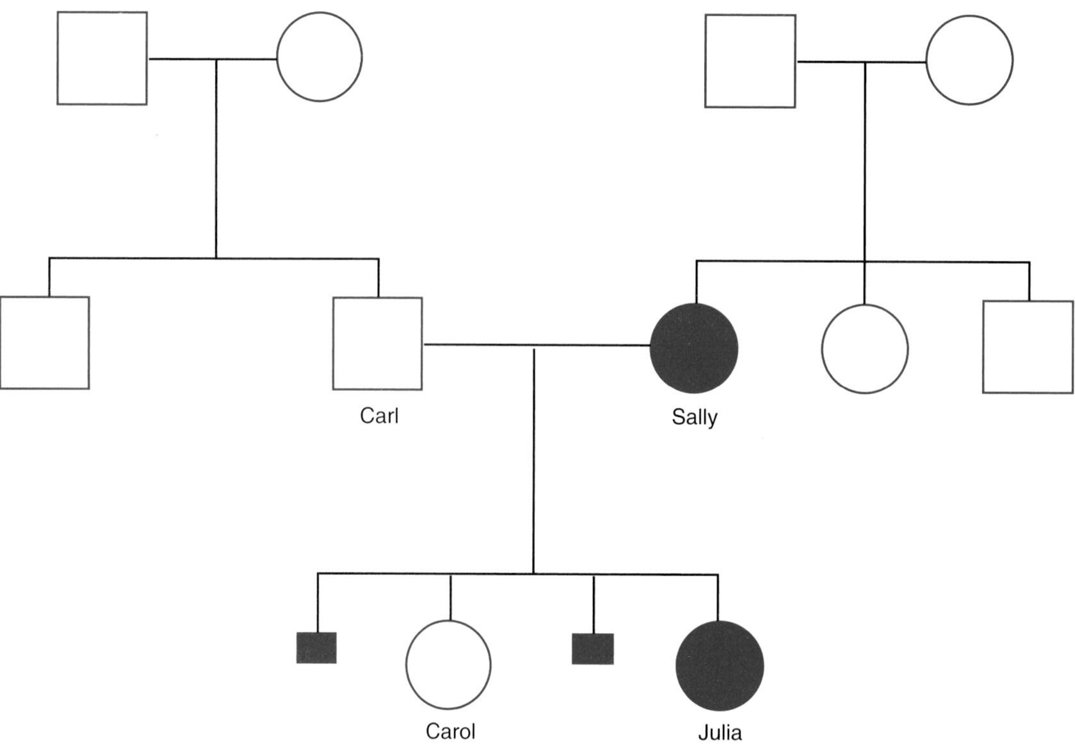

FIGURE 4-41 Pedigree of Sally and Carl's family.

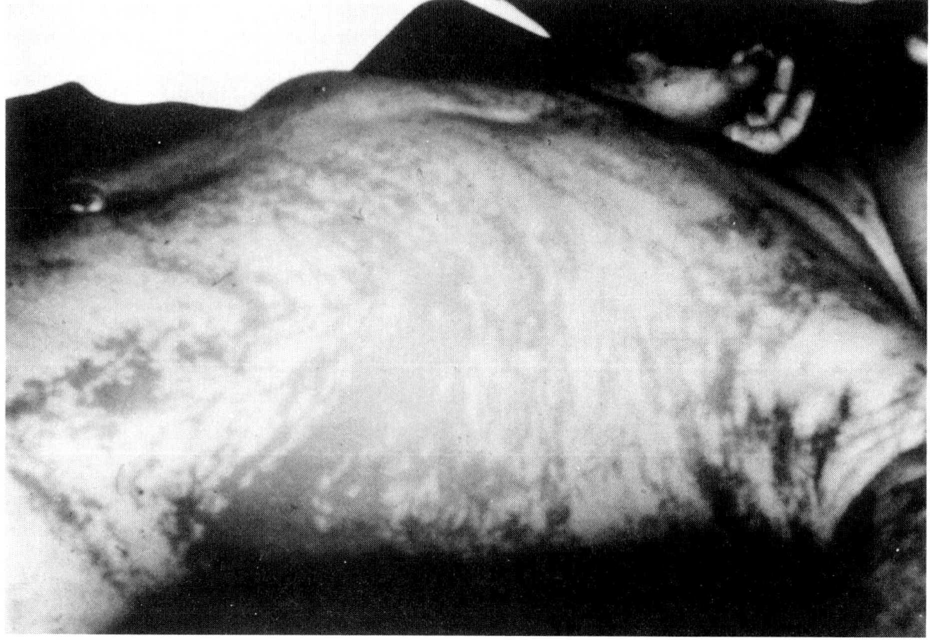

FIGURE 4-42 Streaky hyperpigmentation in child with incontinentia pigmenti. (Courtesy of Dr. Stephen Gellis, Children's Hospital, Boston.)

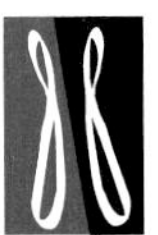

Part VII

Sally is asked to donate a blood sample for a research study on incontinentia pigmenti. DNA is isolated from the lymphocytes and digested with two restriction enzymes, BamHI and HpaII. HpaII cuts only nonmethylated DNA. A Southern blot is made, and the blot is hybridized with a DNA probe for the HPRT gene on the X chromosome. The data, depicted in Figure 4-43, are interpreted as showing evidence of non-random X chromosome inactivation.

Attempts to understand the genetic basis for incontinentia pigmenti have focused on identification of the gene locus or loci. There is evidence for at least two gene loci on the X chromosome that may account for incontinentia pigmenti in different families. Several individuals with incontinentia pigmenti have been described who have had X chromosome rearrangements involving Xp11, a site on the short arm just above the centromere. This has suggested that the responsible gene might reside in that region. In contrast, however, genetic linkage studies indicate that there is an incontinentia pigmenti locus near the tip of the long arm of the X chromosome at Xq28 (see Chapter 5 for chromosome nomenclature).

One possible explanation for these findings is that there are two incontinentia pigmenti loci on X, one at Xp11 and one at Xq28. Alternatively, it has been suggested that the Xp11 chromosome changes might somehow upset regulation of the incontinentia pigmenti locus despite its location at a distant site. Resolution of this issue will likely require identification of the gene or genes.

Another approach to genetic study has focused on a role for X chromosome inactivation. Because the incontinentia pigmenti phenotype is lethal in hemizygous males, it has been hypothesized that there might be selection against cells expressing the mutant gene in heterozygous females. This would predict nonrandom X chromosome inactivation, occurring because those cells in a heterozygous female that express the mutant gene might preferentially die. Moreover, differences in the proportion of cells

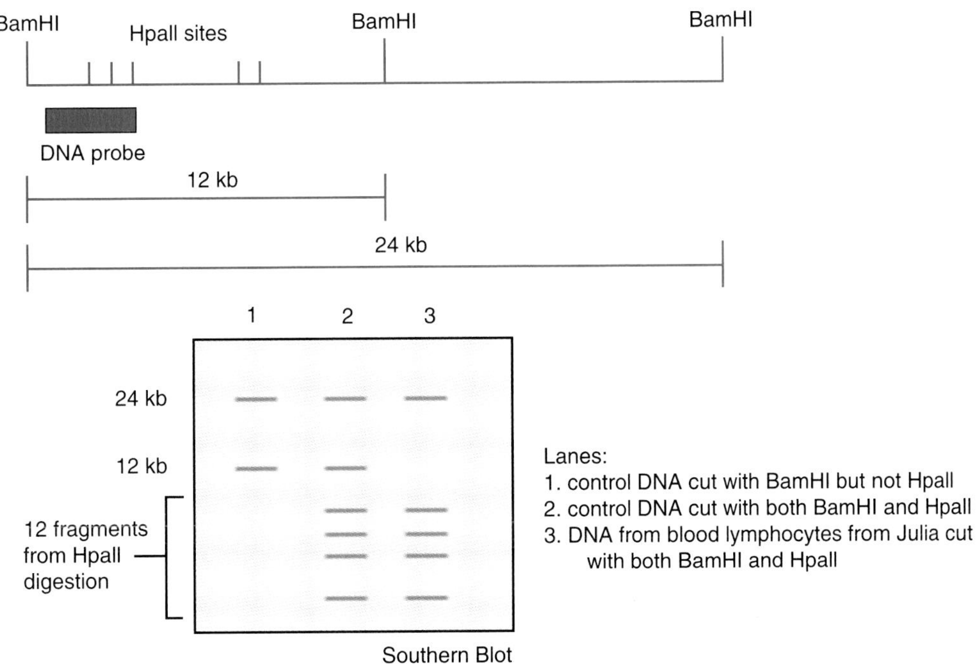

FIGURE 4-43 (Top) Map of region of HPRT gene, showing BamHI and HpaII cutting sites. The middle BamHI site is polymorphic, so the DNA probe hybridizes with either a 12- or a 24-Kb fragment in BamHI-cut DNA. (Bottom) Southern blot, with DNA cut with BamHI alone (lane 1), and with both BamHI and HpaII (lanes 2 and 3). DNA in lane 2 is from a normal control, and that in lane 3 is from a woman with incontinentia pigmenti.

with activity of the normal or mutant X might explain some of the variability of the incontinentia pigmenti phenotype.

Investigators have searched for nonrandom X chromosome inactivation by using RFLPs, taking advantage of methylation-sensitive restriction enzymes. Recall that many genes on the inactivated X chromosome are hypermethylated. Some restriction endonucleases, such as HpaII, are unable to cut methylated DNA. The X-linked gene HPRT encodes the enzyme hypoxanthine phosphoribosyl transferase and is heavily methylated on the inactive X chromosome. There is a BamHI RFLP in the HPRT gene, so that either a 12-Kb or a 24-Kb fragment is seen when a Southern blot of BamHI-cut DNA is hybridized with an HPRT probe. There are also a number of HpaII sites in the gene, which, if unmethylated can be cut with HpaII to result in a number of smaller fragments.

In a female who is heterozygous for the BamHI RFLP, two bands will be seen, a 12-Kb and a 24-Kb, if DNA is not cut with HpaII. If the DNA is cut with BamHI and HpaII, the same two bands will occur but, in addition, several smaller bands will be seen, representing the results of HpaII digestion. Approximately half the time, the 24-Kb BamHI allele will be the active (nonmethylated one) and will be cut with HpaII. The other half of the time, the 12-Kb BamHI allele will be nonmethylated and cut with HpaII. Overall, one expects to see both the 12-Kb and 24-Kb alleles (from cells in which one or the other is methylated and hence not cut with HpaII), as well as smaller fragments from whichever allele was cut with HpaII in a given cell. In a clone of cells, however, one of the X chromosomes will always be active and only one of the two BamHI alleles will be consistently nonmethylated and will always cut with HpaII. This BamHI allele will be cut into smaller fragments by HpaII and so will not appear on the Southern blot. Nonrandom X-inactivation can be inferred if one of the two BamHI alleles tends to be fainter than the other on the Southern blot.

When this assay is used to test females with incontinentia pigmenti, evidence for nonrandom X-inactivation is seen frequently, though not always. Usually peripheral blood lymphocytes or cultured skin fibroblasts are tested, and these tissues may not be representative of other cells in the body. Also, one might expect different levels of nonrandom inactivation to occur in more mildly versus more severely affected individuals. Apparently, though, there is selection against at least some cells that express the X chromosome with the incontinentia pigmenti mutation. It is possible, incidentally, that individuals who have X chromosome rearrangements at Xp11 have nonrandom X-inactivation because of the chromosome rearrangement, leading to expression of incontinentia pigmenti alleles on the other X chromosome, thereby causing a clinically evident incontinentia pigmenti phenotype that would otherwise have gone undetected.

REVIEW QUESTIONS

1. A child is born with a syndrome of multiple congenital anomalies. No etiologic diagnosis is made. How could the pedigree be compatible with:
   a. autosomal dominant transmission?
   b. autosomal recessive transmission?
   c. X-linked transmission?
2. A man with hemophilia A (an X-linked recessive trait) has a daughter who is severely affected. How might this be explained?
3. A genetic linkage study is performed in a family for diagnosis of Duchenne's muscular dystrophy. The polymorphism detects two alleles, *A* and *B*. Is the consultand likely to be a carrier?

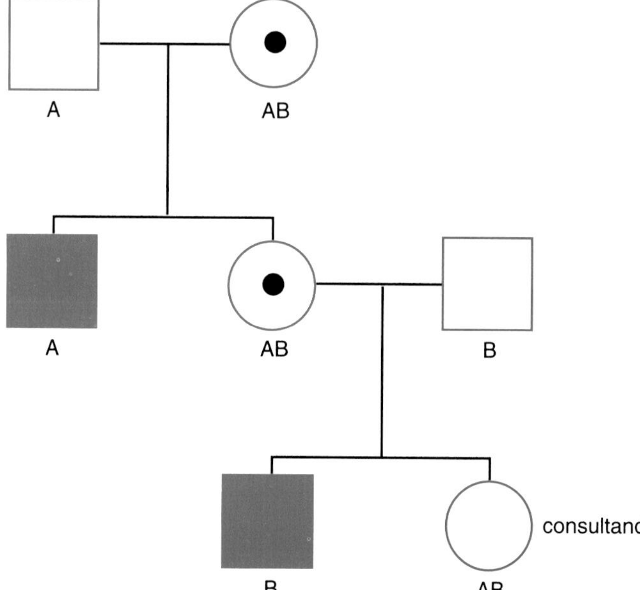

4. When multiplex PCR is used to diagnose Duchenne's muscular dystrophy, a confirmatory Southern blot is done if only a single exon is found to be missing, whereas no such study is done if two or more exons are found to be deleted. Why is this so?

FURTHER READING

4.5 Cloning the Gene for DMD

Kunkel LM, et al. Analysis of deletions in DNA from patients with Becker's and Duchenne's muscular dystrophy. Nature 1986;322:73–77.

Monaco A, Neve RL, Colletti-Feener C, et al. Isolation of candidate cDNAs for portions of the Duchenne's muscular dystrophy gene. Nature 1986;323:646–650.

Monaco AP, Kunkel LM. A giant locus for the Duchenne's and Becker's muscular dystrophy gene. Trends Genet 1987;3:33–37.

4.10 Pathogenesis and Treatment of Muscular Dystrophy

Cox GA, Cole NM, Matsumura K, et al. Overexpression of dystrophin in transgenic mdx mice eliminates dystrophic symptoms without toxicity. Nature 1993;364:725–729.

CASE STUDY

Carney RG. Incontinentia pigmenti. A worldwide statistical analysis. Arch Dermatol 1976;112:535–542.

Crolla JA, et al. Incontinentia pigmenti and X-autosome translocations. Hum Genet 1989;81:269–272.

Harris A, et al. X-inactivation as a mechanism of selection against lethal alleles: further investigation of incontinentia pigmenti and X linked lymphoproliferative disease. J Med Genet 1992;29:608–614.

Landy SJ, Donnai D. Incontinentia pigmenti (Bloch-Sulzberger syndrome). J Med Genet 1993;30:53–59.

Migeon BR, et al. Selection against lethal alleles in females heterozygous for incontinentia pigmenti. Am J Hum Genet 1989;44:100–106.

Sefiani A, et al. Linkage studies do not confirm the cytogenetic location of incontinentia pigmenti on Xp11. Hum Genet 1988;80:282–286.

Sefiani A, et al. The gene for incontinentia pigmenti is assigned to Xq28. Genomics 1989;4:427–429.

5 Chromosomes and Chromosomal Abnormalities

OVERVIEW

Disorders considered thus far, although differing in modes of transmission, have all resulted from changes in single genes. We now turn our attention to the consequences of gain or loss of large groups of genes due to chromosomal changes. Although the genetic basis for these disorders can be complex, involving the interaction of tens or hundreds of genes, from the clinical point of view these are among the most easily recognized genetic conditions. Moreover, collectively, chromosomal disorders are relatively common—affecting perhaps 1 in 200 births. This has led to efforts to carry out population screening to identify pregnancies at risk and to offer prenatal diagnosis.

The spectrum of chromosomal anomalies includes one of the most widely recognized genetic conditions—Down syndrome—as well as exceedingly rare and obscure patterns of congenital anomalies resulting from chromosomal rearrangements that may be unique to particular families. We will begin this chapter with discussion of a well-defined, albeit rare, phenotype, that of Wolf-Hirschhorn syndrome. This results from deletion of material from the short arm of chromosome 4, which was recognized in the early days of clinical cytogenetics. We will see how methods for chromosomal analysis were developed and led to the recognition of major trisomy and monosomy syndromes, and how techniques gradually were refined to allow detection of increasingly subtle chromosomal abnormalities.

Aside from changes of chromosome number and deletions of blocks of genetic material, chromosomal abnormalities include structural rearrangements such as rings, inversions, and translocations. We will explore the consequences of such rearrangements, in terms of both clinical phenotypes and effects on chromosome segregation in meiosis. Approaches to prenatal diagnosis and population screening will also be considered. We will see how some individuals may display mosaicism (i.e., contain a mixture of chromosomally normal and abnormal cells) and how this affects phenotype.

Recent advances in molecular genetics have vastly increased the resolution of cytogenetic analysis, bridging the gap between defects in whole chromosomes and changes in individual genes. We will see how the technique of fluorescence in situ hybridization is providing both new diagnostic tools and new insights into molecular pathogenesis. Finally, new insights into pathogenetic mechanisms will be noted, particularly the phenomenon of genomic imprinting. The Perspective discusses the impact on a family of having a child with a chromosomal abnormality. The chapter closes with the story of a developmentally impaired boy with a chromosomal abnormality resulting from a triplet repeat mutation in a single gene.

5.1 A Newborn with Multiple Congenital Anomalies

Karen was born after a full-term pregnancy. There were no medical problems during the pregnancy, but there was concern toward the end of term because her mother, Deborah, was not gaining weight as fast as expected. No prenatal testing had been done, as Deborah was 27 years old and there is no known family history of genetic problems. An ultrasound examination done in the eighth month, though, showed the fetus to be small for gestational age. At birth, Karen weighs 4 pounds, 10 ounces. She is noted to have a number of unusual features (Figure 5-1), including widely spaced eyes, prominent nose, large ears, and bilateral iris colobomas. She is brought to the special care nursery where, on day 2 of life, she is found to have a heart murmur due to a ventricular septal defect. Karen is examined by a geneticist, who suggests that she might have Wolf-Hirschhorn syndrome.

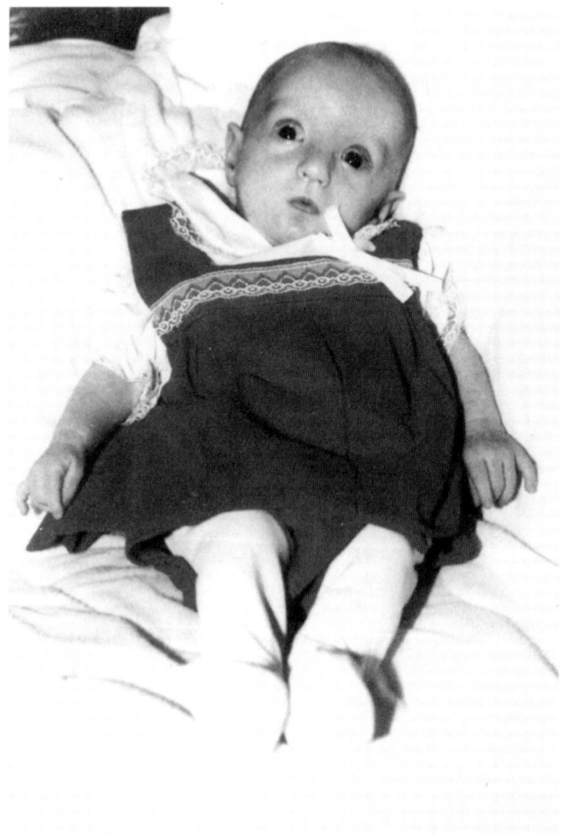

FIGURE 5-1 Child with Wolf-Hirschhorn syndrome.

Approximately 3% of all babies are born with a congenital anomaly. In most cases, these anomalies involve a single part of the body—for example, cleft lip or a heart defect. Usually the problems are compatible with survival and can be surgically repaired. The body is formed normally beyond the one area that is malformed.

There are some infants, however, who have multiple congenital anomalies, indicative of a more generalized disruption of embryonic development. Some of the malformations may be obvious (e.g., facial or limb defects). Internal organs such as the brain, heart, and kidneys often are involved as well. Surgical treatment may be possible for specific malformations, but defects of the central nervous system, which lead to neurologic problems and developmental impairment, cannot be corrected.

Multiple congenital anomalies can occur due to many causes, such as infections acquired in utero, exposure to toxic agents (teratogens), and genetic factors. Genetic problems include single gene defects, chromosomal abnormalities, and multifactorial disorders. The latter involve effects of multiple genes or environmental influences and are the subject of the next chapter.

Definition of the etiology of multiple congenital anomalies can be challenging but is important both for care of the infant and for counseling of the family. Recognition of a pattern of anomalies can lead to the diagnosis of a syndrome—a constellation of distinctive features that is fairly consistent from individual to individual. Knowledge of features associated with a syndrome can alert the clinician to examine the child for internal anomalies that are not otherwise apparent. Some congenital infections can be treated with antibiotics to prevent further damage. The parents can be informed about future prospects for their child, both medically and developmentally. Finally, if a genetic disorder is identified, the family can be counseled about risks of recurrence.

Wolf-Hirschhorn syndrome consists of a set of multiple anomalies with a very distinctive facial appearance. The eyes are set widely apart, and the nose is prominent. A coloboma is a defect in the formation of the iris or retina. There is often a cleft lip and palate. Cardiac defects are common and may cause heart failure in the early days of life. Usually, the brain is severely malformed, resulting in profound developmental impairment. Wolf-Hirschhorn syndrome is due to a chromosomal anomaly, specifically deletion of material from the short arm of chromosome 4.

5.2 Chromosomal Analysis

Because Wolf-Hirschhorn syndrome is suspected, blood is sent to the laboratory for chromosomal analysis. Deborah and her husband Steven are told of the possible diagnosis. They are very anxious about Karen's future. They understand that the chromosomal analysis may take 2 weeks to be completed. Clinically, Karen is doing fairly well; her heart disease does not require treatment. She is being kept in the hospital mainly for nutritional support, until she gains some weight.

The ability to examine the chromosomes for structural or numeric abnormalities sparked a revolution in medical genetics, beginning in the late 1950s. Prior to 1956, the normal human chromosome number was believed to be 48, a misconception attributable to great difficulty visualizing chromosomes. Each chromosome contains thousands of genes and millions of base pairs of DNA. During interphase, the genetic material is unraveled so that individual chromosomes cannot be easily resolved. Chromosomes condense into more compact structures during cell division. Condensation begins in prophase when the nuclear membrane disperses. Chromosomes are most easily visualized at metaphase, when they are lined at the center of the cell (Figure 5-2).

Even at metaphase, though, chromosomes may overlap one another too much to permit detailed study. The breakthrough that enabled chromosomal analysis was the discovery, independently in 1952 by Arthur Hughes and by T. C. Hsu, that cultured cells exposed to hypotonic saline swell, allowing clear

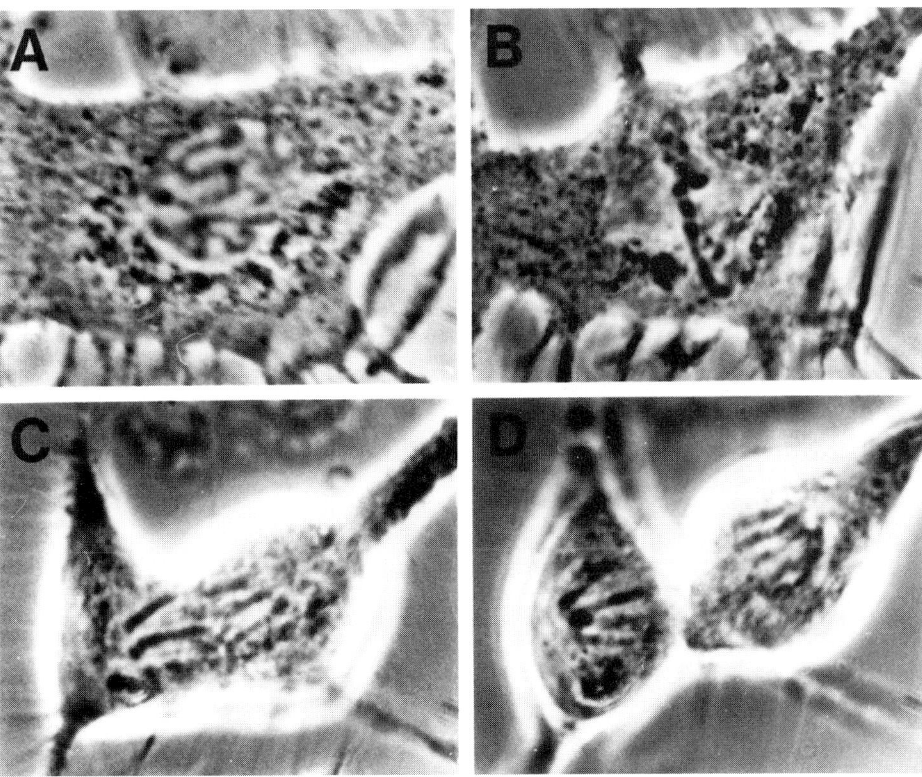

FIGURE 5-2 Stages of mitosis: (**A**) prophase; (**B**) metaphase; (**C**) anaphase; (**D**) telophase.

separation of metaphase chromosomes. It is ironic, however, that Hsu, after demonstrating the ability to study individual chromosomes clearly, did not determine the correct human chromosome number. This discovery fell to Tjio and Levan who, in 1956, determined that the correct human chromosome number was 46.

Development of a practical approach for human chromosomal analysis led to a frenzy of study of individuals with congenital anomalies and delineation of the chromosomal basis of several syndromes. The first was Down syndrome, a relatively common disorder characterized by a distinctive facial appearance (Figure 5-3), decreased muscle tone, developmental impairment and, sometimes, congenital malformation of the heart or other organs. In 1932, a Dutch ophthalmologist, P. J. Waardenburg, took note of the relative consistency of the phenotype of Down syndrome and suggested that it might be due to an abnormality of chromosome number. In 1959, Lejeune et al. found that individuals with Down syndrome have 47 chromosomes, including an extra copy of what appeared to be the smallest chromosome, designated *21*.

Two other chromosomal syndromes were identified in quick succession, both involving abnormalities of sex chromosomes. Females have two X chromosomes and males an X and a Y. Females with short stature, lack of secondary sexual development, and infertility—referred to as Turner's syndrome—were found to have 45 chromosomes, including only a single sex chromosome, an X. Males with Klinefelter's syndrome—male external genitals but slight breast development, tall stature, and infertility—were found to have 47 chromosomes with an XXY karyotype.

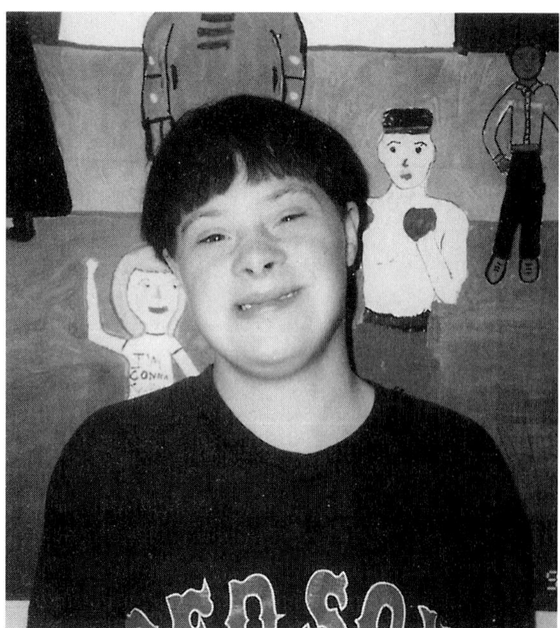

FIGURE 5-3 Child with Down syndrome.

The list of disorders associated with **aneuploidy** (a complete diploid set of chromosomes with one or more extra or missing chromosomes) compatible with live birth was rounded out with the description of trisomies 13 and 18 (Table 5-1), both of which are associated with severe congenital anomalies and cognitive impairment. Affected individuals rarely survive the neonatal period.

By the early 1960s, cytogenetic analysis had been incorporated into routine medical care, and cytogenetics laboratories were established in many hospitals. At first, chromosome studies were done on cultured skin fibroblasts. In 1960, a technique was introduced that made chromosomal analysis simple and rapid. A substance called *phytohemagglutinin*, derived from kidney beans, was found to stimulate the division of peripheral blood T lymphocytes. A few drops of blood could be inoculated into sterile culture medium containing this substance and, over the course of approximately 72 hours, the lymphocytes would divide several times. Addition of a substance such as colchicine inhibits the mitotic spindle, accumulating cells at metaphase. These cells are then swollen with hypotonic solution, fixed onto glass slides, stained, and viewed with the microscope (Figure 5-4).

In 1965, Wolf-Hirschhorn syndrome was identified as being due to a chromosomal abnormality. It was found to be associated with a deletion of material from the short arm of one of the larger chromosomes, eventually designated as *4*. Another chromosome deletion syndrome involving the short arm of chromosome 5 was found to be associated with multiple congenital anomalies, central nervous system impairment, and an abnormal cry likened to the mewing of a cat. It therefore was named *cri du chat syndrome* by its French discoverers.

Chromosome nomenclature was first standardized in 1960 at a conference in Denver. Metaphase chromosomes contain two strands of chromatin, **chromatids**, having replicated their genetic material at interphase (Figure 5-5). The chromatids remain

TABLE 5-1 Major chromosome aneuploidy syndromes compatible with live birth

Syndrome	Chromosomal Abnormality	Major Features
Patau's syndrome	Trisomy 13	Cleft lip and palate, severe central nervous system anomaly, polydactyly
Edwards' syndrome	Trisomy 18	Low birth weight, central nervous system anomalies, heart defects
Down syndrome	Trisomy 21	Hypotonia, characteristic facial features, developmental delay
Turner's syndrome	Monosomy X	Short stature, amenorrhea, lack of secondary sexual development
Klinefelter's syndrome	XXY	Small testes, infertility, tall stature, learning problems
Triple-X	XXX	Learning disabilities, no major physical anomalies
XYY	XYY	Learning and behavioral problems in some individuals

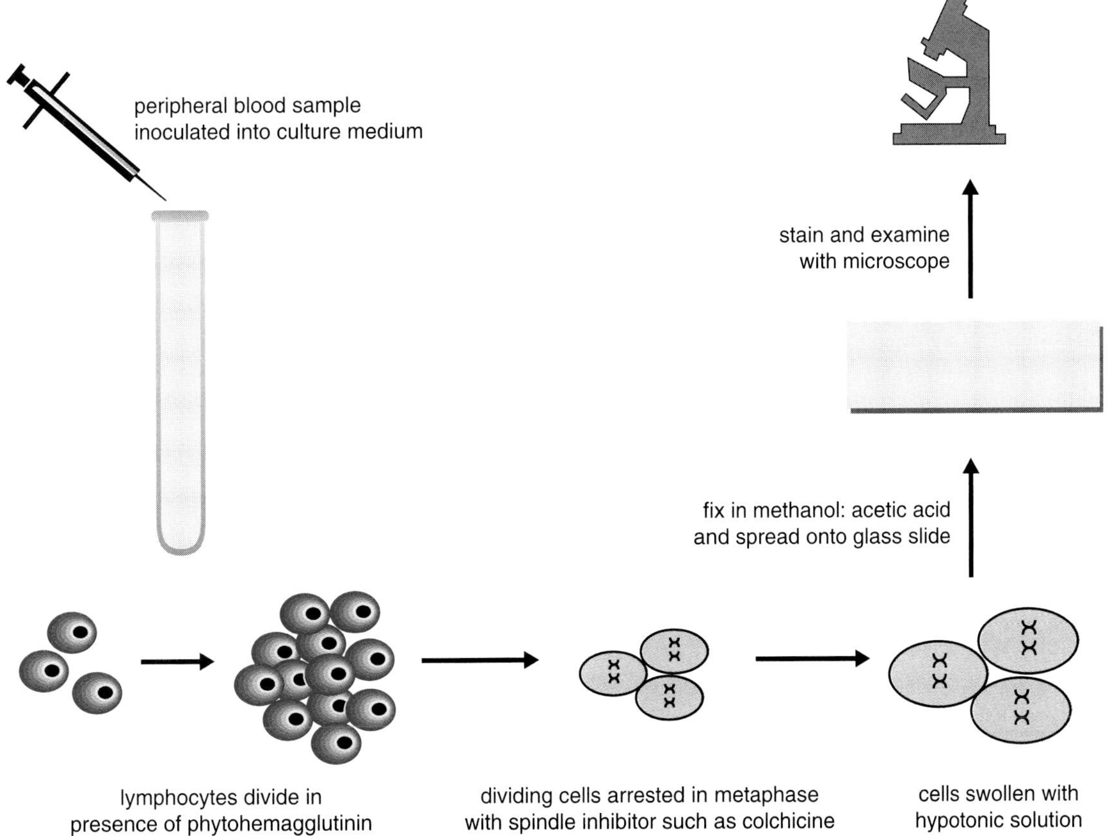

FIGURE 5-4 Steps in chromosomal analysis from peripheral blood. Blood is collected and inoculated into culture medium. Phytohemagglutinin in the medium stimulates T lymphocytes to divide. After 72 hours in culture, dividing cells are collected using a drug such as colchicine, which inhibits the mitotic spindle. The cells are swollen by treatment with hypotonic saline, fixed in a mixture of methanol–acetic acid, collected by centrifugation, and spread onto a glass slide. The slide then is stained, and the chromosomes are examined with a microscope.

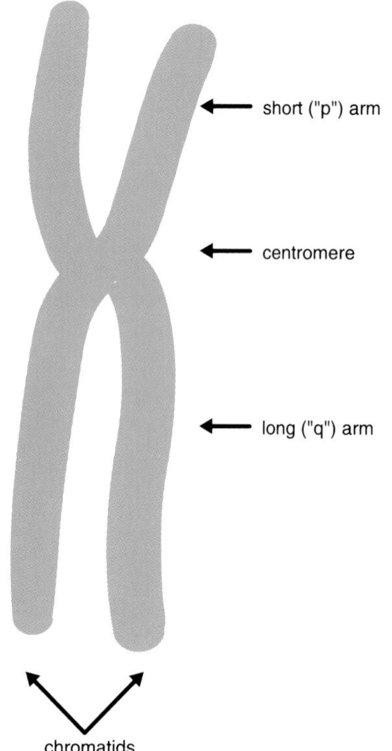

FIGURE 5-5 Structural features of a chromosome, indicating long arm, short arm, centromere, and chromatids.

attached at the **centromere**, which divides most chromosomes into a short arm and a long arm. The short arm is abbreviated *p* (for *petite*, reflecting the influence of French investigators who were among the pioneers in human cytogenetics), and the long arm is abbreviated *q* (because it follows *p*).

When chromosomes were first studied, it was difficult to distinguish some chromosomes from one another (Figure 5-6). Individual chromosomes were numbered from largest to smallest. The three largest were designated *group A*. The first and third have their centromeres near the center—referred to as **metacentric**. The second has a centromere slightly displaced from center and is called **submetacentric**. *Group B* consists of two large chromosomes, each with its centromere located toward the end of the chromosome, also designated as submetacentric. *Group C* includes seven pairs plus the X chromosome. Most members of this group are submetacentric. The three pairs of *group D* chromosomes have centromeres near the end of the chromosome (**acrocentric**). The small amount of short-arm material includes DNA that encodes ribosomal RNA. *Group E* includes three pairs of chromosomes, one metacentric and two submetacentric, and *group F* two pairs of small metacentrics. *Group G* contains two very small acrocentrics, also with ribosomal DNA comprising substantial portions of the short arms; one of these is the extra chromosome of Down syndrome. The Y is a small submetacentric.

5.3 Subtle Chromosomal Abnormalities

Two weeks after Karen's blood sample is drawn, the laboratory reports that she does have a deletion involving material on the short arm of chromosome 4 (Figure 5-7), confirming the clinical diagnosis of Wolf-Hirschhorn syndrome. Deborah and Steven discuss the finding with Karen's pediatrician and with the geneticist who was asked to see her. They are told that Karen has no life-threatening problems but that she will likely have significant developmental impairment. They are not really surprised at this by now, having had several weeks to adjust to Karen's condition. After nearly 3 weeks in the hospital, Karen is now gaining weight, and she is ready to be taken home.

Chromosome aneuploidy syndromes compatible with live birth all were described by the early 1960s, along with several that were associated with loss or gain of smaller amounts of chromosome material. Through most of the 1960s, though, the level of precision of chromosomal analysis was fairly crude. It was even difficult to identify unequivocally all the chromosomes. The group D chromosomes all look alike, for example, as do all the members of groups B, F, and G. Also, loss or gain of a small amount of material from within a chromosome arm would go unnoticed.

In 1968, a second technological revolution—namely, specific staining techniques—advanced the state of clinical cytogenetics. These techniques revealed banding patterns that allowed study of the fine structure of chromosomes. The first technique, which involved staining with the fluorescent dye quinacrine, was introduced by Caspersson et al., who hypothesized that variations of base sequence would exist along the lengths of chromosomes and could be revealed by differential binding of the

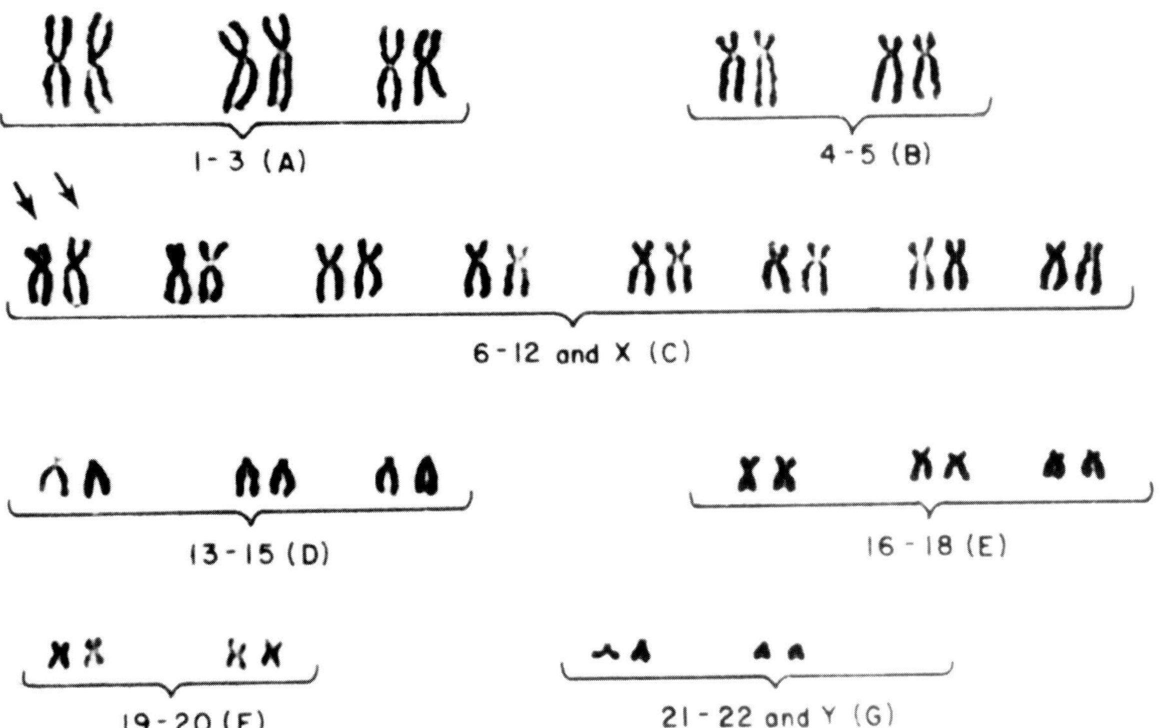

FIGURE 5-6 Normal female karyotype. Arrows indicate X chromosomes.

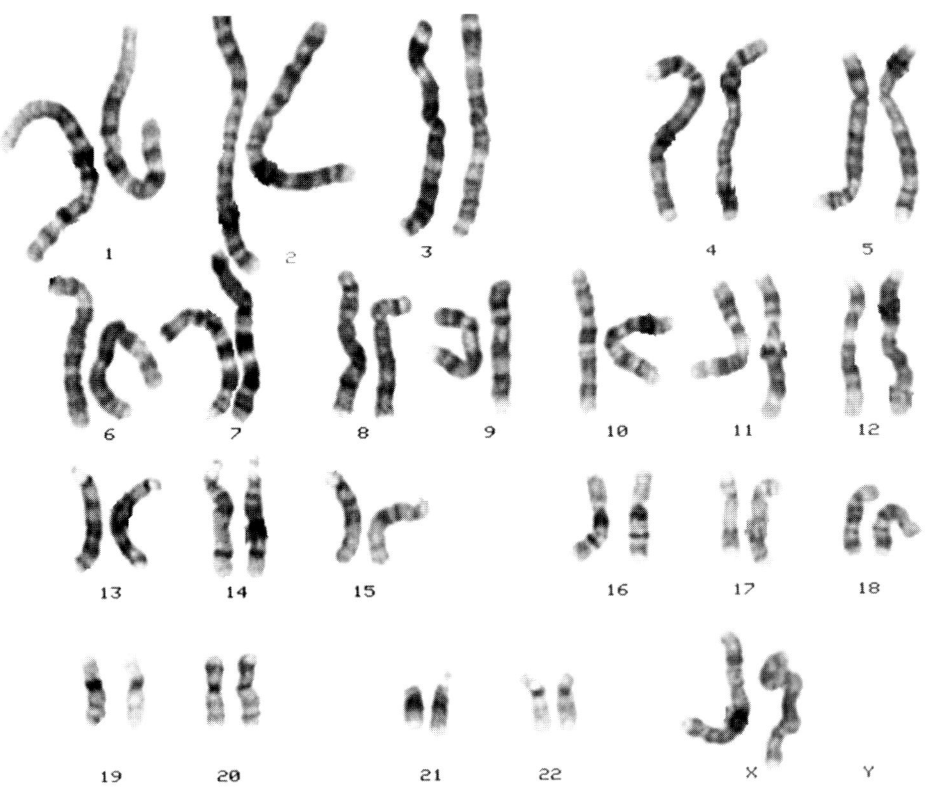

FIGURE 5-7 Karyotype showing evidence of deletion of the short arm of chromosome 4, with abnormal material present on 4p (chromosome 4 on right side of pair).

alkylating agent quinacrine mustard, which reacts with G-C base pairs. It later turned out that even the nonalkylating analog quinacrine dihydrochloride elicits bands of bright and dull fluorescence due to fluorescence enhancement in AT-rich DNA and quenching in GC-rich regions (Figure 5-8a).

The introduction of quinacrine banding was followed quickly by the development of many other banding techniques (Table 5-2). The most useful involved pretreatment of chromosomes with a variety of agents (e.g., the protease trypsin), followed by staining with Giemsa. Essentially the same bands were seen as were elicited by quinacrine; most bright fluorescent bands seen with quinacrine are darkly stained with Giemsa (Figure 5-8b). At a conference in Paris in 1971, quinacrine bands were designated *Q-bands*, and Giemsa bands *G-bands*. In addition, a pattern that is the reverse of Q- or G-banding was produced with another staining technique, designated *R-banding* (Figure 5-8c). An approach to staining condensed chromosome material near the centromeric regions was called *C-banding*. Other

TABLE 5-2 Summary of chromosome banding techniques

Technique	Method	Features
Q-bands	Quinacrine staining	Bright fluorescence bands
G-bands	Giemsa staining after chromosome pretreatment	Dark G-bands correspond with most bright Q-bands
R-bands	Various techniques	Reverse pattern compared with G and Q, useful for defining ends of chromosomes
T-bands	Various techniques	Highlights ends of chromosomes
C-bands	Protein/DNA extraction, staining with Giemsa	Highlights centromeric regions
NOR	Silver staining	Stains nucleolus organizer regions (acrocentric chromosomes in humans)

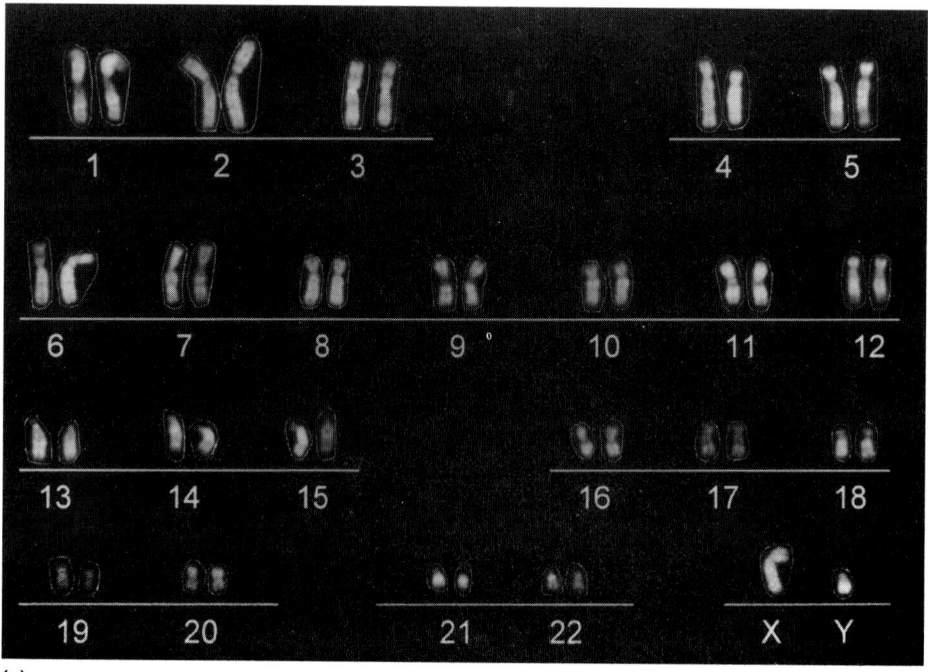

(a)

FIGURE 5-8 (**a**) Q-banded karyotype.

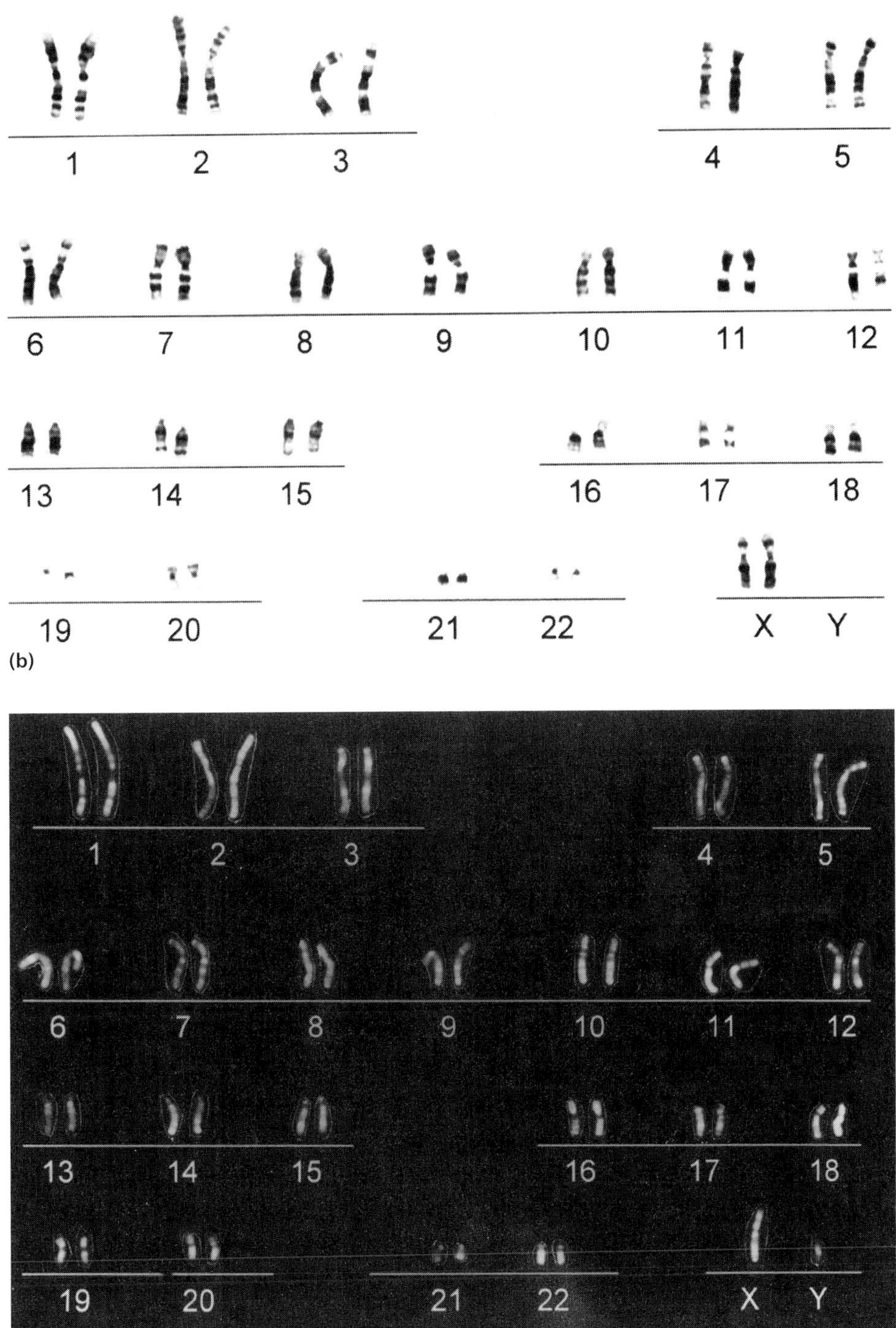

FIGURE 5-8 *(continued)* (**b**) G-banded karyotype; (**c**) R-banded karyotype.

chromosome staining techniques include *T-bands*, which stain the ends of chromosomes, and *NOR staining*, which highlights the nucleolus organizer regions of the acrocentric chromosomes.

The introduction of chromosome banding techniques led to a second wave of identification of clinical syndromes. First, individual chromosomes could be unequivocally identified. Deletion of 4p gives rise to Wolf-Hirschhorn syndrome, whereas deletion 5p results in cri du chat syndrome. Second, some individuals with congenital anomalies but normal numbers of chromosomes were found to have loss or gain of small amounts of genetic material. Many of these anomalies had escaped detection when unbanded chromosomes were studied.

What accounts for the banding patterns? The enormous stretch of DNA that composes each chromosome is actually a highly organized structure. The DNA double helix measures approximately 2 nm in diameter, but DNA does not exist in the nucleus in a "naked" form. It is complexed with a set of lysine- and arginine-rich proteins called *histones*. Two molecules of each of four major histone types—H2A, H2B, H3, and H4—associate together with about every 146 base pairs to form a structure known as the *nucleosome* (Figure 5-9). Successive nucleosomes are separated from one another by up to 80 base pairs, like beads on string. This is more or less the conformation of actively transcribed chromatin but, during periods of inactivity, regions of the genome are more highly compacted. The next level of organization is the coiling of nucleosomes into a solenoidlike structure, held together by another histone, H1, and other nonhistone proteins.

During prophase, chromosomes begin to condense into even more compact structures. This condensation does not occur evenly along the length of the chromosomes. Some regions, particularly those

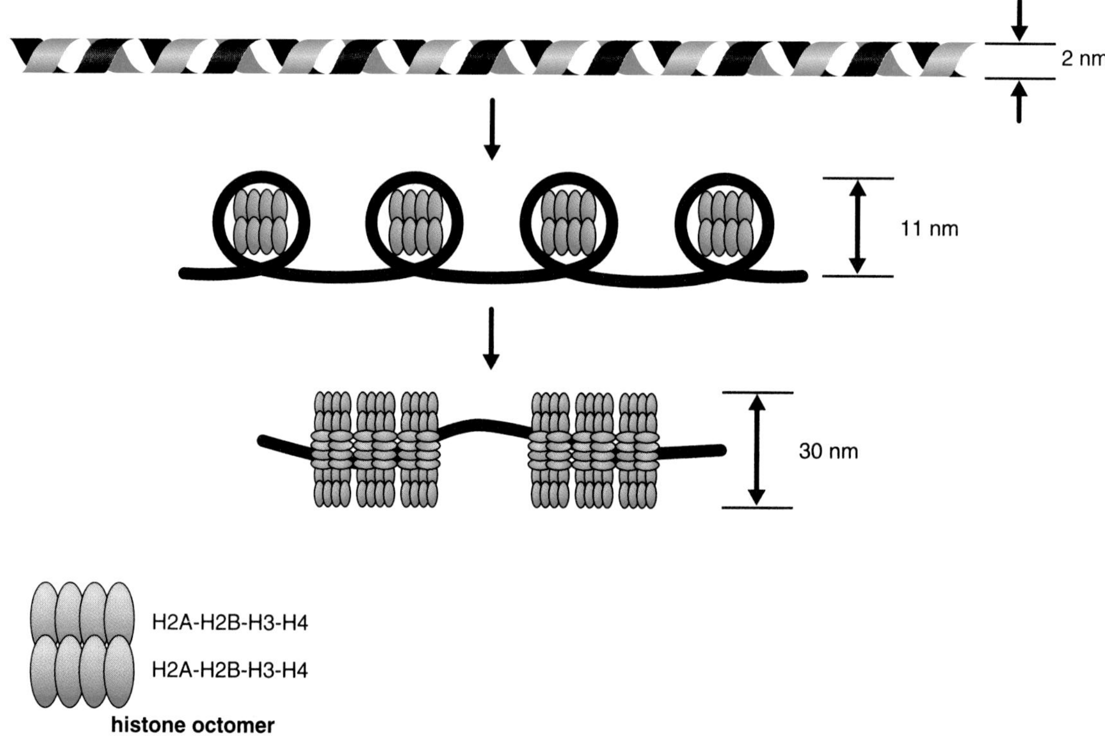

FIGURE 5-9 Levels of organization of chromatin. The DNA double helix has a width of approximately 2 nm. The fundamental unit of chromatin is the nucleosome, which consists of 146 base pairs of DNA wound around a core consisting of two copies of each of the four histone proteins (H2A, H2B, H3, and H4). These are arrayed as beads on a string. The diameter of a nucleosome is 11 nm. The nucleosomes are, in turn, wound into a solenoid-like structure measuring 30 nm. This is further coiled and condensed to compose a metaphase chromosome.

near the centromeres, remain condensed throughout the cell cycle. These regions are referred to as being **heterochromatic**, because they stain differently from other regions even during interphase. It has been found that the DNA in these regions consists of relatively short runs of bases repeated hundreds of thousands of times. Presumably, this DNA plays a structural role and does not encode genes. It is tightly bound to proteins. C-banding procedures involve harsh treatments that extract DNA and proteins from the chromosome, leaving only the tightly compacted centromeric heterochromatin.

Dark G-bands also correspond with regions that are more highly compacted early in mitosis. They are less likely to contain actively transcribed genes and are, on average, richer in AT base pairs. In situ hybridization studies have elucidated further the basis for differences in DNA content between Giemsa light and dark bands (Figure 5-10). The highly repeated simple-sequence DNA found mainly at centromeres is predominantly one type of repeated DNA. There are, in addition, sequences that are somewhat longer and repeated thousands of times. One class of such sequences consists of approximately 300 bases of a specific sequence repeated up to 900,000 times. This type is referred to as *Alu sequence* because it is cleaved by the restriction endonuclease *AluI*. It tends to be located within introns or between genes and is relatively GC-rich. Another class of intermediately repetitive DNA is referred to as *long-interspersed repeated sequences* or *LINE sequences*. The most prevalent LINE sequence is referred to as *L1* and is approximately 6.4 Kb. The 3′ end is repeated 50 to 100,000 times; the 5′ end often is deleted, so is repeated only 4 to 20,000 times. LINEs are AT-rich and tend to be located in regions with fewer active genes.

Chromosome banding techniques elicit patterns reflective of the organization and compaction of DNA sequences along the chromosome. The techniques are used daily in cytogenetics laboratories around the world and permit precise identification of relatively subtle chromosomal anomalies. If chromosomes are studied early in mitosis before they are highly compacted, or if compaction is inhibited (e.g., by treatment with the DNA intercalating agent ethidium bromide), very fine detail can be discerned (Figure 5-11). Some individuals have been found to have chromosomal rearrangements involving loss or gain of very small amounts of chromatin.

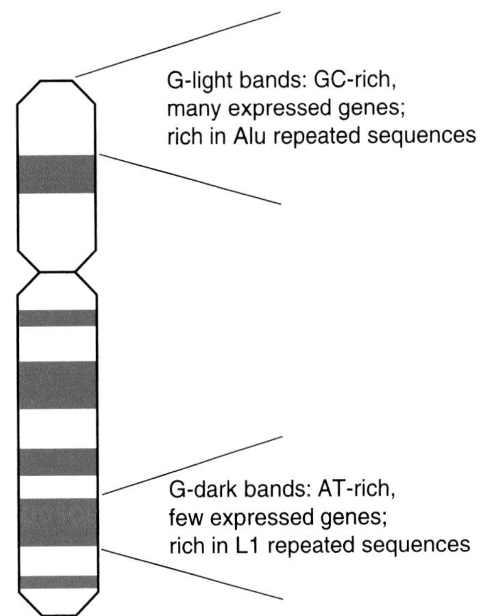

FIGURE 5-10 Giemsa-light bands are rich in expressed sequences and have high GC content. Alu repeated sequences are GC-rich and are found in or near expressed genes. Giemsa-dark bands have fewer expressed genes and a higher AT content. L1 repeated sequences, which are more AT-rich, predominate in these regions. (Data from Korenberg JR, Rykowski MC. Human genome organization: Alu, Lines, and the molecular structure of metaphase chromosome bands. Cell 1988;53:391–400.)

5.4 Chromosomal Rearrangements

Karen is now 2 years old. She did not require surgery for the ventricular septal defect, which closed on its own. She has had innumerable ear infections and has been hospitalized several times for pneumonia. Her parents are now interested in having other children, and speak with the geneticist to review the risk of recurrence of similar problems. It had been recommended that they have chromosomal analysis after Karen's deletion was identified, but they had put this off. Now, both parents have blood drawn. Deborah is found to have an apparently balanced translocation between chromosomes 4 and 9 (Figure 5-12).

The gain of an entire chromosome, such as occurs in Down syndrome, is the result of **nondisjunction**. This involves missegregation of chromosomes at meiosis or mitosis, with two copies of a particular

FIGURE 5-11 Chromosome 4 shown at three different levels of banding resolution from low (left) to high (right).

chromosome going to one cell and no copy to the other cell (Figure 5-13). Fertilization of a disomic germ cell results in a trisomic zygote. Embryos monosomic for an autosome usually do not survive. Most autosomal trisomies are also nonviable, except for trisomies 13, 18, and 21. If nondisjunction occurs during mitosis in the developing embryo, **mosaicism** results, wherein the embryo consists of a mixture of trisomic and normal cells (again the monosomic cells usually die, except for monosomy X) (Figure 5-14).

Changes of chromosome structure can involve single chromosomes or an exchange of material between chromosomes (Figure 5-15). A piece of a chromosome may be lost by deletion or may be duplicated. The former results in monosomy for a group of genes, and the latter in trisomy for the genes. Chromosome segments also can be inverted—

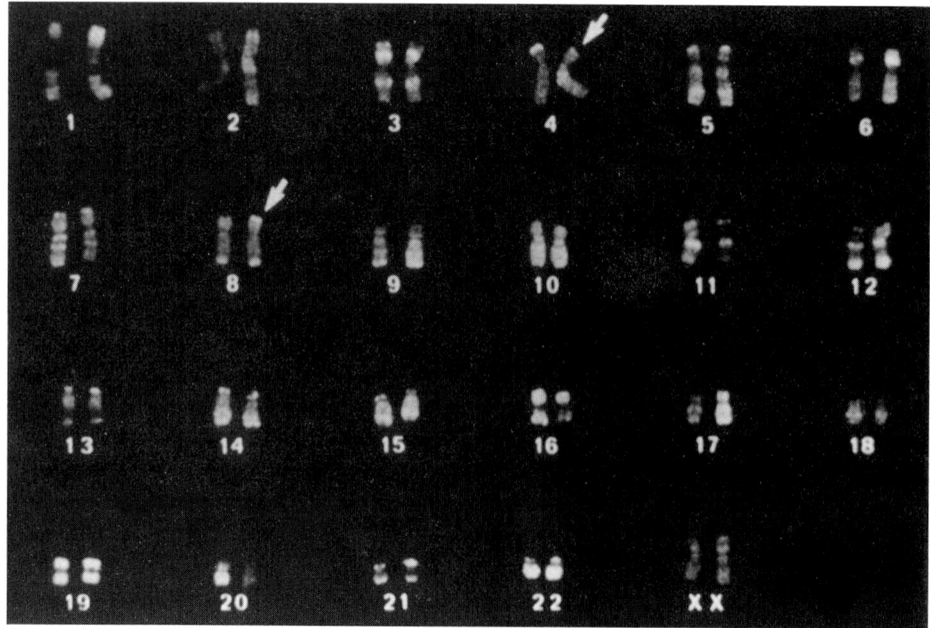

FIGURE 5-12 R-banded karyotype showing balanced translocation between chromosomes 4 and 8 (*arrows*). (Courtesy of Cytogenetics Laboratory, Brigham and Women's Hospital, Boston.)

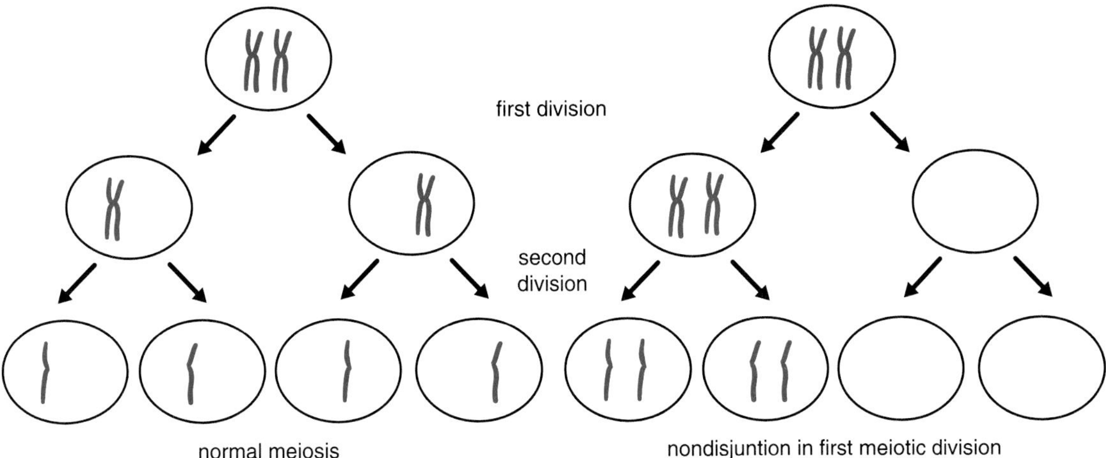

FIGURE 5-13 Nondisjunction in the first meiotic division. Two chromosomes segregate to same cell. On fertilization, gametes with two copies of a chromosome give rise to a trisomic zygote.

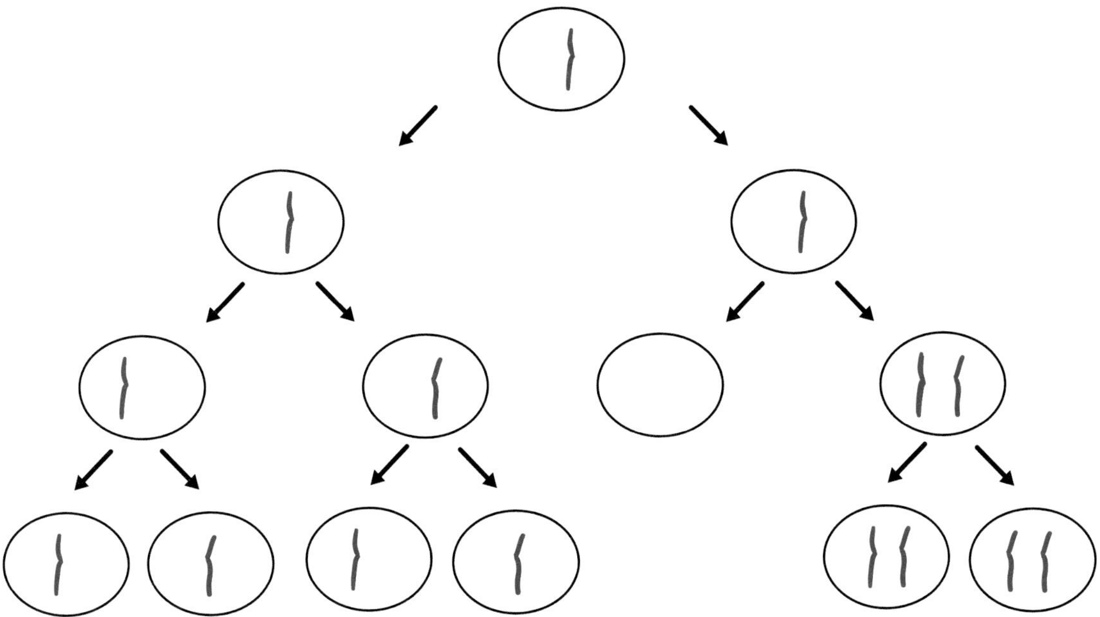

FIGURE 5-14 Mitotic nondisjunction leading to mosaicism. Two chromatids go to same cell during mitotic division. This results in trisomy for one chromosome in one daughter cell and monosomy in the other. The monosomic cell usually dies, but the trisomic cells may survive, leading to a mixture of normal and trisomic cells (mosaicism).

flipped 180 degrees from their normal orientation. If no material is gained or lost, the changes probably will have no phenotypic impact. Rarely, a gene may be disrupted by the chromosome breakage involved in the inversion, but there are vast regions of genetically inert material between groups of genes, so usually these breaks cause no phenotype. As will be seen in the next section, however, such breaks can lead to unbalanced chromosomes after crossing over in meiosis. Another intrachromosomal rearrangement is formation of a ring. This usually arises from breakage of the two ends and their subsequent fusion into

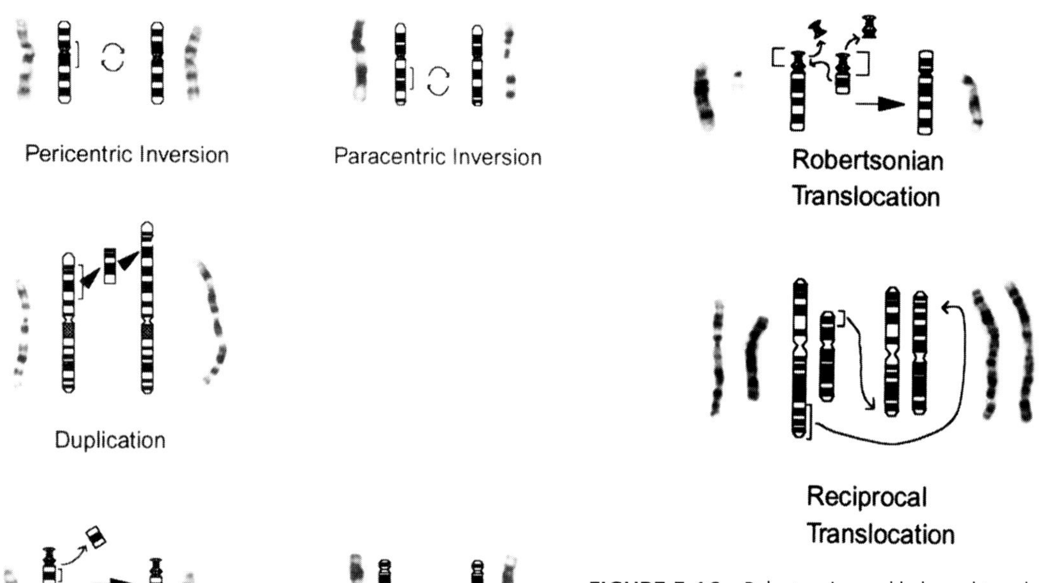

FIGURE 5-16 Robertsonian and balanced translocations. (Reprinted with permission from *Current Protocols in Human Genetics.*)

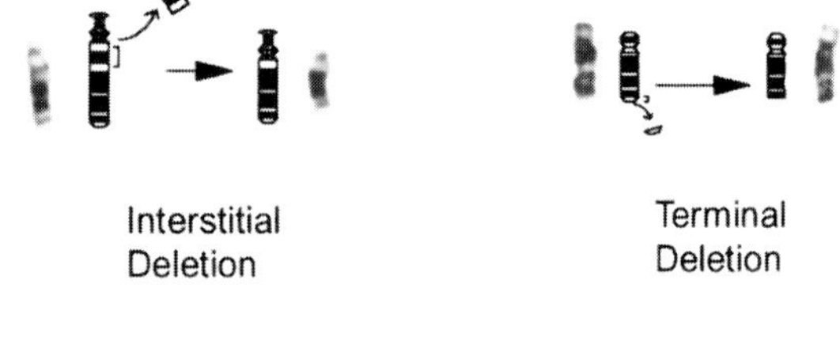

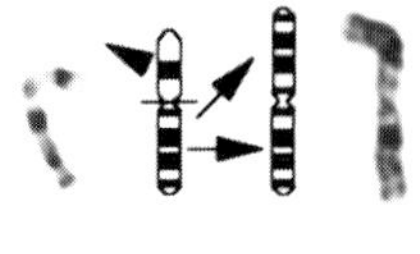

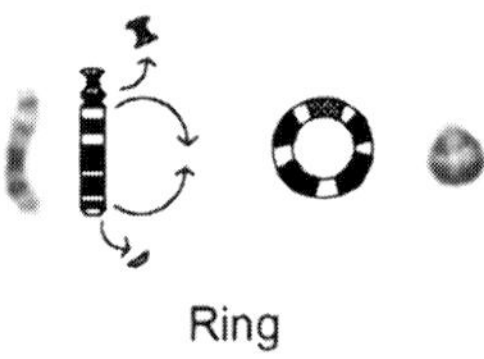

FIGURE 5-15 Types of chromosome abnormalities. (Reprinted by permission from *Current Protocols in Human Genetics.*)

a ring structure. There may be phenotypic consequences from deletion of chromatin from the two ends and also from mitotic instability of rings, resulting in trisomic or monosomic cells.

Translocation involves the exchange of material between chromosomes (Figure 5-16). Usually, translocations arise as apparently reciprocal exchanges. If no material is lost or gained, the translocation is said to be balanced. Balanced translocations—and inversions, for that matter—are occasionally found as variants in the general population. It is estimated that approximately 0.2% of individuals carry an asymptomatic chromosomal rearrangement. If one comes to medical attention, it is usually as a consequence of the generation of unbalanced gametes during meiosis, leading to spontaneous abortion or the birth of a child with congenital anomalies.

Most, if not all, translocations are reciprocal. In the early days of cytogenetics, it was found that at the two ends of a chromosome there is a sort of cap referred to as the **telomere**. Breakage of the chromosome leads to formation of a "sticky end," which tends to be unstable. This is avoided if the rearrangement involves exchange of material between two chromosomes.

Cytogenetic Nomenclature

I. Normal karyotype:
   Male: 46,XY
   Female: 46,XX

II. Aneuploidy:
   Female with Down syndrome: 47,XX, + 21
   Male with Edwards' syndrome: 47,XY, + 18
   Turner's syndrome: 45,X
   Klinefelter's syndrome: 47,XXY

III. Bands:
   Bands are numbered according to landmarks starting from the centromere up the short arm and down the long arm. Chromosomal rearrangements are described by noting the rearrangement and indicating the breakpoint or breakpoints. For example, a female with a deletion of the short arm of chromosome 4 with breakpoint at band p15 has the karyotype: 46,XX,del(4)(p15).

IV. Rearrangements:
   Deletion: 46,XX,del(4)(p15)
   Inversion: 46,XY,inv(2)(p12q12)—inversion of chromosome 2 with breakpoints at p12 and q12

   Ring: 46,XY,r(13)—male with ring chromosome 13
   Translocation: 46,XX,t(3;9)(p14;q21)—translocation between chromosomes 3 and 9, with breakpoints at band p14 on chromosome 3 and q21 on chromosome 9

5.5 Meiotic Segregation of Variant Chromosomes

Deborah and Steven are counseled that there is a risk—in the range of 15%—for an abnormality. Deborah's parents are tested, and her mother is found to carry the same translocation as Deborah. Her mother has a history of three miscarriages but no children born with congenital anomalies. Deborah is an only child. Deborah's mother had two sisters, both of whom are deceased, but each had several children. Deborah's mother is urged to inform these nieces and nephews of their potential for carrying the translocation (Figure 5-17).

Balanced chromosomal rearrangements can spawn gametes with genetic imbalance due to aberrant segregation during meiosis. Meiosis consists of two rounds of cell division and effects a reduction from the diploid to haploid state in germ cells. Homologous chromosomes pair intimately during the first meiotic division, at which time they exchange segments in the process of genetic recombination. The second round of division separates the two chromatids of each replicated chromosome.

Meiotic pairing between chromosomes involved in a balanced translocation requires a complex association of four chromosomes—the two involved in the exchange and the two homologs (Figure 5-18). When first anaphase occurs, these chromosomes can separate in several ways. If the two normal homologs go to one cell and the two involved in the exchange go to another, the resulting gametes will be genetically balanced—either normal or both having rearranged chromosomes. Genetic imbalance will result, however, if germ cells get one normal chromosome and one rearranged chromosome. The translocation complex can also segregate so that three chromosomes go to one cell and only one to the other. Rarely, all four chromosomes can go to the same cell. Obviously, major genetic imbalance results in these instances.

Translocations account for a minority of cases of Down syndrome. Translocations between acrocentric chromosomes in which the long arms fuse at the centromeres are referred to as *Robertsonian translocations* (after the geneticist Robertson, who studied similar rearrangements in mice) (Figure 5-19). A Robertsonian translocation carrier has 45 chromosomes but is phenotypically normal. If, however, both the translocated chromosome and a normal 21 go to the same germ cell at meiosis, fertilization will result in trisomy 21. This mechanism accounts for approximately 5% of cases of Down syndrome; the phenotype is indistinguishable from Down syndrome that ensues from classic trisomy. It is important to identify translocation cases, however, because a carrier is at risk of having additional offspring with trisomy. Robertsonian translocations can be present in many members of a family, all of whom are at risk of having children with Down syndrome.

Similar rules apply to other balanced translocations, although the phenotypic consequences of

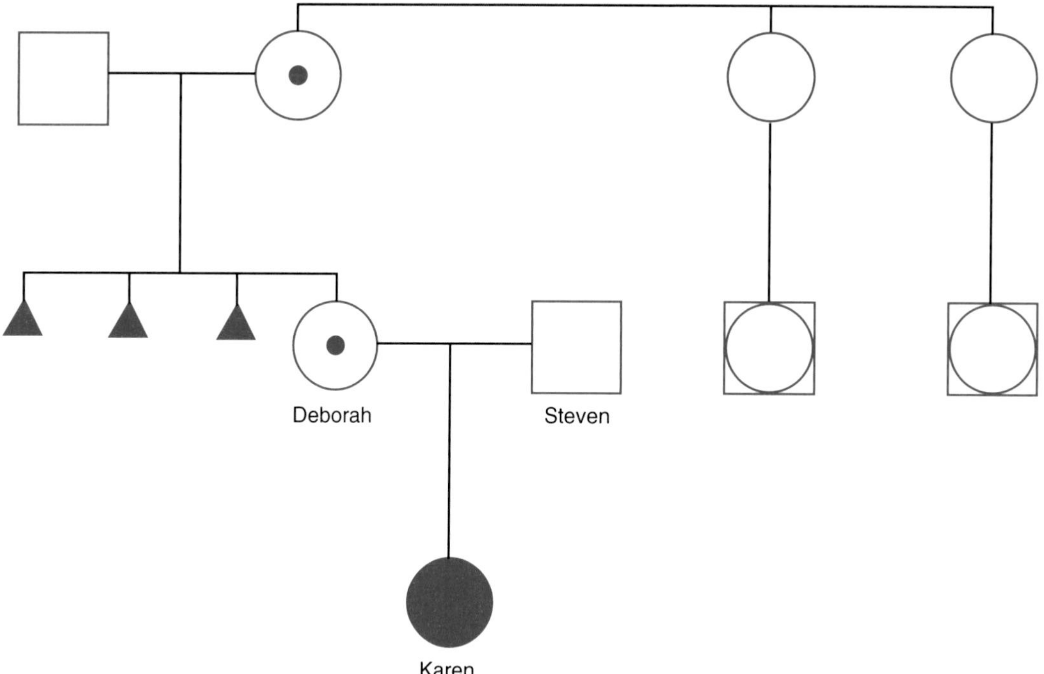

FIGURE 5-17 Pedigree for Deborah and Steven.

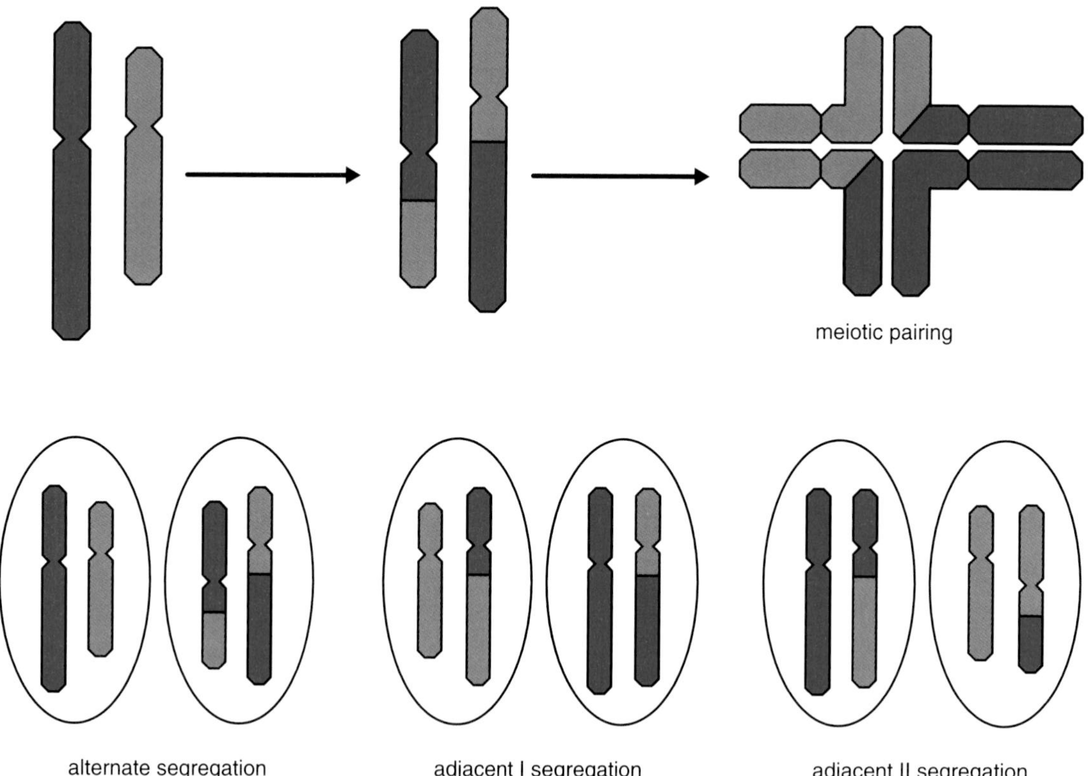

FIGURE 5-18 Meiotic segregation of balanced translocation. The reciprocal translocation is diagrammed at the left. During meiosis, homologous chromosomes pair to create a cross-shaped structure. If alternate chromosomes segregate to the same pole, daughter cells get either the normal chromosomes or the balanced translocation. If homologous centromeres segregate (adjacent I segregation), or nonhomologous centromeres segregate (adjacent II segregation), gametes with unbalanced chromosomes result.

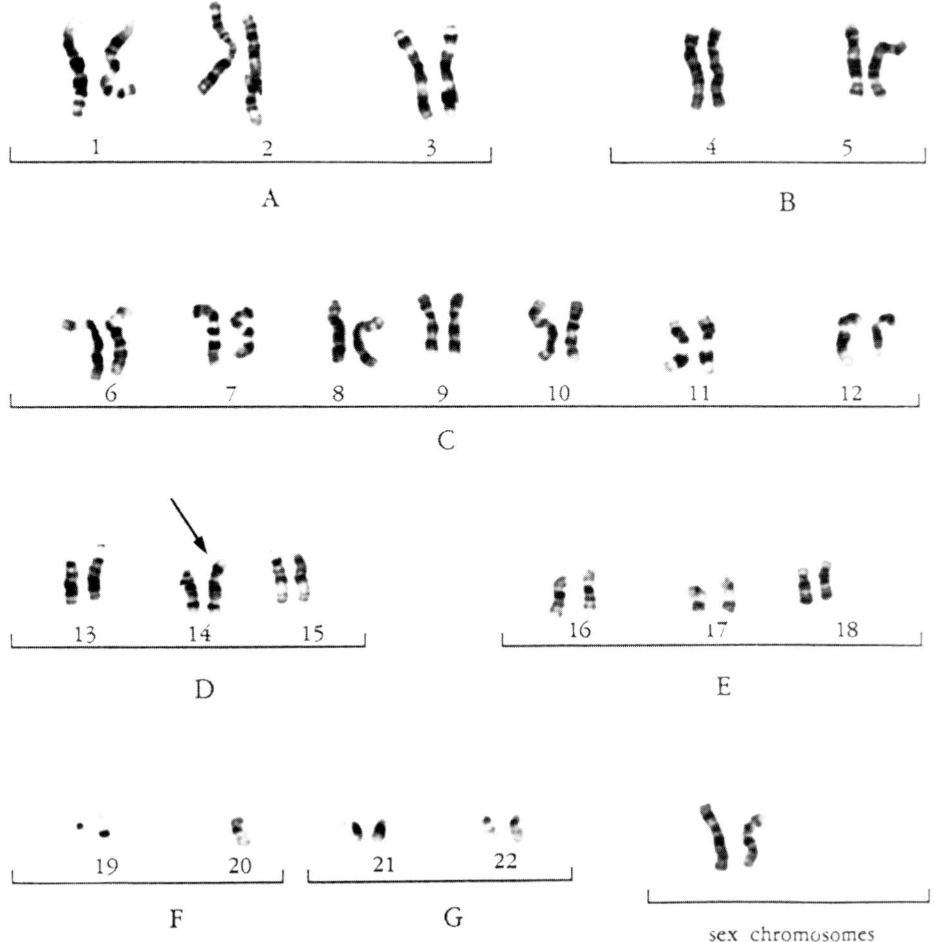

FIGURE 5-19 Karyotype showing Robertsonian translocation between chromosomes 14 and 21 (*arrow*), resulting in Down syndrome.

imbalance are less predictable. Unbalanced offspring are trisomic for one chromosome segment and monosomic for another. Innumerable case reports describe such partial trisomies and monosomies.

Genetic imbalance can also result from meiotic segregation of inverted chromosomes (Figure 5-20). Pairing between homologs where one chromosome is inverted requires formation of a loop. Crossing over within the loop leads to duplication and deficiency of genetic material. If the inversion does not include the centromere, called a **paracentric inversion,** dicentric and acentric chromosomes result. These are usually unstable at mitosis and lead to nonviable phenotypes. Inversions that involve the centromere, called **pericentric inversions,** lead to partial trisomies and monosomies, some of which may be viable. Here, too, balanced rearrangement in a carrier can predispose to unbalanced products in an offspring.

Balanced chromosomal rearrangements usually come to attention through a child who has the rearrangement in an unbalanced form. Sometimes, it is discovered when chromosomal analysis is done as part of an evaluation for recurrent miscarriage. Once found, other relatives may be tested to determine whether they also carry the rearrangement. Carriers are provided with genetic counseling, and prenatal diagnosis can be offered.

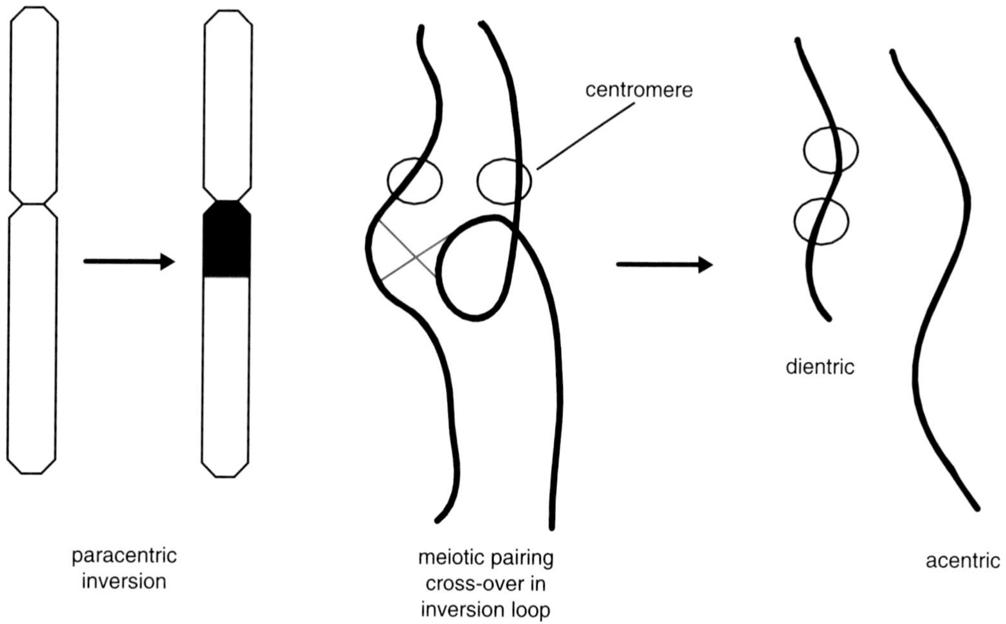

FIGURE 5-20 Paracentric inversion segregates dicentric and acentric fragments after crossing over in meiosis. The homologous chromosomes form a loop at the site of inversion. Dicentric and acentric fragments result from crossing over within the loop. If the inversion is pericentric, the recombinant chromosomes each have a single centromere, but genetic material at each end is duplicated or deleted.

5.6 Prenatal Diagnosis of Chromosomal Abnormalities

Deborah is now pregnant. Amniocentesis is done at 16 weeks and, 2 weeks later, a normal male karyotype is reported: 46,XY. Deborah and Steven are delighted when, 5 months later, a healthy baby boy is born.

Visualization of chromosome structure requires access to dividing cells. These can be obtained from the fetus by chorionic villus biopsy or amniocentesis. Amniotic fluid cells must be grown in culture for several days, after which dividing cells are harvested and studied. Some cytotrophoblast cells obtained with chorionic villus tissue divide so rapidly that they can be harvested immediately for chromosomal analysis. Extraembryonic mesoderm can also be obtained from this tissue and grown in culture.

When a parent is known to carry a balanced rearrangement, it is relatively simple to analyze fetal chromosomes to look for evidence of imbalance. Although it constitutes an important indication for prenatal studies, the presence of a balanced rearrangement in a parent is not the most common reason to undertake a prenatal study. A far more common indication is concern about the risk of trisomy, particularly trisomy 21.

The frequency of trisomy 21 in the general population is approximately 1 in 700 live births. The vast majority of affected individuals have trisomy 21 (i.e., 47 chromosomes with no translocation). This usually occurs unexpectedly in a family without prior history of Down syndrome. Recurrence risk for trisomy is about 1% for future pregnancies. The cause of nondisjunction is unknown, but maternal age has been well documented to predispose to the birth of a child with Down syndrome.

The frequency of birth of offspring with Down syndrome remains relatively flat when plotted as a function of maternal age, until approximately age 30 (Figure 5-21). The curve rises very steeply at around

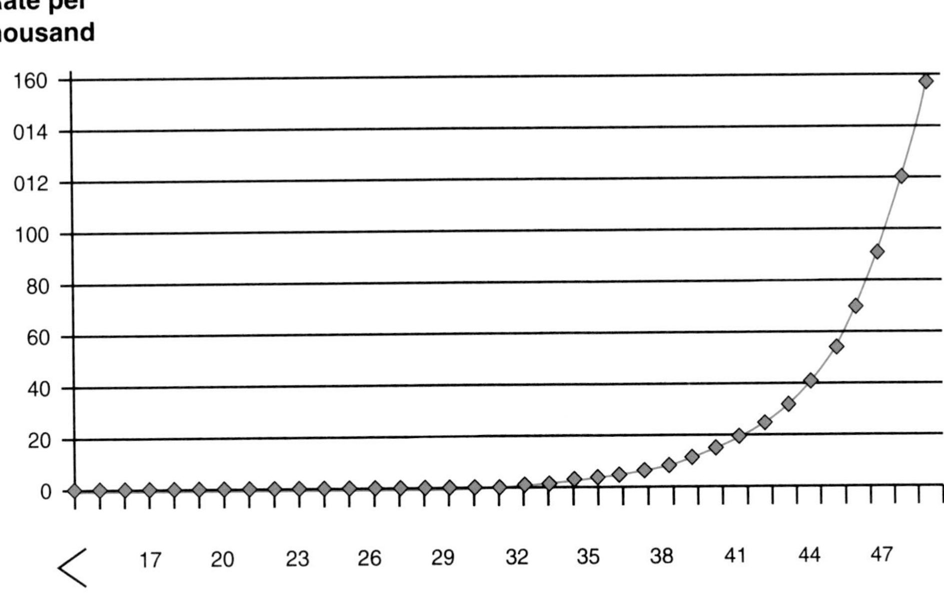

FIGURE 5-21 Frequency of live birth of child with Down syndrome as a function of maternal age. (Data from Hook EB. Rates of chromosome abnormalities at different maternal ages. Obstet Gynecol 1981;58:282–285.)

age 35, with frequencies higher than 1 in 40 after age 40. The maternal age effect is believed to result from the fact that oocytes remain in the first meiotic division from fetal life until the egg cell is ovulated. If homologs drift apart during this long time, nondisjunction may result. Evidence for a paternal age effect is much weaker. This may be due to the fact that in the male, meiosis begins after puberty and lasts only a few weeks. As was noted in Chapter 2, continued mitotic division of spermatogonia may increase the rate of gene mutation, but the rate of nondisjunction is lower than in oogenesis.

The steep rise in rate of nondisjunction in women in their thirties or beyond has led to a standard practice in many countries of offering prenatal diagnosis to women older than 35 years. The rate of complications due to amniocentesis—mostly miscarriage—is approximately 1 in 200; the risk of detecting a fetus with a chromosomal abnormality (any trisomy, not specifically trisomy 21) also is approximately 1 in 200 in women aged 35 years or older.

Although the risk of having a child with trisomy increases with maternal age, most pregnancies occur in younger women, and hence most births of children with trisomy occur to younger women. Efforts have been made to identify those at risk of carrying affected fetuses by means other than assessment of age. Some fetuses with Down syndrome have characteristic features detectable by ultrasonography, but identification of these features is not always possible.

In recent years, several biochemical measurements have been found to be correlated with the occurrence of a fetus with trisomy. Alpha-fetoprotein is a serum protein produced by the fetus. A small quantity is excreted through the urine into amniotic fluid, and some crosses the placenta into the maternal circulation. Congenital malformations that create defects in the fetal body wall lead to large increases in amniotic fluid and maternal serum alpha-fetoprotein levels (see Chapter 6). During the course of screening for such defects, it was noticed that low amniotic fluid levels tend to occur more often if the fetus has a chromosomal abnormality, particularly trisomy 21. It was later found that addition of two other measurements—unconjugated estriol and human chorionic gonadotropin—increased the sensitivity of the screen. Fetuses with Down syndrome are associated with low alpha-fetoprotein and estriol levels but a high human chorionic gonadotropin level in maternal serum. Low values for the three substances indicate a risk of trisomy 18.

It is estimated that applying this "triple test" will detect 60% of fetuses with Down syndrome. There is a high false positive rate: More than 95% of the abnormal serum tests will turn out not to be associated with trisomy. A truly universal prenatal test for trisomy will require a risk-free and inexpensive means of fetal testing. Experiments are in progress with an approach based on analysis of fetal cells in the maternal circulation. Some fetal cells (trophoblast or blood cells) leak across the placenta just as does alpha-fetoprotein. Genetic analysis of these cells may someday actualize the dream of a risk-free test.

5.7 Chromosomal Mosaicism

Karen is now 5 years old, and her brother Daniel is 3. Deborah is pregnant again, and this time chorionic villus sampling is done. Analysis of the direct preparation reveals three cells with trisomy 16. No trisomy 16 cells are found in the cultured preparation from the chorionic villi. Deborah and Steven are counseled that trisomy 16 cells probably are not present in the fetus but represent confined placental mosaicism. The fetus does carry the balanced translocation, but Deborah and Steven understand that this is likely to cause no problems. Five months later a healthy girl, Christine, is born.

When nondisjunction occurs during mitosis rather than meiosis, the result is chromosomal mosaicism. Approximately 2% of individuals with Down syndrome are mosaics. Some may have had complete trisomy 21 at conception and then lost the extra chromosome from some cells; others may have started with 46 chromosomes but then gained a trisomic cell line owing to nondisjunction. Generally, Down syndrome due to trisomy 21 mosaicism is indistinguishable from Down syndrome in general. In some instances, the phenotype may be ameliorated by the presence of nontrisomic cells. If trisomy 21 mosaicism is detected prenatally or at birth, it is difficult to predict whether a full Down syndrome phenotype will occur. The exact proportion of trisomy 21 cells may differ from tissue to tissue. A low percentage of abnormal cells in peripheral blood does not necessarily predict a low proportion in heart or brain.

Mosaicism may occur for any chromosomal trisomy and for some structurally rearranged chromosomes. In the latter case, the mosaicism often results from instability of the abnormal chromosome at mitosis. Some trisomy syndromes are viable only in a mosaic state. Nonmosaic trisomy, for most chromosomes, results in miscarriage during the first trimester. Trisomy 8, for example, has been seen clinically only when it is mosaic, and as such produces a characteristic syndrome. In contrast, complete trisomy 8 results in miscarriage.

Detection of mosaicism by prenatal diagnosis can result in situations in which the phenotypic consequences are hard to predict. Not only is it difficult to know whether the fetus has physical abnormalities, but also the abnormal cells may be not derived from the embryo. Chorionic villus biopsy samples two cell types, neither of which is embryonic. Rapidly dividing cytotrophoblast cells are particularly prone to mitotic nondisjunction, and the mosaic chromosomal abnormalities often are not represented in the fetus. Cultured chorionic villus tissue is derived from extraembryonic mesoderm. Amniotic fluid cells originate from fetal skin and bladder, as well as from the lining of the amnion. The latter, again, are not embryonic.

This has created a dilemma in genetic counseling for many families. Mosaicism noted in uncultured chorionic villus material should be confirmed with cultured material. Mosaicism that is confined to the cytotrophoblast is much less likely to be clinically significant. For amniotic fluid analysis, it is common to grow samples in multiple tissue culture flasks or to grow cells on coverslips in multiple culture dishes (Figure 5-22). The latter permits analysis of colonies derived from individual cells. Sometimes chromosomal mosaicism will arise by nondisjunction in culture. This is referred to as *pseudomosaicism*. If the chromosomal anomaly is not seen in multiple culture flasks or dishes, it is presumed to be pseudomosaic and probably is not clinically significant. On the other hand, if the abnormal cells occur in multiple culture dishes, true mosaicism is present. This is more likely to have phenotypic impact, although in some cases the proportion of affected cells

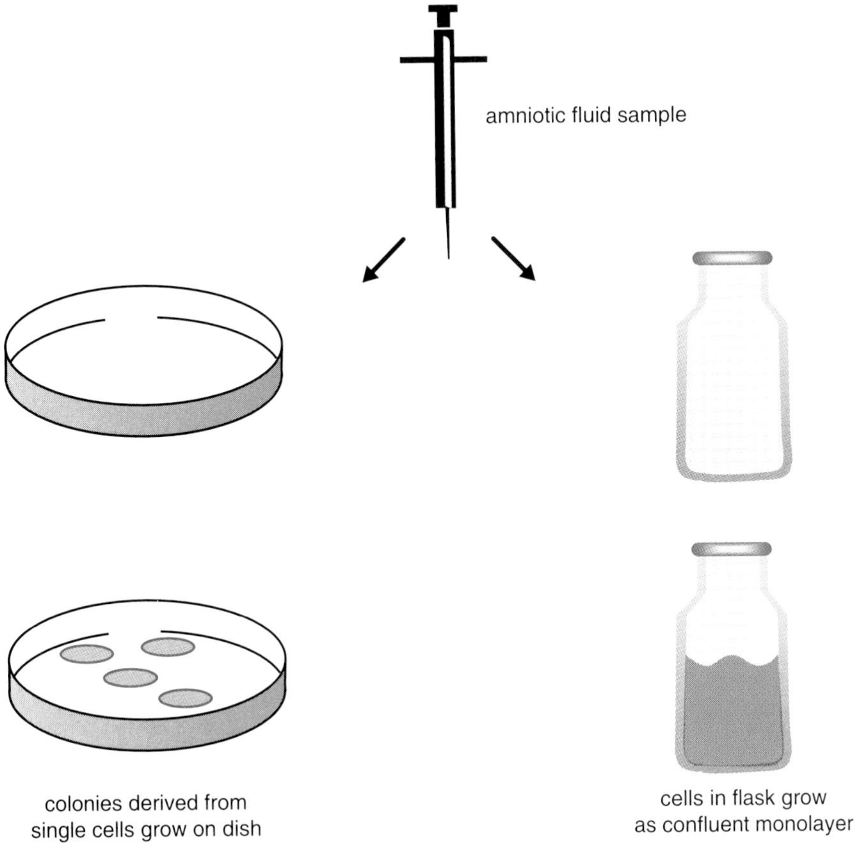

FIGURE 5-22 Amniotic fluid may be inoculated into culture dishes or flasks. In the dishes, single cells grow into discrete colonies. If cells are grown on cover glasses in dishes, they may be harvested in situ and directly examined with the microscope. In this way, cells within individual colonies may be directly examined. Cells that are inoculated into flasks grow as a confluent monolayer. These cells are enzymatically removed from the flask and harvested for chromosomal analysis.

may be low, and the exact phenotypic consequences may be difficult to predict. Many families who undergo prenatal diagnosis are not well prepared for the possibility of receiving results that are of uncertain clinical significance.

5.8 Genotype-Phenotype Correlations

Deborah and Steven are asked to provide a fresh blood sample from Karen for research purposes. Cells are grown for chromosomal analysis. The chromosome preparations are used in fluorescence in situ hybridization studies (Figure 5-23).

Although chromosomal abnormalities are among the easiest of genetic disorders to recognize, we know relatively little about their pathophysiology. Why does it matter to have too many or too few chromosomes? All genes are still represented, albeit in incorrect quantity.

Down syndrome is by far the most common trisomy syndrome. It is remarkable that individuals with Down syndrome may look more like one another than like members of their own families. There must be hundreds—perhaps more than a thousand—of genes on chromosome 21. The means by which they exert their effects to create the Down syndrome phenotype is believed to be increased gene dosage. Levels of transcription one and a half times higher than normal have been demonstrated for

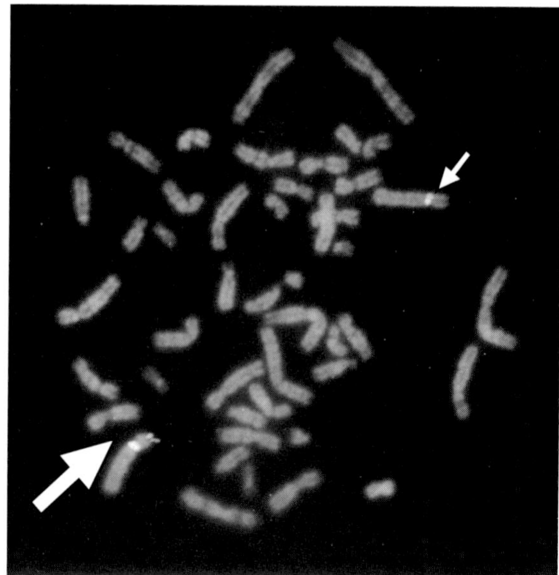

FIGURE 5-23 FISH studies of chromosome 4p. The normal 4 (*large arrow*) hybridizes with a sequence near the tip of the short arm, as well as a control sequence near the centromere. The deleted copy of chromosome 4 only hybridizes with the centromeric sequence (*small arrow*).

several genes on chromosome 21 in trisomic individuals. Apparently, gene dosage is finely regulated, and increases or decreases are not well tolerated.

There is evidence that only a subset of genes on chromosome 21 is responsible for the Down syndrome phenotype (Figure 5-24). Some individuals have been found to have trisomy for only a small region of chromosome 21 due to segregation of an unbalanced translocation. This has led to identification of a so-called critical region of chromosome 21 sufficient to give the stigmata of Down syndrome.

Chromosome deletion syndromes also affect gene dosage. Deletion of large segments or of entire chromosomes usually is poorly tolerated. Except for 45,X Turner's syndrome, there are no other whole-chromosome monosomy syndromes routinely compatible with live birth. Smaller deletions may result in recognizable syndromes, however. Some involve very small segments, sometimes so small as to not be visible with a microscope. It is believed that complex phenotypes result from simultaneous deletion of a group of genes, each of which is responsible for a specific developmental process (Figure 5-25). The exact phenotype may vary according to the extent of the deletion and which genes are lost.

Although the larger chromosome deletions are readily identified by standard cytogenetic analysis, deletions of fewer than a million base pairs or so cannot be resolved with the light microscope. It has long been suspected that submicroscopic deletions might underlie some congenital anomaly syndromes, and recent molecular genetic studies have confirmed this hypothesis. For example, DiGeorge's syndrome consists of a set of congenital anomalies including absence of the parathyroid glands (causing neonatal hypocalcemia), absence of the thymus (leading to immunodeficiency), facial anomalies, and congenital heart disease (particularly anomalies of the great vessels). Rarely, individuals with DiGeorge's syndrome

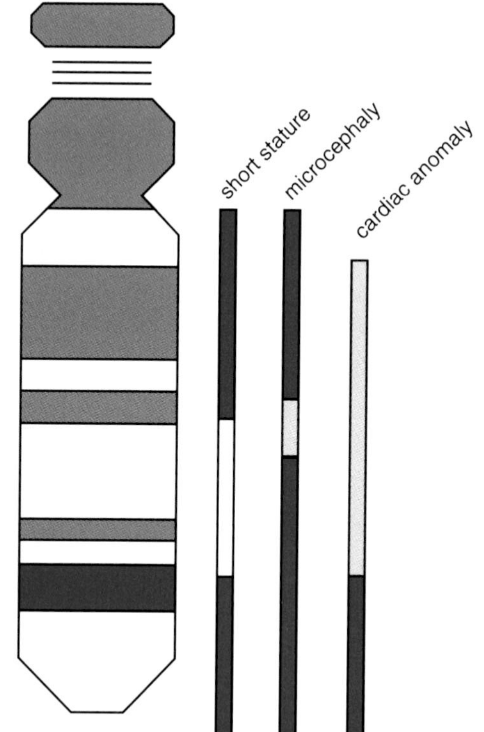

FIGURE 5-24 Map of some features associated with Down syndrome. The *dark boxes* indicate extent of regions in which a single gene or cluster of genes, when duplicated, accounts for the feature. *Light boxes* indicate regions where two genes or gene clusters determine the feature when duplicated. *Clear boxes* indicate regions with three or more genes or clusters. (Modified from Korenberg JR, Chen X-N, Schipper R, et al. Down syndrome phenotypes: the consequences of chromosomal imbalance. Proc Natl Acad Sci USA 1994;91:4997–5001.)

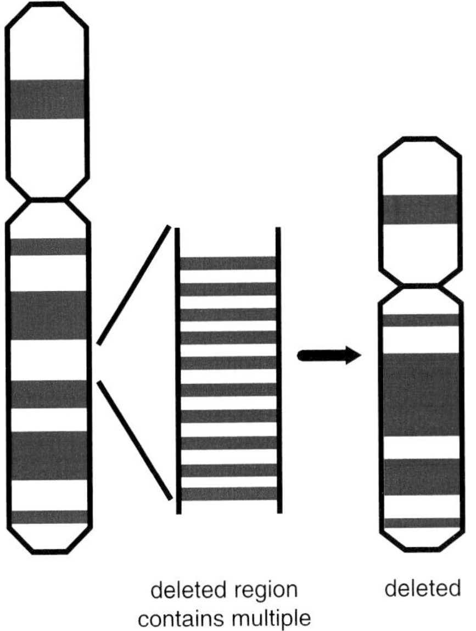

FIGURE 5-25 Deletion of a region that contains multiple contiguous genes. A complex syndrome results from the simultaneous loss of these many genes.

have been found to have visible deletions or other rearrangements of part of chromosome 22, but the majority have normal chromosomes with conventional analysis.

This clue that chromosome 22 might be involved in DiGeorge's syndrome led to efforts to clone DNA from chromosome 22 and determine whether DNA from the region is deleted in some individuals with DiGeorge's syndrome. To detect very small deletions, the technique of fluorescence in situ hybridization (FISH) was used. We have already considered the use of in situ hybridization in gene mapping (see Chapter 2). Early use of this technique was based on labeling of cloned DNA with tritiated thymidine and visualization of areas of hybridization by autoradiography. This approach is time-consuming and has low sensitivity and precision. During the 1980s, it became possible to label cloned DNA sequences with fluorescent tags, which vastly improved the ability to visualize hybridization. The approach works best with very long DNA probes derived from cosmids or yeast artificial chromosomes (YACs), which, because of their length, can be labeled to high specific activity.

When DNA sequences from the region of chromosome 22 involved in DiGeorge's syndrome were labeled and used for FISH, more than three-fourths of affected individuals were found to have deletions. Most deletions were too small to be seen in standard cytogenetic preparations but are easily detectable by the absence of a FISH signal on one homolog (Figure 5-26). FISH has become a standard tool for diagnosis of DiGeorge's syndrome.

The chromosome 22 FISH probes have proved to be valuable not only in the diagnosis of DiGeorge's syndrome but also in analysis of another syndrome, velocardiofacial syndrome (VCFS). VCFS is transmitted as an autosomal dominant trait and consists of cleft lip or palate (or both) and congenital heart defects. In more than two-thirds of individuals with VCFS, deletions are detected with the same

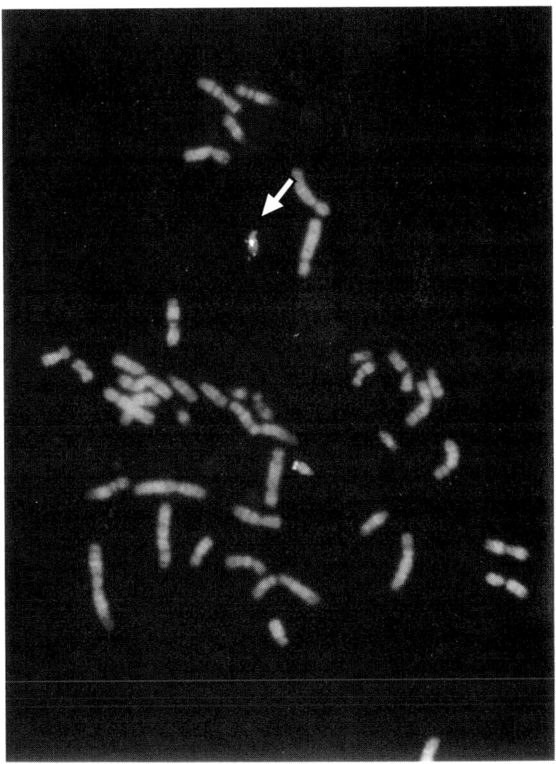

FIGURE 5-26 FISH detection of microdeletion on chromosome 22. Two sets of FISH probes are used: one for the DiGeorge's syndrome region and one that hybridizes with a DNA sequence near the end of chromosome 22. The latter is used to help identify chromosome 22. In this photograph, the normal 22 hybridizes with both sets of probes (*arrow*), whereas the deleted 22 hybridizes only with one set of probes (center). (Courtesy of Dr. Stana Weremowicz, Brigham and Women's Hospital, Boston.)

FISH probes that are used for DiGeorge's syndrome. These two disorders therefore result from chromosome deletions in the same region. Possibly, the deletions responsible for the more severe disorder, DiGeorge's syndrome, are larger and result in deletion of more genes than in VCFS. Testing this hypothesis will require identification of the genes in the region and comparison of the extent of deletions in the two disorders.

Many other microdeletion syndromes have been defined (Table 5-3). Many of these were viewed previously as sporadic entities of unknown cause. Identification of microdeletions has opened new approaches to diagnosis and promises to provide insights into pathogenesis. Even the larger-deletion syndromes, such as Wolf-Hirschhorn, are being subjected to molecular analysis. Affected individuals may have various combinations of congenital anomalies, and efforts are under way to create phenotypic maps that relate deletions of different sizes to particular combinations of anomalies. It is not yet clear that a simple correlation will exist between each deleted gene and specific congenital anomalies. Eventually, a more complete gene map of the deleted regions will be available, and more precise correlations between genotype and phenotype should become possible.

Aside from its use in gene mapping and detection of submicroscopic deletions, FISH has also been applied to identification of subtle chromosomal rearrangements. Sometimes, a translocation may involve a very small chromosome segment, making determination of its source difficult. FISH probes have been developed that consist of a cocktail of chromosome-specific DNA sequences, allowing whole chromosomes to be "painted" (Figure 5-27). This approach allows small pieces of translocated material or small chromosome "markers" to be identified, permitting more precise genetic counseling and prenatal diagnosis.

5.9 Genomic Imprinting

It might be assumed that genes located within a deleted region would be expressed at 50% of normal levels, but sometimes this is not the case. Some genes are not equally expressed from the maternal and paternal alleles. Rather, either the maternal or the paternal allele is expressed preferentially, at least during early development. If the deleted allele is the one that is normally active, the gene will be functionally absent, even though one copy is retained.

This phenomenon is referred to as **genomic imprinting** (Figure 5-28). It was first studied in mammals in experiments with mice, where it was found that embryos derived by duplication of either the

TABLE 5-3 List of chromosome microdeletion syndromes

Syndrome	Features	Deletion
Langer-Giedion syndrome	Sparse hair, bulbous nose, cone-shaped epiphyses, cartilaginous exostoses, MR	8q24.1
WAGR	Wilms' tumor, aniridia, genital and renal anomalies	11p13
Retinoblastoma	Retinoblastoma, MR, dysmorphic facies	13q14.1
Prader-Willi syndrome	Hypotonia, eating disorder, obesity, MR	15q11 (paternal origin)
Angelman's syndrome	Hypotonia, seizures, inappropriate laughter, incoordination	15q11 (maternal origin)
α-Thalassemia and MR	Thalassemia trait, MR	16p13.3
Miller-Dieker syndrome	Lissencephaly, dysmorphic facies	17p13.3
Smith-Magenis syndrome	Characteristic facies, MR	17p11.2
Alagille's syndrome	Biliary dysplasia, pulmonary stenosis, vertebral anomalies, dysmorphic facies	20p11
DiGeorge's syndrome	CHD, hypoplasia of parathyroid and thymus, facial anomalies	22q11
Velocardiofacial syndrome	CHD, palatal anomalies	22q11

MR = mental retardation; CHD = congenital heart defect.

Note: FISH probes are available for Miller-Dieker syndrome, Prader-Willi and Angelman's syndromes, and DiGeorge's and velocardiofacial syndromes.

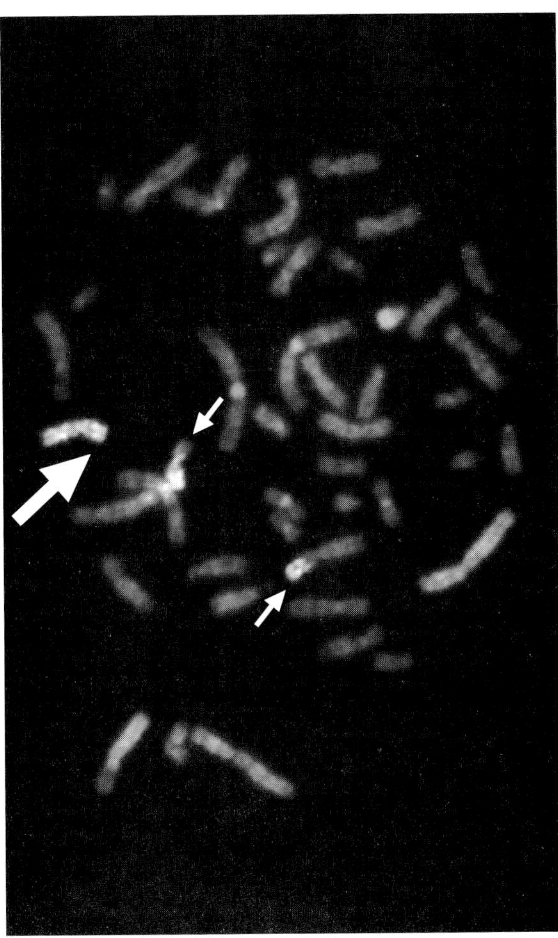

FIGURE 5-27 Chromosome painting of balanced translocation. In this case, DNA probes were used to stain ("paint") chromosome 11 in a metaphase with a translocation between chromosomes 8 and 11. The normal 11 is shown painted at the left (*large arrow*). The two halves of the translocation (the 8;11 and the 11;8) are indicated with *small arrows*. (Courtesy of Dr. Stana Weremowicz, Brigham and Women's Hospital, Boston.)

maternal or paternal chromosome set were not viable (Figure 5-29). This was not due to homozygosity, because the mice came from inbred strains that are normally homozygous. Apparently a set of chromosomes from each parent is required for normal development.

Using a set of balanced translocations, mice have been created that have inherited both copies of just one chromosome, or one chromosome region, from one parent. These experiments have shown that some chromosome regions are imprinted and others are not (Figure 5-30). Another system for the study of imprinting is the transgenic mouse. Foreign DNA sequences may integrate into imprinted regions of mouse chromosomes. Transmission of a transgene from one parent will result in expression of the gene, whereas transmission from the other will not. The imprint is reset in the germ line each generation. Imprinted genes have been found to be methylated at some CpG sites, suggesting that methylation may be involved in the suppression of activity in imprinted regions.

The effects of imprinting first came to light in humans through studies of rare individuals affected with cystic fibrosis who had, in addition, severe growth and developmental delay. They were found to have inherited the cystic fibrosis gene mutation—along with other genes on chromosome 7—from just one parent (Figure 5-31). This is referred to as **uniparental disomy**. Uniparental disomy is believed to be due to loss of one chromosome in a trisomic conceptus. Trisomy 7 would be nonviable, but if, early in development, one of the three copies of chromosome 7 in a trisomic embryo is lost by nondisjunction, the normal chromosome number would be restored. If the remaining copies of chromosome 7 are derived from the same parent, however, uniparental disomy results, which will have phenotypic consequences if the chromosome includes imprinted genes.

The effects of imprinting may also be detected in gene deletion syndromes. Prader-Willi and Angelman's syndromes are distinct disorders that are both associated with deletions of the same region of chromosome 15 (Figure 5-32). If the deletion is of the paternal 15, however, the phenotype is Prader-Willi syndrome, whereas if maternal sequences are deleted, the result is Angelman's syndrome. In some instances of Prader-Willi or Angelman's syndrome, no deletion is seen but, instead there is uniparental disomy for the paternal (giving Angelman's syndrome) or the maternal (giving Prader-Willi syndrome) copy of 15. Presumably, there are imprinted genes in this region, some of which are expressed on the paternal and some on the maternal 15.

The full range of phenotypes that result from imprinting effects has not yet been defined. There is concern that examples of confined placental mosaicism may be due to nondisjunction of a trisomic conceptus that might have phenotypic consequences if uniparental disomy results. Also, instances have

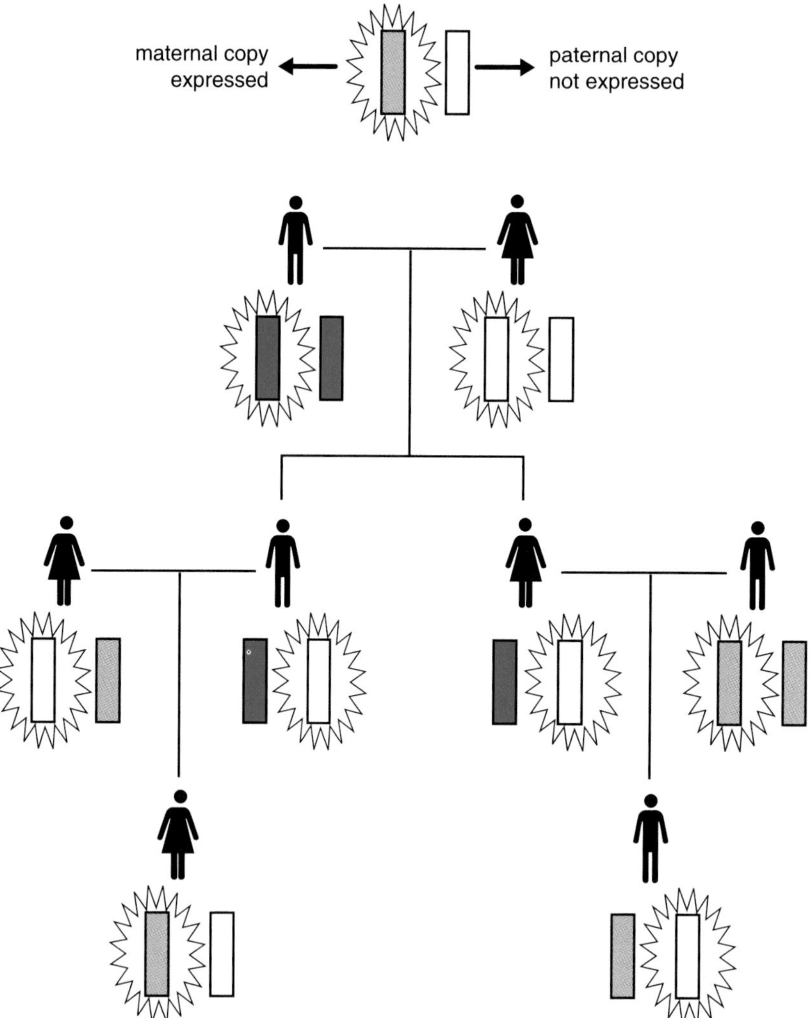

FIGURE 5-28 Concept of genomic imprinting. In this example, the paternally derived copy of a gene is not expressed, whereas the maternally inherited copy is expressed. The imprint is "reset" in the germ line, so that in the next generation, the active copy of the gene depends on the parent of origin, not on whether that copy was active in the parent.

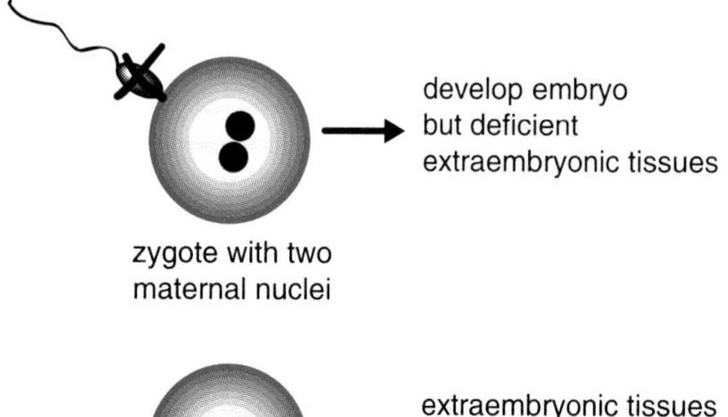

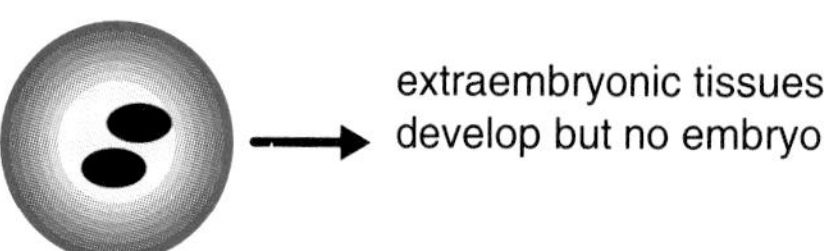

FIGURE 5-29 Using microsurgical techniques of nuclear transplantation, mouse zygotes were created that contained two male or two female pronuclei. Both types of zygotes were incompatible with normal development. Those with two maternal genomes produced embryos without extraembryonic tissues, whereas the opposite was seen in zygotes with two paternal pronuclei. (Data from McGrath J, Solter D. Development of mouse embryogenesis requires both the maternal and paternal genomes. Cell 1984;37:179–183.)

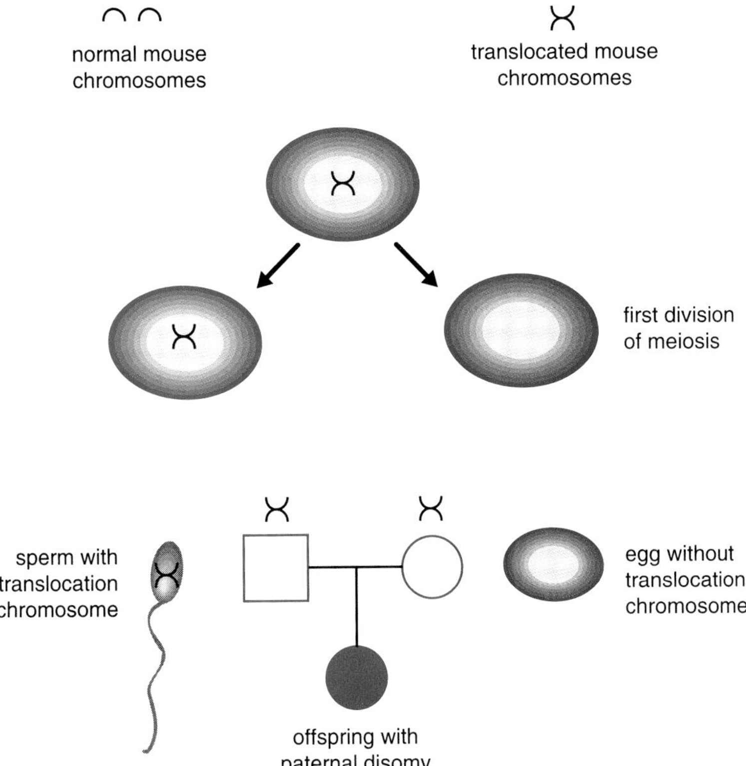

FIGURE 5-30 Imprinted chromosomes in the mouse. Mouse chromosomes are normally telocentric (centromes at end of chromosome). A Robertsonian translocation between the two copies of a homologous pair results in a single metacentric chromosome. During meiosis, one cell receives the translocation chromosome and the other does not. If both parents of a mouse cross have the translocation, some offspring will inherit the translocation chromosome from only one parent and therefore have two copies of the chromosomes that contributed to the translocation from the same parent. This is referred to as *uniparental disomy*. For some chromosomes, this makes no difference to mouse development but, for others, phenotypic changes do result, indicating that some mouse chromosomes have imprinted genes, whereas others do not have imprinted genes with an obvious effect on development. (Data from Cattanach B, Kirk M. Differential activity of maternally and paternally derived chromosome regions in mice. Nature 1985;315:496–498.)

been reported wherein the offspring of a phenotypically normal balanced translocation carrier has an abnormal phenotype in spite of having inherited the same balanced translocation. Some have been found to have inherited from the same parent both the translocation chromosome and one of the normal chromosomes involved in the translocation, resulting in uniparental disomy (see Figures 5-29 and 5-33). Others have been found to have deletions of material not found to be deleted in the parent, suggesting instability of the rearranged chromosome. It is apparent that the full range of phenotypic consequences of chromosomal abnormalities has not yet been discovered.

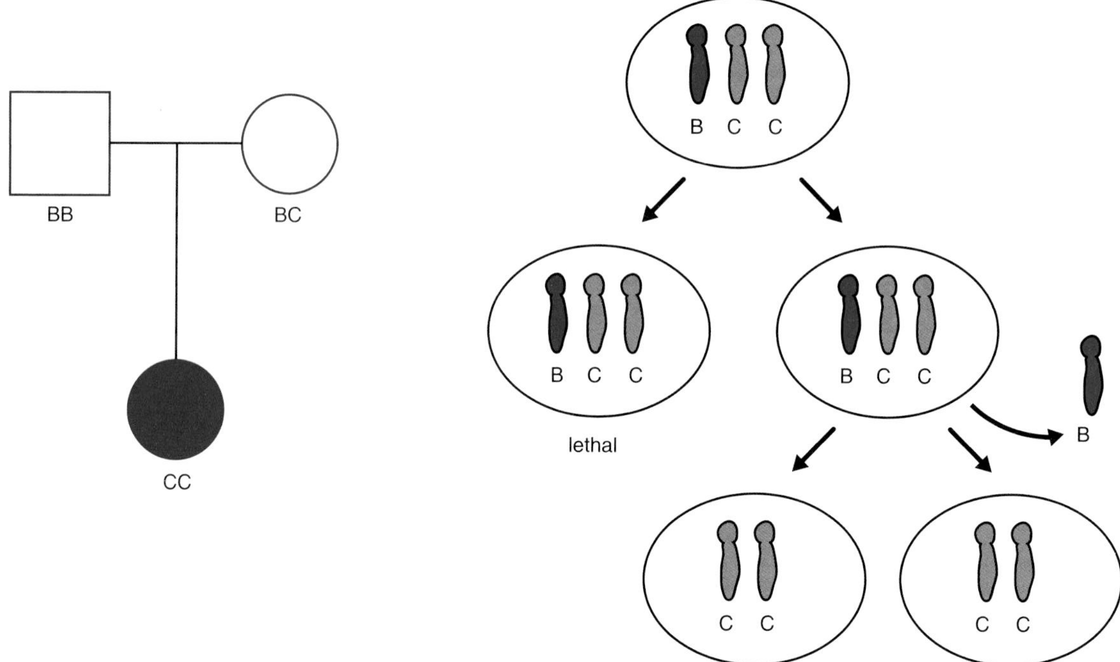

FIGURE 5-31 Maternal isodisomy resulting in cystic fibrosis in a child. Parental haplotypes for a polymorphic marker closely linked to the cystic fibrosis gene are indicated. The affected child has cystic fibrosis along with short stature, inheriting two copies of the marker from her mother, who is a cystic fibrosis carrier. The paternal allele *B* is not represented in the child. As indicated at the right, the child is believed to have started as a trisomy 7 conceptus, owing to nondisjunction in the second division of meiosis in the mother. Trisomy 7 would be lethal but, if a second nondisjunction event early in meiosis resulted in loss of the paternal chromosome 7, disomy would result, allowing the embryo to develop to term. Maternal disomy (isodisomy in this case, because the two copies of chromosome 7 arose from the same homolog) here leads to the child having cystic fibrosis and, due to the effect of unidentified imprinted genes, short stature as well. (Data from Spence JE, Perciaccante RG, Greig GM, et al. Uniparental disomy as a mechanism for human genetic disease. Am J Hum Genet 1988;42:217–226.)

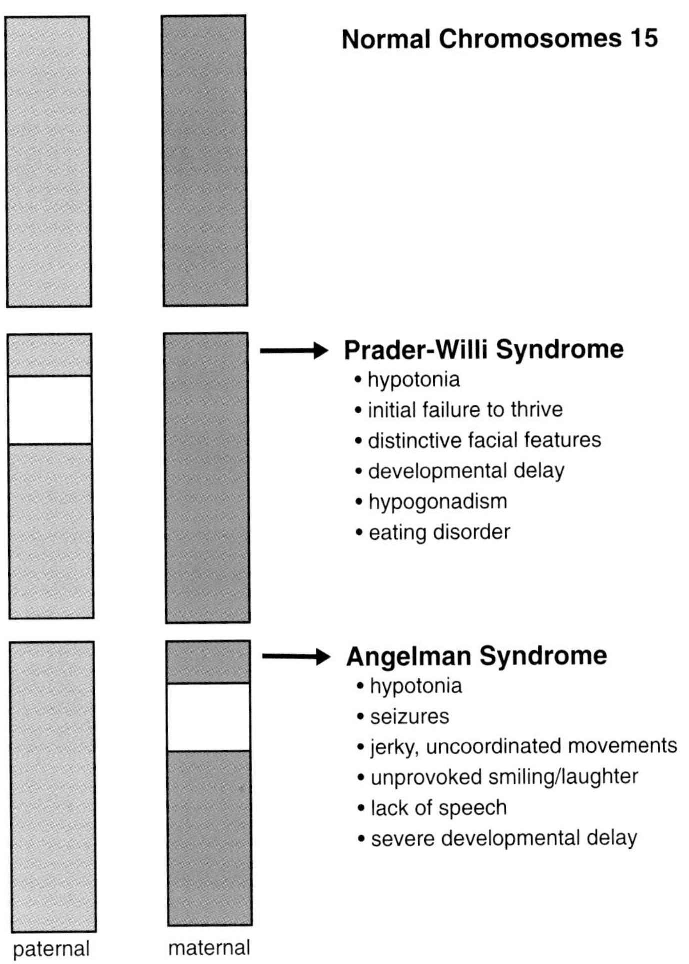

FIGURE 5-32 Deletion of sequences from the paternal copy of chromosome 15 results in Prader-Willi syndrome, whereas deletion from the maternal copy of chromosome 15 leads to Angelman's syndrome. (Data from Knoll JHM, Lalande M. Cytogenetic and molecular studies in the Prader-Willi and Angelman syndromes: an overview. Am J Med Genet 1993;46:2–6.)

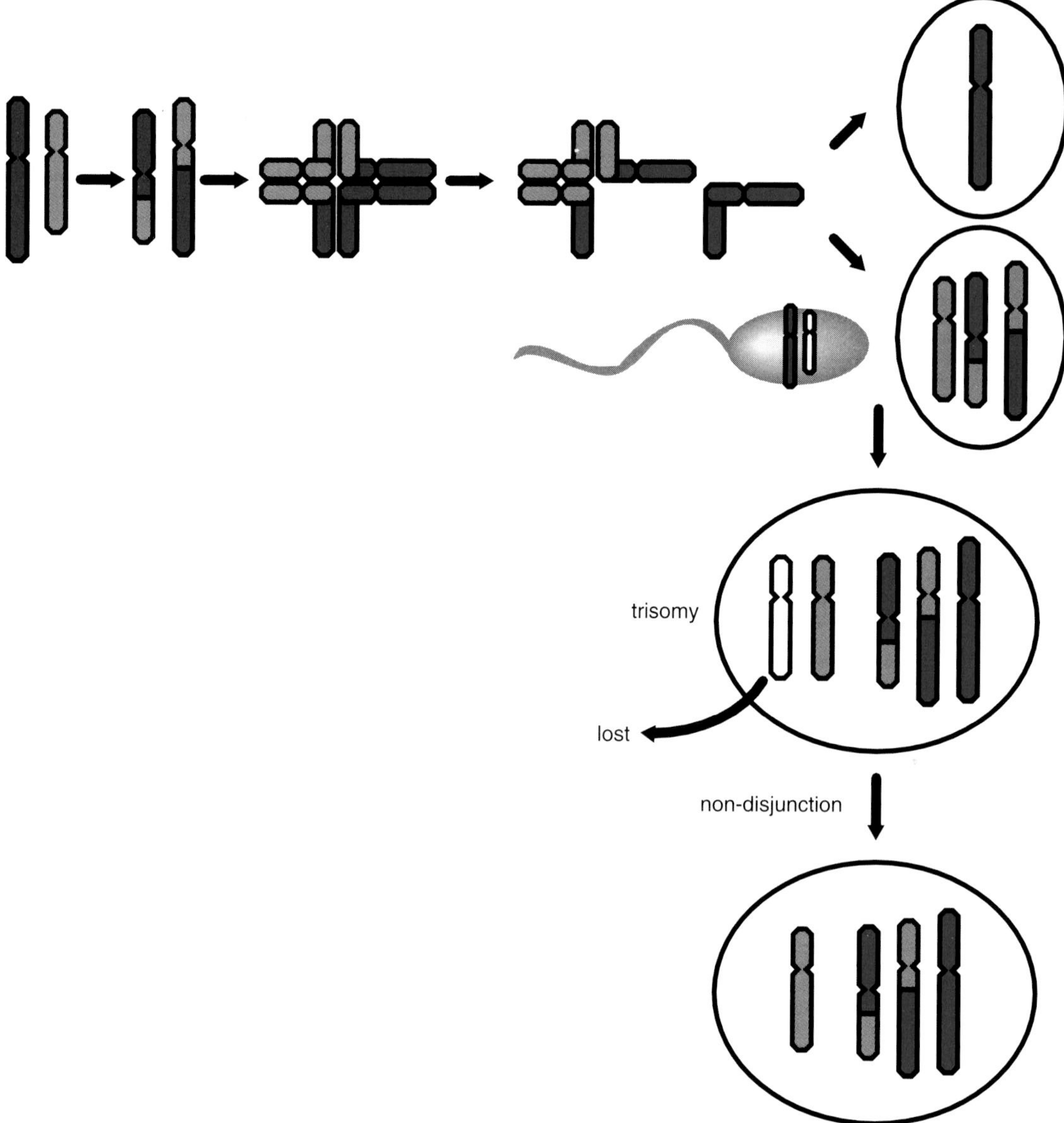

FIGURE 5-33 Uniparental disomy for chromosome involved in balanced translocation. During meiosis, three of the chromosomes segregate to one cell and only one to the other. This is referred to as 3 : 1 segregation and results in trisomy for one chromosome after fertilization. A subsequent nondisjunction event results in uniparental disomy, which may have phenotypic consequences if the chromosome carries imprinted genes.

Perspective

Living with a Child with a Chromosome Deletion

KATHLEEN CALLAHAN

Having never needed any medical attention, I went to a local obstetrician on learning I was pregnant with my first child. I was not gaining much weight, and my doctor, patting me on the back, would say, "Good job." His waiting room was full, and I was always asking questions as he slipped out the door. "This must be what it is like to have a baby," I thought. What little I knew then.

At 33 weeks, I went into labor and was put on an alcohol drip at an area hospital. After spending 24 hours in the dark ages, I was transferred to a major Boston facility. As each shift changed for the next 2 days, so did my doctors. Tired of reciting my story, I wished they would first read my chart. In the stillness of the night, my first child was born in an operating room filled with blue outfits and blank stares. Weighing just over 2 pounds, she was whisked away. I never saw her. Hours later, her father and I learned she was blind.

From that moment on, each day brought new calamities, until we were numb with pain. One team of experts after another came to look at our daughter. All of us were searching for answers. The genetics staff asked to look at our fingers and toes. We heard terms such as *FLK* and *dysmorphic* and were horrified on learning their meaning.

Three months later, with every possible test completed and repeated, we left the neonatal intensive care unit (NICU) with no diagnosis, just a general consensus that our Alissa would never be able to do much of anything. I had never given disabled people much thought before; now I was painfully learning. At that point, I felt any enjoyment in my life was over. How could I take out a dysmorphic baby? And so I hid her under my coat at the pediatrician's office.

Alissa cried most of the time, as did I. I wondered if her life held any chance of happiness. I feared the days ahead with absolute terror.

At 9 months, Alissa started having seizures. On her first birthday, she weighed 8 pounds. During her second year, Alissa cried less. Hope battled with despair. We still searched for answers no one had. We soon welcomed the words "I'm sorry, I don't know," instead of some of the outrageous explanations we received from some professionals.

It was my desire to have another baby. We moved far from the town in which we lived after a toxicologist suggested that we were perhaps exposed to something. I sought the best obstetricians in the city, and in consult after consult, I heard I had nothing to fear. I did not have a medical condition. I finally met the physician of my dreams. I actually jumped in the air after leaving his office. He was knowledgeable and compassionate. He listened to my concerns, recognized my fears. Six weeks into my second pregnancy, I started bleeding. "It is happening again," I thought. My doctor suggested I call the geneticist to lessen my anxiety. On doing so, we learned about a physician who had seen Alissa in the NICU and who had another patient with a syndrome that reminded him of Alissa. In the weeks ahead, a diagnosis of 4p minus syndrome was made in Alissa. My husband and I decided to be tested. We were aware there was no turning back. We now knew the child I was carrying was also at risk. No one ever gets accustomed to waiting for test results. It is pure anguish.

Tremendous joy engulfed us when our second daughter was born fat and healthy. I remember feeling constantly happy that year. Alissa's problems became an element in my life rather than consuming it. I could now have chats with other moms about babies without feeling uncomfortable. Although we still faced special problems, I could hardly believe we had a thriving child. We became alive again.

Case Study

Part I

Kim and Lewis are at their wits' end in trying to deal with their son Jeremy. Jeremy is 8 years old, and he is impossible to manage. He is in constant motion, always getting into things, often getting himself hurt. Jeremy is in first grade, and he is having major problems with schoolwork. It has been difficult to discern whether his school problems reflect inability to do the work or a problem in focusing his attention.

Kim and Lewis will do whatever is necessary to help Jeremy, and themselves. His problems have put a strain on their marriage, in part because of a disagreement over how to discipline him. The strain is also affecting his 10-year-old sister, Kelly. In addition, Kim is getting pressure from her parents, who live nearby and think that she is being too strict with Jeremy. Kim's sister Heidi has a son, Tim, with a temperament just like Jeremy's, which makes for eventful family gatherings.

In desperation, Kim and Lewis bring Jeremy to the pediatrician to discuss what can be done to help him. Jeremy's pediatrician has always been reassuring about his behavioral problems. Even when Jeremy was very young, his parents had concern about what seemed to be slow development, particularly compared with his older sister. He was an irritable baby, which was attributed to colic. He didn't sit on his own until 8 months and began walking at 17 months. There is a range of normal times to achieve these milestones, his parents were told. Speech was particularly slow to come. He only began saying words well into his third year, and was not talking in sentences until 5. Jeremy has been receiving speech therapy and has made considerable progress, although sometimes his speech is still a bit difficult to understand.

The pediatrician cannot find anything "wrong" with Jeremy on examination but, in light of the concerns of his parents and the school, makes a referral to a neurologist. He also arranges for Jeremy to have tests of hearing and vision, both of which are normal.

Jeremy has both behavioral and developmental problems. These are relatively common in the general population and represent a major challenge in diagnosis and management. Jeremy's mother's sister also has a developmentally delayed son (Figure 5-34). X-linked recessive inheritance would be most compatible with the pedigree, although learning disabilities are common, so a nongenetic cause would be most likely.

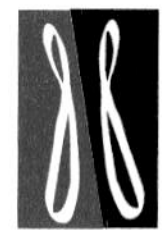

Part II

It takes weeks to get the neurology appointment, and it is necessary for both parents to take the day off from work to get to the visit. Much of the initial part of the evaluation takes place with only Kim in the room, while Lewis tries to park the car. Jeremy is seen first by a medical student, then by a resident, and finally by the neurologist. Jeremy is dismantling the doctor's office and is extremely restless when the time comes to discuss the results of the evaluation.

No specific physical problems have been found, and no neurologic deficits were noted. The diagnosis for

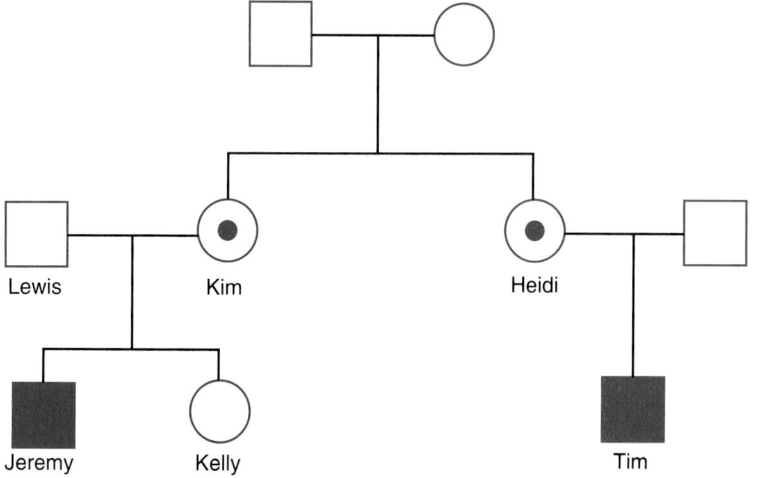

FIGURE 5-34 Pedigree of Jeremy's family.

Jeremy is attention deficit disorder, and it is recommended that Jeremy be started on methylphenidate. He is also referred to a psychologist to set up a behavior modification program.

The neurologist recommends that a number of blood and urine tests be done in an effort to rule out some possible causes of developmental impairment. With all the confusion in the room by the end of the visit, Kim and Lewis cannot remember what these tests were. They bring Jeremy to the blood-drawing laboratory on their way out of the hospital, making for a difficult end to a trying day.

Kim and Lewis are surprised to get a telephone call from the neurologist approximately 8 weeks after the visit. One of the tests done on Jeremy, a chromosome test, was found to be abnormal (Figure 5-35). Arrangements are made for Kim and Lewis to meet with the neurologist the next day. They leave Jeremy with his grandparents for that meeting, so they can concentrate on speaking with the doctor.

The neurologic evaluation was intended to determine whether there was a medical cause for Jeremy's problems and to determine whether medication might be helpful in treatment.

Laboratory evaluation for a child with learning and developmental problems has a low yield of positive findings. If the history and physical examination do not provide clues suggestive of a particular diagnosis, it would be common to check chromosomes, including fragile X, as well as plasma and urine amino acids. Other metabolic tests might be included if the history suggests events such as intermittent metabolic crises or if unusual features, such as motor abnormalities, organomegaly, or funduscopic abnormalities, are found on examination.

Children with attention deficit disorder and hyperactivity may respond to treatment with stimulant medications, such as methylphenidate (Ritalin). It is unclear why stimulants have the paradoxical effect of helping those with hyperactivity calm down and focus their attention, but this is often what happens. Contrary to some negative press, these medications do not sedate the child but rather seem to help the child focus and modulate his or her behavior. Used carefully, medication can be a turning point in the life of a hyperactive child, reducing stress both at home and at school. It should be used as part of an overall treatment plan, including psychological counseling.

The karyotype shows 46 chromosomes, with an XY sex chromosome complement. There are no structural abnormalities, but the X chromosome displays a fragile site at the tip of the long arm. This is the typical cytogenetic finding of fragile X syndrome. The fragile X site is seen only if the cells are grown in culture medium that is deficient in thymidine and folic acid, or in medium supplemented with an agent such as fluorodeoxyuridine (FUdR), an inhibitor of thymidylate synthetase. For some reason, depletion of nucleotide pools stimulates the expression of fragile sites such as fragile X. There are a number of other folate-sensitive fragile sites in the genome, but fragile X is the only one known to be associated with a phenotype. Usually, not all cells of a male with fragile X syndrome display the fragile X chromosome; the proportion may range from approximately 5% to more than 50%. The cytogenetic test now is rendered obsolete by direct DNA testing but, as recently as 1991, cytogenetic testing was the only means available for discerning these chromosomal abnormalities.

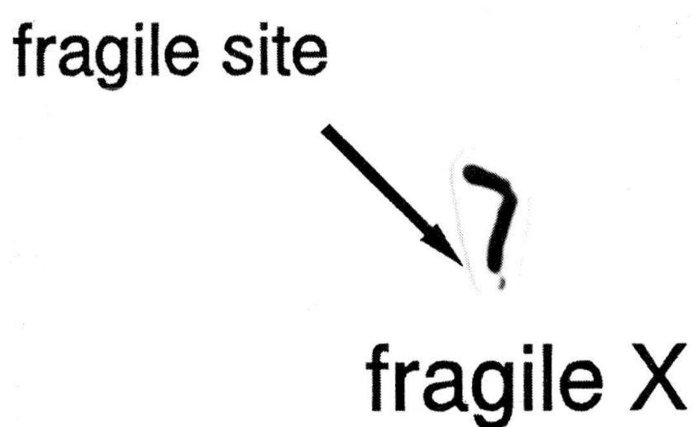

FIGURE 5-35 Fragile site at end of X chromosome.

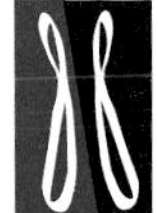

Part III

Kim and Lewis are told that Jeremy has fragile X syndrome, and they are shown a picture of his chromosomes. Both Kim and Lewis have college degrees, but neither has a strong science background. They have heard of chromosomes but never of fragile X. Both had occasionally used marijuana in college and begin to wonder to themselves about chromosome damage from drug use.

The neurologist explains that fragile X syndrome is a genetically determined disorder and that Jeremy has many of the characteristic features. The diagnosis does not alter the management plan that has been put together for Jeremy but does have implications for the family. Referring back to the family tree that had been put together by the medical student during the first visit, the neurologist suggests that Kim's sister have her son tested.

Many questions go through the minds of parents when a child is diagnosed with a disorder such as fragile X: What does it mean to their son, and what can be done to help him? What are their chances of having other affected children? Does this have implications for other members of their immediate family, such as Kelly, and for others in the extended family? How should others in the family be informed? How did this happen, and why? It is common for the parents of a child with a genetic disorder to blame themselves for what has happened. The term *fragile X* implies chromosomal damage—hence the point about drug use. The family history of another child with developmental delay certainly should lead to that child being tested as well.

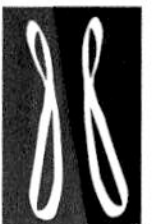

Part IV

Heidi brings Tim to his pediatrician for testing, and he turns out also to have fragile X. Gradually, both families have become very knowledgeable about fragile X syndrome. The Ritalin is working well for Jeremy, helping him to focus his attention, and it has been helpful for Tim as well. Both Jeremy and Tim are provided special help at school and are making encouraging progress. The families have joined a local fragile X support group and have met other families with the disorder. Kim and Lewis ask to have their daughter Kelly tested, and her chromosome study is normal. By now they know, however, that fragile X testing is not as reliable for carrier status in females as it is for affected males.

To the family's surprise, they learn that Kim's father's brother has a daughter who also has a developmentally delayed child, in this case a daughter. She is tested for fragile X and also is found to be positive.

Fragile X syndrome can affect both males and females. Approximately 50% of the male offspring and 30% of the female offspring of a carrier mother will develop fragile X syndrome. Kelly is not developmentally impaired, but she has a 50% risk of being a fragile X carrier. The cytogenetic test for fragile X is not as reliable for females as for males, however. A positive test in a female is significant, but many female carriers for fragile X have normal cytogenetic studies. The reason for this will be apparent later, when the fragile X mutation is discussed.

As word spreads through a family about a diagnosis such as fragile X, it is common to discover other affected individuals. In this case, a fairly distant relative also turns out to be affected. The pedigree can be modified as shown in Figure 5-36.

This presents what might appear to be an unusual dilemma for an X-linked trait—that two males would have to be carriers yet would not manifest signs of the condition. Of course, it is possible that this cousin has fragile X owing to a coincidental independent mutation. This would be a remarkable coincidence. Families like this are seen commonly in fragile X, and this phenomenon of nonmanifesting carrier males is referred to as the *Sherman paradox*. Empirically, it has been found that such transmitting males never have offspring with fragile X syndrome, but their daughters who inherit the mutation are at high risk of passing it on to their offspring. This represents another peculiar aspect to the genetics of fragile X syndrome, which will be explained when the molecular basis of fragile X is considered.

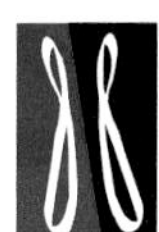

Part V

Approximately 10 years have passed since fragile X syndrome was diagnosed in Jeremy. Although he has required considerable extra help, Jeremy graduated from high school and now is working at a local restaurant clearing tables. Kelly is engaged to be married and remains concerned about the possibility that she is a fragile X carrier. She and her fiancé see a genetic counselor to discuss the current state of fragile X testing. They are told that genetic markers are now available that are located nearby the fragile site on the X chromosome. Genetic testing based on DNA analysis can be done, using genetic linkage to follow the fragile X gene through the family. Kelly and her fiancé believe that they have followed this discussion but definitely understand that the testing will require that blood specimens be collected from Kelly and other members of her family.

The data in Figure 5-37 is obtained and used to provide an estimate of Kelly's risk of being a fragile X carrier.

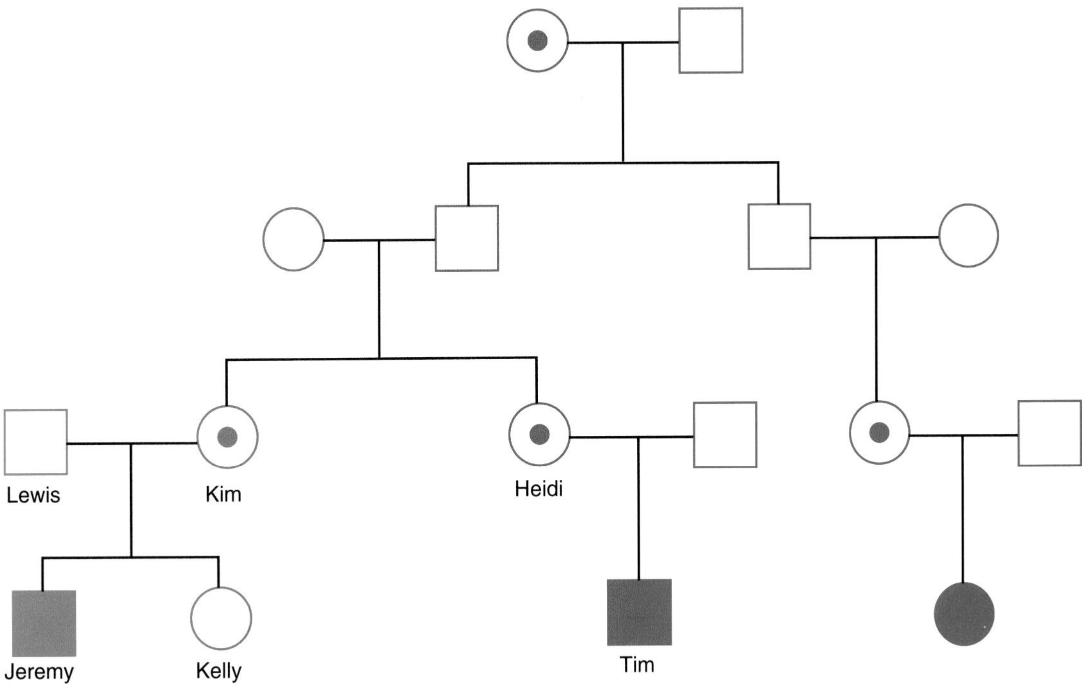

FIGURE 5-36 Modification of pedigree.

Prior to the advent of direct testing for the fragile X mutation, it was possible to use genetic linkage analysis to track the fragile X gene mutation through a family. The map depicted in Figure 5-37 indicates seven polymorphic genetic markers that were used in this family. *F9* is the gene for factor IX, involved in blood clotting. Otherwise, the markers correspond to "anonymous" DNA fragments, of unknown function but useful in linkage analysis. The distance between these markers and the fragile X locus (indicated by *FX*) is shown in the figure. Each marker was tested as a restriction fragment–length polymorphism on a Southern blot. The numbers under each symbol in the pedigree correspond with the band size (in kilobases) of each marker in that individual.

Kim is heterozygous for only one marker, RN1A. The other markers are therefore irrelevant to the determination of Kelly's carrier status. Kelly received the same 1.1-Kb allele as her brother Jeremy, indicating that she is likely to be a fragile X carrier. The distance between that marker and fragile X is 5 cM, which represents 5% recombination. We can calculate the risk that Kelly is a carrier as follows:

Risk =
(probability that the 1.1-Kb allele is in coupling with fragile X in Kim and that Kelly is nonrecombinant) + (probability that the 1.25-Kb allele is in coupling with fragile X in Kim and that Kelly is recombinant)

The probability that the 1.1-Kb allele is in coupling with fragile X in Kim is 95% (she gave this allele to Jeremy with the disorder), and there is a 5% chance that a recombination event occurred going from Kim to Jeremy. The probability that Kelly is nonrecombinant is also 95%, so the probability for the first part of the equation is: (0.95)(0.95) = 0.90.

The probability that the 1.25-Kb allele is in coupling with fragile X in Kim is 5% (because this assumes that Jeremy is recombinant). The probability that Kelly is also recombinant is 5%. Hence, the probability for the second part of the equation is: (0.05)(0.05) = 0.0025.

The total probability is 0.90 + 0.0025 = 0.9025. Kelly therefore has a high risk of being a fragile X carrier. Kelly is heterozygous only for markers on the other side of fragile X, which could be used to offer her prenatal diagnosis.

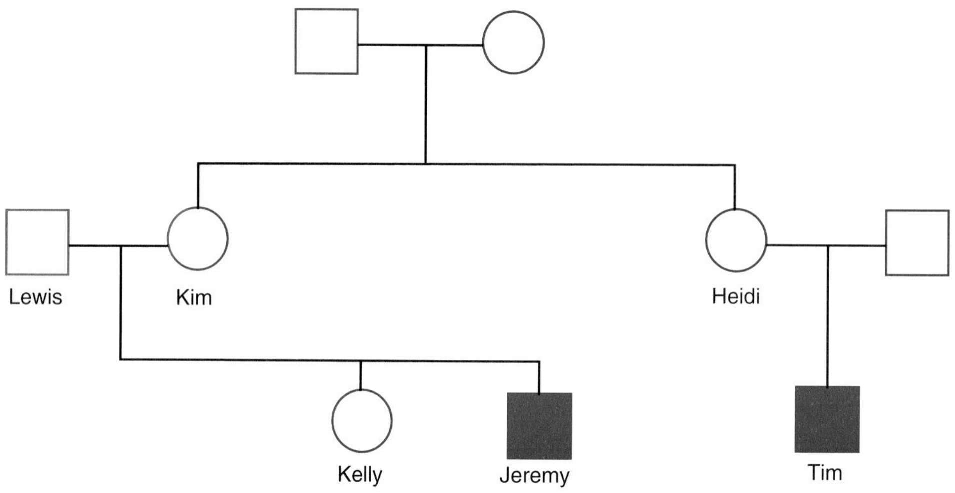

Lewis Kim Heidi

Kelly Jeremy Tim

	Lewis	Kim	Kim	Kelly	Kelly	Jeremy
F9	2.4	2.5	2.5	2.5	2.4	2.5
cX55.7	4.5	4.5	4.5	4.5	4.5	4.5
RN1A	1.1	1.25	1.1	1.1	1.1	1.1
FX						
VK21A	9.9	10.9	10.9	10.9	9.9	10.9
U6.2	7.0	7.0	7.0	7.0	7.0	7.0
U6.2-20E	5.8	9.6	9.6	9.6	5.8	9.6
1A1-1	2.9	5.2	5.2	5.2	2.9	5.2

(5 cM between RN1A and FX)

FIGURE 5-37 Restriction fragment—length polymorphism alleles for family members are indicated below the corresponding pedigree symbols.

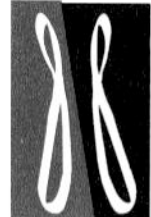

Part VI

Kelly understands that she is probably a fragile X carrier and that she has a risk of passing the trait onto either a son or a daughter. Four years have passed since the genetic linkage testing, and Kelly and her husband are now ready to start a family. Through the fragile X support group, they have heard that the fragile X gene has been identified, so they return to the genetic counselor to discuss whether this will allow more definitive testing of Kelly's carrier status and options for prenatal diagnosis. They are particularly interested in the possibility of preimplantation diagnosis of fragile X syndrome, which seems to them to be far preferable to amniocentesis or chorionic villus biopsy.

Blood is drawn from Kelly, and from other members of her family, to arrange for direct testing of the fragile X mutation (Figure 5-38).

The fragile X gene was isolated by positional cloning and is referred to as *FMR-1*. It is expressed in the central nervous system, but its function is thus far unknown. The gene has a peculiar structure, however, which is responsible for most of the mutations resulting in fragile X syndrome. In the 5′ untranslated region, there is a stretch of CGG bases repeated from approximately 5 to 50 times (Figure 5-39). The exact number of repeats differs in different individuals, representing a genetic polymorphism. When

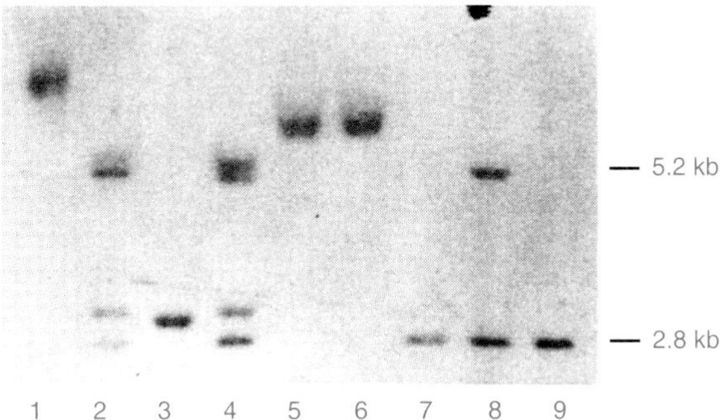

FIGURE 5-38 Southern blot showing results of fragile X analysis. Lane 1, control male with fragile X; lane 2, Kim; lane 3, Kim's father; lane 4, Heidi; lane 5, Jeremy; lane 6, Tim; lane 7, Lewis; lane 8, Kelly; lane 9, control male.

the repeat number is within this range, no signs of fragile X syndrome generally are present, and the number of repeats is stable as the gene is transmitted from generation to generation.

Individuals with fragile X syndrome appear to have enormous expansion of the repeat region, usually more than 200 copies of CGG. This expansion seems to lead to the methylation of the gene and, consequently, its inactivation. Lack of FMR-1 expression is believed to be responsible for the signs of fragile X syndrome.

Fragile X carriers tend to have between approximately 50 and 200 copies of CGG. This is referred to as *premutation*. Premutation does not affect the methylation of the gene but does render it unstable during meiosis. This instability leads to a jump in copy number into the range of greater than 200 copies, accounting for offspring of carriers being at risk for fragile X. The probability that expansion will occur into the fragile X range depends on the number of CGG repeats. The risk is approximately 20% for those with 60 repeats, 75% for those with 80 repeats, and virtually 100% for those with more than 89 repeats.

Males with fragile X premutations are nonmanifesting transmitting males. Expansion of the premutation to full mutation does not occur in male meiosis, but the daughters of such a male will inherit the premutation, and expansion to full mutation may occur in their offspring, which explains the Sherman paradox.

The Southern blot is performed by double digestion of samples with *EagI* and *EcoRI* and hybridization with a DNA probe from the fragile X region (Figure 5-40). *EagI* is a methylation-sensitive enzyme: It will not cut if the recognition site is methylated. In a normal male, *EagI* will cut and a single 2.8-Kb band will be seen. In a normal female, *EagI* will cut those copies of the gene on the active X but

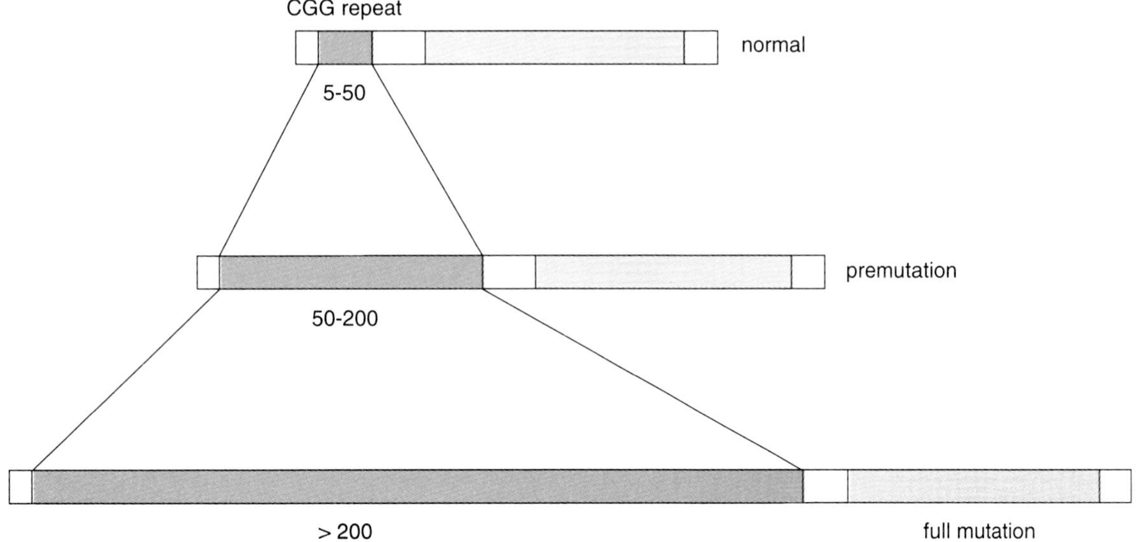

FIGURE 5-39 The CGG repeat region at the 5′ end of the FMR-1 is repeated from 5 to 50 times in the general population. Repeats from 50 to 200 copies are characteristic of premutation, and more than 200 repeats constitute full mutation.

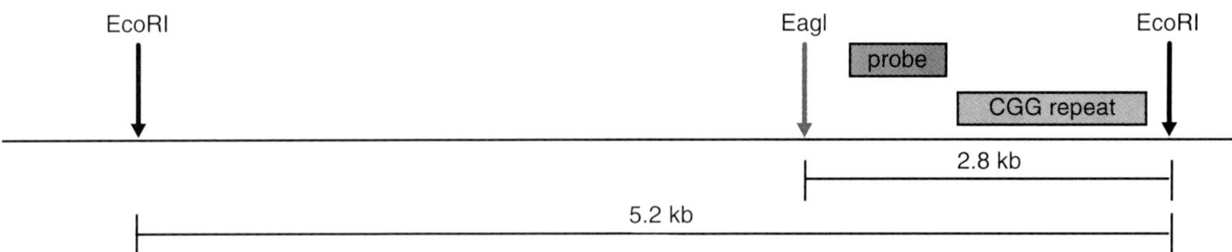

FIGURE 5-40 Map of region of FMR-1 gene showing CGG repeat, EcoRI sites, and EagI site. The approximate location of the DNA probe is indicated.

will not cut the copies on the inactive X chromosome. This is a result of the fact that X chromosome inactivation involves methylation of CpG dinucleotides. As a result, two bands will be seen: a 5.2-Kb band representing the inactive X and a 2.8-Kb band representing the active X.

A male with fragile X syndrome will have no 2.8-Kb band. Instead, there will be a signal at much higher molecular weight, due to expansion of the CGG region. This region also tends to be unstable during mitosis, and therefore there is considerable heterogeneity of its size, which leads to a smear in the high-molecular-weight region of the blot. A female with fragile X syndrome will have a normal pair of bands at 2.8 Kb and 5.2 Kb, representing the unaffected copy of the gene, which will be inactivated 50% of the time. In addition, a smear will be seen above the 5.2-Kb band, representing the expanded CGG, which is always methylated and hence does not cut with EagI.

A male premutation carrier will have a discrete band somewhat higher than the 2.8-Kb band. A female premutation carrier will have normal 2.8- and 5.2-Kb bands, representing the normal allele. In addition, there will be two other bands, just above the normal bands, representing the expanded allele. This allele does cut with EagI if it is on the active X chromosome, because premutation does not affect methylation.

Direct mutation testing can be used as a basis for prenatal diagnosis on chorionic villus biopsy tissue or amniotic fluid cells. Recently, there has been interest in performing preimplantation diagnosis. This involves induction of ovulation of multiple eggs by hormone treatment of the mother, collection of eggs by laparoscopy, and in vitro fertilization using father's sperm. The embryos are allowed to proceed to the eight-cell stage, and one or two blastomeres are collected using a micropipette. This does not damage the embryo, which can be frozen and later thawed and introduced into the uterus. DNA can be isolated from the blastomeres and tested for mutations using the polymerase chain reaction (PCR).

To diagnose fragile X syndrome by PCR, primers are used that flank the CGG repeat region. PCR amplification is done, and the size of the product is measured. This provides a measure of the distance between the PCR primers and, indirectly, of the size of the CGG repeat region. Often, full mutations cause the region to be so large that it cannot be amplified by PCR. In this case, the reaction would produce no product. To be sure that this does not represent failure of the PCR reaction, control primers outside the CGG region would also be used.

Part VII

Kelly and her husband are surprised and delighted to learn that Kelly is not a fragile X carrier after all. Jeremy is now 22 years old (Figure 5-41). He is currently unemployed and lives at home with his parents. He has no medical insurance but fortunately has been in good health. His parents ask about Jeremy's chances of transmitting fragile X to his offspring.

The results obtained with direct testing do not accord with those from the linkage study. Presumably, this indicates that either Kim or Jeremy did represent a crossover between the informative marker and fragile X. It highlights the fact that direct mutation testing is more reliable than linkage-based testing. Kim's risk of having a child with fragile X syndrome is essentially nil.

There is little information available about transmission of fragile X from affected males, few of whom reproduce. One study has shown recently,

FIGURE 5-41 Jeremy at time of high school graduation.

though, that sperm from males with full fragile X mutations do not themselves have full mutations. Thus, it is likely that Jeremy would pass on a premutation if he had children, but probably not a full mutation.

REVIEW QUESTIONS

1. At her first prenatal visit, a 25-year-old woman reports that her partner has a first cousin born with Down syndrome to the partner's aunt when she was 34 years old. The couple is no longer in contact with this branch of the family and does not know whether chromosome studies were performed for the diagnosis.
   a. You arrange to draw blood from the partner for chromosome studies. The couple understands that Down syndrome is due to an extra copy of chromosome 21, and both partners are puzzled about why you would want to test one of them. What outcome are you seeking?
   b. The chromosome study does not reveal anything that would indicate risk of having a child with Down syndrome. Incidentally, though, you discover a balanced translocation between chromosomes 3 and 11. What implications does this finding have for the pregnancy, and what further studies should be offered?
2. Chromosome analysis of a chorionic villus sample reveals mosaic trisomy 15. A subsequent amniotic fluid sample has only cells with normal chromosomes.
   a. How would you explain the discrepancy?
   b. A child is born subsequently, and Prader-Willi syndrome is diagnosed in this child. Is this likely to have been related to the prenatal diagnosis results?
3. A rare individual is found with neurofibromatosis type 1 (NF1), an autosomal dominant condition, and a history of recurrent miscarriages. Chromosome analysis reveals a balanced translocation between chromosomes 17 and 1. The breakpoints on chromosome 17 lie at the locus of the NF1 gene.
   a. How does the chromosome finding explain the miscarriages?
   b. Suppose you have FISH probes for the two ends of the NF1 gene. What might you expect to see if these are used together to stain this person's chromosomes?
4. Women with Turner's syndrome have gonadal dysgenesis, short stature, and sometimes other congenital anomalies, associated with a 45,X karyotype. Some are mosaics with 45,X/46,XX. In contrast, some fetuses recognized as having the same 45,X/46,XX karyotype are born with no discernible phenotype. How might this be explained?

FURTHER READING

5.1 A Newborn with Multiple Congenital Anomalies

Hirschhorn K, et al. Deletion of short arms of chromosome 4-5 in a child with defects of midline fusion. Humangenetik 1965;1:470.

Wolf U, et al. Deletion on short arm of a B chromosome without 'cri du chat' syndrome. Lancet 1965;1:769.

5.3 Subtle Chromosomal Abnormalities

Caspersson T, Farber S, Foley GE, et al. Chemical differentiation along metaphase chromosomes. Exp Cell Res 1968;49:219–222.

CASE STUDY

Caskey CT. Fragile X syndrome: improving understanding and diagnosis. JAMA 1994;271:552–553.

Caskey CT, Pizzuti A, Fu Y-H, et al. Triplet repeat mutations in human disease. Science 1992;256:784–789.

De Boulle K, Verkerk AJMH, Reyniers E, et al. A point mutation in the *FMR-1* gene associated with fragile X mental retardation. Nat Genet 1993;3:31–35.

Fu Y-H, Kuhl DPA, Pizzuti A, et al. Variation of the CGG repeat at the fragile X site results in genetic instability: Resolution of the Sherman paradox. Cell 1991;67:1047–1058.

Hagerman RJ, Jackson C, Amiri K, et al. Girls with fragile X syndrome: physical and neurocognitive status and outcome. Pediatrics 1992;89:395–400.

Hinds HL, Ashley CT, Sutcliffe JS, et al. Tissue specific expression of *FMR-1* provides evidence for a functional role in fragile X syndrome. Nat Genet 1993;3:36–43.

McConkie-Rosell A, Lachiewicz AM, Spiridigliozzi GA, et al. Evidence that methylation of the FMR-1 locus is responsible for variable phenotypic expression of the fragile X syndrome. Am J Hum Genet 1993;53:800–809.

Oudet C, Mornet E, Serre JL, et al. Linkage disequilibrium between the fragile X mutation and two closely linked CA repeats suggests that fragile X chromosomes are derived from a small number of founder chromosomes. Am J Hum Genet 1993;52:297–304.

Reyniers E, Vits L, De Boulle K, et al. The full mutation in the FMR-1 gene of male fragile X patients is absent in their sperm. Nat Genet 1993;4:143–146.

Rousseau F, Heitz D, Biancalana V, et al. Direct diagnosis by DNA analysis of the fragile X syndrome of mental retardation. N Engl J Med 1991;325:1673–1681.

Rousseau F, Heitz D, Tarleton J, et al. A multicenter study on genotype-phenotype correlations in the fragile X syndrome, using direct diagnosis with probe StB12.3: the first 2,253 cases. Am J Hum Genet 1994;55:225–237.

Tarleton JC, Saul RA. Molecular genetic advances in fragile X syndrome. J Pediatr 1993;122:169–185.

Warren ST, Nelson DL. Advances in molecular analysis of fragile X syndrome. JAMA 1994;271:536–542.

6 Multifactorial Inheritance

OVERVIEW

The best-studied genetic traits are not necessarily the ones responsible for the greatest worldwide burden of genetic disease. Single gene disorders tend to be sharply defined traits, for which the genetic contribution lies close to the surface. These were the first disorders to be recognized as inherited and the first to be studied at the biochemical, and then the molecular, level. Chromosomal abnormalities, at the opposite end of the spectrum in genetic complexity, have also been scrutinized in detail, owing to the ability to perform cytologic studies. Genes do not function in isolation, however, nor is the human a closed system isolated from its environment. Therefore, it should be no surprise that among the most common and important of genetic traits are those that are determined by combinations of multiple genes and their interactions with the environment. This is referred to as *multifactorial inheritance*.

The prototype disorder for this chapter is a relatively rare but well-studied congenital anomaly, spina bifida. This is a defect in closure of the neural tube that occurs early in embryonic development. Depending on the location and degree of severity, the malformation may be lethal or may be compatible with a productive life. After providing a brief introduction to the embryology of neural tube defects, we will see how these are etiologically heterogeneous and how empirical data provide the basis for genetic counseling. The general notion of multifactorial inheritance then will be introduced, and the use of twin studies to identify traits with a strong genetic component discussed. We will see that congenital anomalies are best explained by the threshold model of multifactorial inheritance. Neural tube defects are subject to prenatal diagnosis, and we will explore how this is done, both in pregnancies known to be at risk and in the general population via screening. One of the great advances in prevention of this defect in recent years has been recognition of the role of folic acid: Taken prior to conception, folic acid has a protective effect against neural tube defects. We will review the major study that led to this advance and see how such knowledge is being applied to improve prenatal care. If folic acid accounts for one of the major environmental influences on neural tube defects, the genetic factors remain to be elucidated. Animal models may be a key to uncovering these genes, so one such model will be described. Research on neural tube defects extends not only to cause and prevention but also to improved means of medical management. We will see how advances in treatment have improved the lives of those with spina bifida and raised ethical dilemmas regarding those with lethal defects such as anencephaly. The ultimate power of the genetic approach to multifactorial traits will be elucidation of the basis for common disorders, such as cardiovascular disease and cancer. We will look at progress in some of these areas in the final part of this chapter. The Perspective will consider what it is like to grow up with spina bifida, and the chapter closes with a case describing a child with a relatively common medical disorder, the genetic basis of which is gradually coming to light.

6.1 Neural Tube Defects

Alice goes into labor 2 days before her due date following an uneventful pregnancy. Her obstetrician knows that something is wrong with the baby as soon as she is delivered: Both legs are flaccid, and a large mass can be seen protruding from her back in the lumbar region (Figure 6-1). The obstetrician explains to Alice and her husband, Jack, that there is a problem. After Alice and Jack have a brief look at their new baby, she is taken by a pediatrician to the special care nursery. Approximately a half hour later, Jack speaks with the pediatrician, who tells him that their baby has a form of spina bifida called myelomeningocele.

The central nervous system forms during early embryonic life (Figure 6-2). First, the central portion of the embryonic disc thickens, and then a groove forms. Beginning in the third week postconception, this groove closes into a tube, first in the midcervical region and then progressing both rostrally and caudally, like the fastening of a zipper. The anterior neuropore closes at approximately 25 days postconception, and the posterior neuropore 2 days or so later. The most caudal portion of the neural tube forms by a different mechanism. Here, a solid core of tissue becomes canalized due to death of cells at the center and forms a tube.

Failure of closure of the neural tube results in a neural tube defect. Rarely, the entire neural tube remains open, a lethal defect known as **craniorachischisis totalis.** If the anterior-most part of the tube remains open, **anencephaly** results. This, too, is a lethal defect, as none of the essential structures of the forebrain are present. Defect in the region of the hindbrain is referred to as **encephalocele.** This also can be associated with severe neurologic deficits.

The most common type of neural tube defect involves the lumbar region. The defect may involve

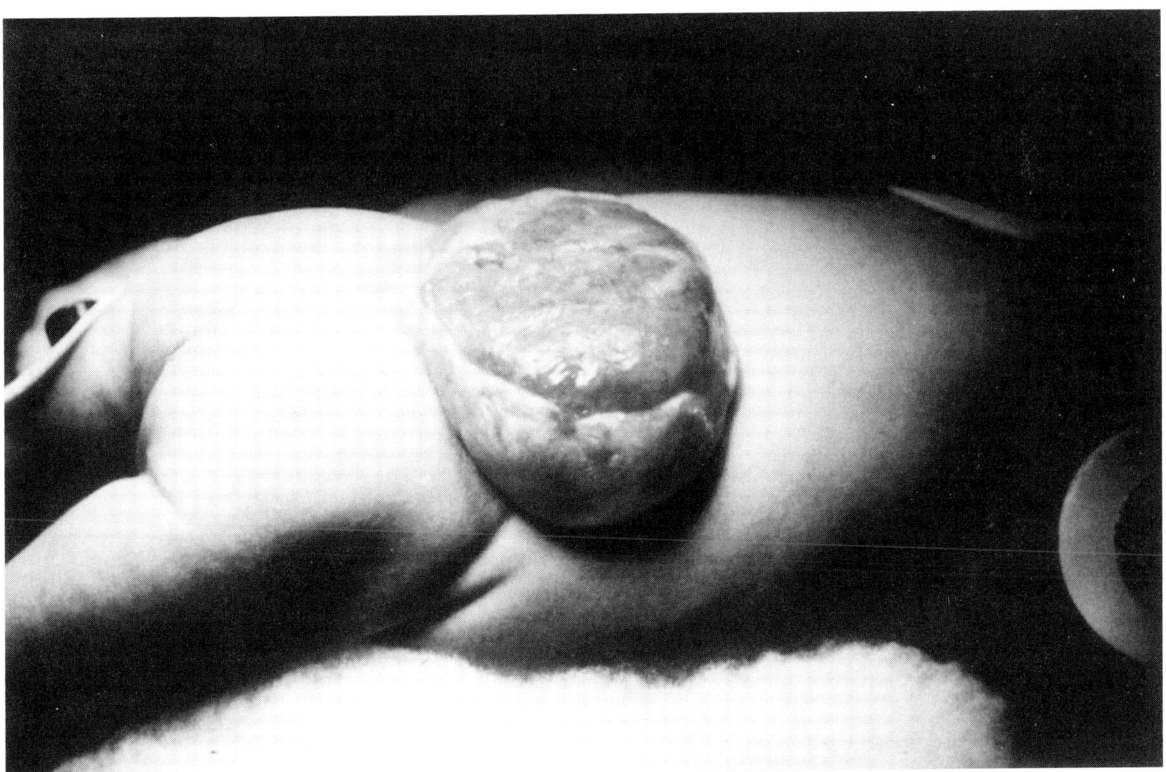

FIGURE 6-1 Child with lumbar myelomeningocele. (Courtesy of Dr. R. Michael Scott, Children's Hospital, Boston.)

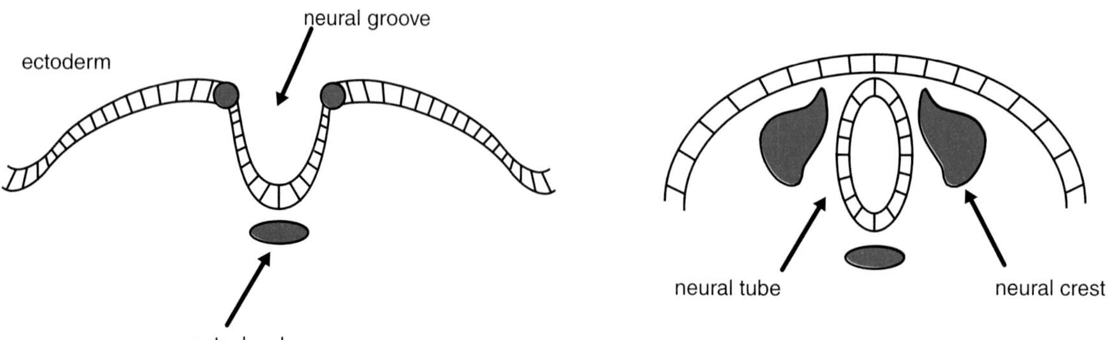

FIGURE 6-2 Diagram of neural tube development. The neural tube begins as a groove in the ectoderm (left), which then folds into a tube (right). Cells at the junction of the neural groove and ectoderm bud off and form neural crest.

the covering of the nervous system, the meninges (**meningocele**), or it may involve the meninges plus part of the spinal cord (**myelomeningocele**). In addition, there is a very mild form in which there is no protrusion of meninges or neural elements but incomplete fusion of the bony arch of a vertebral body. This is a benign condition called **spina bifida occulta.** It is believed that spina bifida occulta may result from failure of canalization rather than incomplete closure of the neural tube.

Sometimes the skin overlying a meningocele or myelomeningocele is broken down, in which case there is high risk of infection unless the baby is treated with antibiotics and the defect is closed soon after birth. Lumbar myelomeningocele results in paralysis of muscles that are innervated from spinal nerves originating from below the level of the defect. A child with a lumbar defect can lead a healthy and productive life but requires special assistance. Physical therapy is needed to maintain joint mobility. If the defect is very low, paralysis may affect only the ankles and walking will be possible using braces to support the feet. Higher defects lead to paralysis of the lower legs or up to the hips, requiring more extensive bracing.

Aside from muscle paralysis, the neural tube defect can also disrupt nerves to the bladder and bowel. The former leads to inability to empty the bladder. Voiding is achieved by catheterization, and the child usually will be taught to self-catheterize as he or she grows older. Lack of bowel innervation leads to chronic constipation, treated with medications that soften stools and help with evacuation.

Alice and Jack name their daughter Laura. Laura is started on antibiotics and she is taken to surgery later in the day for closure of the skin defect. The surgeons try to preserve as much nerve function as possible, but Laura is found to have no movement from the knees down and little movement at the hips. She requires catheterization for emptying her bladder. In addition, a ventricular drain is placed because of hydrocephalus.

Although the defect may be confined to the lower spine, central nervous system problems are common. The body wall grows faster than the nervous system during later development, so there is a relative displacement of the vertebral column in relation to the spinal cord. If, however, the cord is anchored to the body wall by the defect, this movement does not occur. The cord is pulled downward as the body wall grows, exerting traction on the brain stem, pulling it below the level of the foramen magnum, referred to as **Arnold-Chiari malformation** (Figure 6-3). This in turn distorts the spaces in the brain through which spinal fluid flows (ventricles). There is a resulting increase of spinal fluid pressure, or **hydrocephalus,** which leads to lethargy, vomiting, and possibly death, if left unchecked. The treatment is insertion of a catheter into the ventricle to drain fluid into the peritoneal cavity. This relieves the pressure, and function of the brain can be normal.

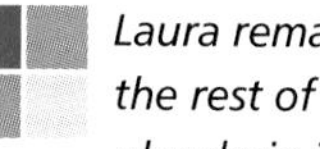

Laura remains in the hospital for 2 weeks, but the rest of her course is uneventful. Her ventricular drain is changed to a ventriculoperitoneal

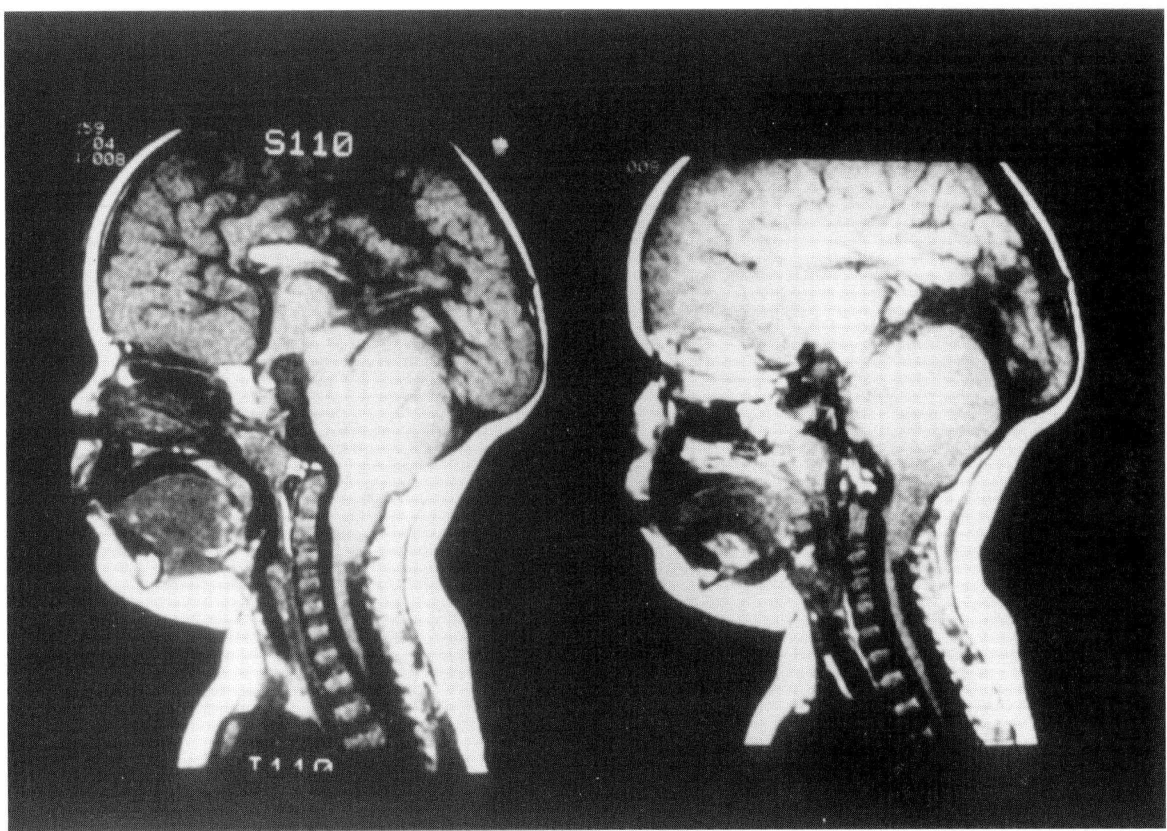

FIGURE 6-3 Magnetic resonance image showing Arnold-Chiari malformation. The brain stem and cerebellum are herniated below the level of the foramen magnum. The two images show different sagittal sections of the same individual.

shunt. She is feeding well and has been alert. The surgical scar is healing nicely. Her parents are taught to carry out bladder catheterization, and Laura is enrolled in an early intervention program with physical therapy.

6.2 Recurrence Risk of Neural Tube Defects

Laura is now 6 months old. She has been making good developmental progress, smiling responsively, reaching for objects, and babbling. She is not yet sitting unsupported, and movements of her legs are minimal. She has had no significant infections or other medical problems. In addition to early intervention, she is followed in a myelodysplasia clinic at a nearby hospital. Alice and Jack speak with a geneticist during one of Laura's follow-up visits. They had briefly met with a geneticist shortly after Laura was born but, in the midst of so much happening, remember little of what was said. Laura and Jack are both in good health, and neither has a family history of myelodysplasia. In fact, the couple had never heard of spina bifida until Laura was born. Alice and Jack are told that their risk of recurrence for future pregnancies is approximately 3%. They learn that another child with a neural tube defect might have myelomeningocele, like Laura, or could have a more severe defect, including anencephaly.

Neural tube defects are etiologically heterogeneous. Some occur in the setting of syndromes of multiple congenital anomalies, due to chromosomal abnormalities such as trisomy 18 or due to single gene disorders. In most cases, however, no underlying syndrome can be identified. Nevertheless, there is a tendency for neural tube defects to cluster in families. For genetic counseling purposes, recurrence risks have been determined empirically (Table 6-1). A couple that has had a child with a neural tube defect with no underlying syndrome faces about a 3% risk of a defect in a subsequent pregnancy. Risk goes down rapidly if only more distant relatives are affected.

Recurrence risks apply to all neural tube defects, irrespective of type or level. That is, a couple with a child with a lumbar defect faces the risk of any kind of neural tube defect in another child, including encephalocele, anencephaly, and the like. Likewise, a couple with a child with anencephaly may have another with a lumbar defect. Whatever factors lead to the recurrence of neural tube defects therefore apply to any kind of defect. There may be exceptions to this rule, however. Individuals with spina bifida occulta probably are not at increased risk of having children with higher defects. This may indicate that canalization defects have an etiology different from tube closure defects. It has been difficult to determine the exact level at which canalization begins, though, so this hypothesis remains untested.

Empirical data are used as the basis for genetic counseling for many other congenital anomalies (Table 6-2). Almost any of these might occur as a component of a chromosome anomaly syndrome, and many occur in single gene disorders. When found in isolation, however, the same tendency to clustering in families is noted as for neural tube defects, and empirical counseling is offered. It seems likely that genetic factors contribute to the etiology of these anomalies yet, unlike single gene disorders, the pattern of inheritance is not simple. Rather it appears that some combination of genetic and environmental factors contributes to their etiology, as will be explored in the next section.

TABLE 6-1 Recurrence risks for neural tube defects according to degree of relationship with proband

Affected Relative	Recurrence Risk (%)
First-degree (parent or sibling)	3.2
Second-degree (niece, nephew, aunt, uncle)	0.5
Third-degree (cousin, great-aunt, great-uncle)	0.17

Source: Data from Toriello HV, Higgins JV. Occurrence of neural tube defects among first-, second-, and third-degree relatives of probands: results of a United States study. Am J Med Genet 1983;15:601–606.

TABLE 6-2 Empirical data on recurrence risk of selected congenital anomalies

Anomaly	Population Incidence	Recurrence Risk in First-Degree Relatives (%)
Cleft lip ± cleft palate	1/1000	4.9
Congenital dislocation of hip	1/1000	3.5
Pyloric stenosis	1/500	3.2
Club foot	1/1000	2–8

Source: Data from Smith D, Aase JM. Polygenic inheritance of certain common malformations. J Pediatr 1970;76:653–659.

Empirical Risk Counseling for Multifactorial Disorders

The use of empirical data for counseling is based on the observation that some disorders tend to cluster in families and may be etiologically heterogeneous. The data must be used with the following cautions:

- The possibility of an underlying single gene or chromosomal disorder must be considered.
- Recurrence risks represent averages. Actual risks may differ in different populations. Also, in a given family, the actual risk may be substantially higher or lower than this average.
- Recurrence risk increases with proximity of relationship with the proband and with the number of affected individuals in the family. There may be gender differences in risk (see section 6.4).

6.3 Multifactorial Inheritance

The model that has been invoked to explain the recurrence of congenital anomalies such as neural tube defects is referred to as **multifactorial inheritance.** The simplest version of the model is one in which multiple genes exert an additive effect on a trait. This is referred to as **polygenic inheritance** and does not take into account a role for nongenetic factors.

As an example, consider a hypothetical genetic system in which two separate loci are involved in determining final adult height (Figure 6-4). There are two alleles at each locus, one dominant, which adds 2 cm to final height, and the other recessive, which adds nothing. Suppose that a person with the genotype *aabb* would be 150 cm tall. Someone with all dominant alleles would then measure 158 cm. A person with the genotype *AaBb* would be 154 cm, as would a person with the genotype *aaBB*. The effects

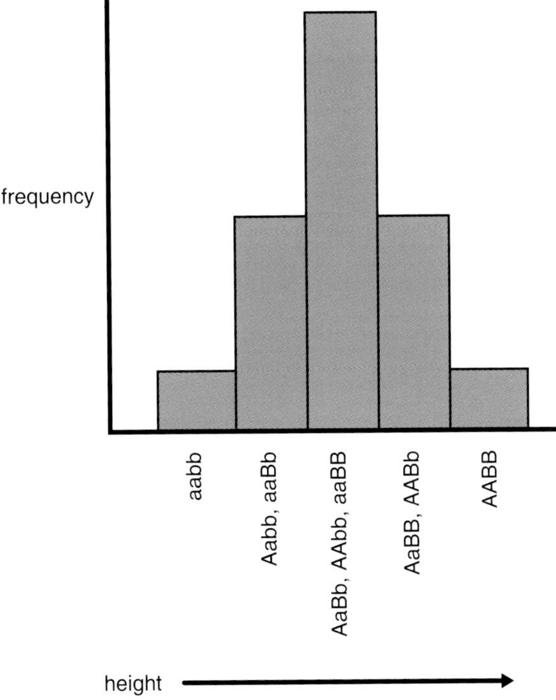

FIGURE 6-4 Simple hypothetical two-locus model for inheritance of height. The dominant alleles each add 2 cm to height and are assumed to be equally distributed in the population, with frequency of each allele being 0.5. According to this model, 1 in 16 individuals will have the genotype *aabb* and be the shortest. Likewise, 1 in 16 will be *AABB* and be tallest. One-fourth will have a single dominant allele or a single recessive allele. The majority will be of average height, having two dominant alleles. Note that, according to this model, it is the number of dominant alleles that determines height, not the specific combination (i.e., *AaBb* and *AAbb* give the same height).

of these genes, then, is additive with respect to height. If alleles are equally distributed, the additive model predicts a normal distribution of the quantitative trait in the population.

Most quantitative traits behave in a more complex manner, however. The trait may be influenced not only by multiple genes but also by environmental factors. There may, in addition, be interactions among genes and between genes and the environment. How, then, can the contribution of genetic factors be dissected out? One of the most powerful tools for doing this is the twin study. Twins can be either identical or nonidentical (fraternal). Identical twins arise by separation of blastomeres early in development or by duplication of the inner cell mass to produce separate embryos. These embryos will have an identical complement of genes. Fraternal twins, in contrast, arise from separate fertilization events. Although fraternal twins develop side by side in the same womb, genetically they are no more alike than any pair of siblings.

Comparison of the concordance of a trait in identical twins with fraternal twins or full siblings is a powerful way to define the degree to which a trait is genetically determined. For a single gene trait with complete penetrance, identical twins will, of course, be fully concordant. Siblings will be concordant less often, depending on whether they both inherit the mutant gene. For a nongenetic trait, concordance will be the same in identical twins or in full siblings and will depend on the degree of similarity of their exposure to environmental or other factors that determine the trait. For a multifactorial trait, concordance in identical twins will be greater than for siblings but not to the extent of a single gene trait with complete penetrance. The degree of concordance in monozygotic twins versus dizygotic twins or full siblings provides a measure of the contribution of genetic factors to the trait. Twin studies have helped to identify genetic contributions not only to congenital anomalies (Table 6-3) but also to common disorders such as hypertension, asthma, and diabetes.

Though powerful, twin studies are limited by the relative scarcity of identical twins who have a trait of interest, a problem that has been partly addressed through the development of twin registries. Another limitation is that twins share not only genetic identity but also some common environmental exposures, beginning with having developed in the same uterine

TABLE 6-3 Concordance rate for common congenital anomalies in identical and nonidentical twins

Trait	Concordance Identical Twins (%)	Concordance Full Siblings (%)
Cleft lip and palate	40	5
Pyloric stenosis	22	4
Clubfoot	32	3
Congenital dislocation of hip	33	4

Source: Data from Smith DW, Aase JM. Polygenic inheritance of certain common malformations. J Pediatr 1970;76:653–659.

environment. Studies of twins separated at birth control for postnatal environmental effects but not for prenatal effects. Another approach has been to compare the offspring of identical twins. These offspring are the equivalent of half-siblings and share half their genes, yet they are born and raised in different environments.

Multifactorial traits have also been studied by analysis of the segregation of these traits through families. Computer models are generated that assume the role of one or more genes, exerting dominant or recessive inheritance, with various levels of penetrance and factoring in interactions with the environment. Data from families then are evaluated, and the relative likelihood that observed data fit with the various models is determined. This *complex segregation analysis* has been used successfully to discern the genetic contribution to a number of common phenotypic traits.

6.4 Threshold Model of Multifactorial Inheritance

Many of the traits that are subject to multifactorial inheritance are quantitative, such as height and blood pressure. It is relatively easy to see how additive effects of multiple genes or environmental factors could determine the values of such traits. Other traits, particularly congenital malformations such as neural tube defects, are less easy to explain by the additive polygenic model. Such all-or-nothing traits are better explained by the **threshold model** of multifactorial inheritance.

The threshold model assumes that there is a "liability" toward development of a disorder that is normally distributed in the general population (Figure 6-5). This liability is composed of contributions from both genetic and environmental factors that can lead to expression of the trait. Individuals will have more or less liability toward the trait, depending on how many of the predisposing genes they have inherited and the degree to which they are exposed to the relevant environmental factors. Up to a point, they will not display signs of the trait. When a threshold of liability is crossed, however, the trait appears. Some will exceed threshold because of having many genetic risk factors and little environmental risk, whereas in others the major contribution to liability may be environmental but, once the threshold is reached, the effect is the same.

An interesting observation has been made in some multifactorial disorders that is handily explained by the threshold model. Some disorders display a gender predilection, affecting either males or females more often. Neural tube defects, for example, tend to occur more often in females. Another example is pyloric stenosis (hypertrophy of the muscle between the stomach and duodenum, leading to obstruction and vomiting), which is more common in males. The recurrence risk of pyloric stenosis is higher for families in which the proband is female than where the proband is male. The reason is that a female, being more rarely affected, is presumed to require a greater liability in order for threshold to be crossed (Figure 6-6). A couple having a daughter with pyloric stenosis is likely to transmit a greater number of genes predisposing to the disorder and therefore has a higher recurrence risk than those who have had a son with the disorder. Recurrence, by the way, is more likely to occur in a son than in a daughter, as the male would require less liability to cross threshold.

The threshold model is based on statistical analysis of the clustering of traits in families. The notion of genetic liability may seem vague. What is really going on? The answer is mostly unknown, as there are few multifactorial traits in which the contributing genetic factors have been identified. It is believed that genetic liability is accounted for by alleles at one or more loci that individually are unable to cause a distinctive phenotype. The phenotype emerges only if an individual has some critical combination of alleles at these loci. Sometimes this is sufficient to

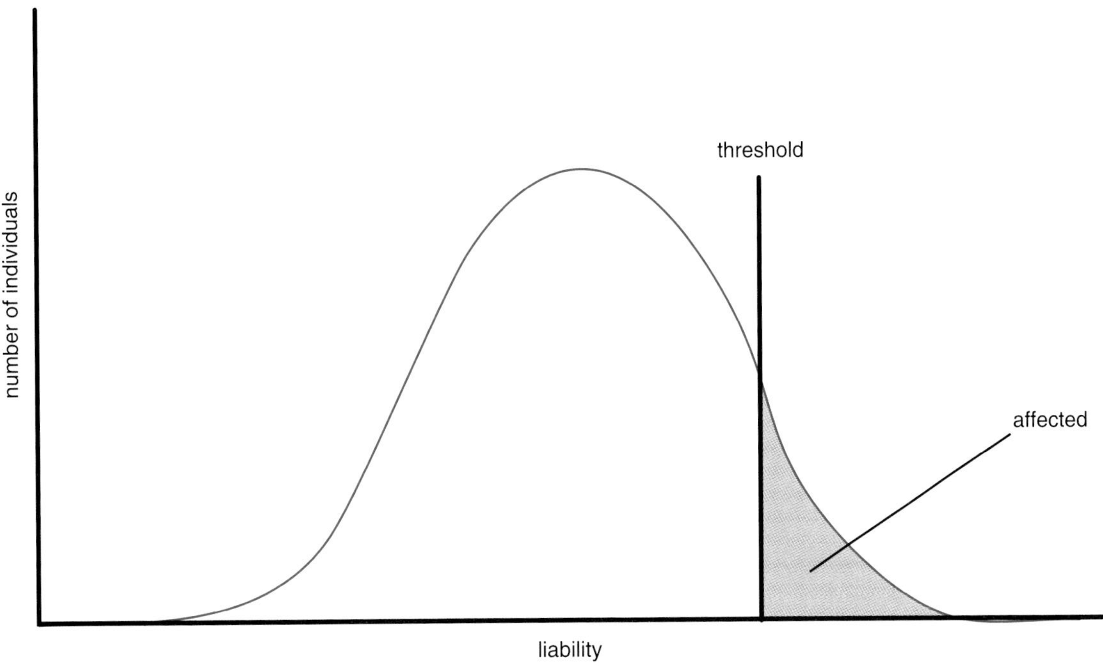

FIGURE 6-5 Threshold model for multifactorial inheritance. There is a liability toward the trait that consists of a combination of genetic and nongenetic factors and is normally distributed in the population. The trait is expressed only in individuals whose liability exceeds a threshold.

produce the phenotype, and other times it just leads to susceptibility to an otherwise harmless environmental agent. We will explore this further in the last section of this chapter, in which will be described some genetic systems that lead to common diseases.

6.5 Prenatal Diagnosis of Neural Tube Defects

Alice now is pregnant. Blood is drawn at 15 weeks for determination of alpha-fetoprotein (AFP), as well as unconjugated estriol and human chorionic gonadotropin. The results indicate a low risk for neural tube defects and Down syndrome. For further reassurance, an amniocentesis is performed at 17 weeks. AFP is found to be normal, and chromosomes are normal as well. A level II ultrasound is performed, and no evidence for neural tube defect or any other congenital anomaly is found (Figure 6-7).

Although the genes responsible for neural tube defects have not yet been identified, prenatal diagnosis is possible in most cases. The most direct approach is examination of the fetus by ultrasonography. A remarkably detailed picture of the fetus can be obtained with modern ultrasound equipment early in the second trimester. Neural tube defects such as anencephaly and encephalocele can be detected easily. Large lumbosacral defects are also easy to see. Smaller lumbosacral defects may be harder to detect unless one is deliberately searching for these (as would be the case if there were a family history of neural tube defect).

Another approach to prenatal diagnosis of neural tube defects involves detection of alpha-fetoprotein (AFP) in maternal serum or amniotic fluid. The use of AFP to screen for trisomies 18 and 21 was described in Chapter 5. Trisomy is associated with low AFP values but, with open neural tube defects, AFP tends to be elevated. The mechanism is probably leakage of AFP from fetal serum into the amniotic fluid and thence across the placenta into the mother's circulation. Levels of maternal serum AFP are

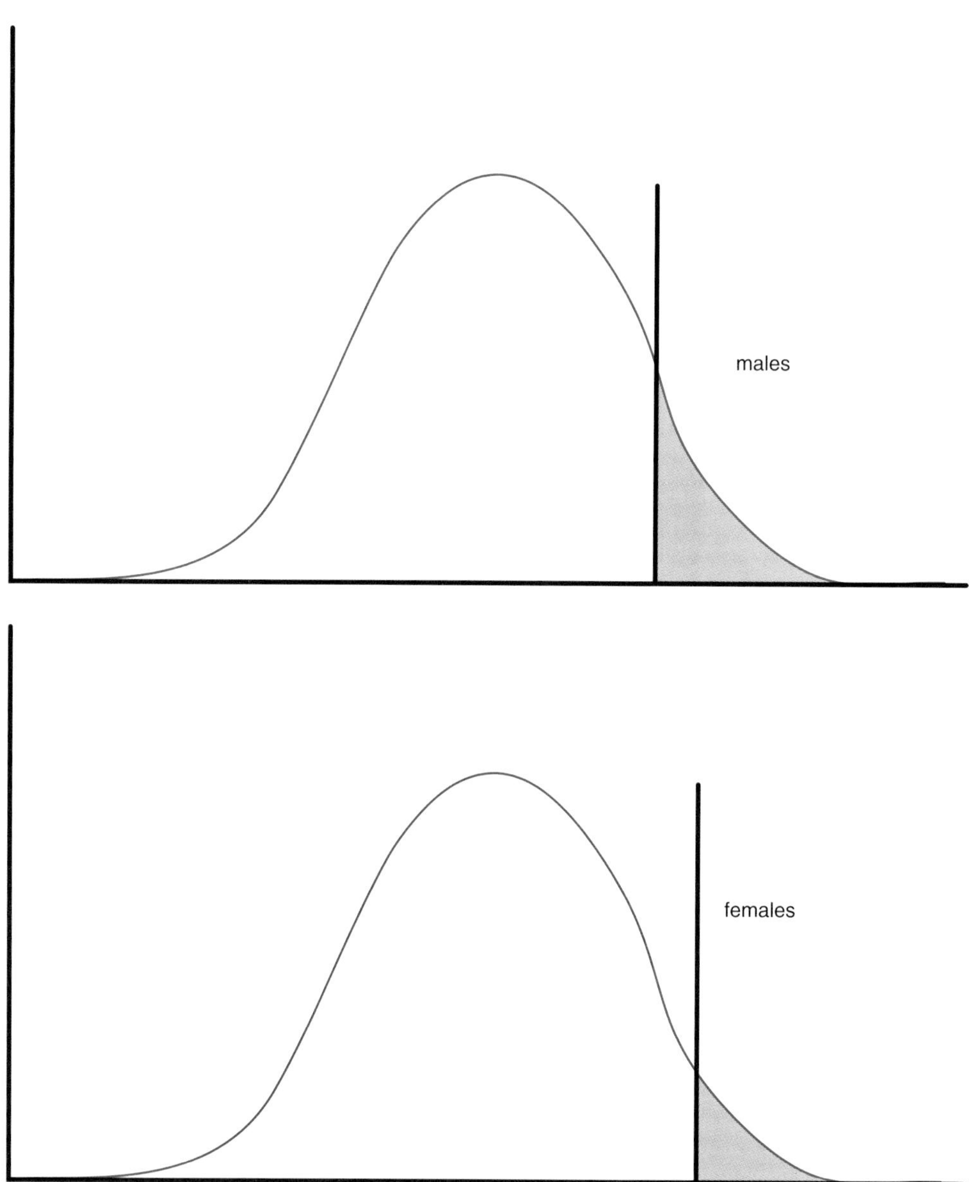

FIGURE 6-6 The threshold for expression of a multifactorial trait may differ in males and females. In this example, the threshold is higher for females. A couple with an affected daughter can be assumed to have a greater overall liability and, therefore, a higher recurrence risk.

elevated in approximately 60% of pregnancies in which the fetus has an open neural tube defect. The detection rate exceeds 90% if amniotic fluid AFP is measured. AFP is not elevated in the presence of closed defects, some of which can have important neurologic consequences. Furthermore, AFP testing is not specific for neural tube defects, as many other factors can lead to elevation of AFP (including other defects in the fetal body wall, fetal hemorrhage, twin pregnancy, renal disease, and incorrect assessment of gestational age). Follow-up study with ultrasonography should be done to confirm abnormal AFP results. Also, another biochemical test can be performed to confirm the presence of an open neural tube defect. This involves assay of the enzyme acetylcholinesterase in amniotic fluid. This enzyme is

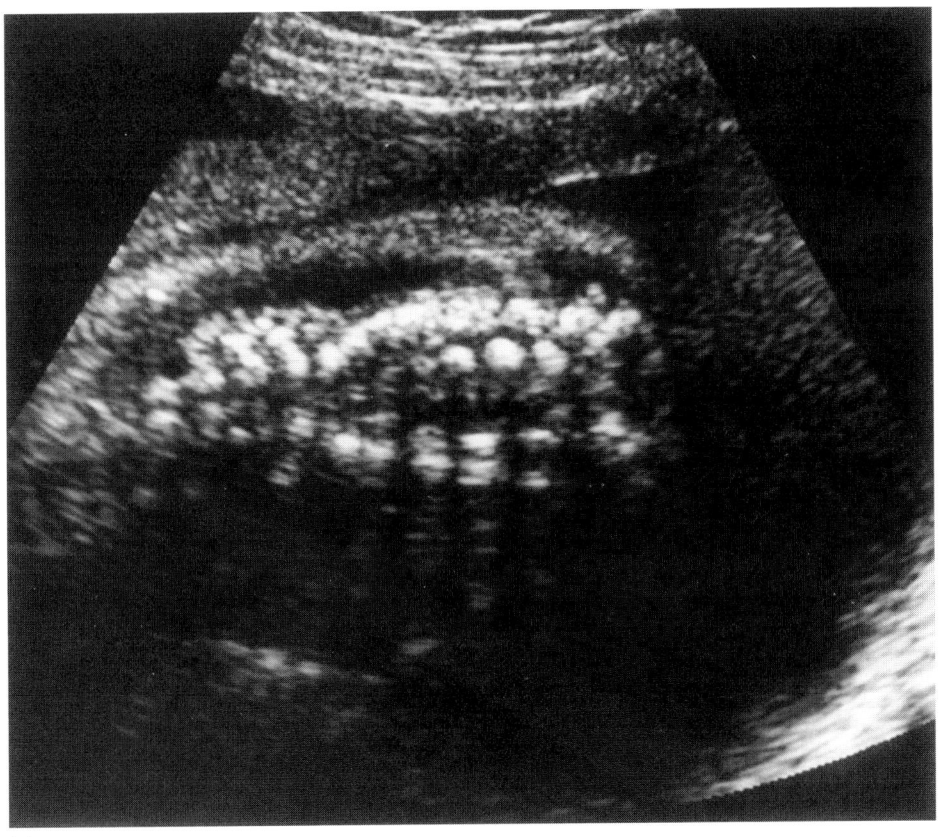

(a)

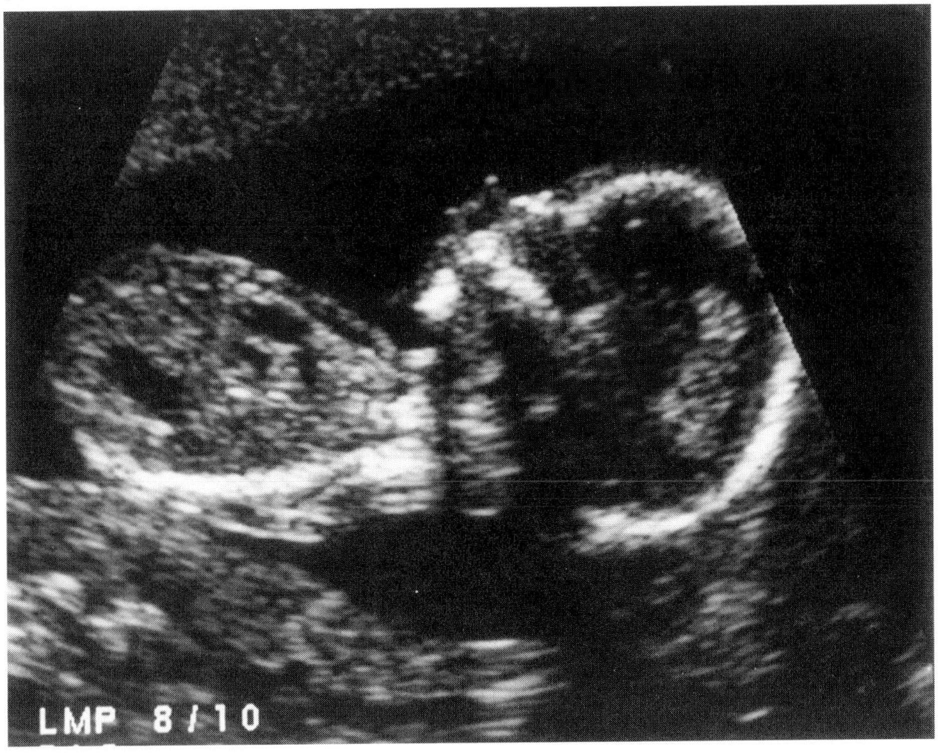

(b)

FIGURE 6.7 (**a**) Ultrasound view showing separation of lumbar vertebrae (looking at back of fetus). The vertebrae are widely separated due to myelodysplasia. (**b**) Sagittal ultrasound view of normal embryo. (Courtesy of Dr. Deborah Levine, Beth Israel Hospital, Boston.)

not normally present in amniotic fluid but is detectable if there is an open neural tube defect.

Prenatal diagnosis is offered routinely to women with a history of a neural tube defect in a first-degree relative. In many regions, all pregnant women are offered serum AFP screening, now most commonly as part of a "triple screen" (see Chapter 5). The initial rationale was detection of open neural tube defects and other congenital anomalies. The potential to detect chromosomal abnormalities was realized only later.

Five months later, Alice gives birth to a healthy baby boy, Joseph. The event is marred only by the fact that, at the same time, Laura is hospitalized to treat a shunt infection. She is discharged from the hospital approximately a week after her brother is.

6.6 Folic Acid and Neural Tube Defects

Alice is thinking about having another child. During a routine gynecologic visit, her physician suggests that she start taking folic acid supplements, 4 mg/day, as a preventive for neural tube defects. On hearing this, Alice points out that she took multiple vitamins during the pregnancy with Laura, although she cannot remember whether these included folic acid. She is told that folic acid can be effective only if treatment begins prior to conception. Indeed, Alice did not start taking the vitamins until nearly 6 weeks after she realized that she was pregnant. Now she feels guilty that this may have caused Laura's problems. Her obstetrician provides a prescription for folic acid, and Alice begins taking the pills that same night.

Studies of the epidemiology of neural tube defects have revealed an interesting pattern. Worldwide, the incidence is approximately 1 per 1000 live births, but there are marked variations in incidence in different populations. The rate has been reported to be as high as 1 in 200 births in the British Isles and 1 in 350 among East Indian Sikhs. Moreover, the frequency of neural tube defects is declining in most parts of the world. In England and Wales, for example, from 1972 to 1981 the rate fell from 1.47 to 0.39 per 1000 persons for anencephaly and from 1.88 to 1.04 per 1000 persons for spina bifida. A study in New York revealed a decline in the incidence of myelomeningocele from 0.4 to 0.25 per 1000 from 1968 to 1979. Clearly, something must be changing in the mix of causes that leads to neural tube defects.

In 1980, Smithells et al. reported a reduction in the incidence of neural tube defects among offspring of women who took a multivitamin supplement prior to conception. During the 1980s, a large multicenter trial was begun involving women in the United Kingdom and several other countries. Women with a history of pregnancy with a neural tube defect were randomized to receive folic acid or vitamins, both, or neither, on a double-blind basis. There was controversy over the inclusion of a group who did not receive vitamins or folic acid, largely because these supplements were viewed as harmless and potentially helpful. Altogether, 1817 women were enrolled in the study. Pregnancy outcomes were monitored for incidence of neural tube defects, and results were monitored on an ongoing basis by a committee. On April 12, 1991, the committee recommended that the study be stopped because data indicated a statistically significant difference in the incidence of neural tube defects in the folic acid–supplemented mothers compared with those who did not receive folic acid. The relative risk of neural tube defects in the folic acid group was 0.28 compared with the control group. No protective effect could be attributed to multivitamin use without folic acid.

Similar results were obtained in a number of additional studies. The United Kingdom study involved the use of folic acid, 4 mg/day, but a protective effect has been seen with much lower doses, in the range of 0.4 mg/day. It now is recommended that all women take folic acid prior to conception and through pregnancy. Women with a history of a pregnancy with a neural tube defect are urged to take 4 mg/day, whereas others take only 0.4 mg/day. The difference is based on the notion that those at higher a priori risk should take the higher dose, because this was the one most definitively shown to be protective and the risk of possible side effects is more justified in this group. (Actually, side effects of folic acid appear to be minimal. The major side effect is masking of the anemia that accompanies chronic vitamin B_{12} deficiency, leading to a delay in diagnosis of that disorder, which causes neurologic deterioration.) The reason for recommending preconceptual use is that the neural tube closes at approximately 4 weeks' post-conception, a time when most women are not aware that they are pregnant. If folic acid supplementation were to begin when pregnancy is recognized, it would be too late to protect against neural tube defects.

It is estimated that more than half of neural tube defects worldwide can be prevented if all women take folic acid. Providing the necessary education and access to the supplements is a formidable problem. Many countries are considering the fortification of cereal grains with folic acid, obviating the need to take supplements.

Alice gives birth to another daughter, Chelsea, who is also healthy. As with Joseph, prenatal testing was normal. Alice took 4 mg folic acid beginning at 6 or so months prior to conception and continuing through the pregnancy.

How does folic acid work? Folic acid is a cofactor for 16 enzymes, serving as a single-carbon donor-recipient. One of these enzymes, methionine synthase, catalyzes the conversion of homocysteine to methioinine (Figure 6-8). This is the only enzyme that uses vitamin B_{12} in addition to folate, and deficiency of vitamin B_{12} has also has been shown to be a risk factor for neural tube defects. In a prospective study of homocysteine levels in pregnant women, elevated values were found in those who gave birth to children with neural tube defects, and this elevation could not be explained by B_{12} or folate deficiency. Some women who give birth to children with neural tube defects may therefore have a deficiency of methionine synthase. This would lead to reduced levels of methionine, which is needed as a methyl donor. Methionine has been shown to be necessary for neural tube closure in rat embryos in culture. In addition, the methionine synthase reaction results in production of tetrahydrofolate from 5-methyltetrahydrofolate. Tetrahydrofolate is required for DNA synthesis. Administration of large amounts of folic acid to women with impaired methionine synthase might stimulate the enzyme, enhancing the amount of product in spite of the deficient enzyme activity. If this is correct, addition of vitamin B_{12} might have similar effects and reduce the dose needed to protect against neural tube defects.

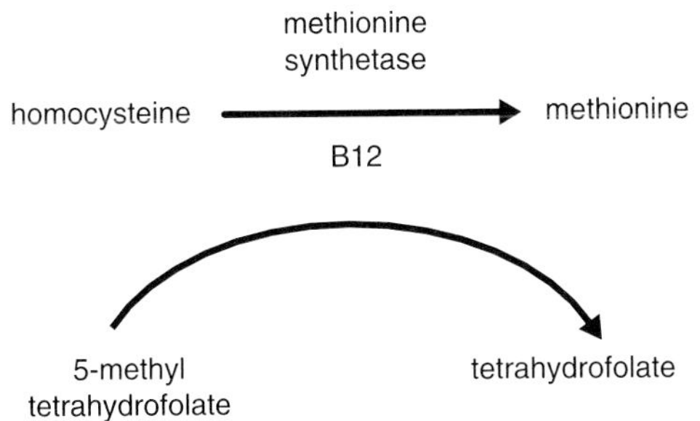

FIGURE 6-8 Conversion of homocysteine to methionine by methionine synthetase, with vitamin B_{12} and methyltetrahydrofolate as cofactors. Deficiency of enzyme activity or of either cofactor leads to accumulation of homocysteine and deficiency of methionine.

This model suggests that a high proportion of neural tube defects may result from deficiency of a single gene, that for methionine synthase. Appearance of the disorder, however, requires an additional factor—namely, low folate or vitamin B_{12} levels. Put differently, most women with impaired methionine synthase have sufficient dietary intake of folic acid and vitamin B_{12} to protect against neural tube defects, and most women with low folate or vitamin B_{12} intake have sufficient methionine synthase activity to protect their fetuses. It is the combination of low enzyme activity with low intake of one or the other cofactor that appears to result in the congenital anomaly.

This hypothesis has not yet been confirmed. Also, there are undoubtedly other genes (including methylene tetrahydrofolate reductase) that contribute to neural tube defects, and there may well be other environmental factors that sometimes act. Nevertheless, a high proportion of cases might be explained by a relatively simple model, involving just one gene and environmental factor. It is possible that other multifactorial traits will be related to the interaction of a single major gene with other genes and environmental factors. The story of neural tube defects shows the power of epidemiologic studies in sorting out such a complex problem.

6.7 Animal Models of Neural Tube Defects

Aside from the major insight provided by studies of the role of folic acid, little else is known about the pathogenesis of neural tube defects. Efforts to identify the gene or genes involved have been thwarted by the relative scarcity of families with more than one affected individual. Because of this, investigators have turned to the study of animal models, particularly the mouse.

In contrast with other conditions we have discussed, here it was not necessary to create an animal model. Mouse colonies are scrutinized routinely for variants that may represent spontaneous mutations and, over the years, many have been found to have neural tube defects. One such mouse is referred to as *curly tail* (Figure 6-9). The phenotype usually is expressed as a kink in the tail, associated with an opening in the lower part of the body near the tail. Like neural tube defects in humans, however, some mice are affected more severely, including a proportion that has exencephaly (i.e., anencephaly). The curly-tail phenotype is genetically determined by a single mendelian locus designated *ct*. Most animals that express the trait are homozygous *ct/ct*, but a small proportion are heterozygotes, indicating that the locus can behave as dominant at times.

The penetrance of the curly-tail phenotype differs in different strains of mice. The original curly-tail mouse arose in a non-inbred strain, but the trait has been introduced into a variety of inbred strains by mating. The frequency of affected animals varies as much as 30-fold in different inbred strains. This indicates that modifier genes at different genetic loci influence the expression of the curly-tail phenotype.

The pathogenesis of the curly-tail phenotype is unknown. It is not responsive to folic acid administration. Experiments have indicated that there may be an asynchrony of growth of endodermally derived gut and the ectodermally derived neural tube. The major progress in understanding curly tail has been made through gene mapping. In the mouse, this is done by selective breeding of curly-tail mice with others that have genetic markers of known location.

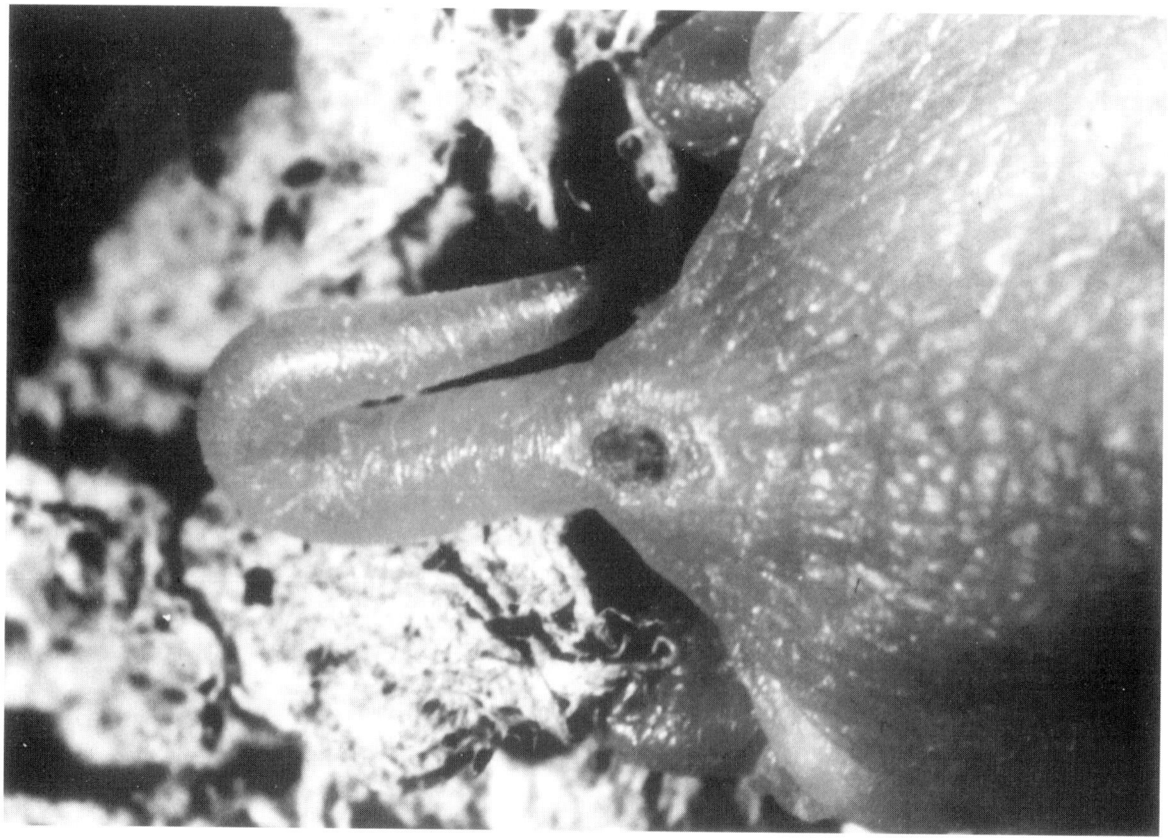

FIGURE 6-9 Lumbosacral defect in curly tail mouse. (Courtesy of Drs. Simon Watkins and Merton Bernfield, Children's Hospital, Boston.)

In this way, it was determined that the curly-tail gene resides on mouse chromosome 4.

Having identified the gene locus, the next step would be to identify the gene itself. Eventually this may be accomplished through a positional cloning strategy, although this can be very laborious and may take years. Alternatively, a candidate gene approach can be used. As the gene map becomes denser, chances improve that there will be many cloned genes located in the region where a trait has been mapped. These genes can be examined as possible candidates for the gene responsible for a trait of interest. Even this approach can be laborious, though, as it may be necessary to find very subtle mutations, even a single base change, in a large gene. Although shortcuts exist to help identify such mutations, the process remains slow. Also, there is no guarantee that the gene being sought has already been cloned.

The curly-tail phenotype is not determined solely by a single gene. Modifying factors, including other genes or environmental effects, must play a role, as not all animals with the curly-tail gene express the phenotype. Identification of the curly-tail gene would, however, represent a major advance in understanding neural tube defects. It would permit studies of the pathogenesis of the disorder and might facilitate the identification of other genes involved in the process. These might be homologous genes located elsewhere in the genome or genes whose products interact with the curly-tail gene product. Moreover, it is likely that the gene or genes involved in other mouse neural tube defects are homologous with genes in humans that have similar effects. Eventually, this may improve further our ability to detect pregnancies at risk or to provide better means of treatment.

6.8 Care of Persons with Neural Tube Defects

Laura is now 9 years old. She attends a public school and is in third grade. She receives special help in reading but otherwise is in regular classes. Her health has generally been good, although she has had to be hospitalized several times due to shunt infections and urinary tract infections. Laura has learned to catheterize herself, which she does several times each day. She spends most of her day in a wheelchair, although with braces and a walker she can get around without the chair. Her family has modified their home to accommodate Laura's needs. Laura is an active child who has many friends. She and her parents know several other children with myelodysplasia, mostly through attending the myelodysplasia clinic and a local support group.

Clinical research on neural tube defects has focused on two areas: prevention and management. Research on prevention has led to methods of prenatal testing and screening to identify pregnancies at risk. It also has revealed the role of folic acid, resulting in efforts to reduce the incidence of neural tube defects. A side benefit is the opportunity to increase public awareness of the importance of prenatal care and maternal-child health in general.

The incidence of neural tube defects was already diminishing before these advances, and it is likely to go down further as new knowledge is incorporated into routine prenatal care. Nevertheless, neural tube defects will not be eliminated. Folic acid supplementation does not protect against all such defects. Undoubtedly, other genetic or environmental factors exist that lead to the disorder in some cases. It also is unlikely that the entire population in any area will be screened for maternal serum AFP. Even if it were, the sensitivity of this screen is only 60%. Finally, many women found to be carrying a fetus with a neural tube defect will not choose to terminate the pregnancy. Indeed, prospective parents will want to know what forms of treatment are available and what the prognosis is.

Most experience with long-term outcome involves lumbosacral myelomeningocele, the most common type of neural tube defect. Just a few years ago, it was not uncommon to offer no treatment for open lumbosacral defects and to let affected infants die of sepsis. With improved techniques of neurosurgical management, it now is common practice to close the primary defect, preserving as much neurologic function as possible. Approaches to the management of hydrocephalus also have improved. Even infants with seemingly massive hydrocephalus can have good cognitive function when early decompression of the ventricular system is performed. Advances in urologic management of bladder immotility include the use of antibiotics to suppress infection.

Self-catheterization has vastly improved patient independence, and effective orthopedic management can help the affected individual maintain mobility. Finally, increased awareness of the needs of physically challenged individuals in society has led to wide-scale integration of those with neural tube defects into all aspects of daily life.

Anencephaly represents the opposite end of the clinical spectrum, and it has raised a provocative ethical dilemma. Anencephaly is invariably lethal, and it is difficult to imagine any advance in medical management that will change this. Anencephalic infants usually are well-formed outside the nervous system, and the question has arisen whether it is ethical to use their organs for transplantation. Donor organs are scarce, but current standards of brain death make it difficult to obtain organs from anencephalic infants. These infants lack a cerebral cortex but generally have brainstem structures and therefore do not meet criteria of brain death. Their natural demise is accompanied by apnea and bradycardia, leading to damage of organs such as heart, kidneys, and liver. Intensive care support preserves the organs but also maintains brainstem function.

There is, as yet, no consensus on how to approach organ donation from anencephalic infants. In some parts of the world, organs have been harvested prior to death on the assumption that the infants are "brain-absent," and hence the usual criteria of brain death do not apply. Hastening the death of a patient, however, is contrary to the generally accepted precept of medical care to place highest priority on the interest of one's own patient rather than sacrificing that patient's interests to the benefit of another person. This issue will likely be the subject of additional research and debate during the next several years.

6.9 Genetics of Common Disease

Geneticists have been accused of focusing on the study of rare traits and ignoring disorders that are the cause of most human suffering. Indeed, progress in understanding the genetic contribution to cardiovascular disease, diabetes, and cancer has lagged behind that of much rarer conditions such as cystic fibrosis or Duchenne's muscular dystrophy. This is a consequence of the enormous complexity of these common disorders as compared with rarer conditions that result from mutations in single genes. Only recently have the tools of molecular genetics begun to penetrate more complex disorders. We will consider the genetics of cancer in a later chapter, but here we will look at genetic approaches to cardiovascular disease.

In the United States, nearly 1 million individuals die each year of complications of cardiovascular disease. Obviously, with such common disorders, multiple members of a family will often be affected owing to chance. Environmental factors such as diet also may play a role. However, genetic factors clearly are involved as well, as evidenced by higher concordance in identical than in fraternal twins.

The leading cause of myocardial infarction is atherosclerosis, the deposition of lipid material within the walls of arteries. Studies of the cause of atherosclerosis have focused on the role of lipoproteins, proteins that transport lipids in the body. The most dramatic genetic cause of premature atherosclerosis is mutation of the gene encoding the low-density-lipoprotein (LDL) receptor. LDLs are the major carriers of cholesterol from the liver to the periphery (Figure 6-10). Specific membrane receptors bind LDLs, which then are internalized into the cell and broken down. Individuals who are heterozygous for LDL receptor mutations have approximately twofold elevations of cholesterol. They develop deposits of lipid along tendons, called *xanthomas*. More importantly, they develop premature atherosclerosis, leading to symptomatic heart disease in their midforties to early fifties. Overall, their risk of coronary artery disease is three to four times higher than in the general population. Familial hypercholesterolemia is an autosomal dominant trait. Rarely, if two heterozygotes have children, an offspring will be homozygous for mutation of the LDL receptor gene, leading to almost complete absence of LDL activity. This is a devastating disorder, characterized by development of xanthomas during childhood and heart disease in the second or third decades of life.

Aside from mutations that lead to deficient LDL receptor activity, there is evidence that different alleles at the LDL receptor locus may account for some of the variability of plasma cholesterol levels in the population. The major protein of LDL is apolipoprotein B (apoB). This is the ligand that binds to the LDL receptor. Polymorphisms of apoB have been demonstrated to account for upward of 3 to 5%

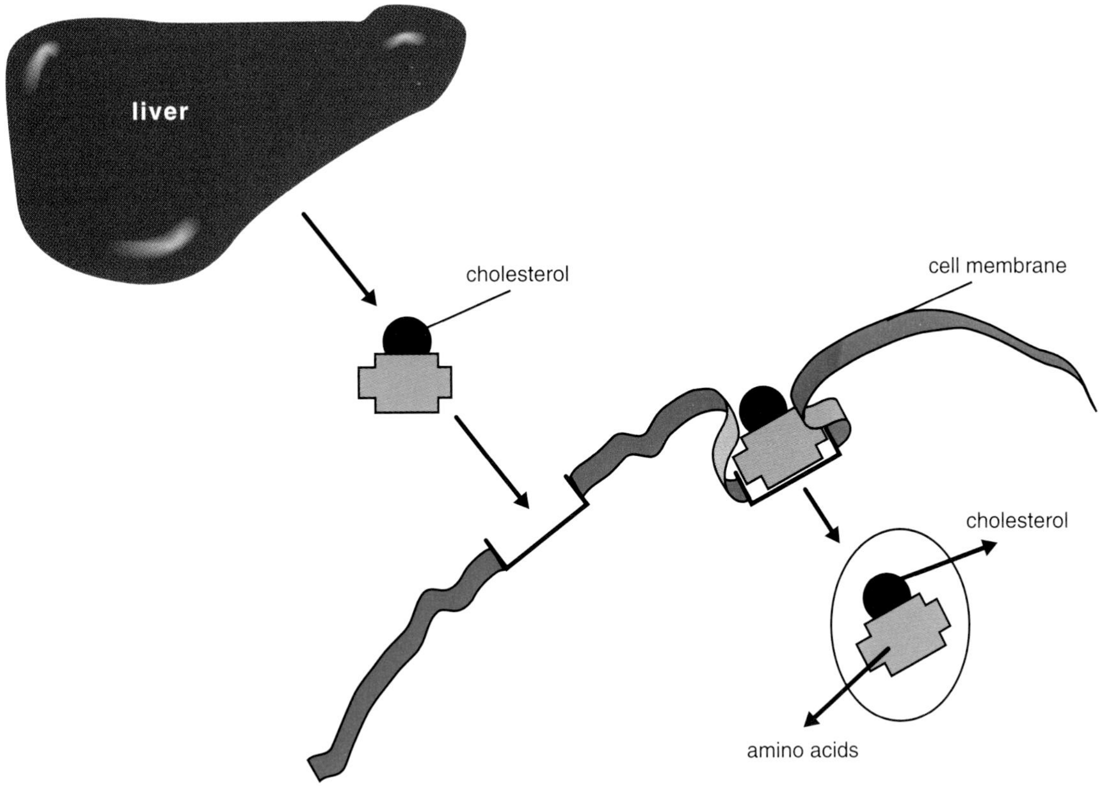

FIGURE 6-10 Low-density lipoproteins (*LDLs*) carry cholesterol from the liver to cells in the periphery. The LDLs enter the cell by endocytosis, binding to a specific membrane receptor. Inside the cell, the proteins are reduced to amino acids, and free cholesterol is released.

of the population variance in plasma cholesterol levels, presumably by altering the efficiency of LDL receptor binding.

Lipids absorbed through the intestine are secreted into the bloodstream as chylomicrons (Figure 6-11). These interact with high-density lipoproteins (HDLs), acquiring apoC and apoE from HDLs and losing apoA I (to HDLs) and apoA IV (which circulates free in plasma). Chylomicron remnants are taken into liver cells via receptor-mediated endocytosis. The ligand in this case is apoE. A similar mechanism occurs in metabolism of very low-density lipoproteins (VLDLs) into LDLs. VLDLs are synthesized by the liver and receive apoC from HDLs. ApoC II activates the enzyme lipoprotein lipase, which leads to loss of triglycerides from VLDLs, converting them to intermediate-density lipoproteins (IDLs). The IDLs also bind to liver cells via apoE and are converted into LDLs. Polymorphisms of apoE therefore influence the efficiency of chylomicron remnant clearance by the liver and of conversion of IDL to LDL. Some apoE alleles increase plasma cholesterol, whereas others lead to a decrease.

Another plasma lipoprotein, Lp(a), has been found to be determined by a single gene. Higher levels of Lp(a) seem to correlate with increased risk of atherosclerosis. It appears, therefore, that a person's risk of atherosclerosis is determined in part by genotype at a large number of loci that encode proteins involved in lipid metabolism. Environmental factors such as diet play a role, and other genetic factors may alter the degree to which dietary factors exert their influence on the risk of atherosclerosis.

Similar complexities apply to the genetics of hypertension. Studies of hypertension have been advanced through the use of an animal model, the spontaneously hypertensive rat. This rat strain was produced by selectively breeding animals that were found to have naturally increased blood pressure,

Perspective

Living with a Child with Spina Bifida

NANCY GOWE

I was 27 years old when our daughter was born with spina bifida. Nicole was our second child. We had a son who was 4 years old. My pregnancy was very uneventful and, at this time, doctors did not routinely do ultrasounds and alpha-fetoprotein testing.

Nicole was delivered by cesarean section, which, as we learned later, is better for children born with an open spine. I think that one of the most frustrating things that we had to deal with from the beginning was misinformation. The pediatrician who examined our daughter immediately following her birth recommended that we let her go quietly. They would place her in a quiet space in the back of the nursery and not feed her. His opinion was that our daughter would never walk, talk, or feed herself and that she would also be mentally impaired. As a result of all this, she would become an emotional and financial burden to our family and this would not be fair to our son.

To this day, we are forever grateful to my obstetrician, who recommended that we transfer our baby to a facility that was used to handling children born with such birth defects. Nicole went by ambulance to Children's Hospital that evening and was evaluated by a neurosurgical team. Medical opinion there was that the problems Nicole had could be dealt with. So began our process of recovery.

I think that right from the beginning you have to learn to take one thing at a time. Our children's care is so complex that it is hard to stay focused. Our first hurdle was for Nicole's back to be closed to cut down the risk of infection. Once that was completed (which, in her case, took 3 weeks of healing), we moved on to dealing with the prospect of hydrocephalus. In our case, it was obvious early on that this would be an issue, so Nicole had a VP shunt inserted as soon as her back was healed.

Nicole was in the hospital for 2 months after she was born. During this period, she was stabilized neurosurgically so that we could move on to deal with the other medical issues facing us. These included bowel and bladder management and orthopedic corrections. As we learned over the course of time, no two children

thereby increasing the frequency of high blood pressure in the strain and allowing studies of numerous candidate genes. A polymorphism in the renin gene has been found to segregate with hypertension in one rat strain with salt-sensitive hypertension. Therefore, in this strain, the renin locus, or another locus in linkage disequilibrium with renin, may contribute to the hypertensive state. Similarly, linkage between the angiotensin-converting enzyme gene and hypertension has been found in another hypertensive strain. Genes that display linkage to the trait in one strain do not necessarily show linkage in another. Hypertension is apparently a complex phenotype that may result from the interaction of genotypes at several distinct loci. In humans, linkage has been found between the angiotensinogen gene to essential hypertension in some families.

Research into the genetic contribution to multifactorial disease will likely yield advances in medical care that will affect large segments of the population. It will be possible to identify individuals at risk of particular diseases through determining their genotypes at specific loci. Meanwhile, as the genetic pathways to these disorders are understood, effective means of prevention or treatment may come to light.

There is concern that genetic testing for disease susceptibility will lead to an erosion of privacy and possible discrimination based on genotype. On the other hand, everyone in the population carries genes that will likely contribute to some chronic and eventually life-threatening illness or another. The challenge will be to use the emerging knowledge of human genetic variation wisely to improve public health without placing physical well-being ahead of human rights.

with spina bifida are the same. Depending on where the lesion is located, one cannot predict what problems a child might encounter.

Nicole is now 13 years old and is preparing to have operation number 18. Surgery is very difficult to deal with because of the disruptions to everyone's life. Nicole has to make special school arrangements to keep up with her classmates and the rigorous schedule of being mainstreamed in regular education. I have to say, though, that our school system has been very supportive. Nicole also finds it very difficult to be away from her friends. By relying on extended family members for support, our son has learned to cope with the many times I have had to be away caring for Nicole in the hospital.

One of the most important factors in being able to coordinate all the care that is involved in Nicole's life is that we have a very supportive pediatrician who oversees everything. Our pediatrician is the individual we can go to when we feel overwhelmed by it all and just need a little guidance. The pediatrician is the person who sees Nicole as a whole person. He can help with all her emotional, medical, and educational needs.

We are very fortunate in that I am able to stay at home and care for my family. It has not always been easy, and one must make some very difficult decisions about whether to go on vacation or to save to buy some new piece of equipment that insurance might not cover. Medical insurance has been a constant worry. It seems that we would just get comfortable with one insurance company and my husband would come home and say his employer was changing insurance plans. No one plan ever covered Nicole completely. By that I mean there would be deductibles, or they would not cover diapers, catheters, or therapies. Over time, we have come to realize that we have to carry a supplemental insurance policy to help defray some of the outside cost. We worry about when Nicole is an adult and must maintain her own health insurance.

We are grateful for the wonderful strides that the medical profession has made in the treatment of persons born with spina bifida. Because of this, Nicole is well on her way to becoming a bright, happy, and productive adult. Nicole is now a very active participant in her medical care, and she lets all her physicians know that she is to be included in the decision-making process.

I hope that I have been able to give you some insight into what our life is like living with a child with spina bifida. I have to tell you, though, that for all the work, worry, and heartache, the rewards of seeing her smiling face and personality every day make it well worth it.

Case Study

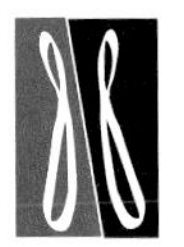

Part I

Elizabeth is a previously healthy 11-year-old. About 4 weeks ago, she had a viral illness characterized by fever, cough, and runny nose. She recovered from that but now has developed a new problem. She is thirsty most of the time, constantly drinking. In addition, she is urinating unusually often and sometimes even wets her bed at night. This is very disturbing to her. She also feels hungry much of the time and is eating a great deal, yet, has been losing weight. At first, her family assumes that this is due to stress, but finally they bring her to her pediatrician, concerned that something may be wrong. A physical examination is done and is entirely normal. Her pediatrician takes a urine sample for testing. He dips a test strip in the urine and finds that it is strongly positive for glucose.

The presence of glucose in the urine is strongly suggestive of diabetes mellitus, and the clinical history is compatible with this diagnosis. Diabetes mellitus results from deficient production of insulin. Insulin is a peptide hormone produced by β cells of the islet of Langerhans in the pancreas. Its secretion is modulated mostly by serum levels of glucose, and its actions are to ehance uptake and utilization of glucose in cells, including muscle, liver, and adipose tissue. Deficiency of insulin leads to decreased glucose transport into cells, high serum levels, and decreased utilization of glucose.

It is not uncommon for diabetes mellitus to be preceded by an apparent viral infection. The basic

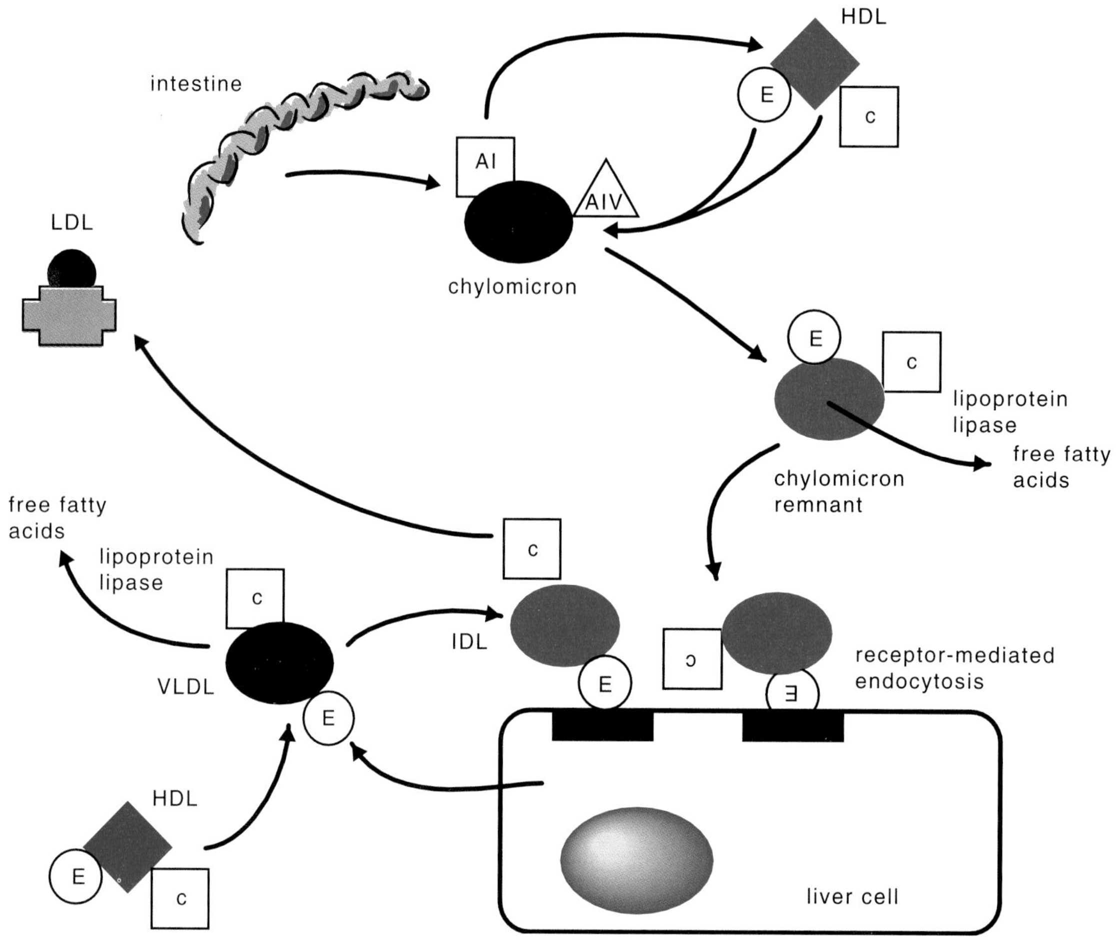

FIGURE 6-11 Lipoprotein pathway (see text). (*HDL* = high-density lipoprotein; *LDL* = low-density lipoprotein; *VLDL* = very low-density lipoprotein; *IDL* = intermediate-density lipoprotein; *AI* = apolipoprotein A I; *AIV* = apolipoprotein A IV; *E* = apolipoprotein E; *C* = apolipoprotein C.)

process, however, appears to be autoimmune destruction of the B cells by T lymphocytes. The symptoms are mainly related to high serum levels of glucose and decreased intracellular utilization. Glucose is excreted in the urine, where it exerts an osmotic effect, causing a large amount of urine production. This, in turn, increases thirst. In the insulin-deficient state, proteins, glycogen, and glucose are produced in the liver in an uncontrolled fashion. This leads to weight loss, despite adequate or more than adequate calorie intake. Eventually, a crisis may occur owing to a combination of dehydration and ketoacidosis due to partial breakdown of fats as an energy source. Often, however, the diagnosis is made before this point, based on the clinical symptoms and demonstration of high levels of glucose in blood and urine.

Part II

Elizabeth is referred to the local hospital emergency room. Her blood sugar concentration is found to be markedly elevated at 350 mg/dl. Diabetes mellitus is diagnosed in Elizabeth, and she is admitted to the hospital that night. She is started on insulin, and her blood sugar rapidly returns to normal. During a 5-day hospitalization, her polyuria and polydipsia resolve. She and her parents are taught to administer insulin by subcutaneous injection, and they are instructed in the new dietary restrictions that Elizabeth will have to follow. Elizabeth is not happy about any of this, and her parents are distressed also. Elizabeth spends several days in the hospital during which she learns about diabetes and its treatment from the diabetes clinic nurse specialist and dietician. She also talks with a counselor about her fears and concern.

Medical management of diabetes mellitus involves administration of insulin to replace the missing hormone. This is a life-long treatment. Aside from restoring normal glucose homeostasis, the initial hospital treatment involves education to teach the patient and/or parents how to use insulin, monitor blood glucose levels, and use a meal plan. Insulin is administered as a subcutaneous injection and usually is given twice or three times a day. Various combinations of short and longer-acting insulin preparations are used. The hormone cannot be administered orally because it would be broken down in the digestive system. Exact dosage tends to vary with dietary carbohydrate intake, level of activity, body size, and pubertal status, so blood glucose levels are monitored regularly, and ketones are measured in urine when blood glucose levels are high. Giving too much insulin can result in dangerous hypoglycemia, whereas too little leads to both the acute symptoms of hyperglycemia, as well as more chronic complications resulting from long-term exposure to hyperglycemia. Simple glucose monitoring systems are available for measuring glucose in blood at home. Dietary modification (a meal plan) is the third component of management, and involves a consistent schedule of meals and snacks to ensure a balance between glucose from dietary sources and the action of injected insulin. This involves limiting the amount of concentrated sugars or other foods that lead to steep rises in serum glucose levels.

It should be clear that management of diabetes mellitus requires major changes of lifestyle, including daily medication, dietary, and activity changes. This is not easy at any age, and may be especially difficult for teens. Much of the focus of early management is, therefore, directed at patient education in self-care and psychosocial support.

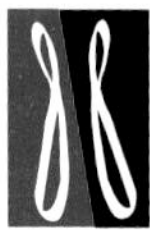

Part III

Three months have passed. Elizabeth has adjusted well to taking insulin and, on some occasions, with supervision by a parent, has administered her own injections. She has been seen several times at the diabetes clinic and has been diligent about monitoring her blood sugar and recording blood glucose data in her diabetes logbook. Elizabeth has a 6-year-old brother, David, who is healthy. Elizabeth's parents ask whether David is at risk of developing diabetes and whether he should have his urine or blood tested. Elizabeth's mother is particularly concerned because her father also has diabetes. In his case, though, the diabetes did not begin until he was 50 years old. He does not take insulin but rather uses a medication to control his blood sugar.

Diabetes mellitus is a term that encompasses two distinct disorders. *Insulin-dependent diabetes mellitus* (IDDM) is the type that presents in childhood. It is caused by destruction of islet cells, leading to permanent loss of ability to secrete insulin and lifelong dependence on insulin injections. IDDM has a population frequency of approximately 1 in 400. The other disorder is *non-insulin-dependent diabetes mellitus* (NIDDM). This is a heterogeneous disorder that usually has onset in adulthood and is commonly associated with obesity. Prevalence varies in different populations from 3% in Europe to as high as 30% in Pima Indians in Arizona. There usually is reduced, but not entirely absent, insulin secretion. Persons with NIDDM may be able to maintain adequate control of blood glucose levels by dietary restriction of glucose intake, weight loss, and exercise. If these measures fail, treatment is oral hypoglycemic agents or insulin.

Both IDDM and NIDDM appear to be multifactorial traits. Concordance in identical twins is approximately 30 to 40% for IDDM and 90% for NIDDM. The risk of IDDM in a sibling of an affected child is 5 to 10%. Family members can be screened for evidence of inflammation in their islets ("insulinitis") by testing for the presence of islet cell antibodies in serum. The presence of islet cell antibodies may precede clinical diabetes by months or years. There is no tendency for IDDM and NIDDM to cluster in the same families, indicating that these are genetically as well as physiologically distinct disorders.

The major genes involved in determining risk of IDDM are those of the HLA complex. The HLA complex on chromosome 6 consists of a number of genes involved in regulation of the immune response (Figure 6-12). These include genes that encode class I and class II HLA membrane proteins. Both class I and class II molecules consist of a dimer with an alpha-chain and a beta-chain. The class I molecules are expressed in all cell types and are encoded by the HLA-A, B, and C loci. The class I alpha-chain is highly polymorphic whereas the beta-chain is invariant. Class II molecules are encoded by three sets of genes in the HLA-D region and are expressed mainly on monocytes, macrophages, and B lymphocytes.

The function of HLA proteins appears to be the presentation of specific peptides at the cell surface. Class I molecules bind peptides of endogenous cellular proteins that are transported from within the cell, whereas class II molecules bind peptides from exogenous proteins that are transported to the cell surface after being degraded in lysosomes. Class I molecules interact with CD8+ cytotoxic T cells, whereas the class II molecules interact with CD4+ helper T cells.

The polymorphic loci of the HLA complex exhibit strong linkage disequilibrium, so particular haplotypes occur together on the chromosome more often than would be predicted by chance. Susceptibility to IDDM has been found to be associated with particular class II alleles encoded by class II loci designated "DQ," e.g., in Caucasians, DQB1-*0302. In blacks, association of DQA1-*0301 and DQB1-*0201 alleles on the same chromosome in coupling leads to an 8- to 12-fold increased risk of IDDM. In white or Japanese IDDM patients, these same alleles are often present, but usually on opposite chromosomes, in repulsion. In contrast, other class II DQ alleles, DQB1-*0602 and DQB1-*0603 confer protection against IDDM. Presumably, specific DQ alleles influence the likelihood that helper T cells will recognize islet cell antigens and lead to destruction of insulin-producing cells. Many other autoimmune diseases have been linked to the HLA complex, and are associated with particular combinations of HLA alleles (Table 6-4).

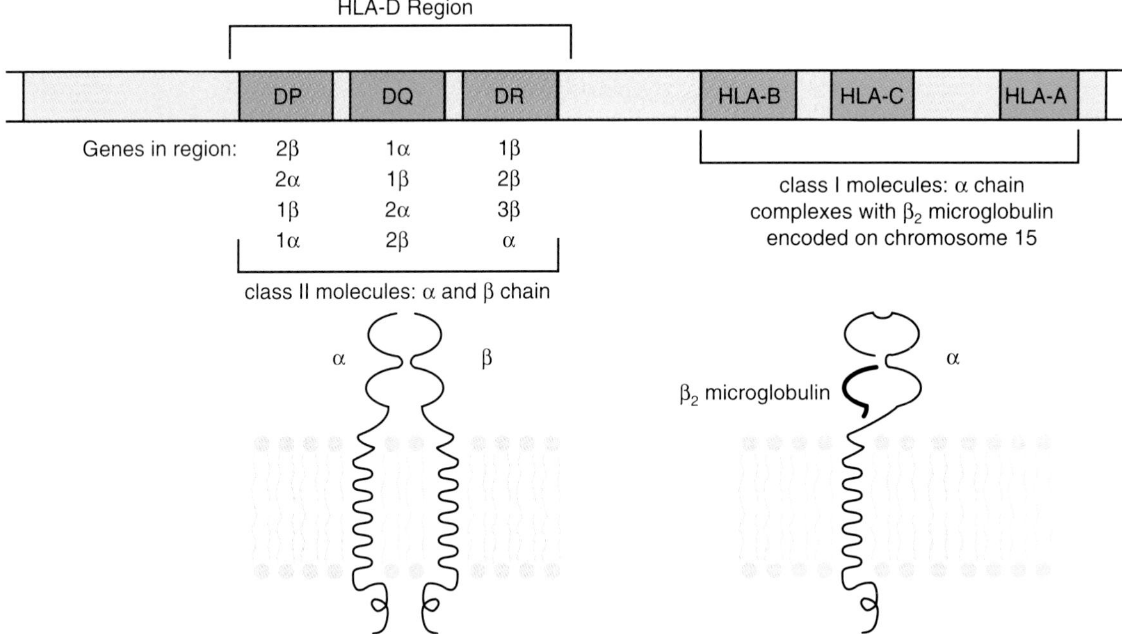

FIGURE 6-12 Human leukocyte antigen (*HLA*) system. Loci of HLA-A, -B, -C, and -D regions on chromosome 6 are shown at the top. HLA-A, -B, and -C are class I molecules in which an alpha-chain encoded in the HLA region binds with a beta$_2$-microglobulin chain. The HLA-D region includes three loci: DP, DQ, and DR. The D region proteins are class II molecules, which consist of an alpha- and beta-chain, both encoded by genes in the D region.

TABLE 6-4 HLA-disease associations

Disorder	Antigen	Relative Risk*
Ankylosing spondylitis	B27	69.1
Juvenile rheumatoid arthritis	B27	3.9
	DR8	3.6
Ulcerative colitis	B5	3.8
Psoriasis	Cw6	7.5
Multiple sclerosis	DR2	6.0
Narcolepsy	DR2	130.0

$$\text{*Relative risk} = \frac{(\%\ \text{antigen} - \text{positive patients})(\%\ \text{antigen} - \text{negative controls})}{(\%\ \text{antigen} - \text{negative patients})(\%\ \text{antigen} - \text{positive controls})}$$

Source: Data from Carpenter CB. The major histocompatibility gene complex. In: Isselbacher K, Braunwald G, Wilson J, et al. Harrison's Principles of Internal Medicine, 13th ed. New York: McGraw-Hill, 1994:380–386.

Genetic studies of NIDDM reveal a very different picture. One variant form that exhibits single gene autosomal dominant transmission is referred to as "MODY"—maturity-onset diabetes of the young. In some families this trait has been found to be linked to a locus on chromosome 20q or another on 12q. In other families, linkage has been found with the glucokinase gene on 7p. In one family, a chain termination mutation has been found in the glucokinase gene. Glucokinase is the first enzyme in glycolysis and plays a crucial role in transport of glucose into the cell. Some Pima Indians with NIDDM are homozygous for a mutation in the beta-adrenergic receptor, which may lead to earlier onset of diabetes.

The processes of secretion of insulin on the one hand, and of cellular responsiveness to insulin on the other, are complex and depend on many genes. In various populations, linkage of genes, including insulin receptor, glucokinase, apolipoprotein D, complement C4B, and other proteins, have been found to contribute to susceptibility to NIDDM. Different genes may have major or minor roles in different populations. It has been hypothesized that genes conferring susceptibility to NIDDM may confer selective advantage to populations that live in harsh conditions where food may be scarce at times. Relative insulin insensitivity will increase the likelihood that excess glucose will be stored as fat rather than glycogen, providing a cache of energy for times of need. In modern urban populations, however, where food is abundant, this is disadvantageous and leads to obesity and high circulating levels of glucose.

Part IV

Elizabeth is now 25 years old. She has had a difficult time with her diabetes, having been hospitalized several times over the years for diabetic ketoacidosis. It has not been easy to maintain an ideal blood sugar level, and her insulin requirement has gradually changed. In addition, during her adolescent years, Elizabeth tended not to adhere to the dietary restrictions. Now Elizabeth is married, and she and her husband are thinking about starting a family. They ask about the risks of having a child with diabetes and whether genetic testing is possible.

The long-term complications of diabetes include a number of problems directly or indirectly related to glucose metabolism. Episodes of hypoglycemia can result from an inappropriately large dose of insulin or failure to consume sufficient carbohydrate, and can lead to loss of consciousness, seizures, and even death or permanent brain damage. In contrast, inadequate insulin can result in an acute crisis called ketoacidosis, in which fatty acids are used as an energy source, leading to the production of keto acids and to metabolic acidosis and dehydration (due to osmotic diuresis of glucose). Cerebral edema is a rare complication that may occur during treatment of diabetic ketoacidosis. Chronic complications of hyperglycemia include damage to small blood vessels in the kidney which may ultimately lead to kidney failure and in the eye can cause blindness. Other long-term effects of diabetes include damage to peripheral nerves and accelerated atherosclerosis which causes stroke, peripheral vascular disease and occlusion of coronary arteries. The risk of developing microvascular complications can be significantly reduced by strict blood glucose control and long-term maintenance of blood glucose levels near normal. Therapeutic research is currently focused on alternative means of treatment that will avoid periods of hyperglycemia, which are inevitable even with perfect compliance.

Genetic counseling for individuals with IDDM is based on empirical data. The recurrence risk in offspring of diabetic mothers is about 4%, of diabetic fathers 5%, 10 to 20 times higher than the general population risk. Although knowledge of HLA associations that increase the risk of diabetes is useful for better understanding of the pathogenesis of IDDM, it is not of practical value in determining genetic transmission in a family. Of greater concern than genetic transmission of diabetes itself, however, is the risk of teratogenic effects of hyperglycemia on the fetus. Infants of diabetic mothers have an increased frequency of macrosomia, and major and minor congenital malformations. The latter include cardiac, brain, renal, and skeletal anomalies. The pathogenesis may be direct effects of hyperglycemia, or fetal hyperinsulinemia resulting from a normal fetal pancreatic response to maternally transmitted hyperglycemia.

Part V

Elizabeth and her husband understand that the risk of diabetes in an offspring is low and that genetic testing is not possible. They are also warned that careful control of blood sugar during pregnancy is important, as exposure of the fetus to high glucose levels can lead to congenital anomalies. Elizabeth is now pregnant, and her pregnancy is being monitored closely.

Strict control of blood sugar during pregnancy and maintenance of near normal blood glucose prior to conception and during pregnancy reduces the risk of fetal malformations to a level comparable to that in the offspring of non-diabetic mothers. However, strict blood glucose control can be difficult to achieve during pregnancy. Some women who are not diabetic have hyperglycemia during pregnancy, referred to as gestational diabetes. It is generally agreed that aggressive control of hyperglycemia with insulin, although it does not guarantee a normal outcome, is the best that can be offered to reduce the likelihood of fetal problems in infants of diabetic mothers.

Part VI

The pregnancy is not easy for Elizabeth. She has a very difficult time maintaining ideal glucose levels and has to be hospitalized twice for adjustment of her insulin dose. The baby is delivered at term and weighs 9 pounds, 5 ounces. The baby girl, Ellen, is delivered by cesarean section because of her large size. Ellen needs to stay in the hospital several days because of episodes of hypoglycemia and jaundice, but otherwise she is well. A heart murmur is heard in the infant during the first few days, but a cardiac evaluation is done, and no congenital heart disease is found. Both head and abdominal ultrasound examinations also are normal. Ellen is finally discharged home and appears to be quite healthy.

The macrosomia is probably related to maternal diabetes. Insulin cross-reacts with insulin-like growth factors to cause fetal overgrowth. This can lead to complications of delivery but usually is not a problem postnatally, because hyperglycemia stops at that time. Some infants of diabetic mothers have episodes of hypoglycemia after birth, owing to high circulating insulin levels resulting from chronic exposure to hyperglycemia in utero. Other problems seen in the neonatal period include polycythemia (increased numbers of red blood cells) and jaundice.

The heart murmur may indicate the presence of a congenital heart defect. Some heart murmurs are noted in infants due to delayed closure of the ductus arteriosus, the connection between the aorta and pulmonary artery that usually closes at the time of birth. These may spontaneously close a few days later and lead to no clinical problems. In general, the congenital anomalies expected in infants of diabetic mothers are readily apparent clinically or by noninvasive testing such as ultrasonography. Their absence in a newborn is therefore very reassuring.

REVIEW QUESTIONS

1. You are asked to provide genetic counseling to a family in which there is a daughter with pyloric stenosis (muscular hypertrophy of the pylorus of the stomach, leading to intractable vomiting in infancy). She is otherwise well and has been successfully treated surgically. There is no other family history of a similar problem.
   a. You explain that pyloric stenosis is a multifactorial trait. You find that the mother is blaming the father for causing this problem, and the father is blaming the mother. How would you address this conflict from the genetic point of view (aside from explaining that assessing blame is not helpful)?
   b. Pyloric stenosis occurs more commonly in males than females. You explain that because a female is affected rather than a male,

the recurrence risk is higher in this family. Why is this so?

c. If the family were to have another affected child, is it more likely for an affected child to be male or female? Why?

2. A couple gives birth to a child with spina bifida. The mother began folic acid supplements 6 weeks after her last period, when she was sure she was pregnant. What advice would you provide for future pregnancies regarding folic acid?
3. Cleft palate is a multifactorial trait. Would you expect the risk of cleft palate in an offspring to be higher if both parents are affected than if one is affected?
4. In rare instances, identical twins are discordant for a chromosome anomaly such as trisomy 21. How might this happen?

FURTHER READING

6.6 Folic Acid and Neural Tube Defects

Hall JG, et al. Clinical, genetic, and epidemiological factors in neural tube defects. Am J Hum Genet 1988;43:827–837.

Mills JL, et al. Homocysteine metabolism in pregnancies complicated by neural-tube defects. Lancet 1995;345:149–151.

MRC Vitamin Study Research Group. Prevention of neural tube defects: results of the Medical Research Council Vitamin Study. Lancet 1991;338:131–137.

Stein SC, et al. Is myelomeningocele a disappearing disease? Pediatrics 1982;60:511–514.

6.7 Animal Models of Neural Tube Defects

Neumann PE, Frankel WN, et al. Multifactorial inheritance of neural tube defects: localization of the major gene and recognition of modifiers in ct mutant mice. Nat Genet 1994;6:357–362.

6.8 Care of Persons with Neural Tube Defects

Committee on Bioethics. Infants with anencephaly as organ sources: ethical considerations. Pediatrics 1992;89:1116–1119.

Peabody J, Emery JR, Ashwal S. Experience with anencephalic infants as prospective organ donors. N Engl J Med 1989;321:344–350.

6.9 Genetics of Common Disease

Berg K. Introductory remarks: risk factor levels and variability. Ann Med 1992;24:343–347.

Caulfield M, et al. Linkage of the angiotensinogen gene to essential hypertension. N Engl J Med 1994;330:1629–1633.

Linkpaintner K. What can molecular genetics of hypertensive rats teach us about the genetics of hypertension in humans? Curr Opin Nephrol Hypertens 1994;3:30–38.

CASE STUDY

Barnett AH, Eff C, Leslie RDG, Pyke DA. Diabetes in identical twins. Diabetologia 1981;20:87–93.

Bell GI, et al. Gene for non-insulin dependent diabetes mellitus (maturity-onset diabetes of the young subtype) is linked to DNA polymorphism on human chromosome 20q. Proc Natl Acad Sci USA 1991;88:1484–1488.

Froguel P, et al. Familial hyperglycemia due to mutations in glucokinase: definition of a subtype of diabetes mellitus. N Engl J Med 1993;328:697–702.

Groop LC, Eriksson JG. The etiology and pathogenesis of non-insulin-dependent diabetes. Ann Med 1992;24:483–489.

Mills JL, Knopp RH, Simpson JL, et al. Lack of relation of increased malformation rates in infants of diabetic mothers to glycemic control during organogenesis. N Engl J Med 1988;318:671–676.

Thorsby E, Rønningen KS. Role of HLA genes in predisposition to develop insulin-dependent diabetes mellitus. Ann Med 1992;24:523–531.

Vaxillaire M, Boccio V, Philippi A, et al. A gene for maturity onset diabetes of the young (MODY) maps to chromosome 12q. Nat Genet 1995;9:418–423.

7 Mitochondrial Inheritance

It used to be thought that the entire genetic constitution of an individual was determined by genes on the 46 chromosomes in the nucleus. Genetic disorders therefore were due to gene mutations, chromosomal abnormalities, or multifactorial traits. Recently, it has become apparent that there is an additional source of genetic information in the cell and, correspondingly, an additional source of potential havoc. Each cell contains hundreds of mitochondria, each of which contains multiple copies of a 16.5-Kb circular DNA molecule. Although most mitochondrial proteins are encoded by nuclear genes, some are encoded by mitochondrial genes, and mutations can lead to energy failure. The mitochondrial genome is subject to a number of peculiarities of inheritance, including maternal transmission and a phenomenon known as *heteroplasmy*, resulting in distinctive patterns of familial disease. Recognition of mitochondrial disorders has opened a new chapter in understanding human pathology.

We will begin our exploration of mitochondrial disorders by consideration of a case history of one such condition, referred to by the acronym *MERRF*. After a review of the features of MERRF, we will see how this disorder is maternally transmitted but how expression can vary widely in a family. The basis for this variation is heteroplasmy, in which different cells in an individual, and different individuals in a family, contain different proportions of mutant and wild-type mitochondria. MERRF is a syndrome of mitochondrial energy failure, and we will see how this was first demonstrated at the biochemical level. We then will turn to the genetic studies of MERRF that resulted in identification of tRNA point mutations in different families. MERRF is but one of a growing class of clinically recognized mitochondrial disorders. Current understanding of some of these other conditions will be summarized. Recognition of mitochondrial disorders has added powerful diagnostic tools, but therapeutic interventions are just being tested. We will next look at some of these efforts. Finally, in addition to the medical advances that have resulted from studies of mitochondrial inheritance, the mitochondrial genome has provided powerful tools for anthropologic studies. One instance of such an application will be briefly described. In the Perspective, we will consider the impact of mitochondrial disease on a family, and the chapter closes with a case history of a man with sudden loss of vision.

7.1 MERRF

Michael is now 12 years old, and he has multiple medical problems. His parents, Sally and Arthur, began to be concerned when he was just 5 years old. At that time, he was noted to have difficulty running and seemed to be unsteady on his feet. Gradually, the symptoms worsened, and Michael has had increasing difficulty keeping up with his peers. Michael began to experience sudden jumps of his muscles at 8 years of age, and a year later he had his first generalized seizure. His seizures have increased in frequency, to the point where Michael has several per day, of various types. Sometimes he will have uncontrollable jerks of the muscles of his hands or legs, causing him to drop things or to fall. Other times, he suddenly loses muscle tone and falls; at other times, he has generalized tonic-clonic seizures. Michael has been treated with many different medications for seizure control. Although he now has fewer seizures, still he has several per day. In addition, since about the time the seizures began, Michael has lost cognitive skills. He was in regular classes in school until fourth grade but has been in special education since then. It is clear to his parents and teachers that Michael has steadily lost intellectual abilities. He has also had difficulty with vision and hearing. Michael's parents have taken him to numerous doctors over the years in an effort to understand the cause of his many problems. Recently, he has had a muscle biopsy. The neurologist who arranged for the biopsy to be done (Figure 7-1) now explains to Sally and Arthur that he thinks he knows what is wrong with Michael: a disorder he refers to as MERRF.

MERRF is an acronym for *myoclonic epilepsy with ragged red fibers*. Myoclonus is a sudden contraction of a group of muscles that causes a brief jerking movement. Generalized myoclonus involves muscles all over the body and can be intermittent or nearly continuous. Myoclonic jerks can interfere with efforts to make ordinary movements, leading one to drop objects or to fall, sometimes causing injury. A person with myoclonic epilepsy may also have generalized seizures associated with body jerks and loss of tone or consciousness. Myoclonic jerks can be difficult to control with antiepileptic medications.

In addition to myoclonus, individuals with MERRF tend to have generalized muscle weakness.

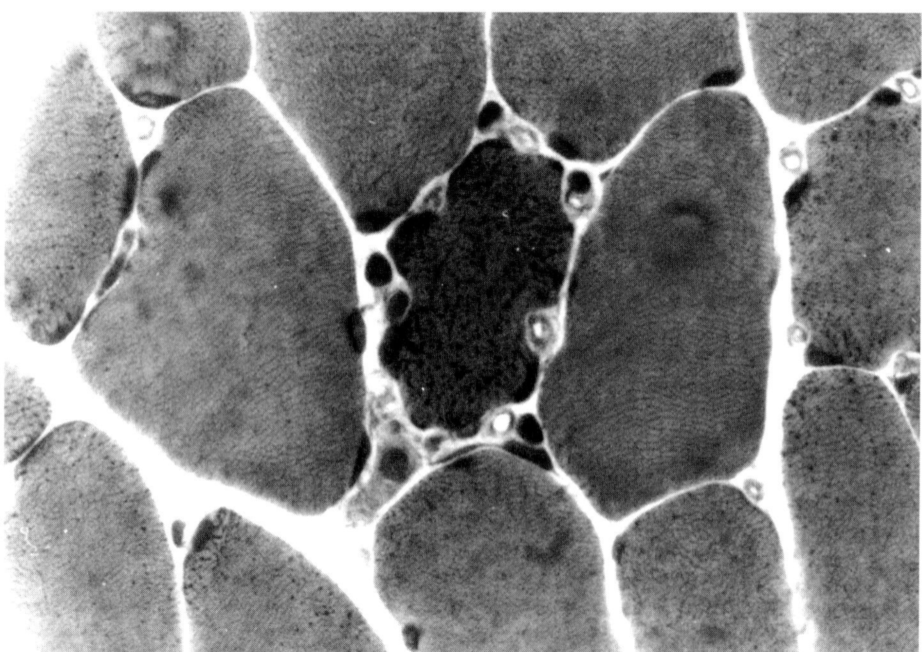

FIGURE 7-1 Muscle biopsy stained with modified Gomori trichrome stain (GTS). The fiber in the center has dark areas of staining at the periphery that correspond with aggregates of mitochondria. These areas stain red with GTS. (Courtesy of Dr. Douglas Anthony, Children's Hospital, Boston.)

The weakness may be noticed first in large muscle groups (e.g., the hip or shoulder girdle). There may be difficulty walking, lifting, or getting up from a lying or sitting position. Gradually, other muscles become involved, including muscles of facial expression, hands, and lower legs. Examination with special stains of a muscle biopsy from an individual with MERRF reveals an appearance known as *ragged red fibers*. These are muscle cells with a ring of red-stained material surrounding the periphery. Through the electron microscope, the red-stained material can be demonstrated to represent a large number of abnormal-appearing mitochondria (Figure 7-2).

Other features of MERRF include ataxia, spasticity, dementia, sensorineural hearing loss, and optic atrophy. *Ataxia* is defined as loss of coordination, leading to difficulty with walking or carrying out coordinated movements of the arms. It may also affect the muscles involved in speech, leading to articulation problems. Spasticity consists of increased muscle tone due to interference with transmission of nerve impulses from the brain to the motor neurons in the spinal cord. This adds to the weakness caused by myopathy and leads to further difficulty walking. There is a gradual loss of cognitive abilities, along with loss of hearing and vision. The cognitive decline may be due in part to visual and hearing deficits, the effects of multiple seizures, and the additional effects of multiple medications. Aside from these problems, however, there appears to be a loss of intellectual abilities due to effects of the disease on the brain itself.

MERRF is a life-long condition that can be rapidly progressive in some people and slowly progressive in others. The major life-threatening complications include respiratory failure due to insufficiency of chest wall muscles and cardiac failure due to involvement of heart muscle.

MERRF is a relatively recently discovered disorder, having been first described as a clinical entity in 1980. The signs may begin subtly, and a large number of neurologic disorders must be considered in the differential diagnosis. It is common, therefore, for

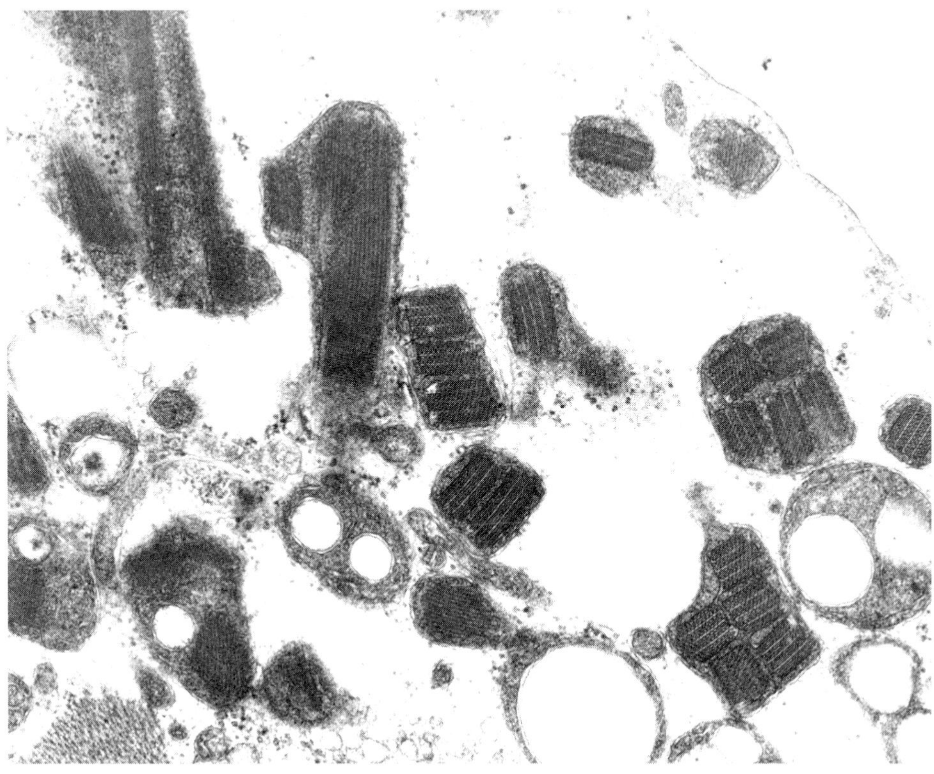

FIGURE 7-2 Electron micrograph of muscle showing abnormal mitochondria. (Courtesy of Dr. Douglas Anthony, Children's Hospital, Boston.)

the diagnosis to be made only after a long and frustrating search for other possible causes. The recognition of MERRF can have important implications both for the affected individual and for other members of the family, as we will now see.

7.2 Maternal Genetic Transmission

Sally and Arthur are devastated by the news of the diagnosis in Michael, but they are also relieved to know that a definite diagnosis has been made. They have a long discussion with the neurologist and, of course, are most interested in knowing whether Michael can be treated to reverse some of his symptoms or, at least, to prevent further progression. They learn that there is no definitive means of treatment, although some experimental approaches are in use. During the course of discussion, the neurologist asks about family history. Michael's parents state firmly that no one else in the family has ever had problems such as this. On further questioning, though, some interesting details emerge. Sally has a history of migraine headaches and hearing loss. The hearing loss is relatively mild, and she does not wear a hearing aid. She also has had some muscle jumps in the past few years, which she has attributed to stress due to her son's illness. Michael has a sister, Judy, who is 16 years old. Judy also has mild hearing loss and had one seizure when she was 10 years old, but she has been well since. Sally had a brother who died when he was 3 years old of some kind of heart problem, but the family does not know more about this. She has one other brother who suffers migraines, hearing loss, and tends to be unsteady on his feet. The family has attributed his problems to alcohol use. Finally, Sally's mother suffers hearing loss and dementia, but these problems did not start until she was in her sixties. Arthur is in good health. He has one sister who is well, and both his parents are alive and well (Figure 7-3).

It is common to discover a family history of multiple neurologic problems when MERRF is diagnosed in an individual. There is a wide range of variable expression of the disorder: Some patients have the full-blown syndrome with myoclonus, myopathy, dementia, and so on, whereas others can have more subtle features. Often, the milder expression of MERRF never comes to medical attention or is attributed to causes such as stress. Typically, the whole picture for the family only comes into focus after a more severely affected relative receives this diagnosis.

The pattern of transmission of MERRF in a family strongly suggests a genetic component but differs from the expectations of mendelian genetics. There appears to be transmission directly from generation

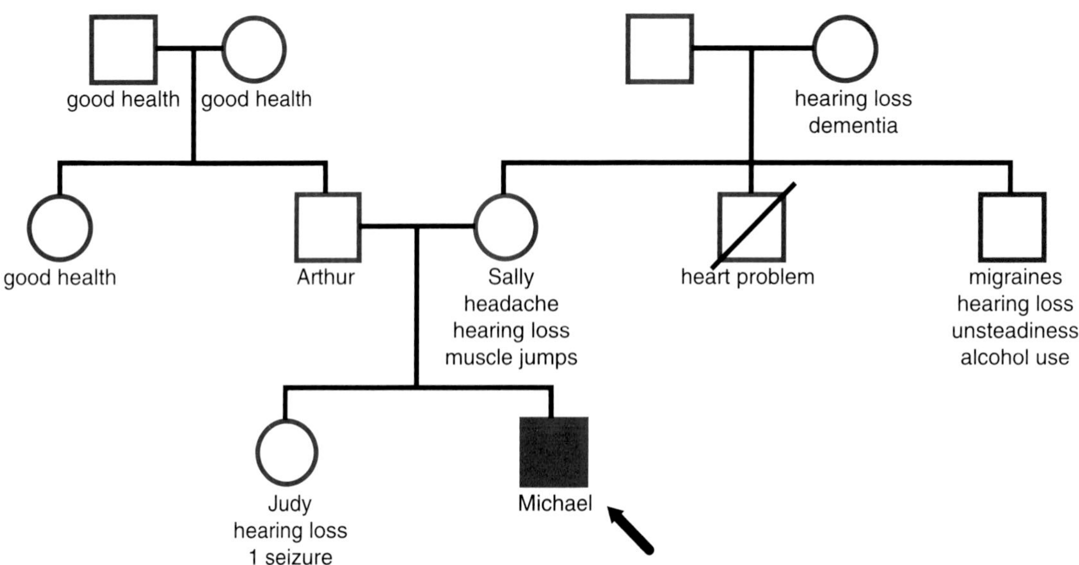

FIGURE 7-3 Family pedigree of Sally and Arthur.

to generation, suggesting dominant inheritance. Both males and females may be affected, but men never transmit the disorder to any of their offspring. Women, on the other hand, pass the trait to all of their children, although expression may be more severe in some than in others (Figure 7-4).

This pattern of inheritance is referred to as **maternal transmission.** It does not fit with the transmission of genes in the nucleus but, over the recent few years, it has become clear that not all the genetic information in a cell resides in the nucleus. DNA is found also within another cellular organelle, the mitochondrion. Mitochondria are responsible for the generation of ATP via aerobic metabolism, and each mitochondrion contains multiple copies of a double-stranded circular DNA molecule of 16,569 base pairs (Figure 7-5). This DNA encodes 13 peptides that are subunits of proteins required for oxidative phosphorylation. In addition, there is a complete set of 22 transfer RNAs and two ribosomal RNAs. These RNAs are involved in translation of mitochondrially encoded proteins within the mitochondrion.

The process of oxidative phosphorylation requires more than 60 proteins. Most are encoded in the nucleus and are transported into the mitochondrion from the cytoplasm. Only a minority of the mitochondrial proteins are encoded in the mitochondrial genome and are synthesized within the mitochondrion. The structure of the mitochondrial genome bears more resemblance to prokaryotic than to eukaryotic genomes. The genes lack introns and are transcribed as polycistronic messages from two promoters. The spaces between genes consist of tRNAs, whose excision from the polycistronic message releases the individual gene transcripts. There are also two sites for initiation of DNA synthesis, one on each strand of the double helix.

The explanation for maternal inheritance lies in the manner by which mitochondrial DNA is passed from generation to generation (Figure 7-6). At the time of fertilization, the sperm sheds its tail, including all its mitochondria. Only the sperm head, containing nuclear DNA, enters the egg. Therefore, all the mitochondria for the next generation are contributed by the egg cell. Hence, mitochondrial genes are exclusively maternally derived, explaining the pattern of maternal transmission.

Maternal Genetic Transmission

- Mother transmits trait to all her offspring.
- Only females transmit the trait.
- This transmission is associated with genes encoded in mitochondrial DNA.

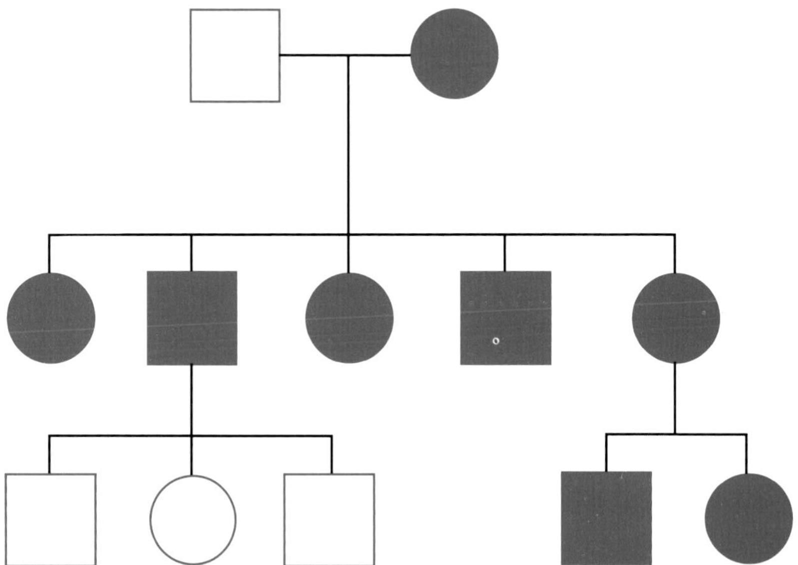

FIGURE 7-4 Maternal genetic transmission. An affected woman transmits the trait to all of her children. Affected men (represented by squares) do not pass the trait to any of their offspring.

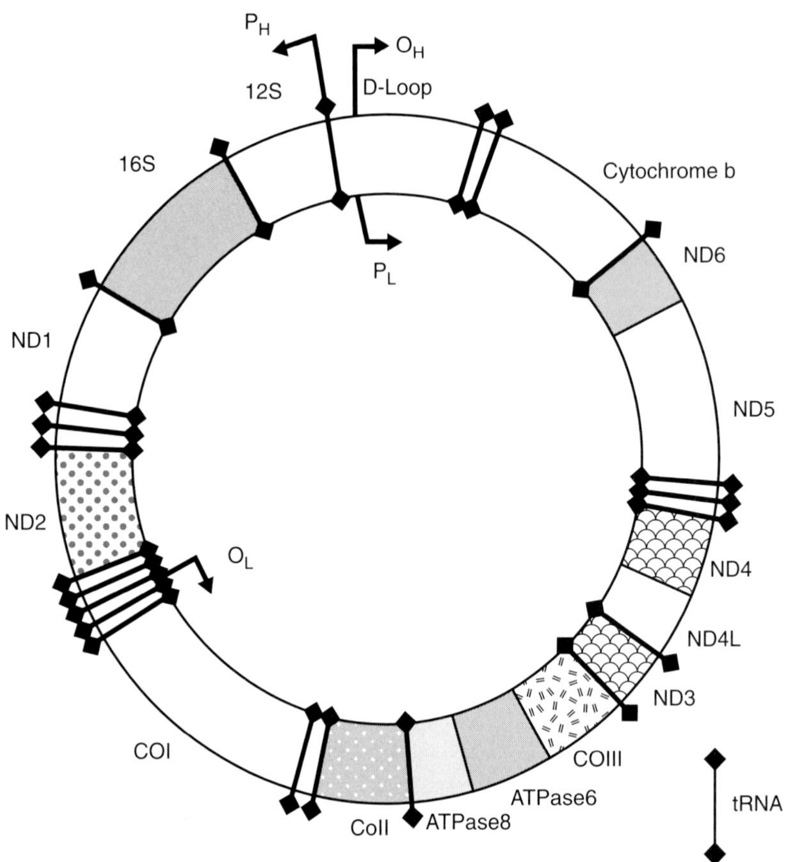

FIGURE 7-5 Map of mitochondrial genome. ND6, ND5, ND4, NDL, ND3, ND2, and ND1 are components of complex I. COIII, COII, and COI are components of complex IV. ATPase6 and ATPase8 are components of complex V. Cyt b is part of complex III. O_H and O_L are the origins of replication of the heavy and light strands, respectively; P_H and P_L are promoters for these strands.

7.3 Mitochondrial Energy Failure in MERRF

News of the diagnosis in Michael sends shock waves through the family. Some refuse to believe that he has a genetic disorder at all, given how different Michael's problems are from anyone else's. Others are afraid that they will develop the same problems or that problems will develop in one of their children. Sally and Arthur make contact with a physician who is an expert in mitochondrial disease, and Michael, Judy, Sally, Arthur, and Sally's mother travel to see this doctor. Another tissue sample for muscle biopsy is obtained from Michael, this time for biochemical studies of mitochondrial function. His mother, sister, and grandmother have studies of vision, hearing, and muscle strength, and an electroencephalogram (brain wave study). In addition, blood samples are obtained from members of the family.

The presence of abnormal-appearing, proliferated mitochondria in muscle cells (ragged red fibers) strongly implicates the mitochondrion as the site of pathology in individuals with MERRF. Maternal transmission also is compatible with this notion. Indeed, the pathogenesis of the disorder can be viewed as a widespread failure of energy production in mitochondria.

Tissues differ in their dependence on oxidative phosphorylation. Some, like red blood cells, are anaerobic. Others, such as neurons and muscle cells,

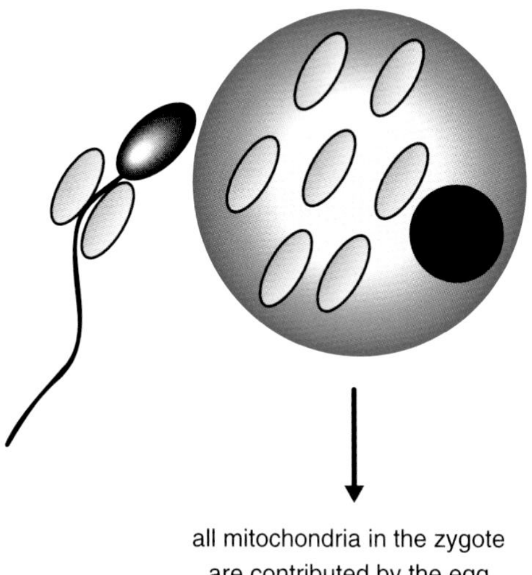

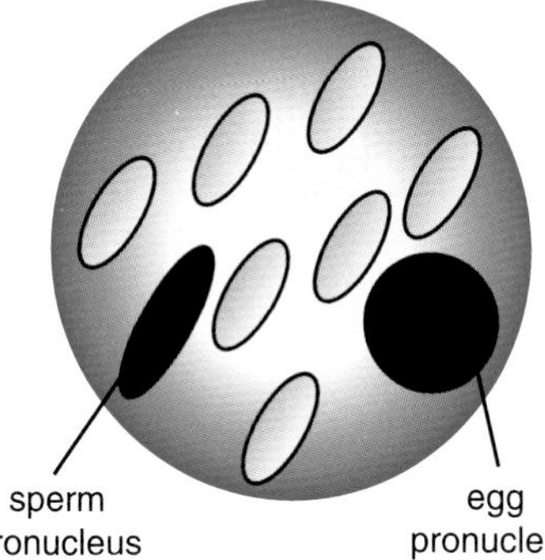

FIGURE 7-6 Mitochondrial inheritance. Sperm mitochondria are shed before entry of the sperm nucleus. All mitochondria in the zygote are contributed by the egg cell.

are highly dependent on oxidative phosphorylation for normal function and survival. One would expect that these latter tissues would be the most sensitive to disruption of mitochondrial function and, indeed, they are the tissues most affected in MERRF.

The process of oxidative phosphorylation involves transfer of electrons through five complexes of proteins (Figure 7-7). Electrons from most carbohydrate and fatty acid oxidation pathways are transferred via nicotinamide adenine dinucleotide-H (NADH) to complex I. Succinate donates its electron to complex II, and other organic acids donate electrons to various other flavoproteins. Coenzyme Q is the ultimate electron acceptor from these flavoproteins or from complexes I and II. Electrons then are transferred to the heme-containing cytochrome complexes III and IV. Each electron transfer through complexes I, III, and IV is associated with transport of H+ across the membrane. Complex V generates adenosine triphosphate from this potential energy. Genes encoded in the mitochondria contribute seven peptides to complex I, one to complex III, three to complex IV, and two to complex V.

Study of one family with MERRF revealed direct evidence of deficient mitochondrial energy production (Figure 7-8). Anaerobic threshold is a measure of the capacity of the muscle to carry out oxidative phosphorylation. Anaerobic threshold in this family was found to decrease with increasing clinical symptoms. Direct measurement of the activity of mitochondrial complexes was done on muscle biopsy specimens. Deficiency of activity of complexes I and IV was demonstrated, in keeping with a greater contribution of mitochondrially encoded peptides to these complexes. Furthermore, it was demonstrated that high-molecular-weight mitochondrial proteins were synthesized at lower levels in cells from these individuals. It remained to be determined, however, how a change in mitochondrial DNA could account for these findings.

The mystery was solved by sequencing the mitochondrial DNA from three unrelated individuals with MERRF. All three were found to have a single base change at position 8344, involving a change from A to G. This occurs in the tRNA for lysine and alters the structure of a critical region of the tRNA. The mutation is not found in the general population and therefore does not represent a polymorphism. Alteration of this tRNA would be expected to have an impact on the synthesis of many mitochondrial proteins, explaining the widespread deficiencies seen in association with MERRF.

Michael is found to have deficient activity in complexes I through V in his muscle. He has the position 8344 mutation in both muscle and blood cells (Figure 7-9).

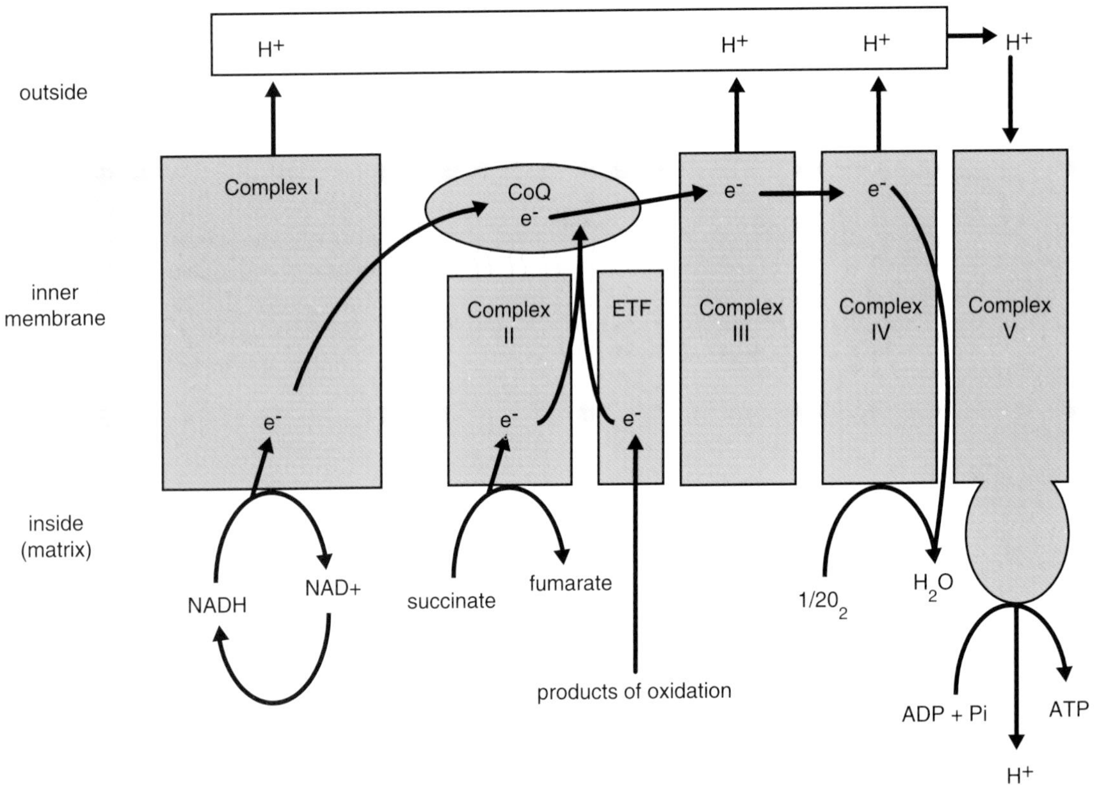

FIGURE 7-7 Oxidative phosphorylation pathway. Pyruvate and most steps of the TCA cycle donate an electron to complex I through NADH. Succinate donates its electron to complex II through $FADH_2$, and other products of oxidation donate electrons to electron transport factors (*ETF*). Coenzyme Q is the ultimate electron acceptor from these pathways, which donates its electron to complex III. This, in turn, passes an electron to complex IV. A proton is pumped from the matrix across the inner membrane along with each electron transfer. The electron then is complexed with oxygen to make water, and each proton is passed through complex V, with the resulting energy release used to make adenosine triphosphate (*ATP*).

7.4 Heteroplasmy

Electroencephalographic abnormalities are found in Michael's mother, sister, and grandmother, despite the fact that only Michael and his sister have ever had a seizure. Hearing loss is present in all of them as well, and Michael, his sister, and his mother have abnormal visually evoked responses, although only Michael has clinically evident visual problems. Blood samples from Michael, his sister, mother, and grandmother all show the presence of a mitochondrial DNA mutation. The proportion of mutant sequence differs from one individual to another in the family, however, correlating more or less with the degree of clinical involvement.

The features of MERRF can vary widely in members of a family. The family usually comes to medical attention through one or a few members with severe or obvious symptoms. Although other siblings of the proband, the proband's mother, and other maternal relatives would be expected to have the same mitochondrial mutation, often these individuals have few or no symptoms of the disorder. Mitochondrial disorders display a wide range of variability of expression, both within and between families.

The source of this variable expression is related to the peculiar manner by which mitochondrial genes are inherited (Figure 7-10). There are hundreds of copies of mitochondrial DNA molecules in each cell, in stark contrast with nuclear genes, where there are only two copies per cell. During cell division,

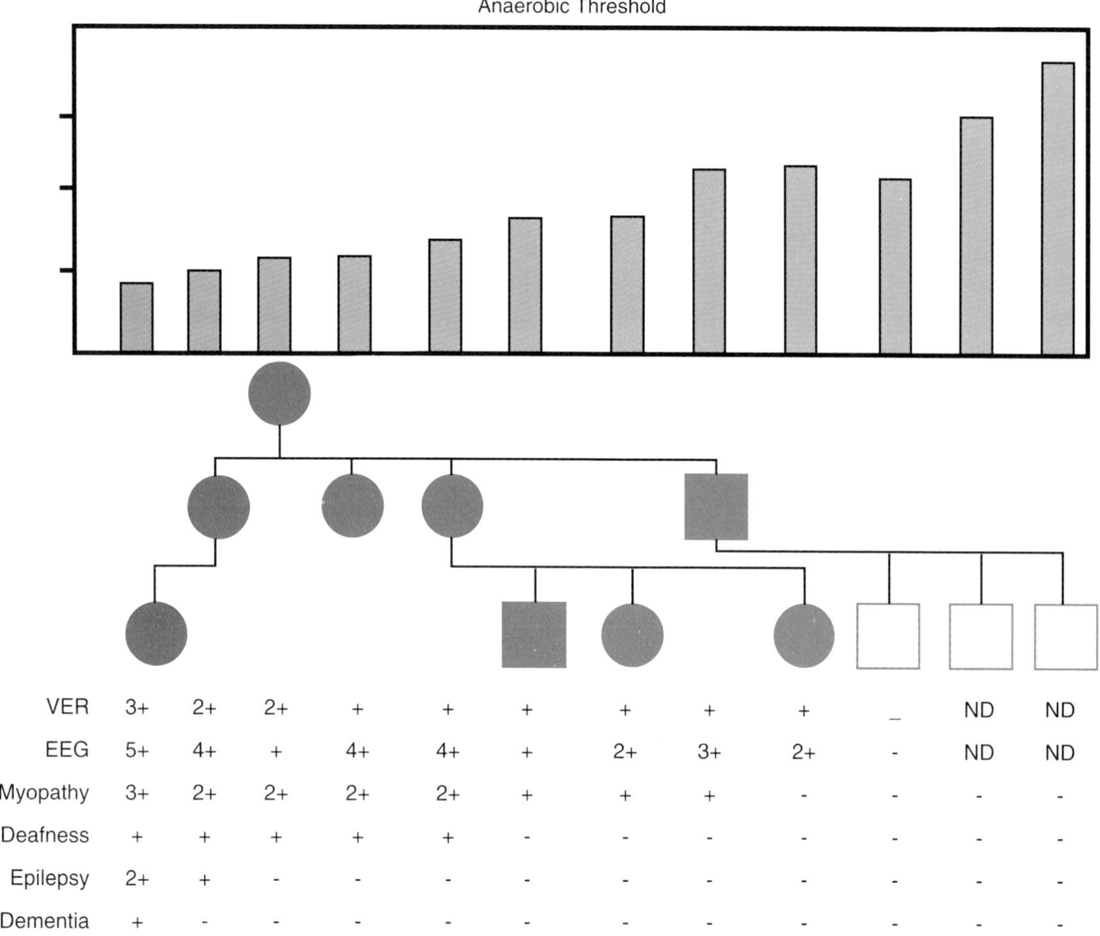

FIGURE 7-8 Clinical features in family with MERRF. The propositus is the individual to the far left. Below the pedigree, results of visual evoked response testing (*VER*), electroencephalography (*EEG*), presence of myopathy, deafness, myoclonic epilepsy, and dementia are shown. (*ND* = not determined; + = abnormal; 2+ = more abnormal; – = normal.) Above the pedigree, the anaerobic threshold is shown for each individual. The propositus was the most dramatically affected, but significant abnormalities were found in many other family members, all of whose mitochondria can be traced to the grandmother. (Redrawn from Wallace DC, Zheng Z, Lott MT, et al. Familial mitochondrial encephalomyopathy (MERRF): genetic, pathophysiological, and biochemical characterization of a mitochondrial DNA disease. Cell 1988;55:601–610.)

each mitochondrial DNA molecule replicates but, unlike nuclear genes, the newly synthesized mitochondrial molecules segregate passively to the daughter cells. A mitochondrial mutation arises as a random change in a single molecule. Over time, chance segregation may lead to proliferation of a mutant molecule in a cell. If, by chance, an egg cell has a large proportion of mutant mitochondria, the resultant child will have a high proportion of mutant mitochondria in all tissues. The same mother may produce another child from an egg with fewer mutant mitochondria, in which case the child will be less severely affected or may not have the disorder at all.

This phenomenon—wherein a cell contains a mixture of mutant and normal mitochondria—is referred to as **heteroplasmy**. Because of the passive segregation of mitochondrial DNA at cell division,

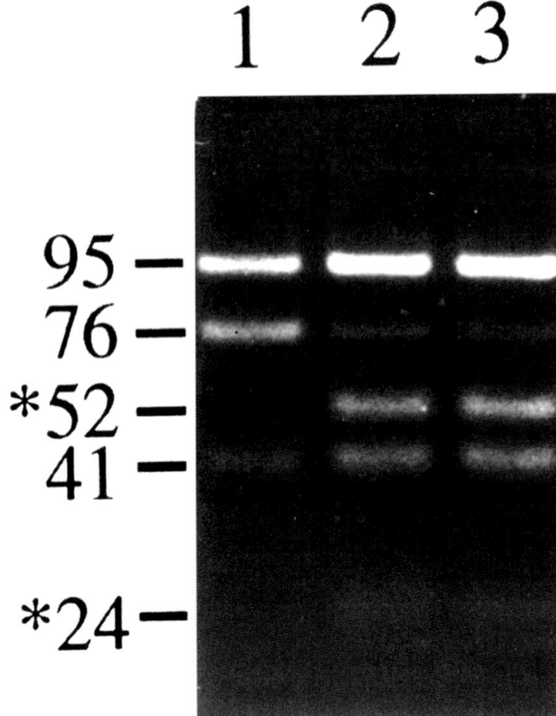

FIGURE 7-9 Analysis of mitochondrial DNA for position 8344 mutation. Since the mutation does not produce or abolish a naturally occurring restriction site, detection of the mutation is performed by introducing a novel nucleotide into the PCR product that creats a BanII restriction site when the 8344 mutation of A to G is present. Digestion of a PCR fragment that does not contain the mutation with BanII produces 95bp, 76bp, and 41bp fragments. When the mutation is present, the 76bp fragment is digested, resulting in 52bp(*) and 24bp(*) fragments. Lane 1 is from a normal control. Lanes 2 and 3 are from individuals who are heteroplasmic for the mutation. (Courtesy of Dr. John Shoffner, Emory University School of Medicine, Atlanta.)

an individual may have different levels of heteroplasmy for a mitochondrial mutation in different tissues. Individuals who carry a MERRF mutation and have symptoms of myopathy have a high proportion of mutant mitochondria in muscle cells. These muscle cells are deficient in the synthesis of mitochondrial proteins. When myoblast cells with different proportions of mutant and wild-type mitochondria are fused in the laboratory, the hybrid cells make mitochondrial proteins in proportion with the relative number of mutant and wild-type mitochondrial DNAs (Figure 7-11).

One of the family's concerns is that Judy eventually will manifest more severe signs of MERRF and that she will transmit the disorder to her offspring. The mutation is found to be highly prevalent in her blood cells, but it is explained that this does not necessarily predict future symptoms. Moreover, although Judy can be assumed to transmit the mutation to her offspring, the specific manifestations or degree of severity cannot be predicted.

Heteroplasmy goes a long way toward explaining the clinical variability of MERRF, while also presenting a diagnostic challenge. Proportions of normal and mutant mitochondria may vary from tissue to tissue, in part explaining differences in expression. Moreover, a paucity of mutant mitochondria in tissues normally used for analysis (e.g., blood or muscle) makes it difficult to confirm the diagnosis. Although in principle a woman who carries a mitochondrial mutation must transmit both normal and mutant mitochondria to all offspring, in practice this may not occur or may be difficult to prove. It has been found that women in whom 35 to 40% of mitochondria in lymphocytes are mutated tend to transmit the mutation to all their children. In contrast, more than 92% of mitochondria in muscle must be affected before biochemical evidence of mitochondrial dysfunction can be measured. This means that clinically unaffected or minimally affected women can have affected offspring, yet the severity of features in an offspring cannot be predicted.

Heteroplasmy

- Presence of mutant and normal mitochondrial DNA molecules in different proportions in different cells.
- Occurs due to passive segregation of mitochondria when cells divide.
- Accounts for variable expression of mitochondrial disorders.

7.5 Other Mitochondrial Disorders

Although mitochondria serve the single function of energy production in the cell, a wide variety of distinct

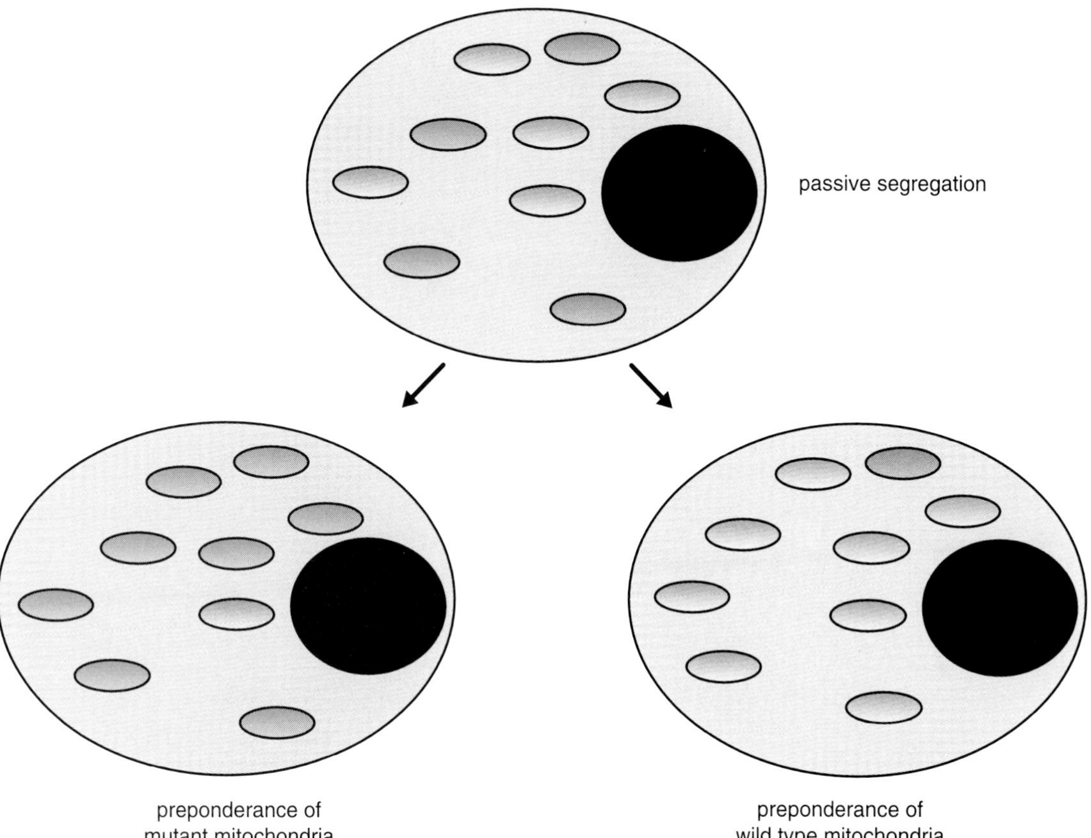

FIGURE 7-10 Concept of heteroplasmy. Both wild-type (blue) and mutant (grey) mitochondria are included in the hundreds of mitochondria in a cell. These mitochondria segregate passively when the cell divides. The proportions of mutant and wild-type mitochondria can change dramatically due to chance segregation. This can lead to variation in the proportion of affected mitochondria in different tissues or different individuals in a family.

disorders have been found to result from different mitochondrial mutations (Table 7-1). These include point mutations, such as the one found in association with MERRF, and large deletions or rearrangements of mitochondrial DNA. The former typically are maternally inherited, whereas the latter usually occur sporadically.

TABLE 7-1 Major syndromes associated with mitochondrial mutations

Syndrome	Clinical Features
Kearns-Sayre	External ophthalmoplegia, pigmentary retinopathy, heart block, ataxia, increased cerebrospinal fluid protein
MERRF	Myoclonic epilepsy, myopathy, dementia
MELAS	Lactic acidosis, strokelike episodes, myopathy, seizures, dementia
Leber's hereditary optic neuropathy	Blindness, cardiac conduction defects
Leigh's syndrome	Movement disorder, respiratory dyskinesia, regression
Others (familial diabetes mellitus; infantile encephalomyopathies)	Varied

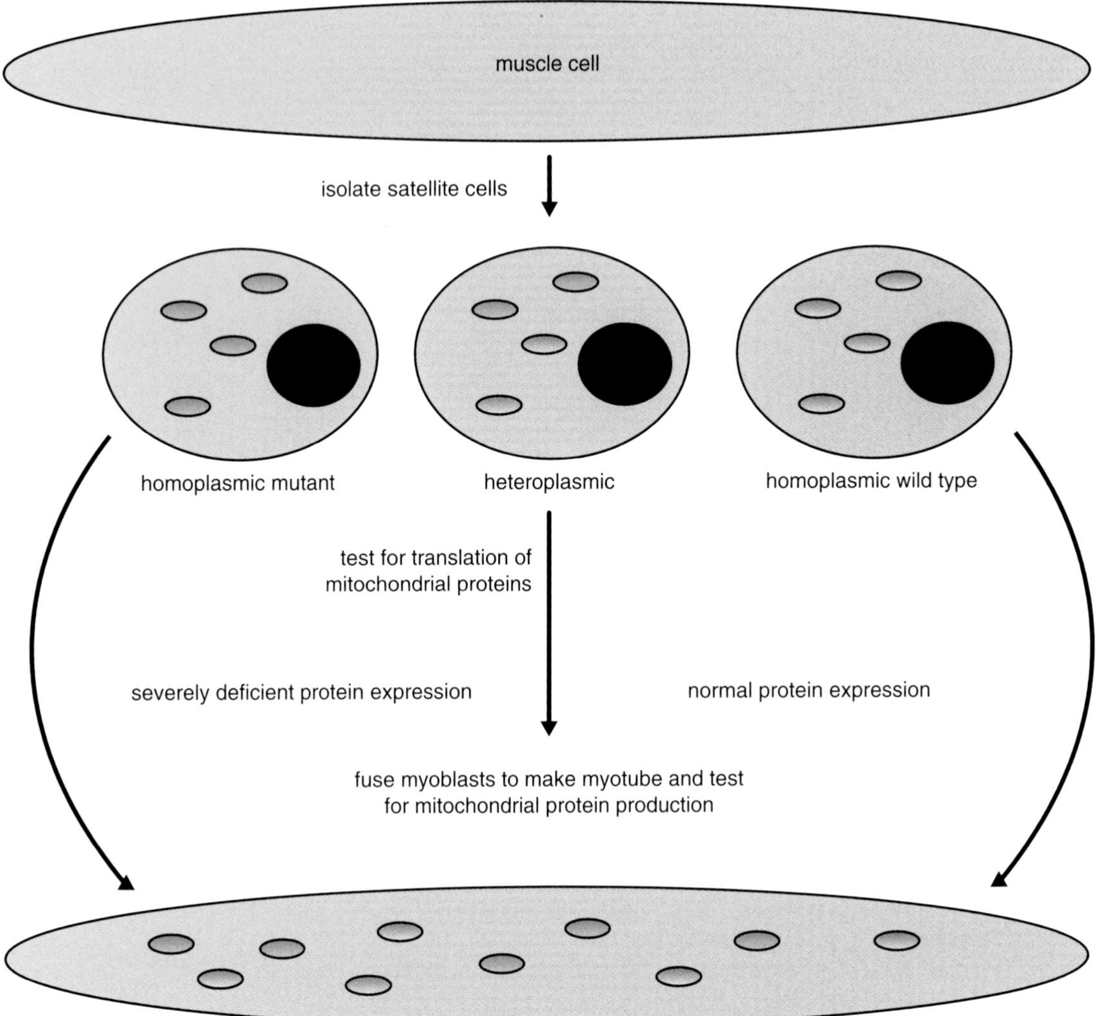

FIGURE 7-11 Myoblasts were isolated from muscle cells obtained from individuals with MERRF. Most of the myoblast clones were found to be nearly homoplasmic for the mutation, but rare heteroplasmic or homoplasmic wild-type cells were found. Cells were labeled with radioactive methionine, and the synthesis of mitochondrial proteins was determined. Myoblasts homoplasmic for the mutation produced little mitochondrial protein, whereas those with the wild-type mitochondria had normal levels. Homoplasmic mutant and wild-type cells then were fused to make myotubes, and mitochondrial protein production was determined. Wild-type levels of protein production occurred with 15 to 16% normal mitochondria. Below this level, there was a sharp drop of protein production and mitochondrial activity in proportion with the number of mutant mitochondria. (Modified from Boulet L, Karpati G, Shoubridge EA. Distribution and threshold expression of the $tRNA^{Lys}$ mutation in skeletal muscle of patients with myoclonic epilepsy and ragged-red fibers (MERRF). Am J Hum Genet 1992;51:1187–1200.)

The first mitochondrial mutations to be described were found in individuals with Kearns-Sayre syndrome. This disorder consists of a progressive external ophthalmoplegia (impairment of eye movements), pigmentary retinopathy (night blindness associated with deposition of pigment in the retina), and at least one of a set of features that include heart block (cardiac conduction defect), cerebellar ataxia, and increased level of protein in cerebrospinal fluid. Onset occurs before age 20, and the course is relentlessly progressive, usually with death before the fifth decade.

Individuals with Kearns-Sayre syndrome usually have deletions in their mitochondrial DNA, although, rarely, duplications are found (Figure 7-12). The deletions tend to be large, involving several thousand base pairs, and remove multiple mitochondrial genes and tRNAs. Kearns-Sayre syndrome usually occurs sporadically, and the deletions tend to be heteroplasmic. In addition to causing Kearns-Sayre syndrome, mitochondrial deletions have been found to occur in chronic progressive ophthalmoplegia, where other features of Kearns-Sayre are absent. The Pearson syndrome of bone marrow, liver, and pancreatic failure also results from mitochondrial DNA deletions. This disorder presents in childhood.

Another disorder associated with mitochondrial DNA deletions which is maternally inherited is characterized by adult-onset diabetes mellitus and deafness. Both deleted and duplicated mitochondrial DNAs have been found in affected individuals, suggesting that a mitochondrial mutation may be inherited that renders the mitochondrial DNA unstable.

Three major syndromes have been defined that are associated with point mutations in mitochondrial DNA. We have already considered MERRF. In addition to the already described tRNA lys mutation at position 8344, another mutation in the same tRNA at position 8356 occurs less frequently in association with MERRF.

The second syndrome is mitochondrial encephalomyopathy, lactic acidosis, and strokelike episodes, or **MELAS.** The strokelike episodes are sudden impairments of neurologic function associated with damage to brain tissue and resembling a stroke. There is episodic lactic acidosis, and ragged red fibers may be present. Clinically, there can be seizures, dementia, and recurrent headache. MELAS is most often associated with point mutations in the $tRNA^{leu}$ gene at positions 3243 or 3771. The disorder is similar to MERRF in that heteroplasmy and consequent variability of expression are common.

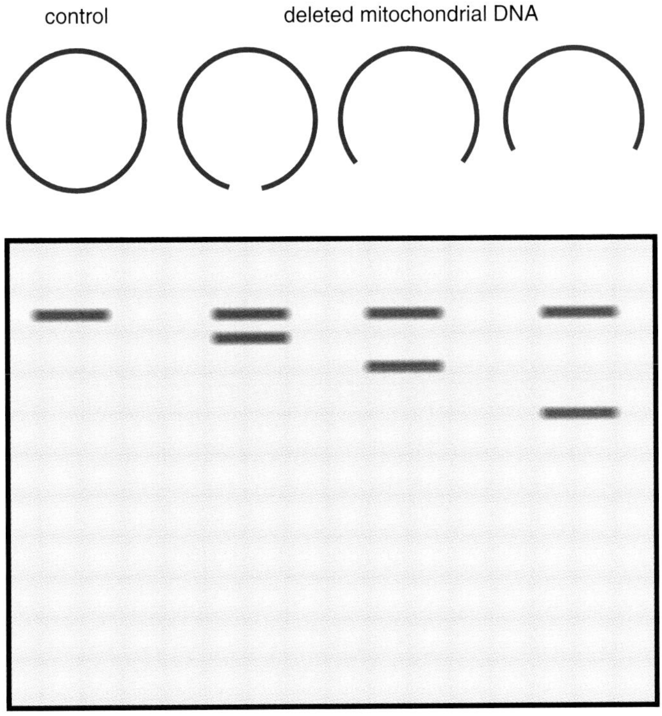

control

deleted mitochondrial DNA

FIGURE 7-12 Deletions of mitochondrial DNA in muscle biopsies from individuals with Kearns-Sayre syndrome. DNA was digested with PvuII, which cuts the mitochondrial genome at one site, resulting in a 16.5-Kb fragment that is detected with a probe to the mitochondrial DNA by Southern analysis. Each individual with the syndrome had two populations of mitochondrial DNA: one of normal size and one of smaller size, the latter corresponding with a deletion. (Modified from Zeviani M, Moraes CT, DiMauro S, et al. Deletions of mitochondrial DNA in Kearns-Sayre syndrome. Neurology 1988;38:1339–1346.)

Another major syndrome is Leber's hereditary optic neuropathy (**LHON**), characterized by sudden loss of vision occurring in the second and third decades. The disorder has been found in association with diverse mitochondrial point mutations. Although LHON mutations are maternally transmitted, symptoms of LHON are far more common in males. This raises the possibility that LHON mutations may interact with other gene products.

Other rare syndromes associated with mitochondrial mutations have been described. These include infantile encephalomyopathies and an early-childhood neurodegenerative disorder, Leigh's syndrome. One family with Leigh's syndrome has been found to have a mutation at position 8993, resulting in deficiency of ATPase 6 proteins.

In addition to maternally inherited mitochondrial mutations, several families have been described with autosomally inherited mitochondrial disorders. Most of the mitochondrial peptides are encoded in the nuclear genome, including those comprising the machinery necessary to replicate mitochondrial DNA. Although the nuclear genes responsible for familial mitochondrial disorders have not been identified, a number of families have been reported with mendelian inheritance of myopathy or encephalomyopathy, in which deficient mitochondrial function can be demonstrated. One interesting family has been described in which chronic progressive external ophthalmoplegia and other neurologic problems are inherited as an autosomal dominant trait. Multiple diverse mitochondrial deletions are found in affected individuals, suggesting that mutation of a nuclear gene leads to unstable mitochondrial DNA replication. Other families have been described in which there are multisystem diseases involving abnormal mitochondrial function associated with depletion of mitochondrial DNA (Figure 7-13). Southern blots of affected tissues reveal little or no hybridization with mitochondrial DNA. Inheritance is autosomal recessive, and it is hypothesized that the defect involves mitochondrial DNA replication.

Given the large number of mitochondrial DNA molecules in each cell, it might be asked whether mitochondrial mutations occur even in healthy individuals. In fact, mitochondrial mutations, particularly deletions, have been found in various tissues as a function of age. This has led to speculation that failure of energy metabolism due to accumulation of mitochondrial DNA damage might underlie some sporadic degenerative disorders, including Parkinson's disease. Further research is needed in this area. Another unanswered question is how such a diverse array of disorders can result from damage to the function of a single cellular organelle.

7.6 Treatment of Mitochondrial Disorders

For all of the insights into Michael's condition that have occurred since the diagnosis was made, little has happened that has directly benefited Michael himself. He has been started on two new treatments: vitamins (E and C) and coenzyme Q_{10}. In spite of this, his seizures continue at about the same rate, and his cognitive skills have not changed. There is some subjective sense that his energy level is improved, but this has been difficult to quantify. His parents have spoken with a number of families coping with mitochondrial disorders and are becoming active in efforts to establish a support group. They remain hopeful that further research will lead to improved means of treating Michael and others with similar problems.

Treatment of mitochondrial disorders has been limited mostly to the management of major symptoms. For example, seizures are treated with antiepileptic medications, hearing aids may be used, and physical therapy might be offered for management of muscle weakness. Unfortunately, there has been little to offer in the way of definitive therapy to prevent progressive neurologic problems such as strokelike episodes or dementia. There has been hope that recognition of the underlying basis for mitochondrial disorders would lead to improved ability to provide specific therapy. Thus far, although there have been some trials of therapeutic agents, none has been consistently successful in clinical trials.

One treatment approach has been to provide an alternative electron acceptor that bypasses complexes I and II. Coenzyme Q_{10} is the natural electron acceptor from these complexes and can be administered orally. There have been reports of clinical improvement in some individuals treated with coenzyme Q_{10}, particularly those with Kearns-Sayre syndrome. The widespread loss of mitochondrial function associated with most mitochondrial disorders makes it difficult to use such therapy, however. Antioxidants,

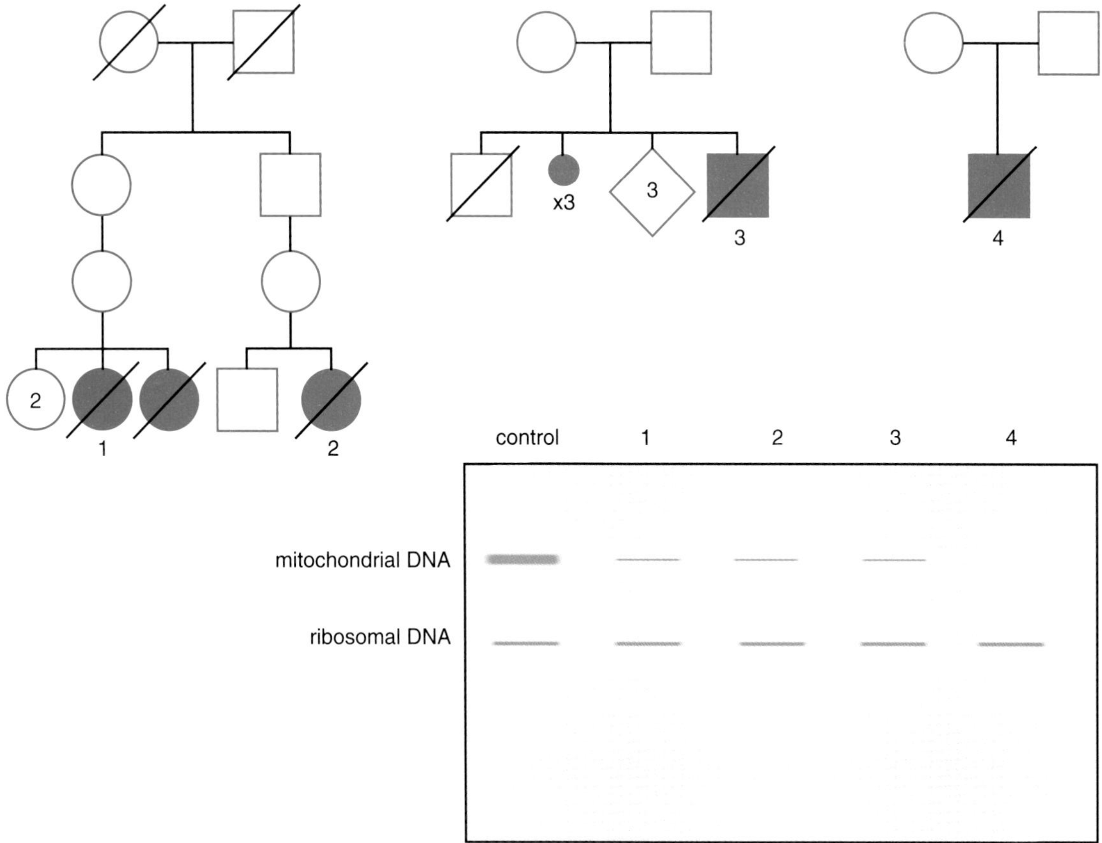

FIGURE 7-13 Three pedigrees of families having infants with fatal mitochondrial disorder. The pedigrees do not show evidence of maternal transmission. DNA from various tissues was cut with PvuII (which cuts mitochondrial DNA at a single site), subjected to electrophoresis, and hybridized on a Southern blot with two DNA probes—one to mitochondrial DNA and one to ribosomal DNA. The ribosomal DNA probe served as a control to demonstrate that equal amounts of DNA were loaded onto the gel. Very faint hybridization occurred with the mitochondrial DNA probe, indicating depletion of mitochondrial DNA in these cells. (Modified from Moraes CT, Shanske S, Tritschler H-J, et al. mtDNA depletion with variable tissue expression: a novel genetic abnormality in mitochondrial diseases. Am J Hum Genet 1991;48:492–501.)

such as ascorbic acid and vitamin E, have also been used as alternative electron acceptors, but there has been little consistent evidence for clinical efficacy.

7.7 Anthropology and Mitochondrial Genetics

Study of mitochondrial genetics has contributed not only to understanding of disease but also to the field of anthropology. Genetic markers have long been used to examine human populations. Specific haplotypes for polymorphic markers may be common in a particular group and help identify subpopulations that have diverged due to migration. Mitochondrial DNA offers many advantages for such studies. The rate of accumulation of DNA changes is estimated to be 5 to 10 times faster for mitochondrial than for nuclear DNA. Also, the absence of recombination makes it simpler to follow inheritance.

Studies of polymorphisms in mitochondrial DNA have revealed a larger degree of variation in African populations than in populations elsewhere in the world. Human mitochondrial DNA variation has

Perspective

Living with a Child with MELAS

EVELYN ARMFIELD

On March 15, 1974 (the Ides of March), our first-born daughter was delivered with no complications. She was healthy and beautiful, and we were overjoyed! She developed on schedule; had memorized, verbatim, *Pinocchio* by age 2½; acted in school plays; took acrobatics, dance lessons, and ice skating lessons; loved school, reading, and life in general. Her sister was born 1½ years later, and her brother 5 years later. They too were born healthy.

At the age of 9, she began "forgetting" multiplication tables. Testing done at our request indicated a math learning disability. At age 10, she started having severe seizures and vomiting, which led to a 2-week hospitalization during which she could no longer see. We were devastated, and her doctors had no idea what was wrong. Over the next 2 years, she experienced a multitude of symptoms, hospitalizations, tests, and consultations. We felt frustrated, as did she, not knowing the cause of her problems.

MELAS, a mitochondrial disease, was finally the diagnosis, with prognosis unknown. We were told that MELAS was extremely rare, with only a dozen cases known at the time, and that it was maternally inherited. There was no printed material available to us and no groups for discussion and support. We felt alone.

As her mother, a very heavy emotional burden disturbs me, even now, 11 years later. Why, with 20 healthy nieces and nephews, had this disease appeared only in my daughter? Why me? Did my mother have the genetic mutation? Did I? Did my siblings? Was there a cure or a treatment? What does the future hold for my two other children? We began to feel that knowing was worse than not knowing. This MELAS might end her life before she had enjoyed many wonderful things, yet there were no answers, not even guesses. "It is impossible to predict," we were told repeatedly.

Our 11-year journey with our oldest daughter seems like a century. From being a creative, bright student, she now is cognitively mentally retarded. She is wheelchair-bound and has hearing and speech losses. She receives her nutrition through a feeding tube. Incontinence, seizures, thyroid problems, diabetes, and mood swings have become a part of her everyday life—and ours. She is totally dependent and has difficulty relating to people, even her siblings. She cannot remember outsiders' names or

been traced to an origin in Africa approximately 150,000 years ago, with subsequent migration and generation of additional polymorphism.

One area in which mitochondrial DNA analysis has been used to advantage is in study of the origins of Amerindian populations. Classic anthropologic studies have indicated that there were three waves of migration from Asia to the Americas, giving rise to Amerindians of North and South America, the Na Dene population in North America, and the Aleut-Eskimos. Analysis of mitochondrial DNA among Amerindians has revealed four haplotypes that are present in 97% of individuals. This has been interpreted as evidence that Amerindian migration occurred in four waves, in each of which a "bottleneck" occurred so that most of the population died and the current population is derived from very few ancestors. This would be an important conclusion, as it would imply a founder effect for Amerindian populations that might also apply to nuclear genes, explaining the high incidence of diseases such as diabetes or hypertension in some populations.

This simple hypothesis has been challenged, however, as further studies have revealed other less common haplotypes. Although it is been proposed that the latter have arisen by random mutation, it has been calculated that this would require an unrealistically high mutation rate. More likely, there were multiple waves of migration or a limited number of major waves with a slow but steady influx of additional migrants. The study of early human colonization of the Americas is made difficult by the loss of remains from early small colonies as well as by forced migrations and extinctions from later European

even relationships within the family.

She attends an adult daycare program for retarded citizens when not hospitalized or home ill. She enjoys music, animals, adaptive bowling, and tabletop activities such as painting. She loves to help with cooking, gardening, and decorating for birthdays and holidays. She spent 15 years in the public school system, the last 10 of which necessitated constant adjustments to her individualized education plans owing to medical changes. Each new brain lesion caused a new side effect, which in turn caused a lessening of her capabilities.

As parents, we realized that there was not much available in sports activities for disabled youngsters in our town, so we began a "challenger league" for bowling and baseball with adaptive equipment. She loved it because she had always been very active in sports. Without realizing it, we became advocates for wheelchair-bound persons—for example, at malls, donut shops, churches, and parking lots. As we experienced difficulties, we complained to government bureaus, businesses, and churches. We could not do anything more medically, having gone to the best doctors and hospitals. What was left? We tried to help research efforts by volunteering our extended family to take part in studies. Financially, we spared nothing in our efforts over 8 to 10 years, forgoing vacations, social engagements, and the like.

Emotionally, we have endured several of our daughter's near-death experiences. We've learned what it means to "feel" another's pain, and we worry constantly for our two other children. We feel helpless in a medical crisis, unable to do anything to stop the course of events. Spiritually, we went through denial, anger, resentment, blaming, questioning, and accepting, and came through still believing, thanks to the wonderful Special Religious Education Program, which gave our daughter and our family dignity, community, caring, and love. We were very lucky to have loyal, loving, helpful, and kind family and friends who have helped us through this experience. The caring and expertise of doctors, clergy, and nurses has eased the burden.

Our eldest daughter has taught us a love of life—moment by moment—and appreciation of small experiences. We focus on what she *can* do instead of what she no longer does. She has touched the lives of family and friends in many ways. We can relate easily to developmentally disabled persons now. Several relatives help with less fortunate persons in our society.

Our daughter has enriched many lives. Her medical status is complicated, fragile, and uncertain, but her smile is contagious. Her speech is limited and repetitive, but the body language of her stories is wonderful. She is still, and always will be, our beautiful first-born daughter.

invaders. The mitochondrial DNA record, though, has been a powerful supplement to more traditional archeologic tools in uncovering the history of human inhabitation of this part of the world.

Case Study

Part I

One day Joseph awakens and notices that he has blurry vision when looking to the left. In addition he sees spots in front of his left eye. There is no associated pain, and he feels well otherwise. At first, he wonders whether this might be an eye infection or whether it might be related to the way he slept. The blurry vision continues through the day and, after lunchtime, Joseph calls his family doctor. He is seen later that same day, and a general physical examination is entirely normal. He is referred to an ophthalmologist, who sees him several days later. The left optic nerve is noted to be slightly swollen, and the right eye is normal. His visual field is constricted in the left eye. An optic neuropathy is diagnosed in Joseph, and a number of studies are arranged, including a brain magnetic resonance imaging (MRI) scan and immunologic tests. Pending the results of these tests, Joseph is not started on any medication.

Sudden onset of blurry vision can occur from a number of causes, including infections, inflammatory diseases, vascular disorders, and other problems. The occurrence of swelling of the optic nerve suggests the likelihood of an inflammatory process, such as optic neuritis. This can occur in isolation or as part of a process such as multiple sclerosis. Swelling of the optic nerve can be a sign of increased intracranial pressure, which might occur in the setting of a brain tumor. It would be unusual for the swelling to be unilateral in such cases, however. Magnetic resonance imaging provides pictures of the inside of the brain. Immunologic tests would be appropriate to consider the possibility of an inflammatory disorder.

Part II

The MRI is read as normal, and the immunologic studies also are normal. Approximately 1 month after the onset of his left visual disturbance, Joseph begins to lose vision in his right eye. He first notes blurry vision in the right eye while reading a book. Within 2 weeks, he finds that he cannot read street signs. Over a 4-week period, his vision deteriorates from 20/15 in the right eye to 20/200. Joseph is in a panic, facing total loss of vision over a period of a month. He is referred to a specialist who performs another ophthalmologic examination. She finds that both optic discs are swollen, and there are bilateral central scotomas. Visually evoked responses are tested and are found to be markedly attenuated.

It would be common for inflammatory optic neuritis to be associated with a normal MRI scan, and immunologic studies also might be normal. It is unusual, though, for isolated optic neuritis to progress so rapidly to involve both eyes. Visually evoked response testing involves measuring electrical activity in the brain resulting from visual stimulation with flashes of light or a changing pattern. An abnormal response indicates a delay in conduction of nerve impulses along the optic nerve or the visual pathways in the brain. Central scotomas are blind spots that correspond to loss of vision surrounding the swollen optic nerves. All these data point to the optic nerves as the site of pathology.

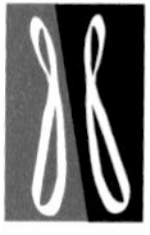

Part III

Joseph receives a diagnosis of Leber's hereditary optic neuropathy. An extensive medical history is obtained. Joseph has been in good health, and he does not smoke. When in tenth grade, Joseph was noted to have palpitations during a routine physical examination. He was hospitalized for a cardiac evaluation, but no specific diagnosis was made. He is a college student and has a night-time job as a cashier.

The clinical data fit with the diagnosis of Leber's hereditary optic neuropathy (LHON). The disorder presents with relatively acute onset of visual loss. It usually begins with blurring of vision in one eye that then rapidly progresses to involve both eyes. There is a rapid decrease of visual acuity, which eventually tends to stabilize. Initially, the optic nerves are swollen and there is a proliferation of small blood vessels. Eventually, the optic nerves appear pale and atrophic, due to loss of nerve fibers. LHON tends to affect men more often than women. Men also have an earlier age of onset (in the second to fourth decades), compared with onset beyond the third decade in women. The major pathology of LHON is confined to the eye. In some individuals, however, there can also be cardiac conduction defects. Usually, these do not cause major clinical problems, although cardiac arrhythmias can occur in some cases.

Part IV

The ophthalmologist explains that Leber's optic neuropathy is a hereditary disorder and questions Joseph about his family history. Joseph is the eldest of three siblings. He has a brother, Tom, who is 16 years old and a sister Ellen, age 14. Both are in good health and have no history of visual problems. Joseph's mother has been treated for breast cancer, and his father is in good health. Neither parent has a history of visual problems. It is arranged for Tom and Ellen to have their eyes examined. Tom has microvascular changes indicative of early Leber's optic neuropathy, but Ellen's eyes are normal.

It has long been known that LHON is a familial disorder, but the pattern of transmission has been enigmatic. More than 75% of affected individuals are male, but males never transmit the disorder, although they are fertile. A high proportion of the sons of female carriers are affected, however. These features are suggestive of maternal transmission and have provided evidence that the basis of the disorder might be mitochondrial dysfunction.

Part V

Three months have passed since Joseph received the diagnosis. His vision has stabilized, but there has been little improvement. He has registered with the state commission for the blind, and he is meeting with a rehabilitation specialist to help him adapt to his visual problem and continue his studies. He is referred to a geneticist who has experience with mitochondrial disorders. A blood sample is drawn and tested for mutations known to be associated with Leber's hereditary optic neuropathy (Figure 7-14).

It is common for visual loss to stabilize eventually in those with LHON and, in fact, some degree of improvement can occur over a long period of time.

The observation of maternal transmission supported the notion that LHON might be a mitochondrial disorder, and this notion was strongly bolstered by the finding of a mitochondrial mutation common to a number of affected individuals from different families. The mutation occurs at position 11,778 of the mitochondrial genome, causing an amino acid substitution in one subunit of complex I. This mutation is not found as a polymorphism in the general population but rather is invariably associated with LHON in a family. Individuals with this mutation may be heteroplasmic, though, perhaps accounting for some variability of expression.

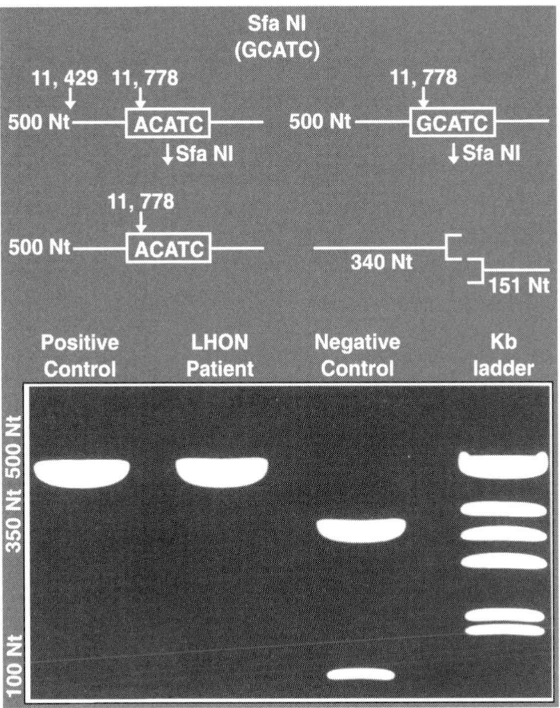

FIGURE 7-14 Polymerase chain reaction assay for nucleotide 11,778 G-to-A mutation responsible for Leber's hereditary optic neuropathy (*LHON*). The mutation creates a new Sfa NI restriction enzyme site. When the region is amplified and cut with Sfa NI, negative control DNA is cut, resulting in two bands (*negative control* lane). DNA from patients with LHON (lanes 1 and 2) are not cut due to the presence of the mutation. (Courtesy of Dr. Donald Johns, Beth Israel Hospital, Boston.)

Several other mutations have been found in other families with LHON. Many are like the 11,778 mutation in that they are found only in those with the disease and not in the general population. Another set of mutations is more difficult to interpret, because those mutations sometimes are found in the general population as polymorphisms. It has been suggested that these changes may act in concert with other mutations to exacerbate the visual loss but would not be sufficient in themselves to cause neuropathy.

The pathogenesis of LHON remains a mystery, despite evidence that mitochondrial mutations are involved. Theories involving toxic effects of tobacco or alcohol exposure or immunologic mechanisms have been proposed. It is possible that the optic nerve suffers from intracellular energy failure, but it is not clear why the effect is so tissue-specific, why it occurs many years after birth, and why it affects men more often than women.

Part VI

The blood test confirms the clinical diagnosis. Tom and Ellen are found also to carry the mutation, as is their mother. Ellen is counseled about her risk of transmitting the disorder to her offspring. Both Joseph and Tom are started on vitamin C, vitamin E, and coenzyme Q_{10}.

In accordance with the expectation of maternal transmission, all the offspring of a carrier female will carry the mitochondrial mutation responsible for LHON. It is difficult to predict the course in an individual, however, although age of onset correlates to some extent with the particular mutation. Unfortunately, there is little to offer to known carriers other than education and close follow-up, as no treatment has been demonstrated to prevent or reverse visual dysfunction. Vitamins E and C have antioxidant properties, serving as surrogate electron acceptors, and coenzyme Q_{10} likewise provides an alternative electron acceptor in the respiratory chain. Although it makes sense to use these medications and they do not appear to have significant side effects, there is relatively little evidence that they are effective.

Epilogue

Two years have passed. There has been some slight improvement in Joseph's vision, but he has learned Braille and is classified as legally blind. He has graduated from college and now is taking graduate courses in engineering. His brother, Tom, recently has experienced blurring of his vision and already is preparing for the possibility of blindness. Ellen remains in good health, with normal vision. Her major concern is the risk of passing the trait on to her children.

REVIEW QUESTIONS

1. A 23-year-old man suffers painless loss of central vision in his right eye. Three months later, he suffers similar visual loss in the left eye. He has no other symptoms and has been healthy throughout his life. After the second eye is affected, you inquire about a family history of sudden visual loss and learn that he has a maternal uncle with a similar problem.

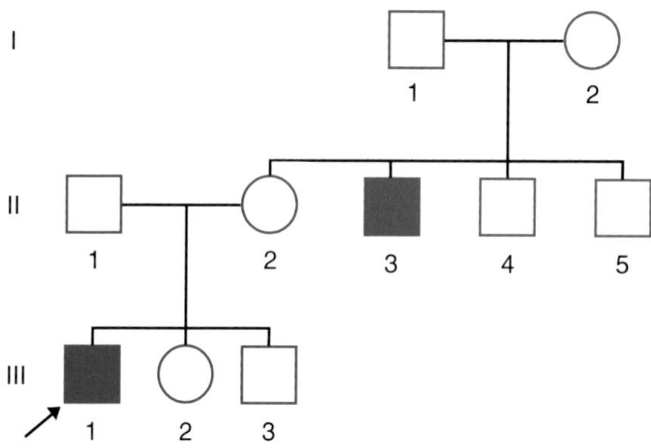

   a. Suspecting Leber's hereditary optic neuropathy, you perform testing and find a mitochondrial mutation. Which family members in the pedigree shown would also be at risk of having this mutation?

   b. How would you counsel your patient about his risk of transmitting the disorder to his offspring?

2. The sister of a boy with a mitochondrial disorder is found not to have the mitochondrial mutation in her blood cells. Is she at risk of having affected children?

3. The following pattern is seen on a Southern blot for restriction enzyme–digested DNA hybridized with a probe that recognizes mitochondrial DNA. The patient has myopathy, paresis of extraocular movements, and hearing block. How would you interpret the pattern?

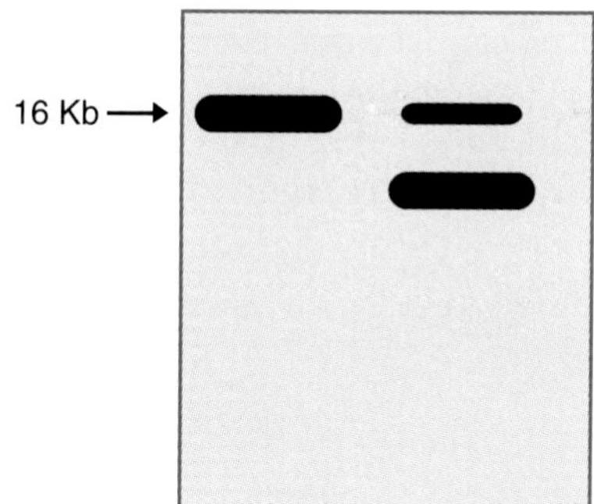

4. You are treating a child with lactic acidosis, encephalomyopathy, and abnormal mitochondria seen on muscle biopsy. His parents are first cousins. Is it possible that the consanguinity is related to this child's problems?

FURTHER READING

7.1 MERRF

Fukuhara N, Tokiguchi S, Shirakawa K, Tsubaki T. Myoclonus epilepsy associated with ragged red fibres (mitochondrial abnormalities): disease entity or a syndrome? Light- and electron-microscopic studies of two cases and review of the literature. J Neurol Sci 1980;47:117–133.

7.4 Heteroplasmy

Larsson N-G, Tulinius MH, Holme E, et al. Segregation and manifestations of the mtDNA $tRNA^{Lys}$ A to

G$^{(8344)}$ mutation of myoclonus epilepsy and ragged-red fibers (MERRF) syndrome. Am J Hum Genet 1992;51:1201–1212.

7.7 Anthropology and Mitochondrial Genetics

Baillet G, Rothhammer F, Carnese FR, et al. Founder mitochondrial haplotypes in Amerindian populations. Am J Hum Genet 1994;54:27–33.

Cann RL. mtDNA and native Americans: a Southern perspective. Am J Hum Genet 1994;55:7–11.

Wallace DC. Mitochondrial DNA variation in human evolution, degenerative disease, and aging. Am J Hum Genet 1995;57:201–223.

CASE STUDY

DiMauro S, Moraes CT. Mitochondrial encephalomyopathies. Arch Neurol 1993;50:1197–1208.

Johns DR, Neufeld MJ. Cytochrome c oxidase mutations in Leber hereditary optic neuropathy. Biochem Biophys Res Commun 1993;196:810–815.

Shapira AHV, DiMauro S. Mitochondrial disorders in neurology. Oxford: Butterworth Heinemann, 1994.

Shoffner JM, Wallace DC. Mitochondrial genetics: principles and practice. Am J Hum Genet 1992;51:1179–1186.

Wallace DC. Mitochondrial DNA variation in human evolution, degenerative disease, and aging. Am J Hum Genet 1995;57:201–223.

Wallace DC, Singh G, Lott MT, et al. Mitochondrial DNA mutations associated with Leber's hereditary optic neuropathy. Science 1988;242:1427–1430.

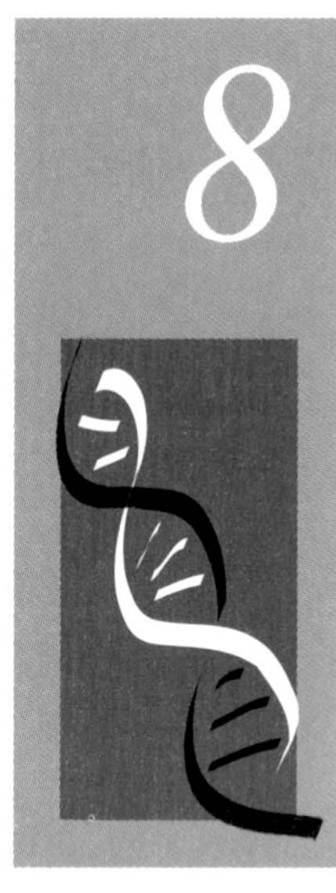

8 Cancer Genetics

OVERVIEW

We have focused thus far on genetic traits present in cells throughout the body, either owing to inheritance from a parent or to new mutation. We have encountered instances in which genetic variation exists from cell to cell in an individual, owing to mosaicism for nuclear genetic traits or heteroplasmy for mitochondrial mutations. We now turn our attention to another form of somatic genetic variation, acquired genetic change leading to malignancy. The twentieth century has witnessed an explosion of knowledge about the genetic basis of human development and disease, beginning with recognition of major patterns of genetic transmission and leading to understanding the gene at the molecular level. There has been a similar blossoming of knowledge about cancer biology. Chromosomal changes in cancer cells were recognized early in the century, and it was later found that environmental agents that cause cancer also are mutagenic. Families at high risk for specific cancers were recognized. All this suggested that genetic change might underlie the pathogenesis of malignancy, a hypothesis that has been overwhelmingly confirmed in recent years. It now is recognized that a tumor arises as a clonal growth, originating from genetic change in a single cell. The properties referred to as *malignancy* represent phenotypic features due to the accumulation of changes in multiple genes. The recognition of these genes, referred to as *oncogenes*, has led to major advances in understanding cancer biology. This, in turn, has led to improvements in diagnostic techniques and the promise of improved methods of treatment for persons with cancer. It has also shed light on mechanisms of normal cell growth and differentiation.

The prototypic disorder for understanding the genetic contribution to cancer is retinoblastoma. We will begin with a case history and introduction to the biology of this tumor. Retinoblastoma was one of the first tumors recognized as having a genetic basis. We will see how susceptibility to retinoblastoma sometimes is transmitted as an autosomal dominant trait and how statistical analysis led to a key hypothesis explaining genetic transmission. This hypothesis, referred to as the *two-hit hypothesis*, formed the framework for understanding the molecular basis for cancer. The gene responsible for predisposition to retinoblastoma was identified by positional cloning. We will trace the developments that led to this accomplishment and see how the gene participates in the two-hit model.

The retinoblastoma gene was the first of a set of *recessive oncogenes* to be discovered. In parallel with the elucidation of this genetic mechanism, an independent set of experiments led to the discovery of a set of *dominant oncogenes*. Recessive oncogenes contribute to the transformation of cells when their function is lost. Dominant oncogenes, in contrast, exert their effects when only one of the two copies of the gene is altered in a cell. We will see how these two sets of genes interact to produce a malignant phenotype and will explore the normal roles of such genes in cell growth and differentiation. Aside from producing major insights into the basic biology of cancer, these discoveries have led to advances in cancer diagnosis and treatment. We will look at current progress in this area and at the promise for the future. The Perspective will present the story of a child with retinoblastoma, and the chapter closes with a case history that illustrates the importance of molecular genetics in the day-to-day management of cancer.

8.1 Retinoblastoma

Seth is being seen for a routine pediatric visit at 2 years of age. His pediatrician becomes concerned when he is unable to see a red reflex in the back of Seth's right eye. After several attempts, he decides that Seth should be referred to a pediatric ophthalmologist. It is not easy to know what to say to Seth's parents, Rita and Arthur. Seth seems to be a healthy and active little boy. He is meeting all his normal developmental milestones and has not manifested obvious visual problems. Actually, Rita has noted that Seth sometimes turns his right eye out, but this happens only when he is tired, and she hasn't made much of it. The pediatrician explains that he is having trouble seeing well into the back of one of Seth's eyes and that, just to be safe, he would like an eye doctor to take a look. When asked what the eye doctor might be looking for, the pediatrician is vague, so when Rita and Arthur take Seth to a pediatric ophthalmologist 3 days later and are told that Seth probably has retinoblastoma (Figure 8-1), they are shocked.

Retinoblastoma is a malignant tumor of the retina. It occurs in young children, usually in the first few years of life. The tumor often comes to attention as an area of pallor at the back of the eye. This may be noticed during an eye examination as part of routine well-child care. Sometimes, family members will notice it when a photograph is taken and the annoying phenomenon of "red eye" is not seen in one or both of the child's eyes. Another sign of retinoblastoma is strabismus, or crossing of the eyes.

Seth is brought into the hospital for a day, and a detailed eye examination is performed with general anesthesia. He actually has two separate tumors in the right eye, and a very small tumor is also found in his left eye. Magnetic resonance imaging of the head reveals no evidence for tumor spread in the orbit. He has other x-rays of the body, likewise indicating no tumor metastasis. Seth's parents meet with an oncologist, who explains the plan to treat him with radiation. After a period of planning, Seth begins a course of radiation therapy. This is a very stressful time for his family, including his parents and 4-year-old sister, Diane.

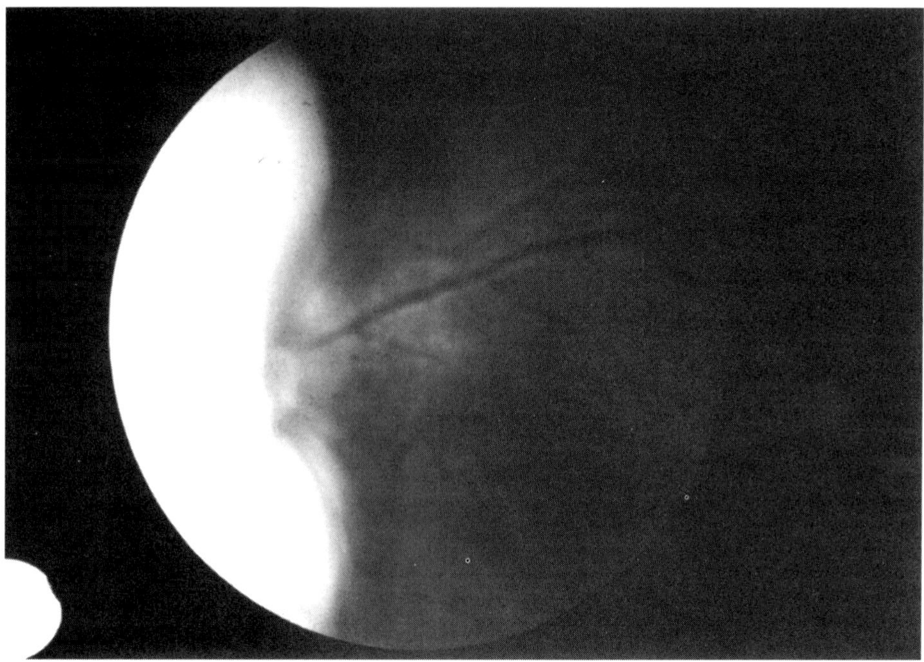

FIGURE 8-1 Photograph of retina in child with retinoblastoma. The left side of the retina is obscured by tumor. (Courtesy of Dr. Robert Petersen, Children's Hospital, Boston.)

Early recognition of retinoblastoma is essential for effective treatment. Retinoblastoma cells can detach from the retina and seed the vitreous of the eye. Spread also can occur back in the orbit, into the brain, and elsewhere in the body. Left untreated, the tumor is invariably fatal, but early recognition not only can be life-saving but also can preserve vision. Treatment typically consists of radiation therapy, applied over several weeks. This usually is effective in dealing with tumors localized to the eye. Systemic chemotherapy is used if there is evidence of tumor spread. Other modes of treatment may be used for very small intraocular tumors. These include photocoagulation (using a light to burn the tumor cells through the lens of the eye), cryotherapy (freezing tumor cells through the eye), or implantation of radioactive substances into the eye. Advanced tumors that have destroyed vision may be treated in part by removal of the affected eye.

Prognosis in children treated for retinoblastoma has improved markedly over the years. Long-term survival now exceeds 90% for tumors recognized before systemic spread. Vision often can be preserved in the affected eye, and there may be minimal complications. For many, the greatest risk is posed by the possibility of development of a second independent tumor at a different site in the same eye, in the other eye, or elsewhere in the body.

8.2 Genetics of Retinoblastoma

The course of radiation therapy is completed after several weeks. Seth seems to have done well, and the ophthalmologist is encouraged that the tumors seem to have shrunk. Meanwhile, Rita and Arthur have done a lot of reading about retinoblastoma, and they have spoken at length with both Seth's ophthalmologist and his oncologist. One issue that has arisen involves family history. Seth's parents are surprised to learn that retinoblastoma can run in families. In fact, there is no family history of retinoblastoma of which either Rita or Arthur is aware. Rita's father died of colon cancer, and Arthur's mother has been treated for breast cancer, but nobody on either side ever has had an eye tumor.

Approximately 10% of children with retinoblastoma have a family history of the disorder (Figure 8-2). Retinoblastoma occurs in these families as an autosomal dominant trait, with approximately 90% penetrance. Among those with no family history, nearly 30% turn out to have a germinal mutation leading to familial retinoblastoma. The remainder have sporadic retinoblastoma with no genetic predisposition.

The major distinguishing features of sporadic versus hereditary retinoblastoma are age at onset and number of tumors. Hereditary cases often are diagnosed within the first year of life, whereas sporadic tumors may be recognized as late as age 7 to 10 years. This is not due entirely to more careful scrutiny of children in hereditary cases, especially as the majority are the result of new mutations. Hereditary cases also tend to be multifocal (occurring at several places in an eye) or in both eyes. Approximately two-thirds of those with hereditary retinoblastoma have multifocal involvement. Aside from multifocal eye involvement, there is a risk later in life of nonocular tumors, such as osteosarcoma or breast cancer.

In 1971, Alfred Knudson reviewed records of children with either unilateral or bilateral retinoblastoma. At the time retinoblastoma was first diagnosed, children who eventually developed bilateral tumors were younger than children whose tumors remained unilateral. The rate of diagnosis of bilateral cases increased exponentially with time, whereas the rate of diagnosis for unilateral cases increased much more slowly (Figure 8-3). Knudson proposed a model (Figure 8-4) in which retinoblastoma formation requires the occurrence of two separate mutation events in a retinal cell lineage. Individuals with hereditary retinoblastoma have inherited one of these mutations, and therefore all their retinal cells carry the mutation. Only one additional event needs to occur to produce a tumor. In contrast, sporadic retinoblastoma only occurs when two independent mutational events occur in the same cell lineage. This would be expected to be much rarer, and so age at onset is later and tumors are invariably unilateral.

Knudson's hypothesis has come to be called the *two-hit hypothesis*. It has been the cornerstone for understanding hereditary predisposition to malignancy. Almost two decades passed after it was proposed, however, before the hypothesis could be verified at the molecular level.

The multifocal nature of Seth's tumors suggests that he has a hereditary form of retinoblastoma. The lack of family history indicates that he

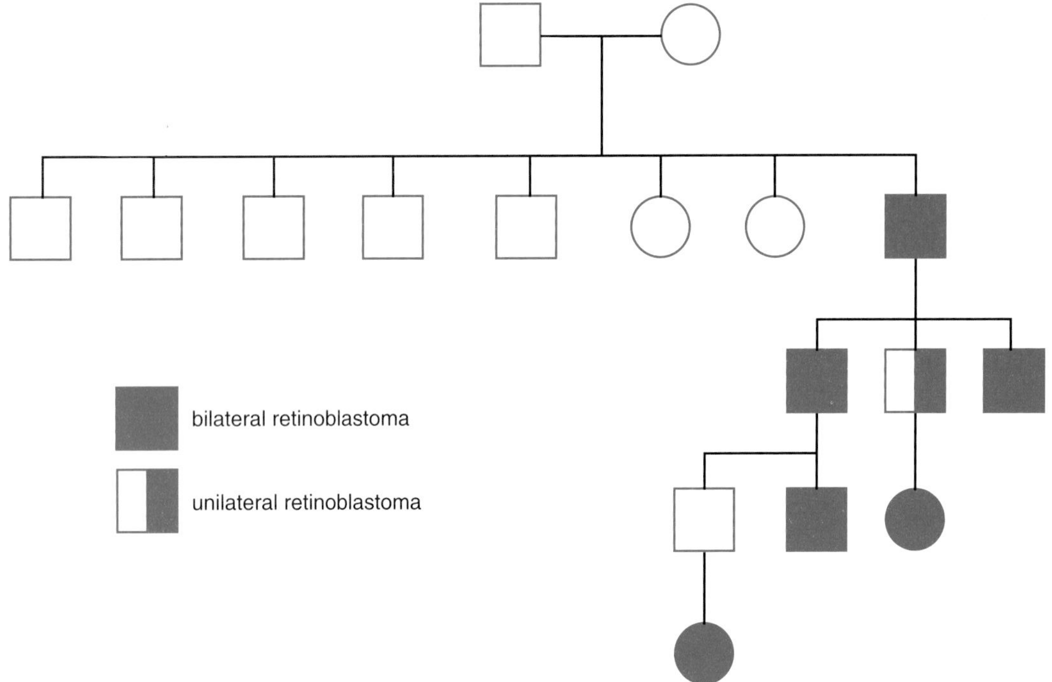

FIGURE 8-2 Pedigree showing dominant transmission of retinoblastoma. Some individuals have unilateral tumors, whereas others have bilateral retinoblastoma. The father of the youngest affected female is himself unaffected, indicative of nonpenetrance. (Redrawn by permission from Vogel F. Genetics of retinoblastoma. Hum Genet 1979;52:1–54.)

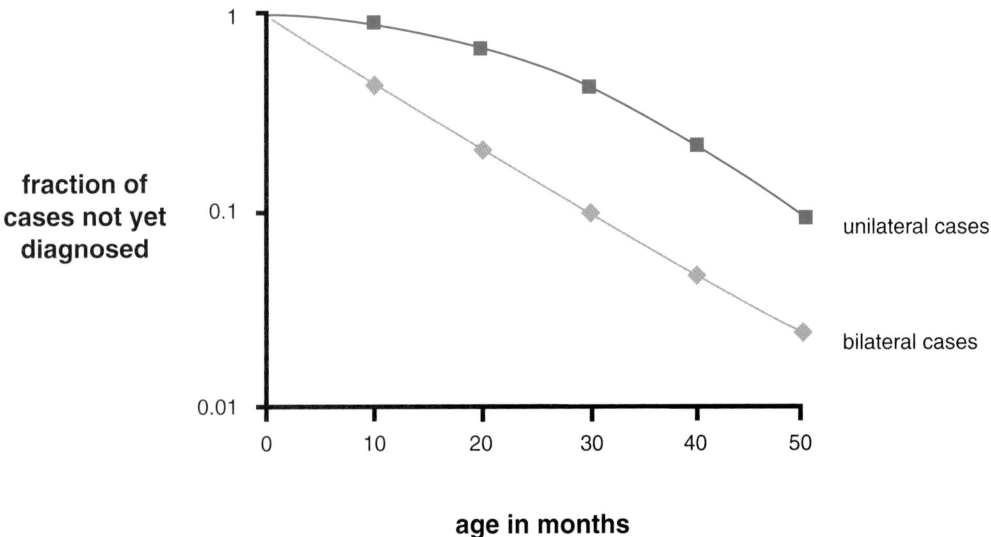

FIGURE 8-3 Graph of fraction of cases of retinoblastoma not yet diagnosed as function of age for individuals who developed unilateral or bilateral retinoblastomas. The rate of accumulation of cases of bilateral retinoblastoma was logarithmic, whereas the unilateral cases increased more slowly. (Data from Knudson AG Jr. Mutation and cancer: statistical study of retinoblastoma. Proc Natl Acad Sci USA 1971;68:820–823.)

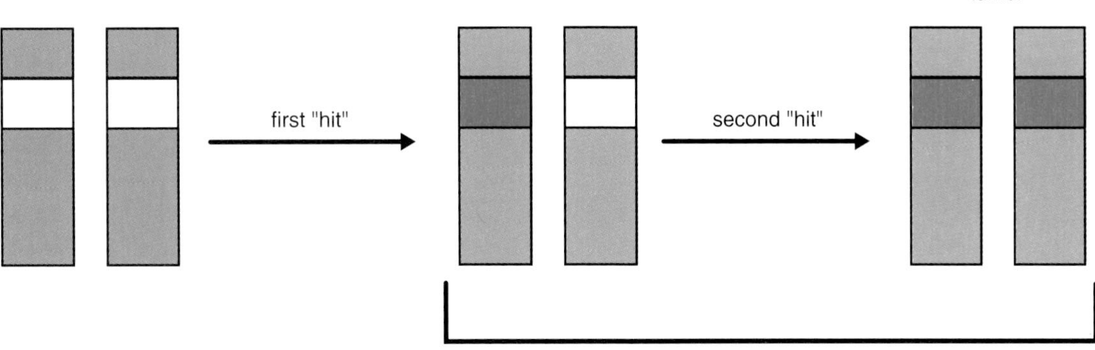

FIGURE 8-4 Two-hit model. A tumor ensues when two events have occurred in the same cell lineage that knock out both copies of a gene. Two rare events must occur to produce a sporadic tumor. A person who is born with the first mutation, however, needs to acquire only one additional mutation to develop a tumor.

probably is affected due to a new mutation. The ophthalmologist explains that Diane, too, might be at risk for having inherited a retinoblastoma mutation, because not all gene carriers develop tumors and therefore one parent could silently pass the gene on to more than one child. An eye examination performed in Diane under general anesthesia reveals no tumor. Rita and Arthur are told that Diane will need to have this test done every 6 months, at least through her early childhood years.

8.3 The Retinoblastoma Gene

Approximately 6 months after completing his course of radiation therapy, Seth is found to have recurrence of the large tumor in his right eye. His vision remains good, but his parents are told that he will need to be treated by a different method this time. This involves cryotherapy, in which tumor mass is destroyed by touching an instrument cooled to very low temperature to the outside of the eyeball. The procedure is done under general anesthesia. Whether the therapy has been successful will not be known immediately. Diane has been through one other eye examination, which again is negative for tumor. Her parents are told that recent developments now permit genetic linkage studies, which might determine whether Diane has inherited the same retinoblastoma mutation as her brother. Blood is obtained from each member of the family and submitted for testing. A single informative marker with five alleles is tested. The results are shown in Figure 8-5.

The first clue to the location of the retinoblastoma gene was provided by rare individuals who developed bilateral retinoblastomas and had multiple congenital anomalies and developmental impairment. These children do not have a family history of retinoblastoma, in spite of having bilateral disease. The occurrence of other congenital anomalies suggested the possibility that they might have a chromosomal abnormality. Indeed, analysis revealed that these children have deletions involving the long arm

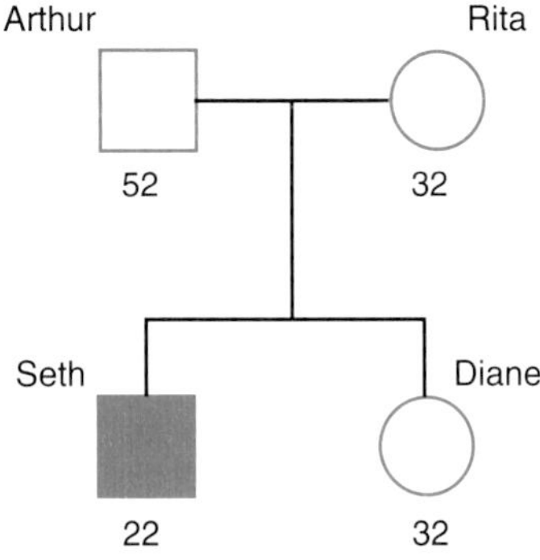

FIGURE 8-5 Seth's family pedigree. Numbers below symbols indicate alleles for the marker gene.

of chromosome 13 (Figure 8-6). The size of the deletion differed from child to child, as did the extent and nature of the congenital anomalies. All had in common, however, deletion of bands 13q13–q14, suggesting that a gene predisposing to retinoblastoma resides in that region.

It also was found that these children had a deletion of one copy of the gene encoding the enzyme esterase D. Although the biologic function of esterase D was unknown, the enzyme could be easily assayed and was found to display a polymorphism detectable by protein electrophoresis, with two alleles. Children with a deletion involving 13q14 expressed half the normal quantity of esterase D and invariably expressed only one allele.

The esterase D polymorphism opened the way to testing the hypothesis that the gene on chromosome 13 that was lost in children with chromosome deletion was the same gene involved in familial retinoblastoma. This was done using genetic linkage analysis, looking at the segregation of esterase D polymorphisms in families with hereditary retinoblastoma (Figure 8-7). Here, chromosome 13 deletions were not present, so individuals expressed both esterase D alleles. It was found, however, that the esterase D alleles segregated in the families along with the retinoblastoma trait. In one study of three families, a maximum logarithm of the odds (lod) score of greater than 3 was achieved in three families. (The actual calculation of lod scores depended on various assumptions for the penetrance of the retinoblastoma mutation and was estimated at 2.64–3.50.) This provided strong evidence that the gene predisposing to retinoblastoma in these families was located at 13q14.

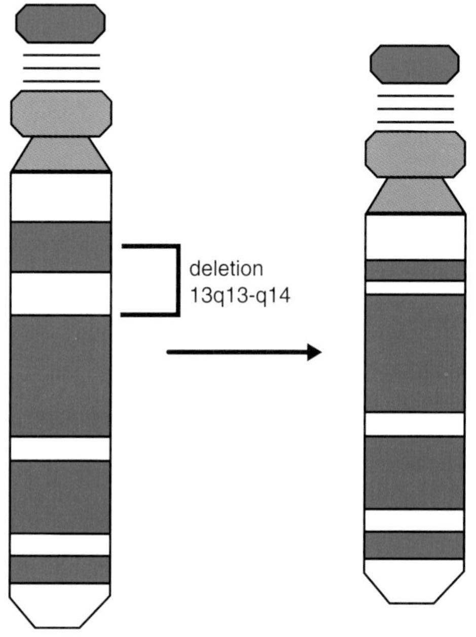

FIGURE 8-6 Diagram of chromosome 13 deletion seen in child with multiple congenital anomalies and retinoblastoma.

It appeared, therefore, that mutation of a gene on chromosome 13 was responsible for the predisposition to tumor formation in familial retinoblastoma. This would correspond to the first hit in Knudson's hypothesis. Although the second hit could, in principle, be accounted for by many possible mechanisms, it was speculated that loss of the remaining normal copy of this gene might be the second hit. Again, esterase D was used to test this hypothesis. Esterase D is expressed in retinoblastoma cells, and a number of tumors were examined to determine which alleles were expressed. All the tumors had been obtained from individuals who were heterozygous for the esterase D polymorphism, and none were from individuals with constitutional deletion of chromosome 13. The tumors, therefore, should have expressed both alleles just as did other cells from these individuals. In four of six tumors, however, only one of the two esterase D alleles was expressed. This indicated that, in a tumor, one of the esterase D alleles had been lost or damaged. The second hit in these cells was probably a deletion that included both the esterase D locus and the retinoblastoma locus.

By the early 1980s, molecular genetic techniques were being applied to the problem (Figure 8-8). Chromosome 13–specific DNA was cloned from a human-rodent somatic-cell hybrid that had lost most human chromosomes but retained chromosome 13. Some of these cloned sequences detected restriction fragment–length polymorphisms. Sequences near 13q14 frequently were homozygous in retinoblastomas, even if they were heterozygous in nontumor DNA from the same individual. Sometimes, only sequences immediately near 13q14 were affected, whereas other times homozygosity extended throughout chromosome 13 (Figure 8-9). Other cases could be explained by loss of the entire chromosome, sometimes with reduplication of the remaining chromosome 13.

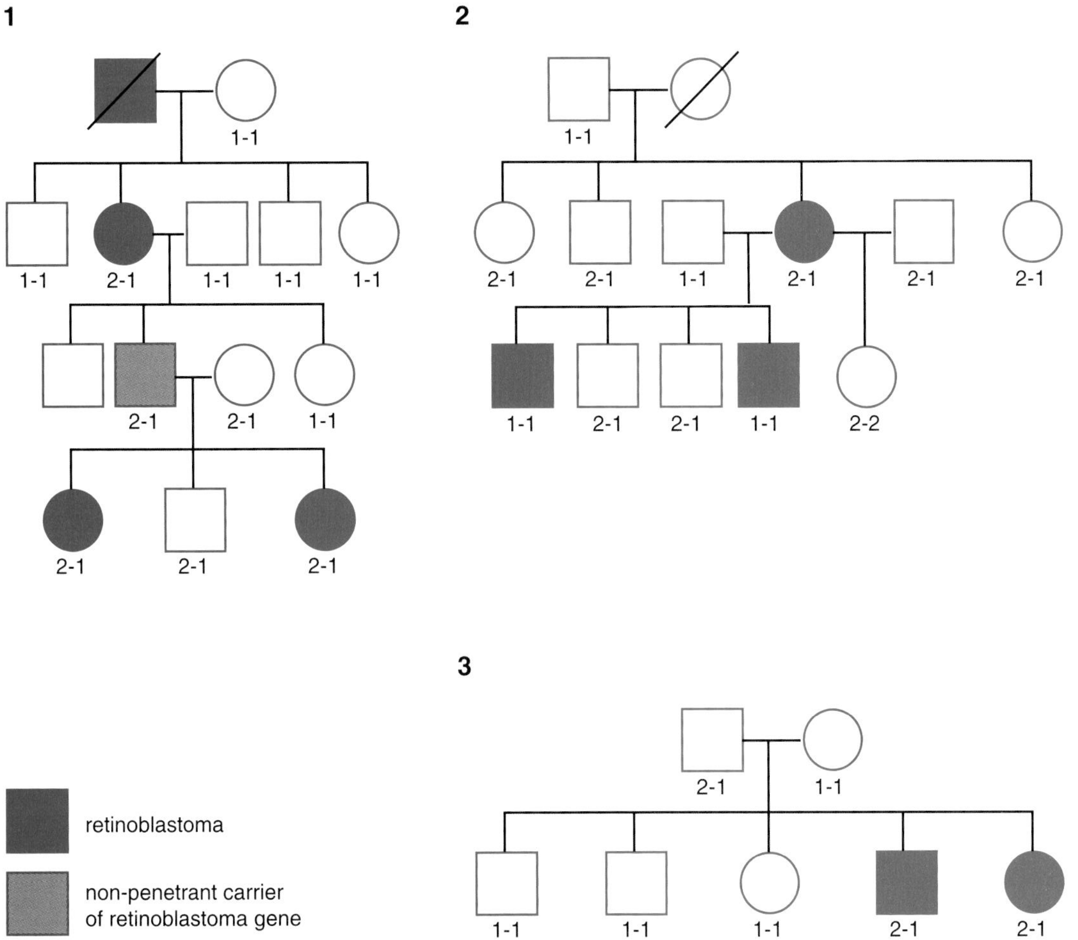

FIGURE 8-7 Segregation of esterase D alleles (alleles indicated as *1* and *2*) in three families with hereditary retinoblastoma. The *2* allele segregates with retinoblastoma in families 1 and 3, whereas the *1* allele is in coupling with retinoblastoma in family 2. (Redrawn from Sparkes RS, Murphree AL, Lingua RW, et al. Gene for hereditary retinoblastoma assigned to human chromosome 13 by linkage to esterase D. Science 1983;219:971–973.)

These findings applied to tumors whether they were familial or sporadic, suggesting that the events that lead to tumor formation involve a locus on chromosome 13 in either case. This is compatible with Knudson's hypothesis. In familial cases, the alleles retained in the tumor invariably were those inherited from the affected parent. This also is consistent with the hypothesis, as it should be the normal allele that is lost in the tumor if this is the second hit. A very satisfying story was coming together, but ultimate proof of the two-hit hypothesis required confirmation at the molecular level.

The molecular data reveal that Diane has inherited from her mother a different retinoblastoma allele than that found in Seth, but the same allele from her father. The family is told that the risk of either parent being a silent carrier is approximately 5%, so Diane's risk of retinoblastoma is estimated to be in that range as well. She will therefore need to continue to have regular ophthalmologic examinations.

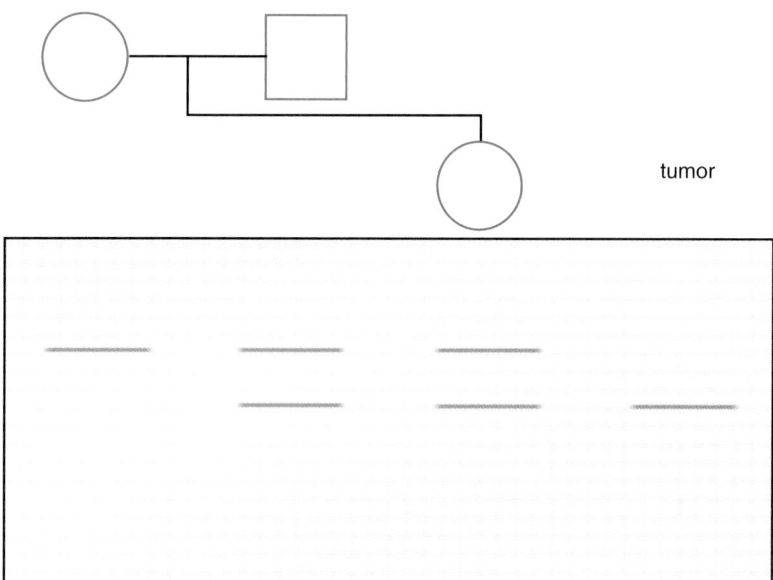

FIGURE 8-8 Loss of heterozygosity for polymorphic marker on chromosome 13. DNA from the mother, father, and daughter, as well as the daughter's retinoblastoma tumor, was digested with *HindIII* and hybridized on a Southern blot with a DNA sequence from chromosome 13q14. The mother is homozygous for the upper allele and the father and affected daughter are heterozygous. DNA from the retinoblastoma, however, shows only the lower allele, indicating loss of the upper allele in the tumor. (Modified from Dryja TP, Cavenee W, White R, et al. Homozygosity of chromosome 13 in retinoblastoma. N Engl J Med 1984;310:550–553.)

8.4 Cloning the Retinoblastoma Gene

Although he remains bright and active, Seth's difficulties continue. After 3 months, it is clear that his tumor still has not responded. Yet another mode of treatment is instituted, this time involving surgical implantation of a radioactive cobalt source directly into the tumor mass.

The search for the retinoblastoma gene focused on band q14 of chromosome 13. The strategy was to clone DNA in the region and search for sequences expressed in normal retinal cells and mutated in persons with familial retinoblastoma. The DNA was cloned by gene walking, but a gene walk requires a starting point. This starting point was provided by the discovery of a set of tumors that were homozygously deleted for a cloned DNA sequence from 13q14.

The cloned sequence in this case was isolated from a flow-sorted chromosome library (see Chapter 4). Clones in the 13q14 region were identified by Southern hybridization with human-rodent cell lines with various chromosome 13 rearrangements and by in situ hybridization. Phage clones then were tested by Southern analysis for whether they were deleted in retinoblastomas. One clone, designated H3–8, was homozygously deleted in two tumors from unrelated individuals, one of whom had bilateral (hence hereditary) retinoblastoma. This individual was heterozygous for the deletion in somatic tissue. The other tumor was a sporadic unilateral retinoblastoma. Constitutional DNA in this case did not carry a deletion, indicating that the homozygous deletions occurred as acquired events in the retinal cells. A human phage DNA library was screened with H3–8, and gene walking was done for a region of approximately 29 Kb. All the cloned DNA was found to be homozygously deleted in both tumors. Other cloned DNA sequences from the 13q14 region were not deleted. Therefore, H3–8 was judged to be very near the retinoblastoma gene.

Subclones of the phage surrounding the H3–8 region then were tested for species conservation. One clone, designated *p7H30.7R*, was found to be present in both human and mouse and hybridized with RNA from human retinal cells. A cDNA library was made from human retinal cells, and a cDNA clone was isolated that was homologous with p7H30.7R. The transcript size was 4.7 Kb. It was present in normal retinal cells but was not found in any of four retinoblastoma cell lines or one osteosarcoma cell line. It also was expressed in spleen, liver, and many tumor types from various tissues.

A subclone of the cDNA was hybridized with DNA from retinoblastoma tumors on Southern blots. Aberrant fragments were found in 30% of

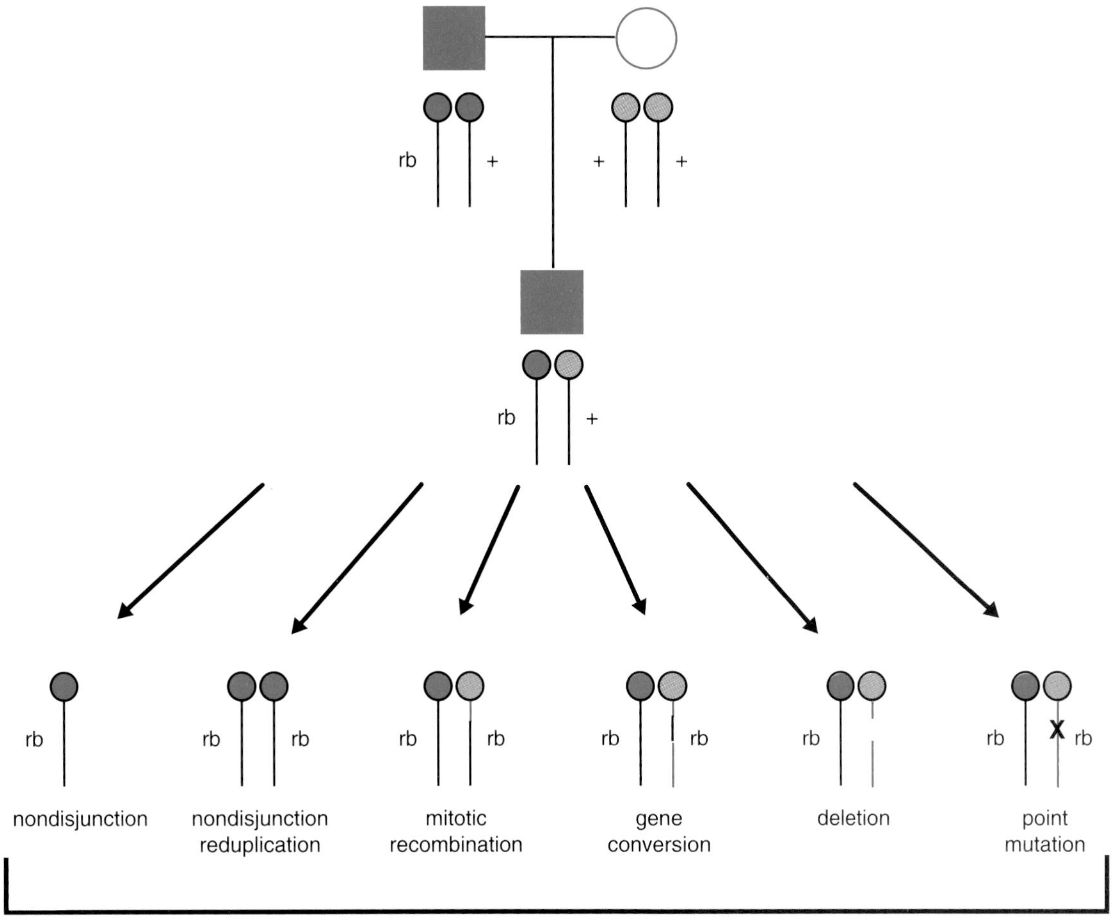

FIGURE 8-9 Mechanisms of acquisition of second hit in retinoblastoma tumors. The child with retinoblastoma has inherited one mutant allele (*rb*) but is heterozygous. Various possible genotypes in tumor tissue that result in loss of the wild-type allele are shown at the bottom. The normal chromosome may be lost by nondisjunction, with or without reduplication of the mutation-bearing chromosome. A recombination event can take place between *mitotic* chromosomes, placing the mutant allele on both homologs. Gene conversion represents a change in sequence due to copying of the mutant sequence onto the other chromosome. The wild-type allele also may be deleted from the chromosome or damaged by mutation. (Redrawn by permission from Cavenee WK, Dryja TP, Philips RA, et al. Expression of recessive alleles by chromosomal mechanisms in retinoblastoma. Nature 1983;305:779–784.)

50 tumors examined, including retinoblastomas and osteosarcomas. Many had homozygous deletions of part of the gene. It appeared highly likely that this was the retinoblastoma gene. This discovery was announced in the same week as was the cloning of the gene for Duchenne's muscular dystrophy, representing the second and third genes to be identified by a positional cloning strategy (after the gene for chronic granulomatous disease).

Subsequently, the same sequence was cloned by other groups and found to be abnormally expressed in retinoblastoma cells: Either no transcript was present or the transcript was of abnormal size. In many cases, structural changes of the gene, including homozygous deletions, were observed. Patients with hereditary retinoblastoma who had tumors with homozygous deletions had heterozygous deletions in constitutional tissue. This was in accord with the two-hit hypothesis. Sequence analysis of the gene, referred to as *Rb*, revealed an open reading frame of 816 amino acids with no homology to known proteins.

Ultimate proof that Rb was the genuine retinoblastoma susceptibility gene was provided by demonstration that introduction of the intact gene into retinoblastoma or osteosarcoma cells suppressed the neoplastic phenotype (Figure 8-10). A vector was constructed, placing the Rb gene under control of the long terminal repeat sequences of a retrovirus, which stimulate transcription. The vector was packaged into virus particles with a helper virus, and cultured retinoblastoma or osteosarcoma cells were infected. Although these cells lacked natural Rb expression, expression was restored by the inserted gene. There are a number of tests for malignant properties of cultured cells, such as cell morphology (neoplastic cells tend to grow in clumps, whereas nontransformed cells remain attached to the culture dish), ability to grow in soft agar (only transformed cells are capable of growing without attaching to a solid substrate), and ability to grow in nude mice (nude mice lack an immune system; tumor cells tend to grow when injected into nude mice). By all these measures, the cells with restored Rb expression had lost their transformed phenotype.

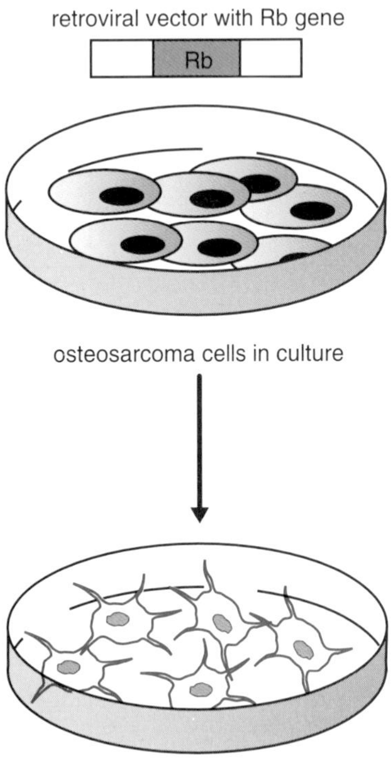

FIGURE 8-10 Suppression of malignant phenotype in cultured osteosarcoma cells by introduction of an intact Rb gene using a retroviral vector. (Data from Huang H-JS, Yee J-K, et al. Suppression of the neoplastic phenotype by replacement of the RB gene in human cancer cells. Science 1988;242:1563–1566.)

Molecular analysis reveals that Seth has a one–base pair deletion in exon 19 of the retinoblastoma gene. This leads to a frameshift, resulting in a truncated protein. Neither of his parents is found to carry this mutation, and it also is not found in Diane. Rita and Arthur are delighted to learn that Diane will no longer require ophthalmologic examinations for retinoblastoma.

The isolation of the Rb gene opened new avenues of genetic diagnosis for families with retinoblastoma. In most cases, though, the mutations are not deletions or large-scale disruptions of the gene that can be detected by Southern analysis. Instead, most are subtle point mutations. One strategy for detection of these mutations is sequencing each exon after amplification by polymerase chain reaction (PCR). This allows unequivocal detection of any mutation. Molecular analysis has obvious utility for families with known hereditary retinoblastoma. A child at risk of inheriting the gene must be followed closely for the development of retinoblastoma, because early detection can vastly improve outcome. This involves frequent ophthalmologic examinations, which must be done very thoroughly to ensure scrutiny of the entire retina. The examinations are performed under general anesthesia in young children, entailing some risk and great discomfort. Molecular testing allows those determined not to have a mutant gene to be freed from regular eye examinations. In addition, molecular testing has proved helpful in sporadic unilateral cases. Some of these represent new mutations of the retinoblastoma gene, and affected persons are at risk of transmitting the mutation. Molecular testing can be helpful for genetic counseling.

The retinoblastoma gene was the first of a series of genes to be identified that are associated with familial predisposition to cancer. Other examples are listed in Table 8-1. These syndromes are dominantly

TABLE 8-1 List of some tumor suppressor genes, associated syndromes resulting from germ-line mutations, and characteristic tumors

Gene	Syndrome	Major Tumors
APC	Familial adenomatous polyposis	Bowel carcinoma
VHL	Von Hippel–Lindau syndrome	Hemangioblastoma, pheochromocytoma
p53	Li-Fraumeni syndrome	Soft-tissue sarcoma, glioma, leukemia
BRCA1 BRCA2	Familial breast, ovarian cancer	Breast and ovarian carcinoma
NF1	Neurofibromatosis type 1	Neurofibroma, astrocytoma
NF2	Neurofibromatosis type 2	Schwannoma, meningioma
RB	Retinoblastoma	Retinoblastoma, osteosarcoma
TSC1	Tuberous sclerosis	Cortical dysplasia, renal angiomyolipoma
WT1	WAGR-Wilms' tumor aniridia, genitourinary anomalies, growth retardation	Wilms' tumor
p16	Familial melanoma	Melanoma, glioma, carcinoma
hMSH2 hMLH1 hPMS1 hPMS2 GTBP	DNA repair genes Hereditary non-polyposis colorectal cancer	Colon carcinoma

inherited. Tumors that occur in affected persons are seen also in the general population, but those with the hereditary form develop them earlier in life and often have multiple independent cancers. The genes have come to be referred to as *tumor suppressor genes* or *recessive oncogenes*. They behave in a recessive manner at the cellular level, so that both copies of the gene must be inactivated in order for a tumor to occur. Within a family, however, they act as dominant traits. What is transmitted from generation to generation, though, is not the malignancy itself but rather the risk of development of malignancy. This is conveyed by transmission of one mutant allele at the tumor suppressor locus, so that only one event need occur—mutation of the remaining normal copy of the gene—for a cell to become a tumor. For these same tumors to occur in the general population requires two rare events: mutation of the first copy of the gene and then mutation of the second copy in the same cell lineage. This does occur but only rarely.

The term *tumor suppressor gene* is apt, though in one sense it is misleading. It implies that the normal function of the gene is to prevent the cell from becoming a tumor. Is this truly the normal role of these genes? Understanding the function of tumor suppressor genes and the way that loss of this function results in tumor formation requires a more complete unveiling of the various genetic changes that accompany cancer.

8.5 Oncogenes

Seth is now nearly 4 years old. He has lost most vision in his right eye and, in spite of multiple modes of therapy, one of the tumors in that eye continues to grow. His left eye has normal vision, however, and there is no evidence for active tumor in that eye. It is decided, after considerable agonizing, that it would be best for Seth's right eye to be removed. This is done, and Seth is fitted with a prosthetic eye. The tumor in his right eye is sent to a research laboratory and grown in culture. Chromosomal analysis (Figure 8-11) reveals multiple genetic changes.

Loss of the Rb gene must not be the only change in retinoblastoma cells, for if it were, there would be no difference in behavior from tumor to tumor. Some tumors grow faster and spread more aggressively than others. Analysis of chromosomes in retinoblastoma tumor cells has indeed revealed a plethora of genetic changes. Chromosome 13 rarely is visibly changed, yet a wide variety of chromosome abnormalities are apparent. It is typical for cancer cells to harbor many chromosomal changes, including aneuploidies, deletions, duplications, and translocations. As far back as 1914, the pathologist Theodor Boveri

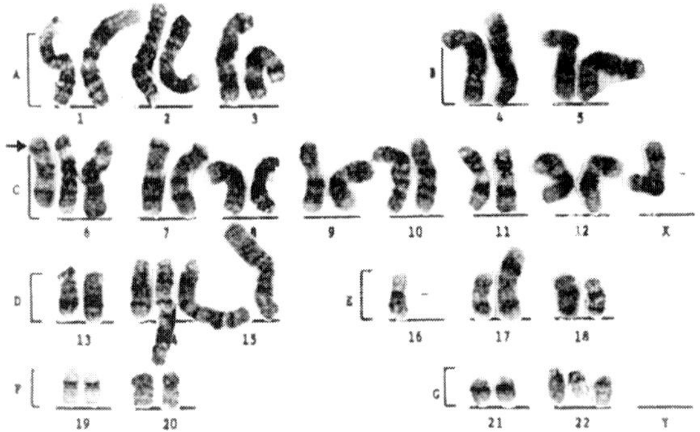

FIGURE 8-11 G-banded karotype from retinoblastoma. Numerous chromosome abnormalities are present including isochromosome for the short arm of chromosome G (arrow). Courtesy of Dr. Brenda Gallie, Hospital for Sick Children, Toronto.

recognized that the nuclei of cancer cells were abnormal in appearance and proposed that chromosomal changes might underlie malignant transformation. It has taken most of the twentieth century to unravel the puzzle, but the elements that underlie the process of malignancy are coming into focus.

The discovery of the first cancer gene, or *oncogene*, was based on work begun long before DNA was known to be the genetic material. Peyton Rous, working at the Rockefeller Institute for Medical Research in New York in 1909, began a series of experiments that started with a chicken that had a lump on its leg. The lump was a soft-tissue sarcoma. When Rous ground up some of this tumor and injected it into other chickens they, too, developed sarcomas. The active agent was identified as a virus—in fact, a retrovirus—and was called *Rous sarcoma virus*. Decades later, it was found that of the four genes in this virus, one, referred to as *src*, was responsible for the transforming properties. When *src* was lost or mutated, the virus was no longer oncogenic.

By the 1980s, approximately 20 different retroviruses, each containing a distinct oncogene, were known to be associated with cancer. These oncogenes were identified by three-letter abbreviations, usually named for tumor types they caused (e.g., *erb*-B for erythroblastosis). The breakthrough in understanding nonretroviral malignancies was recognition that viral oncogenes are homologous with normal eukaryotic genes. For example, the viral *src* gene, or v-*src*, is homologous to a eukaryotic *src* gene, referred to as c-*src* (for cellular *src*). These normal cellular genes are referred to as **proto-oncogenes**. Apparently, the overexpression of a proto-oncogene due to regulation by the viral genome is responsible for its transforming properties.

When a retrovirus containing an oncogene infects an animal, there is usually a latent period of 2 to 3 weeks before a tumor grows. Some, however, are oncogenic with a longer latency of many months. When retroviruses infect cells, the cDNA copy of the viral genome integrates into the host cell genome. Longer-latency retroviruses do not carry oncogenes. Examination of the DNA from tumors in infected animals, however, indicates that the virus has integrated, by chance, adjacent to a cellular proto-oncogene (Figure 8-12). This places the cellular gene under the influence of the active retroviral promoter, activating the proto-oncogene and causing a tumor. Rarely, an aberrant recombination event causes the cellular proto-oncogene to incorporate into the viral genome. This is how oncogenes are "captured" from the eukaryotic genome by retroviruses.

Not all proto-oncogenes were discovered from studies with retroviruses. Oncogenes exert a dominant effect—that is, overexpression of one copy contributes at least some transformed properties to the cell. An assay was developed whereby DNA was isolated from tumor cells, broken into pieces, and then introduced into nontransformed cells (Figure 8-13). In the presence of calcium phosphate, the isolated DNA is taken up into the cells by a process known as **transfection**. Some DNA integrates into the recipient cell genome and is expressed. Using a mouse fibroblast cell line as recipient, it was found that transfection of tumor cell DNA resulted in isolation of fibroblast clones with some properties of transformation, such as ability to grow in soft agar. When human tumor cells were used as the donor, the transfected DNA could be identified in the mouse cells because of the presence of repeated sequences found in the human but not the mouse genome. It was found that the transforming genes corresponded in some cases with oncogenes already known to be involved in retroviral-mediated oncogenesis. Moreover, it was found that specific oncogenes tended to be involved in certain tumor types.

Oncogenes, then, are normal cellular genes that, when their patterns of expression are changed,

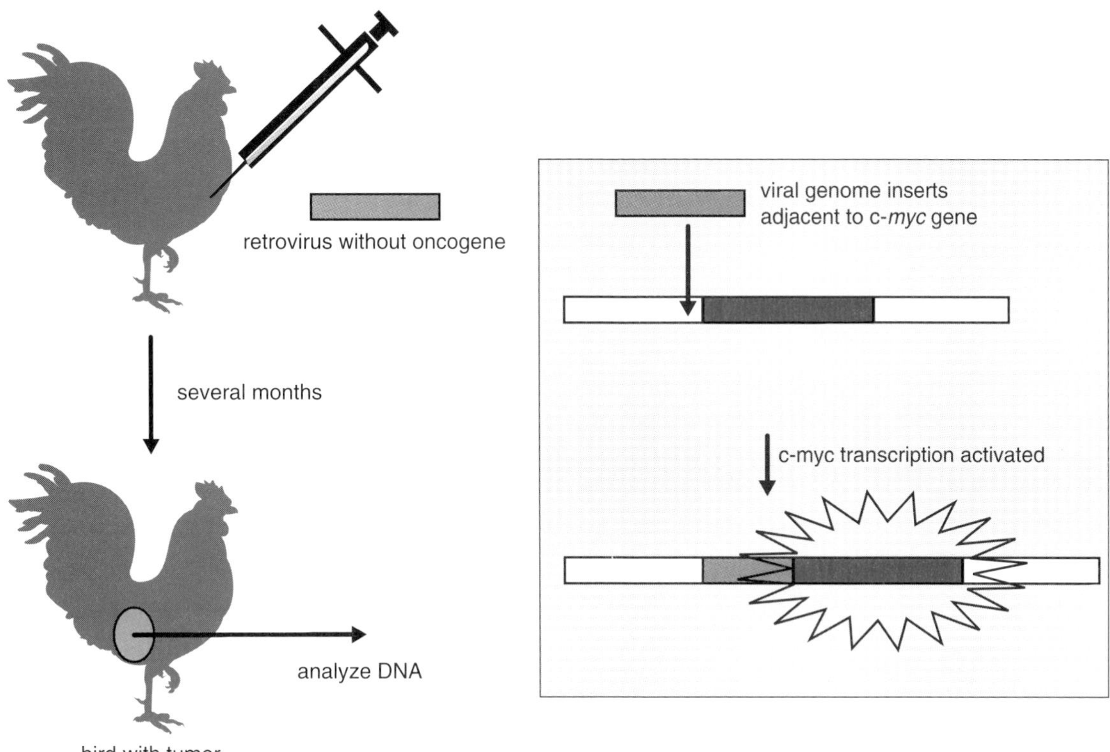

FIGURE 8-12 Insertion of avian leukemia retrovirus lacking oncogene into a site in avian genome adjacent to c-*myc* gene. Activation of c-*myc* transcription results in tumor formation. (Modified from Hayward WS, Neel BG, Astrin SM. Activation of a cellular onc gene by promoter insertion in ALV-induced lymphoid leukosis. Nature 1981;290:475–480.)

confer on the cell some neoplastic properties. There are many routes to oncogene activation. Some have already been mentioned—namely incorporation into a virus or insertion of a retrovirus adjacent to a proto-oncogene. Most mechanisms do not rely on viruses, though. The most direct route to proto-oncogene activation is gene amplification, wherein a block of DNA, including an oncogene and some neighboring genes, is replicated tens or hundreds of times in the cell (Figure 8-14). This probably occurs by random errors of DNA replication but, when it occurs, a selective advantage is conferred on the cell, which proliferates faster than other cells in the tissue. Tumor cells with gene amplification contain tiny objects referred to as *double minute chromatin bodies*, which harbor the amplified DNA.

Another route to proto-oncogene activation is through chromosome rearrangement. This was first demonstrated in the tumor Burkitt's lymphoma (Figure 8-15), which commonly includes a translocation between chromosomes 8 and 14. The breakpoint on chromosome 14 corresponds with the immunoglobulin heavy-chain locus, and that on chromosome 8 with a proto-oncogene, c-*myc*. The immunoglobulin heavy-chain gene tends to undergo rearrangement during the process of lymphocyte maturation (see Case Study at end of chapter) and, on rare occasions, this rearrangement process goes awry. If an aberrant rearrangement juxtaposes the heavy-chain gene with the *myc* oncogene, altered expression of the oncogene ensues and contributes to tumor formation.

In most cases, activation of a proto-oncogene is not associated with a visible change in chromosome structure. The transforming gene isolated from transfection of DNA from bladder carcinoma cells is the *ras* oncogene, also a retroviral oncogene. In bladder carcinomas, the *ras* gene is found to be mutated, a single base substitution leading to an altered amino acid. The mutations tend to occur at characteristic

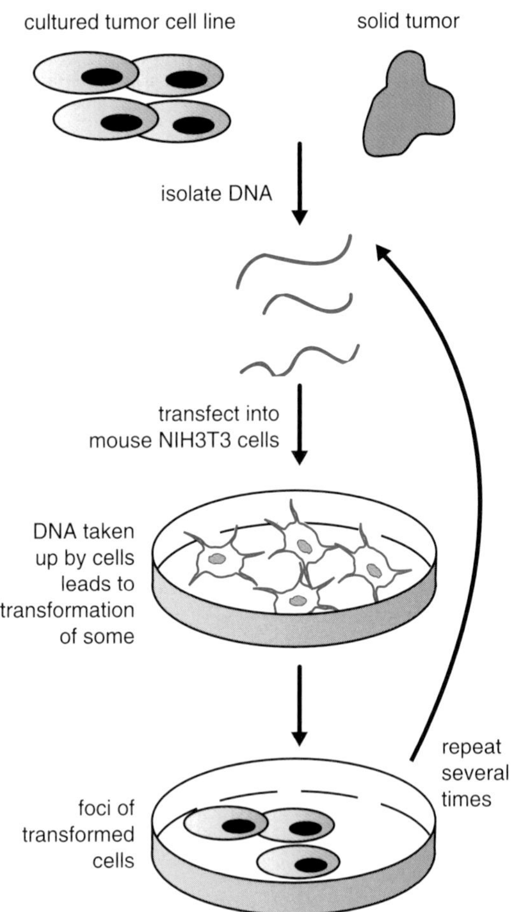

FIGURE 8-13 Identification of transforming genes by transfection of tumor DNA into mouse fibroblasts. DNA was isolated from tumor cell lines or from solid tumors. This was added to cultures of mouse NIH 3T3 fibroblasts, which grow indefinitely in culture but otherwise do not display a transformed phenotype. Some of the human DNA is taken up by the mouse cells and, in rare instances, results in transformation of the cells. These transformed cells had abnormal appearance, grew to high density, and could be cultured in soft agar (indicating lack of requirement for solid surface on which to grow). The process was repeated for several iterations, using DNA from the transformed cells, effectively purifying the transforming genes from the rest of the human DNA. Analysis of the human DNA in the transformed cells revealed homology with known viral oncogenes. (Modified from Krontiris TG, Cooper GM. Transforming activity of human tumor DNAs. Proc Natl Acad Sci USA 1981;78:1181–1184; Shih C, Padhy LC, Murray M, Weinberg RA. Transforming genes of carcinomas and neuroblastomas introduced into mouse fibroblasts. Nature 1981;290:261–264; Pulciani S, Santos E, Lauver AV, et al. Oncogenes in solid human tumors. Nature 1982;300:539–542; Parada LF, Tabin CJ, Shih C, Weinberg RA. Human EJ bladder carcinoma oncogene is homologue of Harvey sarcoma virus *ras* gene. Nature 1982;297:474–478.)

sites in the gene and lead to an altered protein that causes transformation of the cell.

Decades of observations of virally induced malignancies and chromosome changes in cancers converged on the discovery of cellular proto-oncogenes and diverse mechanisms of oncogene activation. An understanding of the normal role of these genes in the cell would seem critical to uncovering their roles in cancer.

8.6 The Normal Roles of Proto-Oncogenes

The discovery of oncogenes and tumor suppressor genes has revealed the major cast of characters responsible for the growth of neoplastic cells. It remains for the tools of molecular biology to reveal their roles. Although this story still is unfolding, major advances have been made relatively quickly.

Dominant proto-oncogenes sort into four classes of molecules, all of which are involved in the control of cell differentiation and proliferation (Figure 8-16). The first class encodes growth factors. The prototype is the proto-oncogene c-*sis*, which encodes the beta-chain of platelet-derived growth factor. Such growth factors are able to stimulate the proliferation of certain types of cells. Overexpression, or expression of an aberrant protein, leads to enhanced cell proliferation. Tumor cells that secrete such factors and simultaneously respond to them are subject to autocrine growth control.

Growth factors must interact with membrane receptors to exert their activity, and growth factor receptors comprise the second class of proto-oncogene

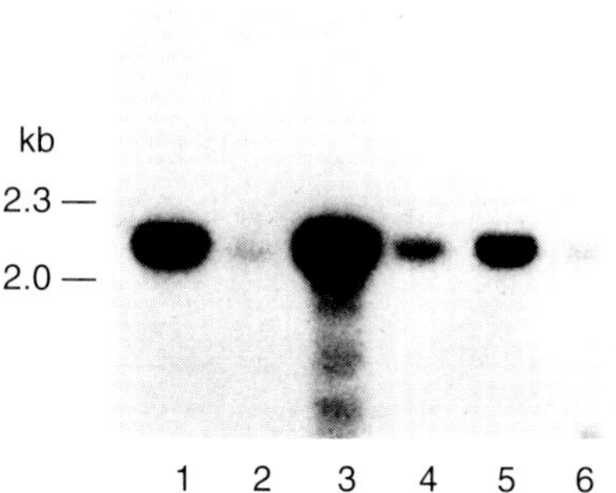

FIGURE 8-14 N-*myc* amplification. Equal amounts of DNA from restriction enzyme–digested cells were subjected to electrophoresis, Southern blotting, and hybridization with a DNA probe to N-*myc*. The IMR-32 cell line (lane 1) and neuroblastoma tumors in lanes 3–5 have N-*myc* amplification, resulting in heavy hybridization bands. The control (lane 2) and the neuroblastoma tumor in lane 6 do not amplify N-*myc* and have fainter hybridization bands.

products. A mutation that renders a receptor active even in the absence of ligand, or that binds ligand abnormally, might render the cell independent of growth factors for control of proliferation. The proto-oncogene c-*erb-B* corresponds with the epidermal growth factor receptor. The viral v-*erb-B* is a truncated form of this receptor with abnormal activity. Binding of a growth factor leads to dimerization of the receptor and activation of tyrosine kinase domains, which leads to phosphorylation of tyrosine residues on the receptor as well as other membrane proteins. Phosphorylation serves as an activating signal that leads the proteins to interact with others. Some of these other proteins also have tyrosine kinase activity and transmit an activation signal to target proteins by phosphorylation. An example of such a downstream target is the *src* gene product.

The third class of membrane-associated proteins has the ability to phosphorylate serine or threonine residues. These include the product of the *raf* proto-oncogene. Another group consists of proteins that bind guanosine triphosphate (GTP), of which *ras* is the prototype. *Ras* is activated by signals transmitted from membrane receptors and, in turn, transmits a signal by modification of other proteins. Oncogenic *ras* mutations lead to overexpression of the activated form of the protein or insensitivity to other proteins that regulate *ras* activity.

The final class of oncogenes is the group that controls transcription in the nucleus. These include *myc*, *jun*, and *fos*. Direct activation of these factors can lead to abnormal cell proliferation or differentiation in the absence of appropriate signals transmitted across the cell membrane and through the cytoplasm. It is here that the Rb gene product is active.

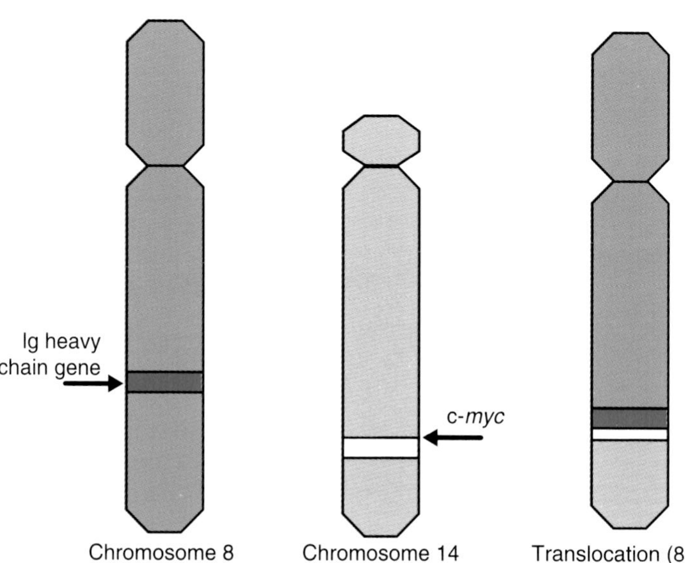

FIGURE 8-15 Translocation between chromosomes 8 and 14 in Burkitt's lymphoma, juxtaposing the c-*myc* proto-oncogene into the immunoglobulin heavy-chain locus. This alters the regulation of c-*myc* expression, contributing to the malignant phenotype. (Modified from Taub R, Morton C, Lenoir G, et al. Translocation of the c-*myc* gene into the immunoglobulin heavy chain locus in human Burkitt lymphoma and murine plasmacytoma cells. Proc Natl Acad Sci USA 1982;79:7837–7841.)

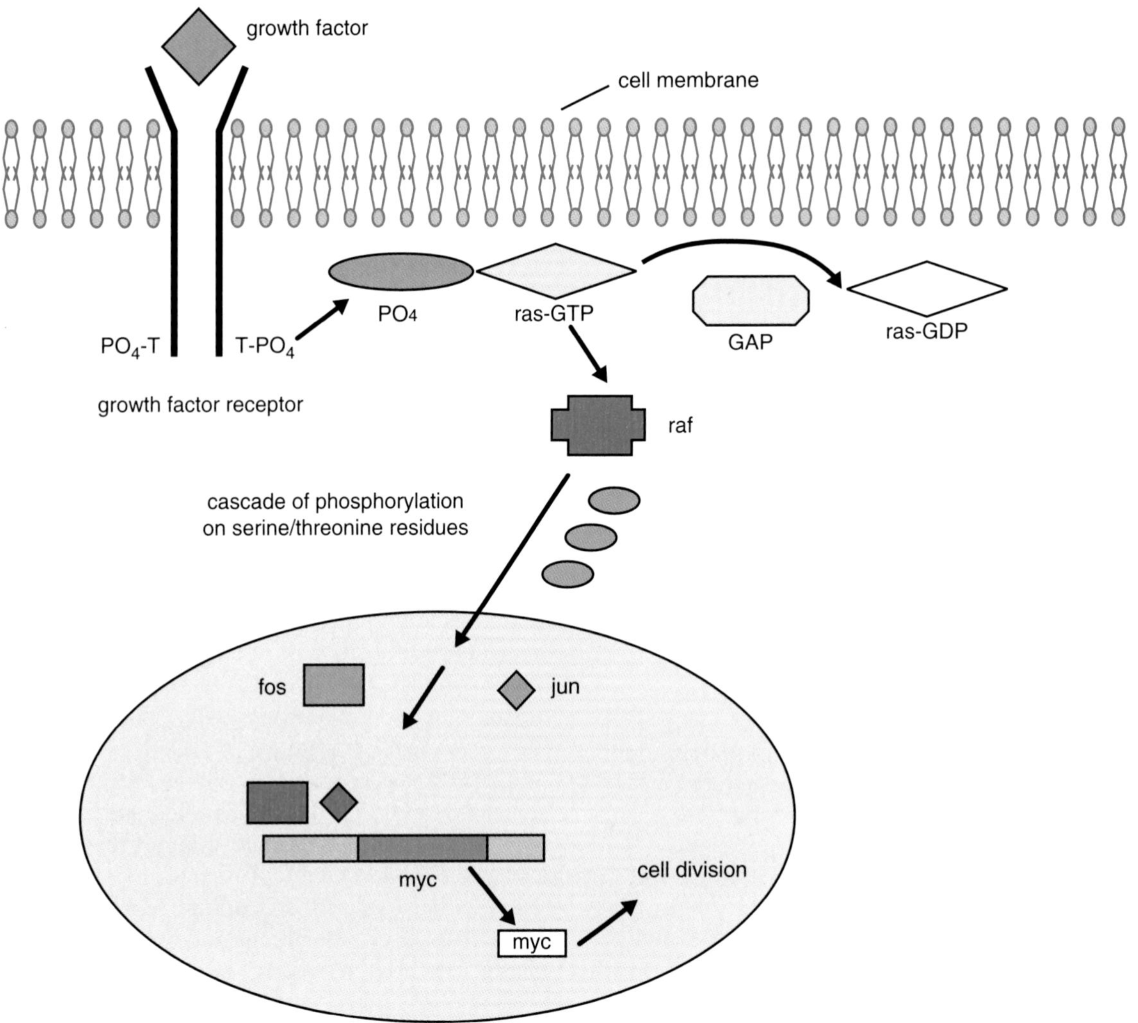

FIGURE 8-16 Cascade pathway of activation of cell division involving oncogene products. A growth factor binds to a specific membrane receptor, which dimerizes and activates tyrosine kinase activity. This leads to autophosphorylation of the receptor, as well as phosphorylation of other membrane proteins at tyrosine (*T*) residues. The oncogene product *ras* then is activated by binding to guanosine triphosphate (*GTP*), which in turn leads to activation of the product of the *raf* oncogene. GTPase-activating proteins (*GAP*) convert the active *ras*-GTP into an inactive *ras*-GDP complex, limiting the duration of *ras* activation. The raf protein sets into motion a series of phosphorylation events involving other proteins, which then enter the nucleus and activate the transcription factors *fos* and *jun*. These then bind to the promoter of the *myc* gene, activating transcription. The *myc* gene product, in turn, activates other genes involved in stimulation of cell division. GDP = guanosine diphosphate. (Redrawn by permission from Krontiris T. Oncogenes. N Engl J Med 1995;333:303–306.)

The Rb protein is phosphorylated as the cell transits from G1 to S phase and plays a role in the transit of the cell from interphase to DNA synthesis. Loss of Rb activity removes this control point. The Rb protein binds to three proteins that were identified initially as oncogenes from three DNA-containing tumor viruses. These oncogenes are referred to as *E1A* (an adenovirus oncogene), *SV40 large T antigen*, and *papillomavirus E7 protein*. It has been hypothesized that overexpression of these proteins in virally transformed cells binds the Rb protein and thus inhibits it, effecting the same outcome as if Rb were homozygously mutated. In fact, expression of SV40 large T antigen in the retina of transgenic mice results in retinoblastoma formation.

The mechanism of action of other recessively acting oncogenes is also gradually coming to light. The NF1 gene product, which is mutated in individuals with neurofibromatosis type 1 (see Chapter 2) is a GTPase-activating protein (GAP), which regulates *ras* activity by stimulating GTPase activity. Loss of the NF1 gene product is believed to lead to unimpeded ras activity. The p53 gene is lost in a wide variety of tumor types and is involved in a familial cancer syndrome, Li-Fraumeni syndrome, in which sarcomas, brain tumors, and leukemias occur with high frequency. The p53 gene appears to be involved in the process of programmed cell death, or **apoptosis**, in some cells. Loss of this gene may render the cells immune to this process, increasing cell survival and accounting for resistance to chemotherapeutic agents. It may also be involved in causing the cell to pause before DNA synthesis to repair DNA damage. Loss of p53 activity would allow mutations to go unrepaired and thus increase the likelihood of survival of cells with genetic alterations that may contribute to malignancy.

A different type of genetic defect has been found to underlie the syndrome of hereditary nonpolyposis colon cancer. Familial adenomatous polyposis is characterized by formation of intestinal polyps, which are at risk for malignant change, and is associated with a tumor suppressor gene (APC) on chromosome 5 (see next section). In familial nonpolyposis colon cancer, malignancies form without preexisting polyps. The mechanism in this case is mutation of one of several genes involved in repair of mutations in DNA. One of these is a gene known as *GTBP*, for G-T binding protein. Homozygosity for a GTBP mutation results in hypermutability of the cell. It has been found that the gene for the type II transforming growth factor–beta (TGF-β) receptor is inactivated by mutation in tumor cells from affected individuals. TGF-β inhibits epithelial cell growth, so loss of the receptor may lead to unimpeded growth and consequent tumor formation.

The oncogenes—both the dominantly acting oncogenes and the tumor suppressor genes such as Rb (sometimes also referred to as *recessive oncogenes*)—are part of a network of proteins that control cell growth. They represent steps in the pathway from the cell surface to the nucleus. A defect anywhere along the chain can render the cell incapable of responding to signals to stop dividing and differentiate or can trick the cell into a cycle of unstoppable proliferation.

8.7 The Molecular Basis of Oncogenesis

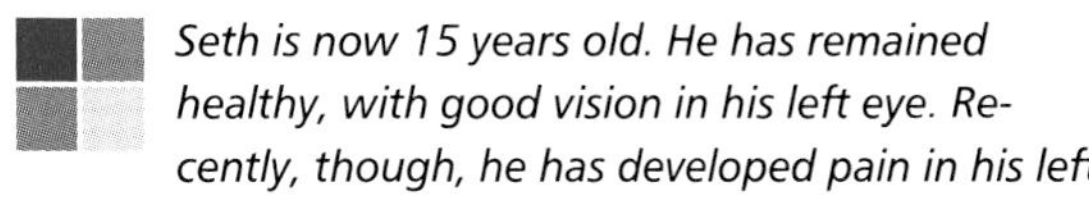

Seth is now 15 years old. He has remained healthy, with good vision in his left eye. Recently, though, he has developed pain in his left leg and noted a swelling near the knee. An x-ray reveals a mass in the bone. Seth is seen in the oncology clinic, and he and his parents are told that the mass is likely to be an osteosarcoma. He is taken to surgery, where a biopsy confirms that suspicion. The leg is amputated, and a program of chemotherapy is begun.

For those individuals who have Rb mutations and survive retinoblastoma, another challenge may lie ahead. Retinoblastoma is a tumor of primitive retinoblasts, which are normally present only in the early years of life. That is why retinoblastomas occur in young children. Later in life, carriers of Rb mutations face an increased risk of other tumors, particularly osteosarcomas and breast cancers. The underlying mechanism is the same as for retinoblastoma—homozygous loss of function of Rb in the tumor cells—but the latency is much longer.

Rb is only one of several genes that, when both copies are inactivated, lead to tumor formation. Other tumor suppressor genes are listed in Table 8-1. Each is responsible for a distinctive syndrome of increased risk of particular types of cancer. Affected individuals tend to develop these malignancies at an early age and are at risk of developing multiple tumors during their lifetimes. Although these syndromes are rare, the same genes also are involved in cancers that occur commonly in the general population. Rb mutations, for example, may be seen in breast carcinomas in women who do not carry germline Rb mutations. Mutations of p53 are found commonly in a wide variety of tumors, particularly sarcomas, in individuals with cancer, not just those with the rare Li-Fraumeni syndrome. The tumors in these cases are homozygously mutated, just like those in persons with the hereditary syndromes. If the germline copies of the gene are wild-type, though, both mutations must be acquired during the course of tumor formation.

The path from a normal to a malignant cell comprises a series of genetic changes involving both dominant oncogenes and tumor suppressor genes. In the familial cancer syndromes, and perhaps in sporadic cancers as well, loss of a tumor suppressor gene appears to be rate-limiting: It is the change that sets into motion a cascade of events wherein genetic damage accumulates, leading to a complex set of abnormal properties. Moreover, there can be a snowballing of genetic change. As the tumor cells experience genetic damage and cell division becomes more rapid and disordered, additional damage occurs. Other tumor suppressor genes may be mutated or lost. Dominant oncogenes are activated. Novel genes may be created by translocation of two genes into one. The fusion proteins encoded by these new genes may confer new, abnormal properties to the cell. Some changes are deleterious and lead to cell death. If, however, a change occurs that causes the cell to grow faster or to be freed from dependence on a growth factor, that cell will have an advantage over others and may become the predominant cell type in the tumor. Tumor growth, then, is a process of natural selection. When the tumor finally is diagnosed, the tumor cells are highly evolved products of hundreds of generations of mutation and selection.

The multistep process of oncogenesis has been well studied in human colorectal tumors (Figure 8-17). *Ras* mutations are found in 40% to 50% of colorectal carcinomas. Further, a tumor suppressor gene on chromosome 5, called *APC*, is associated with a familial syndrome that predisposes to adenomatous polyps. Loss of alleles on chromosomes 17 and 18 also are commonly found in colorectal carcinomas. Vogelstein et al. examined colorectal cancers at various stages, from benign adenomas to metastatic cancers. *Ras* mutations were found in the larger adenomas and carcinomas but not in the smaller adenomas. Deletions of the APC gene were found commonly in adenomas and carcinomas. Loss of sequences on chromosome 18 occurred in advanced adenomas and carcinomas, whereas loss of sequences on 17p were found only in carcinomas. In

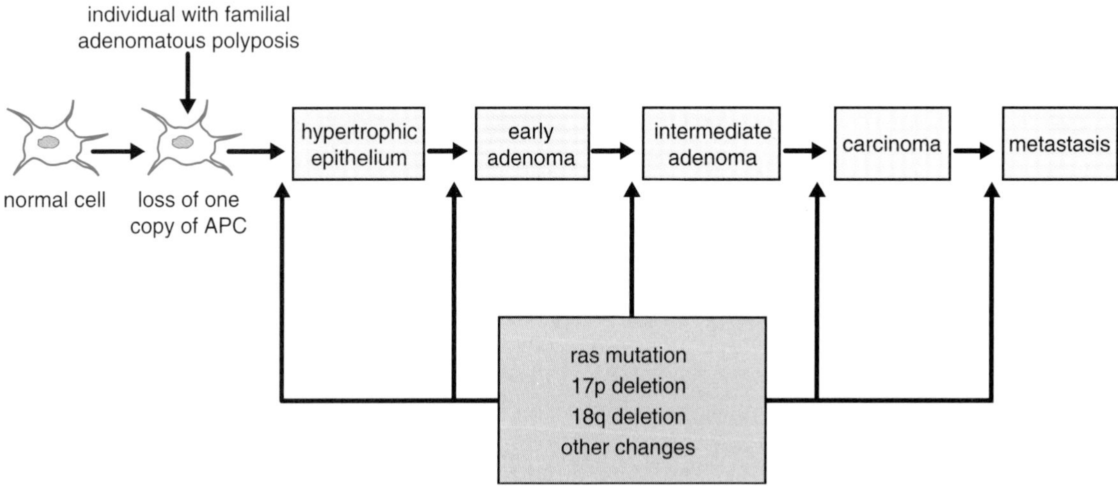

FIGURE 8-17 Multistep pathway from normal cell to metastatic colon carcinoma. The rate-limiting step is postulated to be loss of the APC gene on chromosome 5. Individuals with familial adenomatous polyposis (*fap*) inherit the first mutation (APC gene) and need acquire only a second mutation. Other individuals need to acquire both mutations in the same cell lineage. This leads to a hypertrophic epithelium, however. Additional genetic damage is accumulated, including loss of sequences on chromosomes 17p and 18q as well as other changes. These can occur in any order but lead to progressively abnormal cells including, eventually, cells capable of metastasis. (Redrawn from Fearon ER, Vogelstein B. A genetic model for colorectal tumorigenesis. Cell 1990;61:759–767.)

some cases, individual tumor samples from the same patient differed in terms of allele loss, showing independent evolution of different lineages in the tumor.

The gene on chromosome 17p was eventually found to be p53 (this, in fact, is how p53 was first discovered), and the one on chromosome 18 is called *DCC*. Aside from the insights into the biology of neoplasms gained from this work, a practical advance has been made that provides prognostic information about individuals afflicted with colorectal cancers. When chromosome 18q sequences in tumors were analyzed using microsatellite polymorphisms, it was found that loss of DCC was associated with poor outcome. Patients with stage II tumors that extend through the bowel wall but do not involve lymph nodes have a more favorable prognosis (70% 5-year survival) than those with stage III tumors that do involve lymph nodes (40–50% 5-year survival). More aggressive chemotherapy therefore is offered to stage III patients. Jen et al. have found, however, that stage II patients whose tumors show loss of chromosome 18q sequences have outcomes more like those of stage III patients. Hence, these individuals should be offered the same treatment as stage III patients, despite the absence of lymph node involvement.

In addition to the hereditary disorders due to germ-line mutation of a tumor suppressor gene, there is another class of disorders that leads to hereditary predisposition to malignancy. These disorders represent defects in the mechanisms that underlie replication of DNA and repair of DNA damage and are summarized in Table 8-2. These are recessively inherited, and increased risk of malignancy probably is due to an accelerated rate of accumulation of mutations, some of which activate dominant oncogenes or inactivate tumor suppressor genes.

Knowledge of genetic changes in malignancy has significantly changed the approach to cancer diagnosis and management. In the leukemias, cytogenetic analysis has long been used for diagnostic purposes. The first tumor-specific chromosome change to be identified was the Philadelphia chromosome in chronic myelogenous leukemia (Figure 8-18). Although originally believed to be a deleted chromosome 22, this eventually was found to represent a translocation of chromosomes 9 and 22, juxtaposing the proto-oncogene *abl* to a gene called *bcr* to generate a novel fusion gene. Detection of the Philadelphia chromosome is used in the diagnosis of leukemia and in following response to treatment. The first appearance of Philadelphia chromosome–positive cells in the bone marrow of patients after treatment is often the first sign of relapse. Sensitive detection methods based on PCR now provide early detection of such relapse, allowing treatment to be re-initiated before tumor burden becomes substantial.

Genetic testing can be used not only to diagnose and monitor cancer in affected persons but also to identify some individuals at risk of developing malignancy in the first place. At present, this can be done for members of families with one of the syndromes listed in Table 8-1. Those who inherit a p53 mutation are at risk of developing sarcoma, brain tumor, or leukemia, characteristic of Li-Fraumeni syndrome (Figure 8-19). If the pathogenic mutation in the family is known, then any family member can be offered testing. As is increasingly true in genetics, however, knowing *how* is easier than knowing *why*. Early diagnosis of cancer usually is assumed to lead to a better outcome of treatment, but effective treatments do not exist for some of the cancers of Li-Fraumeni syndrome. The problem is especially difficult for young children. Is it justified to submit

TABLE 8-2 Major DNA repair defects that lead to increased risk of malignancy

Disorder	Features	Molecular Basis
Ataxia-telangiectasia	Ataxia, telangiectasia, lymphoid malignancies	Cell cycle control defect
Bloom syndrome	Short stature, sun-sensitive rash, various malignancies, increased chromosome breakage	Helicase deficiency enzyme involved in DNA replication
Fanconi's anemia	Congenital anomalies, aplastic anemia	Genetically heterogeneous
Xeroderma pigmentosum	Sun sensitivity, skin cancer	Ultraviolet light DNA repair defect

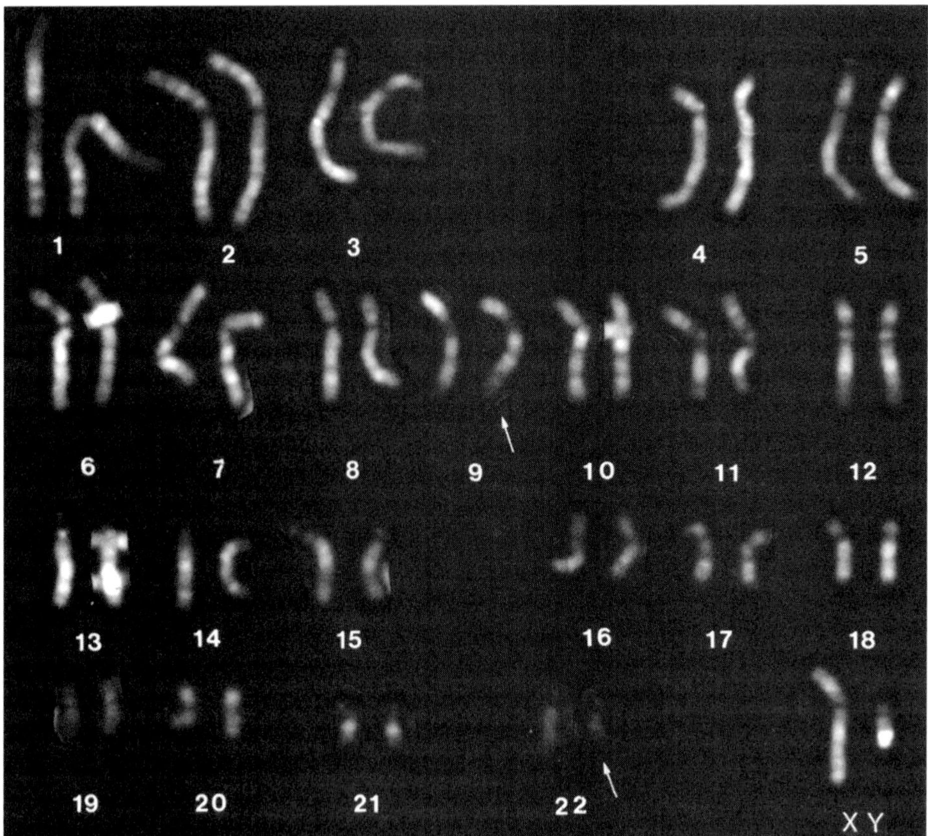

FIGURE 8-18 Karyotype showing Philadelphia chromosome, consisting of a translocation between chromosomes 9 and 22 (*arrows*). (Courtesy of Dr. Ramana Tantravahi, Dana Farber Cancer Institute, Boston.)

children who cannot give informed consent to a lifetime of cancer surveillance, the benefit of which is unknown? Will persons found to have a p53 mutation be denied health insurance or employment? What are the psychological effects of being discovered to carry a mutation, and what is the impact of being found not to carry the mutation when other relatives are not as fortunate?

These questions are being investigated in the context of rare syndromes, such as Li-Fraumeni syndrome. Yet the issues are not confined to these exceptionally affected families. The discovery of a gene for breast and ovarian cancer, BRCA1, raises the same questions for a much larger segment of the population. Is it beneficial for a woman to know, years in advance of signs or symptoms, that she is likely to develop breast cancer? Some have elected bilateral mastectomies, whereas others at risk chose not to know their fate. The BRCA1 gene is involved in a proportion of familial breast cancers and is but one of a group of genes that influence risk of cancer. The proportion of malignancy in the population determined by inherited predisposition remains unknown. Our concept of cancer risk and prevention will change dramatically in the coming years, and the immediacy of the ethical debate will increase accordingly.

Cancer is indeed a genetic disease, sometimes inherited but always showing somatically acquired genetic damage. The discovery of specific genes that underlie the transformation and progression of malignant cells is one of the great achievements in biology during the twentieth century. Cancer diagnosis has been revolutionized, yet therapy still depends on the use of surgery or nonspecific killing of dividing cells with toxic drugs or radiation. Can knowledge of the genetic basis of cancer teach us not only to recognize the disease but also to treat it?

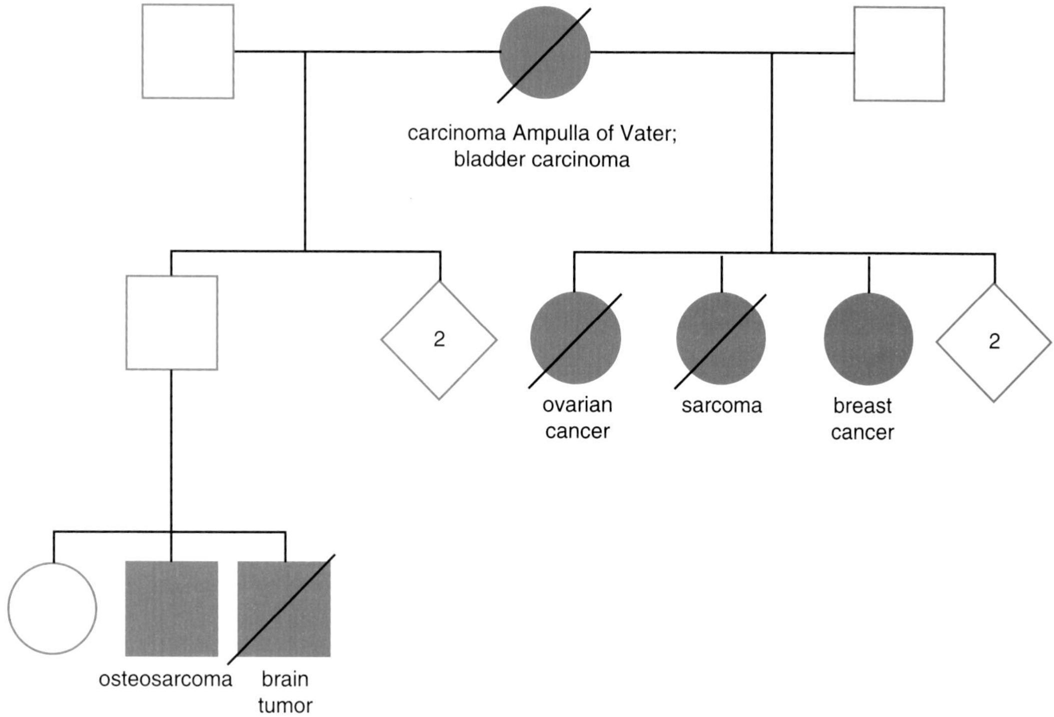

FIGURE 8-19 Pedigree of family with Li-Fraumeni syndrome. (Redrawn from Malkin D, Li FP, Strong LC, et al. Germ line p53 mutations in a familial syndrome of breast cancer, sarcomas, and other neoplasms. Science 1990;250:1233–1238.)

8.8 Genetics and the Treatment of Cancer

The twentieth century has seen great advances in cancer therapy, just as it has seen increases in our understanding of the basis of cancer. Until recently, however, these advances were at best loosely coupled, or perhaps not linked at all. The most dramatic improvements in therapy have been the result of trial-and-error development of new drugs and methods of surgery or radiation therapy. The era of devising treatments based on knowledge of cancer pathogenesis is just beginning.

The trial-and-error approach is credited with making one cancer, childhood acute lymphocytic leukemia (ALL), essentially a curable disease. In the 1950s, virtually all children with ALL died. Over the past four decades, however, therapeutic advances gradually have increased survival. Chemotherapeutic drugs that showed promise were given in combination, sometimes working synergistically. It was discovered that relapse often occurred in the central nervous system, so radiation of the neuraxis was added. Children with some forms of ALL now have an excellent chance of long-term survival. The lives of such children are saved thanks to the thousands who died over the years as therapeutic measures were being developed and tested.

Such success comes at a price, however. The drugs used are highly toxic and therefore are very unpleasant to take. Some children die during therapy from problems such as sepsis, due to suppression of the bone marrow during treatment. Long-term consequences include learning disabilities due to radiation treatment of the developing nervous system. The cure surely is not worse than the disease, but it takes a toll nevertheless. Moreover, the success in childhood leukemia has not been paralleled by such dramatic improvements in other types of cancer. Lung cancers, melanomas, brain tumors, and many others remain deadly. Here, progress has been measured, in many cases, by the addition of months, not years, of survival.

Perspective

Living with a Child with Retinoblastoma

PETER HOMANS

At 9 months of age, my son received a diagnosis of bilateral retinoblastoma, a cancer of the retina, involving both eyes. To say this was like being hit in the stomach with a baseball bat is a profound understatement. This changed my wife's and my lives forever. We and our son spent the next 3½ years going to the hospital approximately every 5 weeks in the hope of discovering that the most recent treatment, which was meant to be the last, had been successful. Every treatment involved putting my son under general anesthesia in order to probe the eye with a stiff metal prod under bright lights and magnification, so by the age of 4 he had been under general anesthesiasia perhaps 50 times, not including the 30 consecutive days under general anesthesia during radiation, when the condition was first discovered.

The problem was that every one of the treatments had a high probability of success—on the order of 95%—and in every case, they were unsuccessful. It was like flipping a coin and having it come up on one side 300 times in a row. So much for probability!

We consoled ourselves with the notion that if you are going to get cancer, this sure as heck is the one to get because, unlike most other forms of cancer, metastases don't seem to occur as a matter of course, and the only path through which any stray cells may pass is the optic nerve into the brain. And it was possible, at each of these monthly trips to the operating room, to see whether or not the tumors were involving the optic nerve. So, essentially, we knew that if we could keep the cancer controlled within the eye, Alex would lead a long and healthy life, with or without both eyes. If the cancer ever spread through the optic nerve, he would die—simple! It's amazing how notions that would horrify most people are available as comfort to victims of illness such as this.

Probably the most difficult and important role for a parent in a medical situation such as this is as the child's protector or representative vis-à-vis the medical community. I know that doctors

Perhaps the trial-and-error approach will lead to major advances in other cancers, too, but now there is hope that new knowledge of the genetic mechanisms that underlie cancer may lead to novel approaches. A step in this direction has already occurred through the development of more precise means of genetic diagnosis. As already noted, therapy for colon cancer may be determined in part by whether deletion of chromosome 18q has occurred, which would indicate that a tumor may be more advanced than its spread may imply. Early diagnosis of relapsing chronic myelogenous leukemia by detection of the Philadelphia chromosome is another example. Genetic markers have also been used to distinguish histologically similar tumors that may respond differently to different means of therapy. An example is the distinction of Ewing's sarcoma from neuroblastoma. These are histologically similar small-cell soft-tissue tumors, yet they are treated differently. Although classic pathologic markers may not distinguish these tumors, Ewing's sarcomas typically have an 11;22 translocation, whereas neuroblastomas are characterized by deletion of chromosome 1p.

Advances in therapy, though, are at a more primitive stage. Conventional modes of treatment are designed to kill all dividing cells. Side effects result from killing of nontumor cells, and treatment failures are due to failure to kill all tumor cells. Sometimes, in fact, the tumor cells that survive develop drug resistance and may be more virulent than their predecessors. Research in cancer therapy therefore is directed toward the development of agents that are more selective for tumor cells. This has been approached by taking advantage of cell surface markers that are unique to tumor cells or by developing means of altering the activity of genes involved in oncogenesis.

The latter has taken advantage of new knowledge of the molecular genetics of cancer. The same and nurses wouldn't have chosen their professions unless somewhere in their decision-making process they deeply desired to alleviate the pains and troubles of others. However, in the course of a medical provider's day, there are many, many patients, too many distractions from insurers, administrators, and the like, and too much suffering. So these caregivers frequently forget to slow down and deal with each patient with the full understanding that this is an immensely scary situation for both patients and parents—certainly the most terrifying experience of their lives. We found ourselves in any number of situations where anesthesiologists or lab technicians or whomever would whisk into Alex's room without any more than a "Good morning, I'm so-and-so and I'm here to take your vitals. Would you mind raising your shirt, please?" If you are a 2- or 3-year-old boy or girl, it's scary enough being in the room in the first place, wondering who is coming in next, without having to be approached totally impersonally and merely instructed to lift this, fold that, move that, cough now, and so on. Thirty seconds on the caregiver's part to say "Hi! What's going on? This is sort of confusing isn't it?" and merely to pass the time a little is worth years of future counseling to a kid who suffers all this.

In addition, one of the most unnerving discoveries we made in the whole process was the insufficiency of the counseling provided by doctors vis-à-vis probabilities. As I mentioned previously, in the early stages of treatment, we were fed lots of quotations of probabilities, such as the high likelihood of success of cryotherapy, the number of patients a particular doctor had whose cancer had reappeared in the knee or some other part of the body in later years, and so forth. And because these probabilities were very favorable to our son's prognosis, we glommed onto them like those little fish that hitch rides on sperm whales. The problem was that we soon discovered that it really couldn't have mattered less what the probabilities were because, if the odds are 1 in 100 and you are the one, that's the story! The probabilities are 100%. You're cooked! So I think it's very important for caregivers, who must be familiar with all the statistics involving these conditions, to remain conscious of how little odds mean if your patient is not beating them and of how cynical those patients and their families become every time those numbers are repeated to them if they are not able to get out in front.

complexity of the system that regulates cell growth, which makes the cell vulnerable to genetic changes that lead to malignancy, also offers many possible sites of intervention. Drugs are being developed that interact either with oncogenes or with other cellular proteins that interact with oncogenes. For example, the oncogene *ras* binds to the cell membrane via an enzyme-mediated reaction called *farnesylation*. Farnesylase inhibitors have been devised that interfere with *ras* action (Figure 8-20). In vitro experiments indicate that these inhibitors can slow the growth of *ras*-transformed cells in vitro. Another approach has been to inhibit the action of oncogenes directly using antisense oligonucleotides, which interfere with the action of oncogenes by binding to RNA transcripts (Figure 8-21). Here, too, promising results have been obtained with in vitro experiments.

Aside from these efforts at drug design, other experiments are aimed at genetic manipulation of tumor cells. The strategy has been to increase the immune response in a tumor by introducing into some of the tumor cells a gene that makes them more antigenic or otherwise attractive to inflammatory cells. In one study, the histocompatibility antigen B27 was introduced into melanoma cells using liposomes. The tumor patients from whom the cells were obtained were HLA-B27-negative, so this produced a new foreign antigen on the tumor cells. The altered cells then were injected back into the patient. No ill effects of the injections could be identified, and an immune response was seen in the tumors. Although the response was not sufficient to change the course of the disease, the results were encouraging in that the approach at least does not cause major side effects.

It is premature to predict the impact of molecular genetic approaches on the treatment of cancer. With the rapidly increasing insights into pathogenesis, though, it is unlikely that the trial-and-error approach that has prevailed to date will continue to set the standard for new advances in cancer treatment.

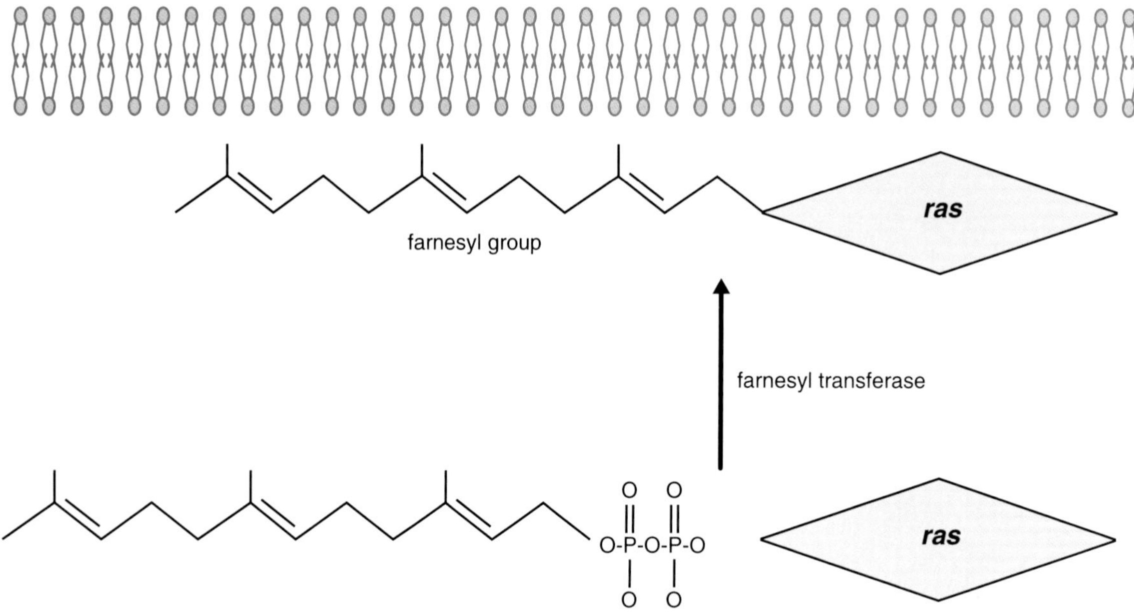

FIGURE 8-20 A farnesyl group is covalently added to the C terminus of the *ras* protein during posttranslational modification by the enzyme farnesyl transferase. The farnesyl group is required to bind *ras* to the cell membrane, its site of action. Farnesyl transferase inhibitors can interfere with this reaction and are being developed as a means to inhibit the growth of cancer cells with activated *ras*. (Modified from Gibbs JB, Kohl NE. Farnesyltransferase inhibitors: *ras* research yields a potential cancer therapeutic. Cell 1994;77:175–178.)

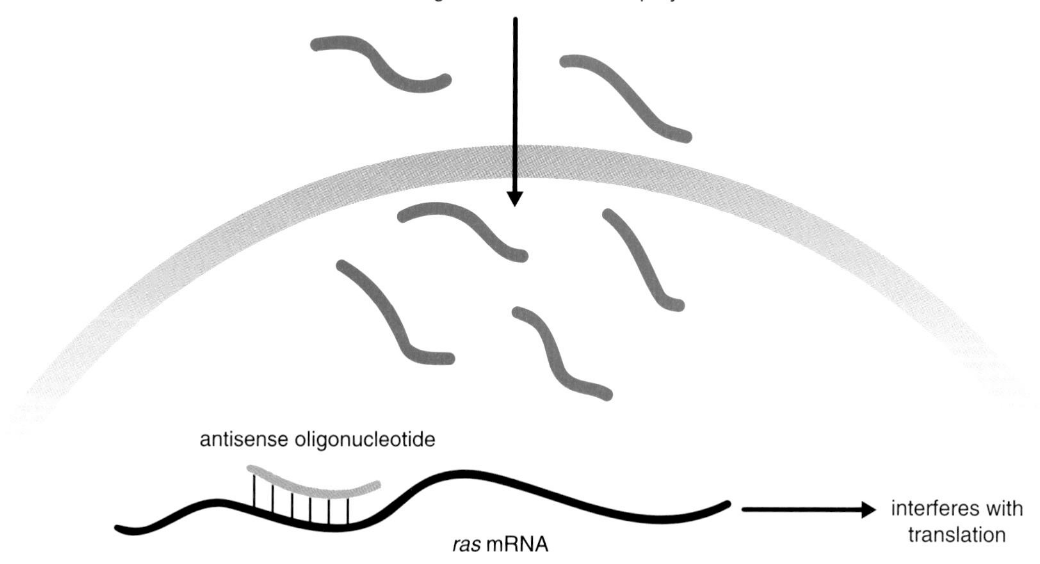

FIGURE 8-21 Oligonucleotides homologous with part of the *ras* mRNA are added to the culture medium and taken up by the cells. In the cell, these bind with the *ras* mRNA to form small double-stranded regions. These interfere with translation, perhaps by rendering the double-stranded regions susceptible to RNase cleavage. Treatment of transformed cells having activated *ras* genes has been shown to result in growth inhibition. (Modified from Schwab G, Chavany C, et al. Antisense oligonucleotides adsorbed to polyalkylcyanoacrylate nanoparticles specifically inhibit mutated Ha-*ras*-mediated cell proliferation and tumorigenicity in nude mice. Proc Natl Acad Sci USA 1994;91:10460–10464.)

Case Study

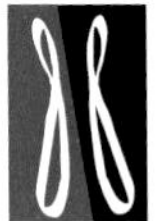

Part I

Tom is in the shower one day when he notices a lump in his left groin. It doesn't hurt and doesn't appear inflamed. Tom is puzzled and a little worried. Was it there before? He doesn't think so. He hasn't been sick lately. It doesn't feel like a hernia. He tries to put it out of his mind; it will probably go away. Nearly a week later, though, it hasn't changed. He mentions it casually to his wife, Jean. She is more concerned than Tom is. She arranges for him to see his doctor, because she knows that Tom won't set up the appointment himself.

The occurrence of a mass in the groin can indicate a number of possible problems. The mass is most likely an enlarged lymph node. An inguinal hernia would be painful, and the mass could be reduced with pressure. There are many possible causes for an enlarged lymph node. These include bacterial, viral, or parasitic infections, and malignancy.

Part II

It has been a long time since Tom has been to the doctor. He is 32 years old and works as a lawyer in a large firm. His health has generally been good, except for a bout of infectious mononucleosis when he was a law student. He takes no medications. The doctor asks whether Tom has experienced fever, chills, weight loss, or night sweats recently. Tom has had none of these symptoms. He asks whether Tom has any pets. Tom does have a cat, who roams the neighborhood much of the time and sometimes does scratch Tom or Jean. A family history is obtained. Tom has one brother, who is 29 years old and is in good health. Both his parents are alive, but his mother received a diagnosis of breast cancer at age 55. His maternal grandmother died of breast cancer. A physical examination is performed and is found to be unremarkable except for a 2-cm, firm, painless mass in the left inguinal region. Several other shotty inguinal and cervical lymph nodes are found. There is no redness over the site of the mass. Tom had expected the visit to be reassuring. He is surprised, therefore, that the doctor refers him to a surgeon to have the mass biopsied.

Tom has not exhibited any symptoms that would suggest infection. He has a history of infectious mononucleosis, which is caused by Epstein-Barr virus. This leads to diffuse lymph node enlargement but generally resolves without long-term effects. Cat-scratch disease is an infection transmitted from a cat scratch. It is associated with painful lymph node enlargement, generally in the region of the scratch. This seems an unlikely cause of Tom's painless enlargement of a single inguinal node. Therefore, malignancy seems the most likely diagnosis. Tom is at an age where malignant lymphomas are the most common kind of neoplasm.

Lymphomas usually do not display a striking familial predisposition. The relationship with Tom's mother's and maternal grandmother's history of breast cancer is therefore likely to be coincidental. Autosomal dominant transmission of predisposition to breast cancer has been well documented in some families. One gene on chromosome 17, referred to as *BRCA1*, has been cloned, and is involved in some families with breast cancer. This gene appears to function as a tumor suppressor gene but accounts for only a small proportion of familial breast cancer. Other genes, yet to be identified, must account for the balance.

Part III

The biopsy is scheduled for a week later, and neither Tom nor Jean can sleep in the days between. The biopsy itself is a quick and fairly painless procedure, done in the office (Figure 8-22). Tom has done some reading and is sure that he has cancer. He actually feels well but now interprets every minor symptom, every headache, every stomach pain, as probably being related to his "disease." His colleagues at work have noticed that he appears distracted, but he doesn't tell anyone what is going on.

Histologic analysis of the lymph node biopsy indicates a pathologic picture referred to as *follicular lymphoma*, predominantly consisting of small cleaved cells. This is also referred to as *nodular poorly differentiated lymphocytic lymphoma*. It is considered to be a low-grade lymphoma that is slow to progress and may be associated with a course extending over many years. It is common for this kind of lymphoma to come to attention through a

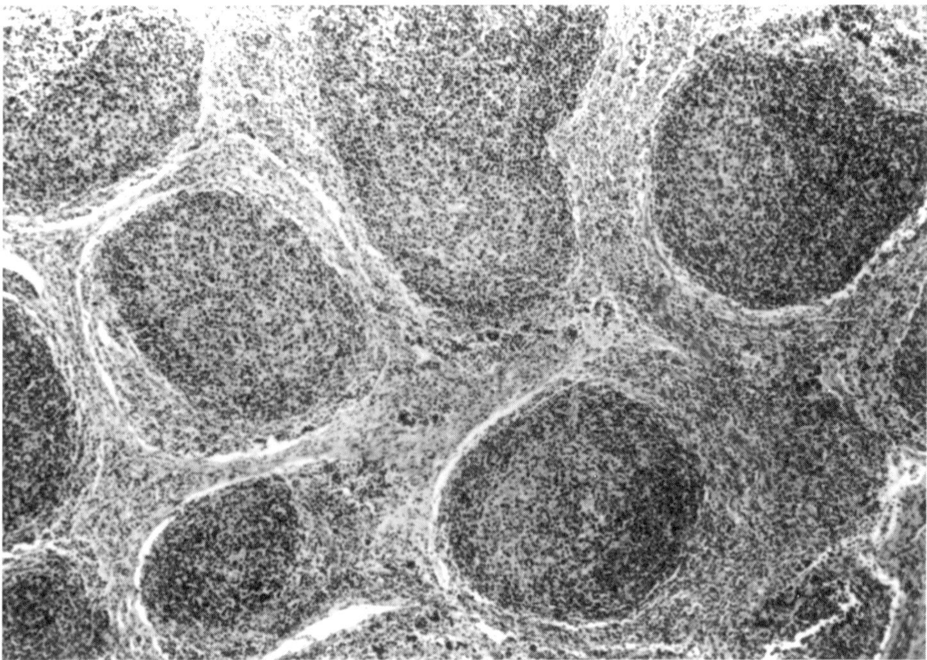

FIGURE 8-22 Photomicrograph of follicular lymphoma showing islands of tumor cells surrounded by stroma. (Courtesy of Dr. Marshall Kadin, Beth Israel Hospital, Boston.)

single painless enlarged lymph node and not to be associated with other symptoms.

Lymphocytes consist of two major types: B cells and T cells. B cells develop mainly in the bone marrow and produce immunoglobulins. T cells develop in the thymus and are involved in cell-mediated immunity. Both cell types contain receptors that recognize specific antigens. These receptors are IgM molecules in the case of B cells and the T-cell receptor molecules for T cells. Each primitive B or T cell has a receptor with distinct specificity that remains the same for progeny of that cell. The specificity of the immune response is based on selective proliferation of those cells that bind to an antigen. Once an antigen has been encountered, the expanded clone of reactive cells remains ready to mount a rapid response in the case of repeated exposure. The basis for antigenic specificity of a receptor molecule is the structure of the molecule and the unique manner in which it is assembled.

Antibody molecules consist of two pairs of identical light and heavy chains (Figure 8-23). There are two kinds of light chains, kappa and lambda; only one type will occur in a given antibody molecule. Heavy chains include mu-, delta-, gamma-, epsilon-, and alpha-chains, which comprise the heavy chains of IgM, IgD, IgG, IgE, and IgA, respectively. Each chain, moreover, includes a variable region, which is responsible for antibody specificity, and a constant region, which is characteristic of the type of heavy or light chain. Antigenic specificity is due to the amino acids that compose the variable region, which consists of three hypervariable regions connected by a framework of more constant amino acids. Noncovalent interactions between the amino acids in the variable regions of the heavy and light chains with epitopes on the antigen lead to binding of antibody to antigens.

Both the light chains and heavy chains are assembled by a process of site-specific DNA recombination. Kappa-chain genes are located on human chromosome 2, lambda on chromosome 22. The heavy-chain genes are all located on chromosome 14. The light-chain genes consist of a cluster of hundreds of different variable (V) segments, four joining, or J, segments, and a single kappa or lambda constant (C) segment. The heavy-chain genes consist of hundreds of V segments, four J segments, a dozen or more diversity, or D, segments, and a C segment corresponding with mu-, delta-, gamma-, epsilon-, and alpha-chains. An antibody-producing gene is assembled early in B-cell

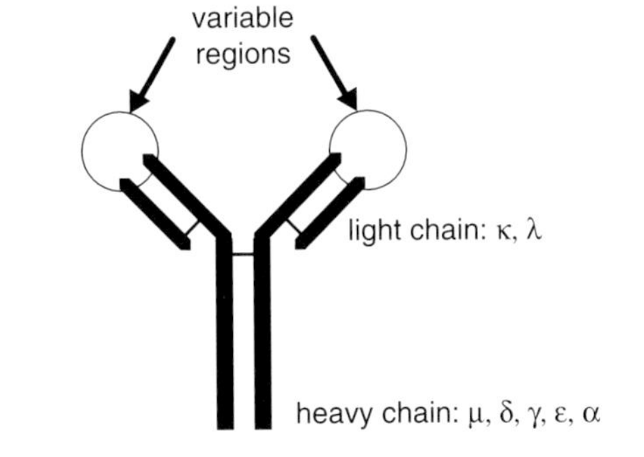

immunoglobulin molecule

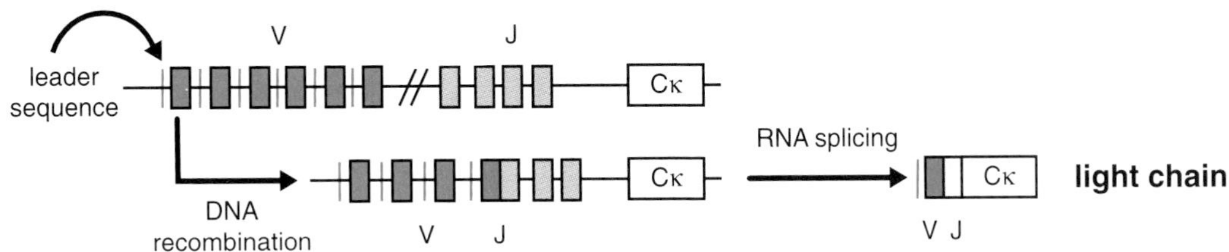

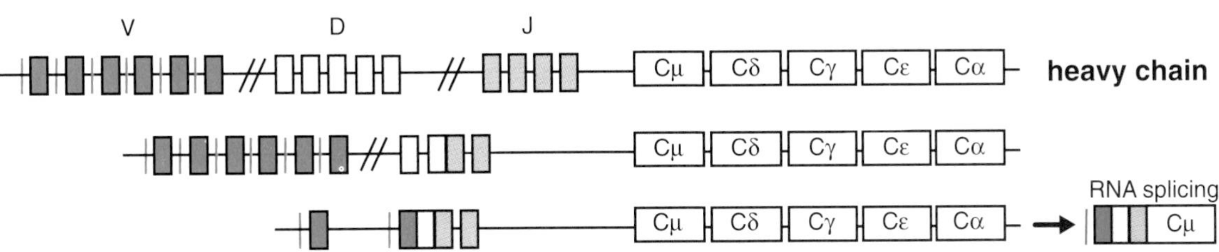

FIGURE 8-23 An immunoglobulin consists of a pair of light chains (kappa or lambda) and a pair of heavy chains (mu, gamma, delta, epsilon, or alpha). Both the light and heavy chains have variable regions (circled) which bind to the antigen and convey antigen specificity. The light-chain regions have a set of multiple V genes (each associated with a leader sequence), four J segments, and a single constant sequence. To construct a light chain, first a V sequence is spliced adjacent to one of the J sequences. This DNA is transcribed, and remaining J sequences and the intron between the J and the C are spliced out in the RNA. To construct a heavy-chain transcript, first a D recombines with a J and then a V recombines with the D-J segment. RNA splicing then produces an IgM molecule, removing extra J segments and the intron between J and C.

development by splicing, at the DNA level, V-J-C segments to produce a light chain and V-J-D-C segments to produce a heavy chain. Once rearrangement has occurred for one allele and for one light chain, the other allele or chain is not expressed, so that each cell produces only a single kind of light chain and heavy chain.

Antibody diversity is the consequence, in part, of the great number of possible chain sequences, depending on the random association of a specific V, J, and C segment for a light chain, or V, J, D, and C segment for a heavy chain, and a given kappa– or lambda–light chain with a heavy chain. Specific DNA sequences flanking the segments ensure that a

V will recombine only with a J, and a J with a D, and so on, but there may be gain or loss of a few bases at the site of joining, adding to immunoglobulin diversity. Furthermore, there may be additional mutation of V-region sequences during B-cell maturation, creating still more diversity.

There is a sequence of antibody formation and secretion during the ontogeny of B cells. The mature B cell produces mostly IgM (and sometimes IgD) on its cell surface. When an antibody molecule binds antigen, a signal is transduced through the membrane that leads to cell proliferation, expanding the clone of antigen-specific cells. Some of the resulting cells produce IgM that is secreted from the cell. Others switch to production of IgA, IgG, or IgE. This class switch involves splicing out of the intervening constant regions to produce a new heavy-chain gene, leaving the same V-D-J segment, preserving specificity.

The 14;18 translocation characteristic of follicular lymphomas is thought to arise due to an aberrant recombination event during B-cell ontogeny. This translocation juxtaposes the immunoglobulin heavy-chain gene on chromosome 14 with the gene bcl-2 on chromosome 18. The bcl-2 gene is involved in the process of programmed cell death, referred to as *apoptosis*. The translocation leads to aberrant expression of bcl-2, which may interfere with a normal process of apoptosis of lymphocytes, leading to their long-term survival and continued proliferation.

Part IV

The pathologic studies are completed within less than a week. Tom's doctor calls him back in to explain that Tom indeed has a cancer called follicular lymphoma. *Laboratory studies are performed, including a complete blood cell count, liver function tests, renal function tests, uric acid and calcium levels, and serum protein electrophoresis. A chest x-ray is performed, as well as computed tomographic scanning of the abdomen and pelvis and magnetic resonance imaging of the brain. Finally, and most painfully, Tom undergoes a bone marrow biopsy.*

It is common for individuals with low-grade follicular lymphomas to have widely disseminated disease at the time of diagnosis. These tumors are very slow-growing and indolent and may involve multiple sites without producing symptoms. Several years can pass between diagnosis and occurrence of clinical symptoms. X-ray and computed tomographic (CT) or magnetic resonance imaging (MRI) studies are performed to determine the sites of enlarged lymph nodes or involvement of the central nervous system. Biochemical tests determine liver and kidney function, which may be impaired by tumor infiltration. Bone marrow biopsy identifies tumor cells at this site.

Part V

The staging indicates that Tom has evidence for involvement of the spleen, liver, and bone marrow. Cytogenetic studies reveal cells with a chromosome 14;18 translocation in the bone marrow (Figure 8-24). An MRI scan of the brain shows no evidence for involvement of the central nervous system. Tom is told that he has disseminated disease. He is asymptomatic now and might remain so for several years. He is told that there is a good chance that chemotherapy now will lead to remission of his disease, but there is also a high likelihood of relapse. Overall, his expected survival is 7 to 10 years, although it could be longer or shorter. Tom is willing to endure whatever is necessary to prolong his life if there is any hope of benefit. He therefore begins a regimen of aggressive chemotherapy using cyclophosphamide, vincristine, prednisone, and procarbazine (a regimen called C-MOPP).

Although follicular lymphomas may be slow-growing, inevitably they do progress to cause symptoms and eventually, death. Most treatment protocols involve the use of chemotherapy, based on the fact that tumors usually are widely disseminated. In some instances, chemotherapy may be delayed until the onset of symptoms, as outcome of treatment with standard regimens has not been demonstrated to be better if treatment is begun at diagnosis rather than later, at onset of symptoms. More recently, though, there has been a tendency to use more aggressive chemotherapy protocols that may prolong overall survival. Generally, these involve cytotoxic drugs that have been empirically determined to be effective. In most cases, remission occurs, but relapse is common.

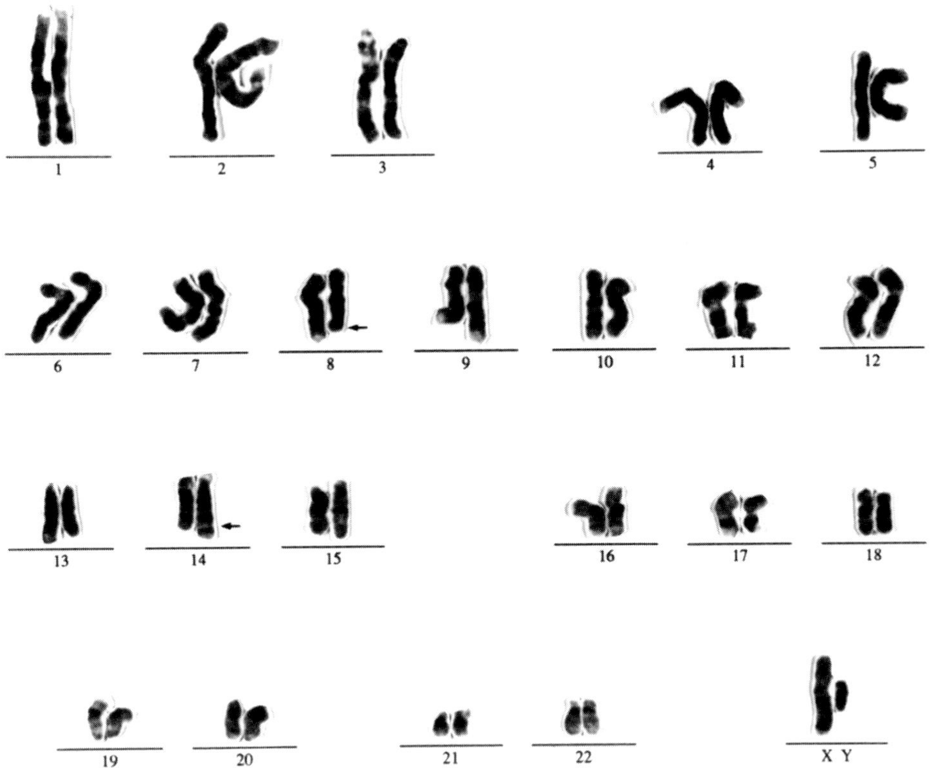

FIGURE 8-24 Karyotype showing translocation between chromosomes 8 and 14 (arrows). (Courtesy of Dr. Jonathan Fletcher, Brigham and Women's Hospital, Boston.)

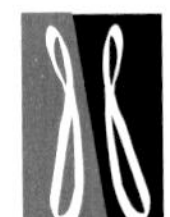

Part VI

The chemotherapy regimen is extremely uncomfortable. Tom is hospitalized at one point for treatment of possible infection when his white blood cell count is low and he has a fever. After several months, however, the treatment is completed. CT scanning reveals no evidence of enlarged abdominal lymph nodes. His liver function tests have returned to normal. A bone marrow biopsy is done. No cancer cells are seen by histologic examination, and molecular studies reveal no evidence of the chromosome 14;18 translocation (Figure 8-25). Tom resumes work, and he and Jean try to pick up their previous lives.

Detection of the 14;18 translocation is a defining feature of follicular lymphomas and therefore helps to confirm the diagnosis. The translocation can be detected in nondividing cells by a molecular genetic assay. The site of translocation is consistent, and PCR primers have been made that recognize flanking sequences in the immunoglobulin heavy chain and the bcl-2 gene (Figure 8-26). When used together in a PCR reaction, product will be seen only if the two genes are juxtaposed by translocation. Normally, of course, the genes are on separate chromosomes, and therefore no PCR product would occur. The PCR reaction provides a very sensitive assay for the presence of cells with the 14;18 translocation and therefore is useful in following response to therapy.

Part VII

Tom is followed by his oncologist every several months, and he does well for approximately 2 years. Finally, however, a CT scan reveals evidence of enlarged lymph nodes. Another bone marrow biopsy is done, and this time cells with the 14;18 translocation are found. Tom has had a relapse. He and Jean knew that this might happen and have already decided what they will do next: Tom will have a bone marrow transplant, following a course of extremely aggressive therapy. It has already been determined that

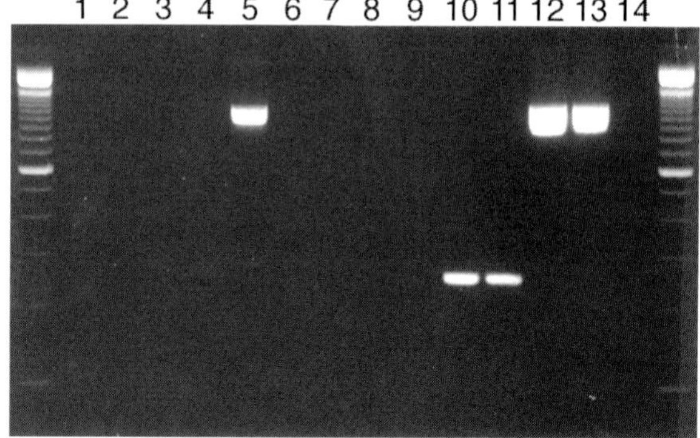

FIGURE 8-25 Polymerase chain reaction (PCR) detection of chromosome 14;18 translocation using PCR primers to the immunoglobulin heavy-chain region and bcl. Lanes 1–8: Patient tissue sample DNA. Lane 5 contains the t(14;18) in minor breakpoint region of bcl-2. DNA is from a follicular center-wall lymphoma from a 20-year-old man. Lane 9 is a negative control containing only water (no DNA). Lanes 10 and 11 are positive controls for the major breakpoint region of bcl-2, and lanes 12 and 13 are positive controls for the minor breakpoint region of bcl-2. These are two sites within bcl-2 where breakpoints can occur and give rise to different-size PCR products. (Courtesy of Drs. Janina Longtine and Jeffrey Sklar, Brigham and Women's Hospital, Boston.)

Tom's brother Mark is a suitably compatible donor. Tom receives the transplantation following high-dose chemotherapy and radiation therapy, which kills his own bone marrow. Hematopoietic growth factors are used to speed the recovery of the transplanted marrow. Tom must use an immunosuppressive drug, cyclosporin, to prevent a rejection phenomenon of graft-versus-host disease. After several weeks in a germ-free environment in the hospital, Tom is discharged, again with no evidence of active tumor.

When relapse occurs, the treatment alternatives include use of different chemotherapeutic regimens and, increasingly, bone marrow transplantation. The rationale of the latter approach is that very aggressive, toxic combinations of chemotherapy and radiation therapy can be administered that are lethal not only to cancer cells but also to normal bone marrow cells. This can be tolerated if the patient is maintained in a germ-free environment during the time when bone marrow is destroyed and then bone marrow is restored by transplantation. The source of transplanted marrow can be an HLA-matched donor (allogeneic) or one's own marrow maintained frozen (autologous), if it is tumor-free.

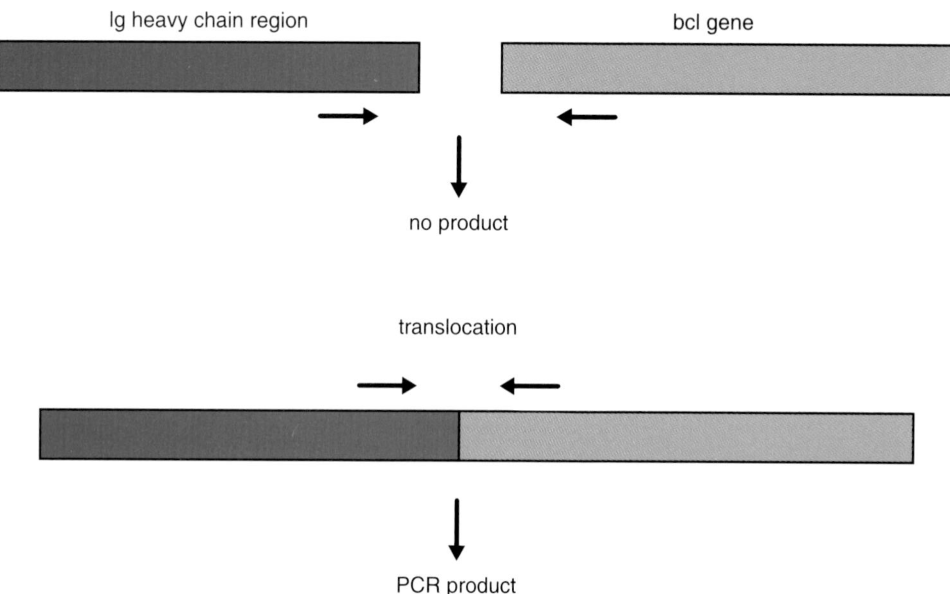

FIGURE 8-26 Detection of t(8;14) by polymerase chain reaction. One half the primer pair is homologous to a sequence in the immunoglobulin heavy chain and the other to a sequence in the bcl gene. A PCR product results only if there is a translocation that juxtaposes the two sequences.

The major complications of bone marrow transplantation are related to the high toxicity of chemotherapeutic drugs, infection, bleeding, and, in the case of allogeneic transplants, graft-versus-host disease (GVHD). The period of susceptibility to infection has been reduced, along with the need for hospitalization in a germ-free environment, by the use of growth factors that stimulate rapid recovery of transplanted marrow cells. GVHD represents a form of cell-mediated immunologic rejection of the host by the transplanted marrow. It is treated by immune suppression, mostly using a drug called *cyclosporin*. This is continued through several months to a year after transplantation, after which the period of vulnerability to GVHD may pass.

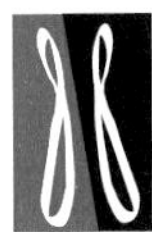

Part VIII

Five years have passed since Tom had his bone marrow transplant. He is now in good health and no longer requires cyclosporin therapy. He is working full-time as an attorney in the same firm, and he and Jean have just adopted a baby girl.

Although it is a relatively new mode of treatment, bone marrow transplantation offers to many individuals with lymphoma the possibility of long-term remission or even cure. It is increasingly being incorporated into the therapeutic regimen in young persons with lymphoma who are otherwise in good health.

REVIEW QUESTIONS

1. Bloom syndrome is an autosomal recessive disorder that is associated with a high frequency of spontaneous chromosome breakage due to a defect in a gene involved in DNA replication. There is an increased frequency of malignancy associated with this syndrome. Why do you think this is so?
2. A vestibular schwannoma is surgically removed from a person with neurofibromatosis type 2 (NF2). DNA is isolated from the tumor, cut with a restriction enzyme, and hybridized with a polymorphic DNA probe closely linked to the NF2 locus. The following Southern blot pattern is seen, along with the pattern from analysis of blood DNA from the same person.

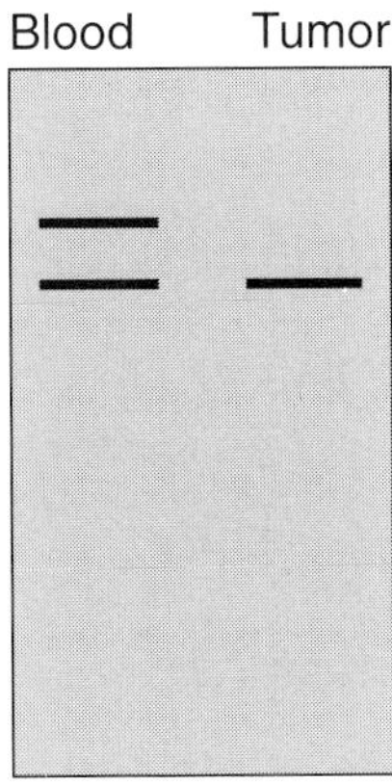

   a. What would you conclude about the pathogenesis of vestibular schwannoma in this disorder?
   b. If the father of this person also has NF2, would you expect the father's or mother's allele to have been the one lost in the tumor?
3. Loss of one or both copies of the retinoblastoma gene is found in many malignant tumors other than retinoblastoma, in individuals with no family history of retinoblastoma. How might this be explained?
4. Multiple myeloma is a malignancy arising from antibody-producing cells. Generally, a single type of antibody is produced by a given tumor, rather than a mixture of antibodies of various specificities. Why is this?

FURTHER READING

8.3 The Retinoblastoma Gene

Godbout R, Dryja TP, Squire J, et al. Somatic inactivation of genes on chromosome 13 is a common event in retinoblastoma. Nature 1983;304:451–453.

Sparkes RS, Sparkes MC, Wilson MG, et al. Regional assignment of genes for human esterase D and retinoblastoma to chromosome band 13q14. Science 1980;208:1042–1044.

8.4 Cloning the Retinoblastoma Gene

Dryja TP, Rapaport JM, Joyce JM, Peterson RA. Molecular detection of deletions involving band q14 of chromosome 13 in retinoblastomas. Proc Natl Acad Sci USA 1986;83:7391–7394.

Friend SH, Bernards R, et al. A human DNA segment that predisposes to retinoblastoma and osteosarcoma. Nature 1986;323:643–646.

8.5 Oncogenes

Varmus HE. The molecular genetics of cellular oncogenes. Annu Rev Genet 1984;18:553–612.

8.7 The Molecular Basis of Oncogenesis

Jen J, et al. Allelic loss of chromosome 18q and prognosis in colorectal cancer. N Engl J Med 1994;331:213–221.

Vogelstein B, Fearon ER, et al. Genetic alterations during colorectal-tumor development. N Engl J Med 1988;319:525–532.

8.8 Genetics and the Treatment of Cancer

Nabel GJ, Nabel EG, et al. Direct gene transfer with DNA-liposome complexes in melanoma: expression, biologic activity, and lack of toxicity in humans. Proc Natl Acad Sci USA 1993;90:11307–11311.

CASE STUDY

Verdonck LF, van Putten WLJ, Hagenbeek A, et al. Comparison of CHOP chemotherapy with autologous bone marrow transplantation for slowly responding patients with aggressive non-Hodgkin's lymphoma. N Engl J Med 1995;332:1045–1051.

Weiss LM, Warnke RA, Sklar J, Cleary ML. Molecular analysis of the t(14;18) chromosomal translocation in malignant lymphomas. N Engl J Med 1987;317:1185–1189.

Yunis JJ, Frizzera G, Oken MM, et al. Multiple recurrent genomic defects in follicular lymphoma. N Engl J Med 1987;316:79–84.

Yunis JJ, Oken MM, Kaplan ME, et al. Distinctive chromosomal abnormalities in histologic subtypes of non-Hodgkin's lymphoma. N Engl J Med 1982;307:1231–1236.

9 Developmental Genetics

OVERVIEW

Identification of a disease gene is only the prelude to understanding the pathogenesis of a disease process. The study of the human genome presents an unprecedented opportunity for understanding normal human development and physiology. Embryonic development involves the orderly unfolding of a genetic program. Genes are activated that control processes of cell division and differentiation. Elucidation of this program is one of the central themes of twentieth-century genetics.

Syndromes of abnormal development have provided both impetus and opportunity for the study of developmental genetics. In this chapter, we will trace the steps that led to the discovery of the genetic basis for one such disorder, Waardenburg's syndrome. This autosomal dominant condition is characterized by deafness, patchy hypopigmentation, and distinctive facial features. We will see how the gene locus was identified by linkage analysis, aided by finding a rare affected individual with a visible chromosome rearrangement. Although the gene would have yielded to a positional cloning effort, identification occurred much more quickly using a candidate gene approach, which involves testing of genes residing at or near the locus for mutations in affected individuals. The candidate gene for Waardenburg's syndrome was pinpointed first from studies of a naturally occurring mouse model of the human disorder. The list of genes known to be involved in development is expanding rapidly, and the candidate gene approach is an increasingly productive way to identify disease genes. The result of these efforts is a gradual elucidation of the basis for human developmental syndromes whose pathogenesis has long been a mystery. In the Perspective, we will explore life with Waardenburg's syndrome, and the chapter closes with the story of a child with multiple congenital anomalies, showing how drastically new knowledge of genetics can alter counseling.

9.1 Waardenburg's Syndrome

Jean, an engaging 3½-year-old girl, is brought by her mother Karen for an assessment at a local hospital-based program for the deaf. Karen was not surprised when Jean was found to be hard of hearing, because both Karen and Jean's father Eric are hard of hearing. Karen is able to speak and can communicate in American Sign Language (ASL). ASL is Jean's primary language, and she can already communicate quite effectively. An evaluation is set up to be sure that Jean is provided with an appropriate school program to take advantage of her excellent cognitive abilities. There has already been a dispute with the school system over the most appropriate class placement and services for Jean. Karen and Eric are divorced, and Jean lives with her mother and two siblings. Her brother Aaron, age 6, has the same parents and hears normally. Her half-sister, Ellen, is 1 year old and also hears normally. Ellen has the same mother but a different father. Karen does not have a history of hearing loss on her side of the family. Eric is not available to provide family history information but, according to Karen, his mother and his brother also have hearing loss. Eric has another sibling, a sister, who can hear normally, as can her three children. Jean is examined by a pediatrician as part of the evaluation, and she is noted to have features of Waardenburg's syndrome (Figure 9-1).

FIGURE 9-1 Child with Waardenburg's syndrome.

There are diverse causes of congenital hearing loss, involving both genetic and nongenetic factors (Figure 9-2). Approximately 30% of individuals with hearing loss are affected on a genetic basis. Nearly one-third of the genetically based hearing losses are associated with syndromes in which deafness is but a component of a more complex disorder. The remaining two-thirds of genetic hearing losses are undifferentiated, in which there are no distinctive features other than hearing loss and no way to discern the specific mechanism of hearing impairment. Among these cases of undifferentiated deafness, nearly a third are inherited as autosomal dominant, almost two-thirds are autosomal recessive, and a small proportion are X-linked.

Waardenburg's syndrome is a dominantly inherited disorder with pleiotropic effects. The most prominent feature is hearing loss. The deafness is *sensorineural*, meaning that it affects hearing at the level of the cochlea or the acoustic nerve. This is distinguished from conductive deafness, in which middle-ear structures such as the tympanic membrane or ossicles are affected. Waardenburg's syndrome is responsible for 2% to 3% of cases of congenital deafness. The hearing loss is present from birth and tends to be nonprogressive. The degree of impairment differs from one person to another even within the same family. It may range from total deafness to milder hearing loss manageable with hearing aids.

Another feature of Waardenburg's syndrome is pigmentary dysplasia. This is most obvious on the head where there is often a patch of white hair in the frontal region. In addition, there can be patches of hypopigmented skin scattered along the body. The irises of the two eyes may be of different colors, one brown or green and the other blue. Blue eyes lack iris pigment, so this is another example of patchy hypopigmentation.

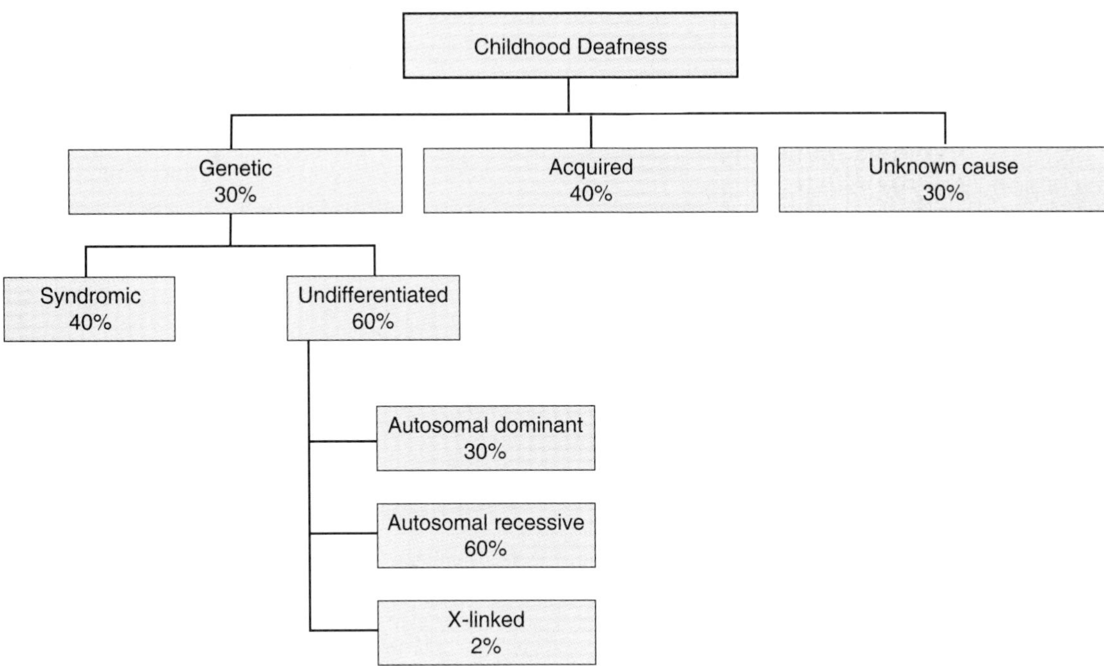

FIGURE 9-2 Causes of childhood deafness.

A third defining feature of Waardenburg's syndrome is dystopia canthorum. The eyes appear to be placed widely apart due to lateral displacement of the inner canthi. This does not affect vision and usually does not cause cosmetic problems. It is detected by measurement of facial features.

Three types of Waardenburg's syndrome have been distinguished clinically (Table 9-1). Type 1 includes all the features already noted. Type 2 includes sensorineural hearing loss and pigmentary dysplasia but not dystopia canthorum. Type 3 includes the same features as type 1 but with the addition of limb malformations. Although the expression of Waardenburg's syndrome varies widely within a family, the basic features of types 1, 2, or 3 tend to be consistent within a family. This suggests that different mutations, perhaps at different gene loci, account for these different forms of Waardenburg's syndrome.

TABLE 9-1 Classification of Waardenburg's syndrome

Type	Features
1	Hearing loss, pigmentary dysplasia, dystopia canthorum
2	Hearing loss, pigmentary dysplasia
3	Hearing loss, pigmentary dysplasia, dystopia canthorum, limb deformities

9.2 Mapping the Gene for Waardenburg's Syndrome

The pediatrician explains to Karen that Jean has Waardenburg's syndrome. Jean finds this interesting but learns that this diagnosis does not have much of an effect on Jean's care. She understands that Jean has a 50% chance of passing the trait on to any of her offspring, which Jean will need to be told when she is older. It is recommended that Jean attend a preschool to help her develop her sense of identity as a deaf person. A preschool for deaf children is found where ASL is used, and Jean is enrolled. Karen is generally satisfied that Jean's needs have been addressed. Several months after the assessment, the audiologist working with the group is contacted by a geneticist who is seeking families with Waardenburg's syndrome for study. After contacting Karen herself, the audiologist puts the geneticist in touch with the family. Blood samples are obtained from Karen, Jean, Eric, Eric's mother and father, and Eric's sister, who also is affected with Waardenburg's syndrome (Figure 9-3).

The genes responsible for the various types of Waardenburg's syndrome have been pursued vigorously because of the opportunity to gain insight into a major cause of genetically determined hearing loss. Initial efforts were focused on genetic linkage studies, with the goal of applying a positional cloning strategy.

When the search for linkage began in the 1970s, the only available markers were protein polymorphisms, including blood group antigens. Although there was evidence at one point of possible linkage with the ABO blood group gene on chromosome 9, more complete analysis of data from four families revealed a peak logarithm of the odds (lod) score of only 1.6 at 20% recombination. By 1990, using a set of both protein and DNA polymorphisms, 23% of the genome had been excluded as showing significant linkage to the Waardenburg's syndrome locus. That left more than three-fourths of the genome to search.

An important clue that narrowed the search was reported in 1979. A child was identified in Japan who had hearing loss, dystopia canthorum, heterochromia iridis, and a patch of hypopigmented skin. Although this child had many features of Waardenburg's syndrome, no relative was affected. This is an unusual finding for Waardenburg's syndrome, and it prompted practitioners to perform chromosomal analysis. The child was found to have an inversion of chromosome 2 involving the region 2q35–q37 (Figure 9-4). Neither parent carried the inversion, indicating that it arose de novo in the child. This raised the possibility that a gene or genes responsible for Waardenburg's syndrome might reside in this region of chromosome 2. As in the case of retinoblastoma, a rare chromosomal abnormality helped point the way to a disease gene locus.

There was another clue that directed attention to chromosome 2q. For a long time, breeders have scrutinized mice for mutant phenotypes. One of these,

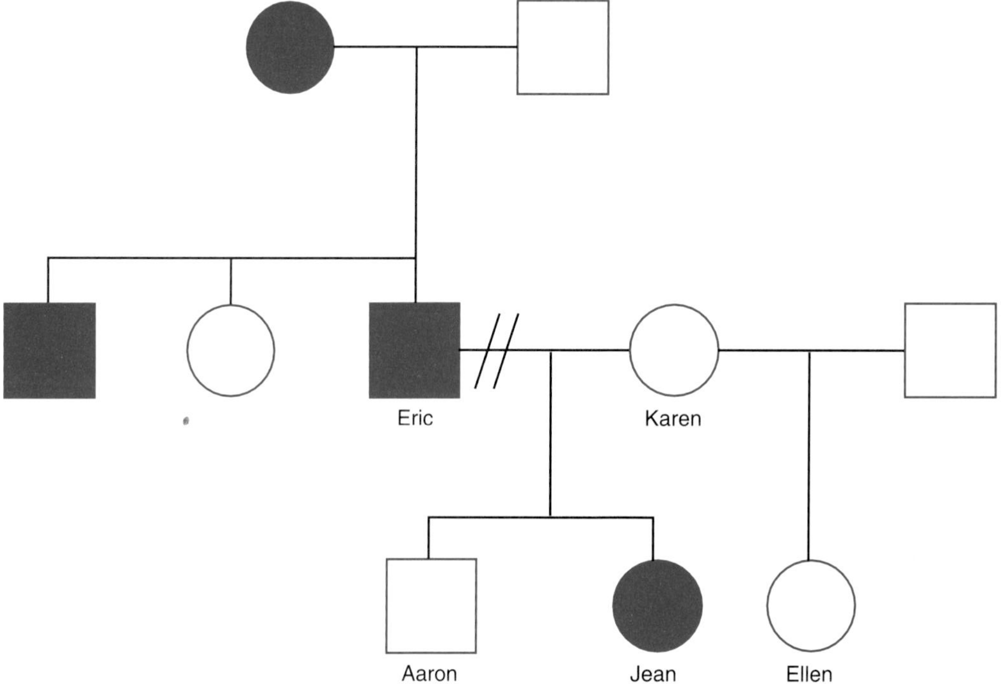

FIGURE 9-3 Pedigree for Jean's family.

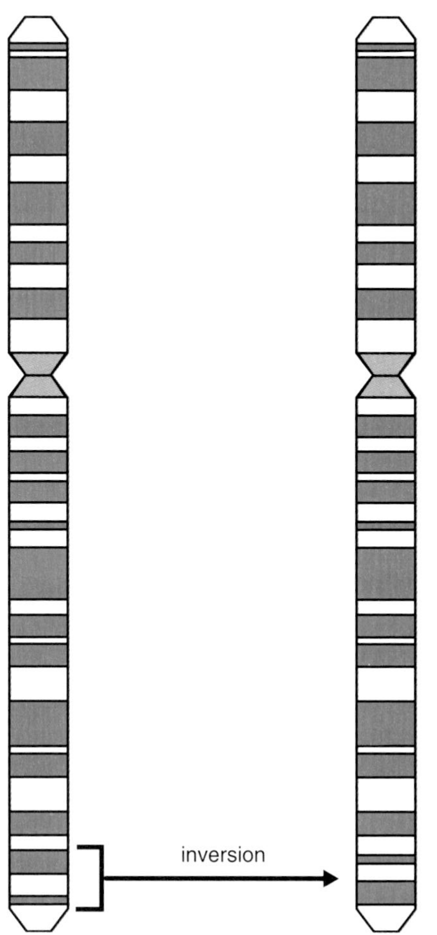

FIGURE 9-4 Diagram of chromosome 2 inversion in child with features of Waardenburg's syndrome. (Modified from Ishikiriyama S, et al. Waardenburg syndrome type I in a child with de novo inversion (2)(235q3y.3). Am J Med Genet 1989;33:506–507.)

designated *splotch*, has been considered a possible homolog to Waardenburg's syndrome. *Splotch* heterozygotes have patches of white spots but have normal hearing. Homozygotes have malformations of the organ of Corti in the cochlea, however. They also develop neural tube defects. Through breeding experiments the *splotch* locus was mapped to mouse chromosome 1.

As gene mapping has proceeded in both humans and mice, it has become apparent that there are groups of genes that are closely linked to one another in both species. This indicates that the reshuffling of the genome during evolution is not complete; blocks of genetic material remain intact even as multiple chromosome rearrangements occur (Figure 9-5). Islands of homology have been found, allowing human chromosome regions that correspond with mouse chromosome regions to be identified. The human chromosome region corresponding with the area of mouse chromosome 1 where *splotch* maps is on 2q.

Armed with these leads, investigators focused attention on 2q35–q37 as a region where the Waardenburg's syndrome gene might reside. Analysis of

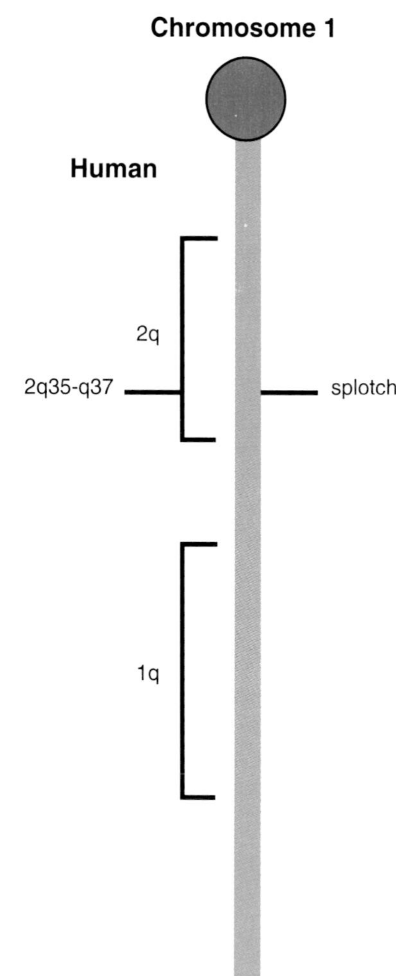

FIGURE 9-5 Diagram of mouse chromosome 1, indicating blocks of homology with human chromosomes 2q and 1q. The mouse *splotch* locus maps to a region that is homologous with human chromosome 2q35–q37.

five families with a restriction fragment–length polymorphism (RFLP) for the gene encoding placental alkaline phosphatase–1 (ALP-1) revealed a peak lod score of 4.767 at $\theta = 0.023$. This provided compelling evidence that the Waardenburg's syndrome gene resides in this area (Table 9-2).

9.3 Identification of Candidate Genes

With gene linkage established and a chromosome rearrangement from an affected individual in hand, the stage seemed set for positional cloning of the Waardenburg's syndrome gene. As we have seen with cystic fibrosis, though, this can be a slow and laborious enterprise. Fortunately, a shortcut led to identification of the gene within 2 years of finding linkage. The critical clue came not from the mouse but actually began with a species much further down the evolutionary path—the fruit fly *Drosophila*.

Observations of *Drosophila* for mutations date back well before work with mice. The *Drosophila* genome began to be mapped in the 1920s, principally by Thomas Hunt Morgan and his group. These early efforts laid much of the groundwork for our current understanding of genome organization in eukaryotes.

Over the years, a curious set of *Drosophila* mutants have been studied. *Drosophila* species have the typical arthropod body plan of three segments—head, thorax, and abdomen. Rare mutants, referred to as **homeotic mutants,** have distinctive and often bizarre disruptions of body segmentation. The mutant *Antennapedia,* for example, has legs protruding from the head where antennae should be. Other mutants are characterized by segmentation defects (e.g., conversion of a legless abdominal segment into a thoracic segment with legs).

Many of the genes responsible for these phenotypes have been cloned. It has been found that they share a region of homology that has come to be called the **homeobox** or **Hox**, which consists of approximately 60 amino acids (corresponding with 180 base pairs of DNA) that has DNA-binding properties. Other genes involved in the regulation of development share a different DNA-binding domain called the **paired box** or **Pax.** The Pax box consists of 128 amino acids. In *Drosophila,* there are five Pax genes, most of which appear to be involved in body segmentation.

There are several groups of Hox genes in *Drosophila* that share sequence homology. One group, referred to as the **homeotic complex (HOM-C),** consists of eight genes. These genes are expressed in an anterior-posterior pattern in the *Drosophila* embryo, in a spatial order that is the same as their arrangement on the chromosome (Figure 9-6). The gene at the 3′ end of the complex is expressed in the head region, and loss of function of this gene causes disruption of development of head structures. Genes located in a 5′ direction are expressed in progressively more posterior segments. Areas of expression tend to overlap, however. The overlapping expression explains the transformation of one body segment into another in a homeotic mutant: When one Hox gene is not expressed in a segment, persistent expression of another leads to transformation of the segment. Pax genes also are expressed in specific segments in the *Drosophila* embryo. Both Hox and

TABLE 9-2 Locating the gene for Waardenburg's syndrome

	Recombination Fraction (θ)					
Family	0.00	0.05	0.10	0.20	0.30	0.40
1	2.26	2.06	1.86	1.42	0.93	0.42
2	0.96	0.88	0.80	0.62	0.44	0.23
3	1.15	1.00	0.84	0.53	0.24	0.05
4	1.45	1.28	1.10	0.73	0.36	0.09
5	−1.26	−0.55	−0.33	−0.14	−0.05	−0.01
Total	4.56	4.68	4.27	3.16	1.91	0.77

Source: Data from Foy C, Newton V, Wellesley D, Harris R, Read AP. Assignment of the locus for Waardenburg syndrome type I to human chromosome 2q37 and possible homology to the splotch mouse. Am J Hum Genet 1990;46:1017–1023.

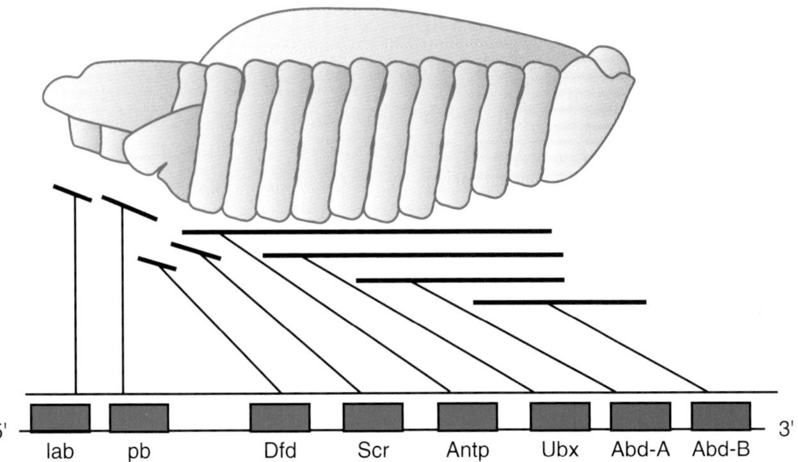

FIGURE 9-6 Domains of expression of genes of the HOM-C complex in *Drosophila*, with map of region shown at bottom. (Redrawn by permission from McGinnis W, Krumlauf R. Homeobox genes and axial patterning. Cell 1992;68:283–302.)

Pax proteins bind to DNA and activate the transcription of other genes. They are believed to act as switches that invoke tissue-specific developmental programs.

The discovery of Hox and Pax boxes prompted a search for homologous genes in higher eukaryotes, including humans. Using DNA probes for Hox or Pax boxes, regions of homology were indeed found. Moreover, mammalian Hox genes, like their *Drosophila* counterparts, are arranged in tandem arrays and are expressed in an anterior-posterior direction that mirrors their order on the chromosome (Figure 9-7). There are nine Pax genes in the human genome, seven of which also include Hox box sequences.

There are no known natural Hox mutants in the mouse, but study of Hox function has been accomplished by targeted disruption of these genes. A vector containing a mutant Hox gene has been introduced into mouse embryonic stem cells. The disrupted segment recombined with the wild-type gene and introduced the mutation into the mouse gene. The embryonic stem cells then were injected into mouse blastocysts, which in turn were implanted into a pseudopregnant uterus and brought to term. Breeding of these chimeric mice resulted in fully heterozygous animals. Mice with Hox mutations tend to have congenital anomalies in the body

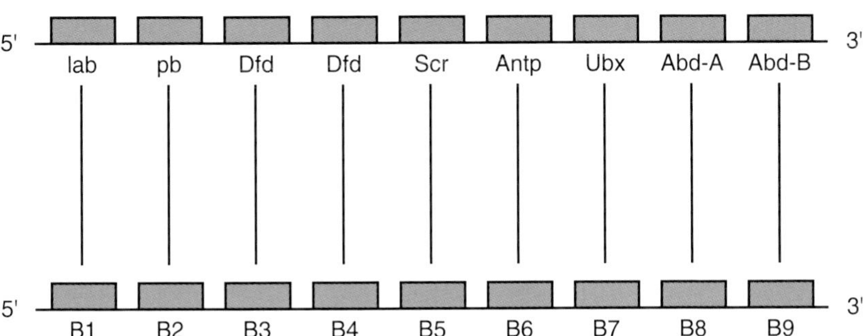

FIGURE 9-7 Genes of HOM-C complex in *Drosophila* (top) and homologous genes of Hox-2 cluster in mouse (bottom).

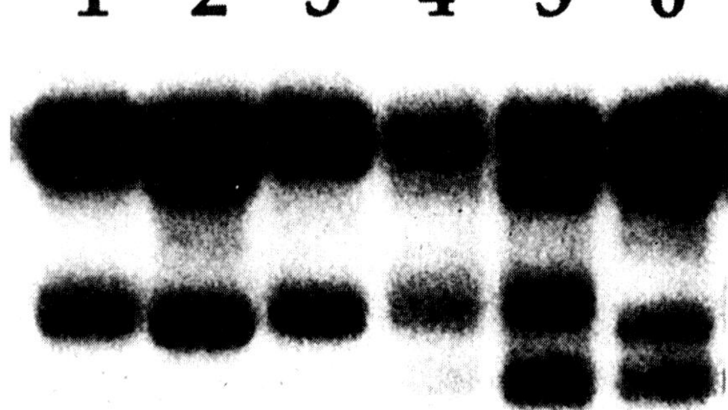

FIGURE 9-8 Single-strand conformational polymorphism autoradiogram showing PAX3 mutation in exon 5. Lanes 1–3 are from controls. Lane 4 is from Karen, lane 5 from Eric, and lane 6 from Jean. (Courtesy of Dr. Clint Baldwin, Boston University Center for Human Genetics.)

region where the mutant gene is expressed, indicating that the mammalian Hox genes, like their *Drosophila* counterparts, function to regulate the proper morphogenesis of different anterior-posterior regions of the embryo.

In contrast to Hox, three natural mouse Pax mutations have been identified. One is called *undulated,* the phenotype of which is skeletal deformities. It is due to a single base change in the Pax1 gene. The second mouse mutant is *splotch,* associated with Pax3 mutations. *Splotch* has arisen naturally several times, and different *splotch* alleles have been characterized. The third mouse mutant is referred to as *small-eye*, due to Pax6 mutations. Finding the gene that encodes for *splotch,* the mouse homolog to human Waardenburg's syndrome, opened the way to finding the Waardenburg's syndrome gene in humans.

9.4 The Gene for Waardenburg's Syndrome

Further linkage studies confirm that the locus for Waardenburg's syndrome type 1 maps to chromosome 2q37. Recently, it has been found that a gene called Pax3 *maps to the region and is mutated in* splotch *mice. The geneticist studying Jean's family decides to examine the human homolog of Pax3, designated* HuP2, *for mutations in individuals with Waardenburg's syndrome. Single-strand conformational polymorphism analysis is performed, and the data shown in Figure 9-8 are obtained.*

The human homolog to Pax3 was found to map to chromosome 2q35–q37 by fluorescence in situ hybridization (Figure 9-9). Families with Waardenburg's syndrome then were tested for Pax3 mutations. Polymerase chain reaction (PCR) primers were synthesized for regions flanking exons, and PCR products were analyzed by the technique of single-strand conformational polymorphism (SSCP) analysis (see Chapter 2, Case Study). SSCP can be highly sensitive and will reveal a variety of kinds of mutation—base pair changes, insertions, and deletions. Some mutations will be missed, however. For example, conformation may not be altered sufficiently to distinguish mutant from wild-type strands. This can

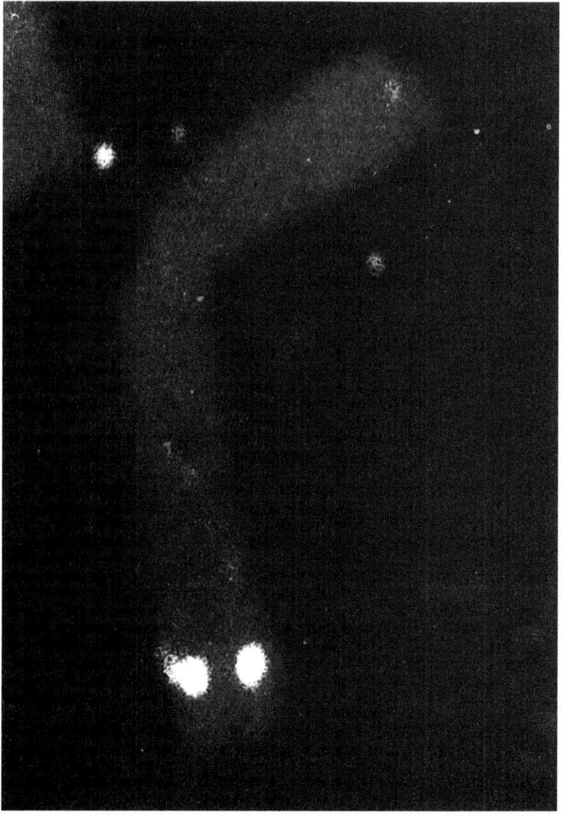

FIGURE 9-9 Localization of human PAX3 gene to chromosome 2q by fluorescence in situ hybridization. (Modified by permission from Wu B-L, Milunsky A, Wyandt H, et al. In situ hybridization applied to Waardenburg syndrome. Cytogenet Cell Genet 1993;63:29–32.)

be mitigated if small PCR fragments are tested, in which case a subtle mutation may have more of an impact on conformation.

SSCP testing with exon 2 of human Pax3 in one family with Waardenburg's syndrome revealed an abnormal band that segregated with the disorder. The exon was sequenced, and a single base change—G to A—was found, leading to a change from proline to leucine in the protein. This is a site near one of the DNA-binding domains of Pax3 and is highly conserved in other species, making it likely to be pathogenic.

Another approach that has been used to identify Pax3 mutations is heteroduplex analysis. After performance of PCR, amplified DNA fragments corresponding with exons are separated into single strands and are allowed to re-anneal. In a heterozygous individual, four kinds of double-stranded molecules may result. Wild-type strands may anneal with one another or mutant strands with one another. Alternatively, the sense mutant strand may anneal with the antisense wild-type strand, or vice versa. The latter two are referred to as **heteroduplexes** and will have a mismatched region at the site of mutation. This mismatch changes conformation of the heteroduplex and slows its migration through an agarose gel. Exons found to have heteroduplexes then are sequenced to identify the mutations. Using this method, several other Pax3 mutations were found in Waardenburg's syndrome families.

Three classes of mutation have been found. One involves mutation of the paired box domain, which probably disrupts DNA binding. A second class of mutation is exemplified by a two-base deletion near a homeodomain, also found in this gene, that leads to frameshift. The third class leads to stop codons or frameshifts near the N terminus, causing truncation of the protein and probably degradation of the truncated product.

Thus far it has been difficult to determine whether there are genotype-phenotype correlations. The initial work was done mostly with type 1 Waardenburg's syndrome families, but mutations have also been found in some type 3 families. A separate gene, referred to as *micro-ophthalmia,* has been found to be responsible for Waardenburg's syndrome 2.

All of the Pax3 mutations would be expected to decrease or abolish function of one allele. An intact allele would be present in heterozygotes. Presumably, the loss of one functional allele results in reduced gene dosage, and this is insufficient for normal function. In *splotch* mice, homozygotes are much more severely affected than heterozygotes, indicating that some Pax3 function is necessary for development.

Pathologically, both Waardenburg's syndrome and *splotch* represent abnormalities of neural crest development. Neural crest is derived from ectodermal tissue at the edge of the folding neural tube in the early embryo. It migrates laterally and gives rise to diverse tissues, including melanocytes, schwann cells, sympathetic and sensory neurons, calcitonin-releasing cells in the thyroid, adrenal medulla, facial bones, and the aortic outflow tract of the heart, among others. In the mouse, Pax3 is expressed in the upper part of the neural tube and in the neural crest. Both tissues are abnormal in homozygous *splotch* animals, but only the neural crest is abnormal in heterozygotes. Apparently, there are different thresholds for expression of the mutant phenotype in these different tissues.

Loss-of-function mutations appear to be much less common in Hox genes. This may indicate lethality of the phenotype or, conversely, less sensitivity to decrease in gene dosage. A human Hox gene mutation has been found in one family with craniosynostosis—premature fusion of sutures in the skull. The gene is called *msx2*. The pathogenesis of this phenotype is not yet known, but introducing a **gain-of-function** change in mice causes craniosynostosis.

The SSCP picture shows two normal bands, corresponding with a control. In lane 2, there are three bands. The top and bottom are the two normal bands; the middle band represents a polymorphism. This polymorphism leads to a secondary structure for one strand that is distinguishable, but the other strand migrates at the same rate as one of the other bands. This explains why three bands, not four, are seen. Jean's mother (lane 3) has the normal pattern. Her father has the polymorphism but also a new band, found at the bottom, representing a mutation. Here, too, the mutation results in one new band, not two. Jean has the same pattern as her father. Sequencing (Figure 9-10) reveals a mutation that substitutes an A for a G in the antisense strand (note both an A and a G in the mutant sequence at this position, instead of only a G). This causes an arginine codon to become a chain-termination codon.

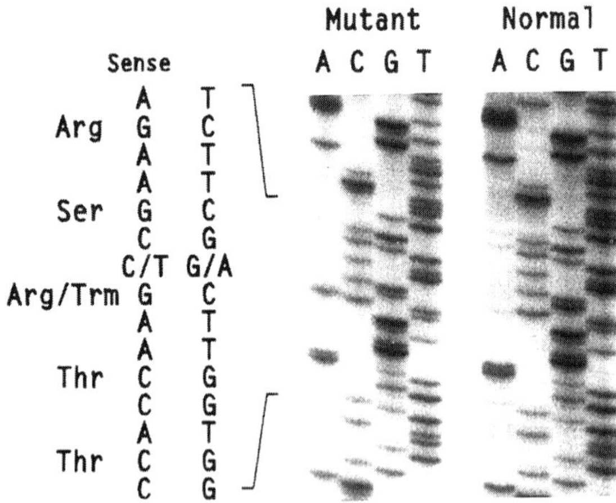

FIGURE 9-10 Sequence analysis of Jean's exon 5, showing PAX3 mutation. The antisense strand was sequenced. (Courtesy of Dr. Clint Baldwin, Boston University Center for Human Genetics.)

9.5 Inducers of Differentiation

Hox and Pax genes regulate the expression of genes in a region-specific and tissue-specific pattern, but if these genes set into motion programs of differentiation, what tells the cell to express one Hox gene or another? Studies in embryology have examined the signals that control the anterior-posterior (i.e., head-to-tail) as well as the dorsal-ventral differentiation. This work has shown that specific parts of the embryo are required to induce the development of other parts.

In the frog *Xenopus*, for example, the dorsal (animal) pole of the egg gives rise to ectodermal structures—skin and nervous system—whereas the ventral (vegetal) pole gives rise to endodermal tissues. Reconstitution of frog embryos has indicated that signaling molecules in the animal or vegetal poles control this dorsal-ventral distinction (Figure 9-11). The RNA encoding Vg1, homologous to a family of growth factors designated *transforming growth factor–beta* (TGF-β), is at first uniformly distributed in the oocytes but later is transported to the vegetal pole where the protein is made.

In *Drosophila*, a pair of proteins, bicoid and nanos, are concentrated at the anterior and posterior ends of the early embryo. RNAs encoding these proteins are synthesized by nurse cells at the anterior end of the embryo and then are transported into the embryo. Bicoid RNA remains in the anterior region, whereas nanos is transported posteriorly due to signals in the 3′ untranslated region of the message. Other proteins are concentrated at the anterior or posterior, dorsal or ventral end of the embryo. It appears that these form concentration gradients that direct tissue-specific gene expression.

The situation in the whole embryo is mimicked in the developing limb. The anterior-posterior axis determines the formation of the first through fifth digits. The proximal-distal axis determines formation of the limb from humerus (or femur) to digits. The anterior-posterior axis is controlled by cells in the posterior mesenchyme that secrete a signal known as the *zone of polarizing activity* (ZPA). A gradient of this signal's activity is established, and cells nearest the source become the fifth digit, those farther away

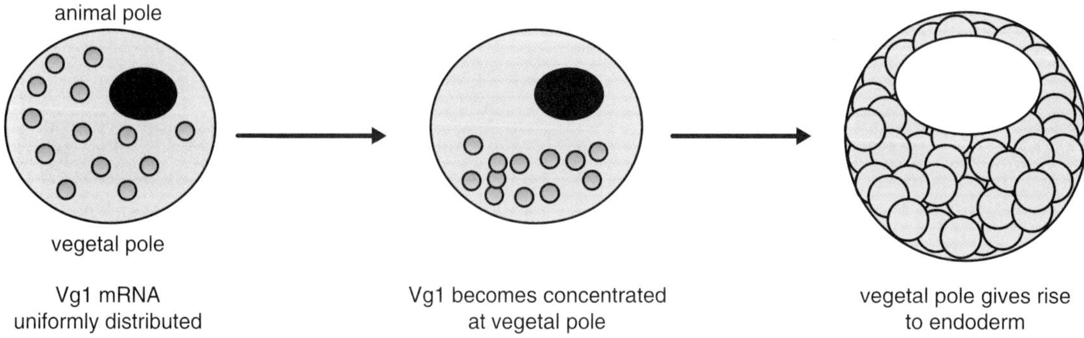

FIGURE 9-11 Establishment of dorsal-ventral axis in amphibian. Vg1 mRNA is initially uniformly distributed in the oocyte, but it is later concentrated at the vegetal pole, which eventually will give rise to endoderm.

progressively the fourth through first (Figure 9-12). If the ZPA is transplanted to the anterior end of the developing limb, a mirror-image duplication results, with a little finger on both sides.

A gene has been identified that appears to represent the ZPA signaling molecule. Referred to as *Sonic hedgehog*, it is secreted by cells in the ZPA region. A set of Hox genes are expressed in the developing limb. Like those that are expressed in the nervous system, these, too, are clustered in the genome and are expressed in an anterior-posterior order that mirrors their position on the chromosome. *Sonic hedgehog* can induce expression of these Hox genes, and it is believed that differential Hox gene expression due to a concentration gradient of *Sonic hedgehog*, perhaps acting through other intermediates, ultimately is involved in determining the formation of individual digits.

The proximal-distal axis of the limb originates from an apical ectodermal ridge, which also secretes inducer molecules. These include a protein in the fibroblast growth factor family (FGF4), which is secreted from the posterior part of the apical ridge, and FGF2 and FGF8, secreted from the entire ridge. Another cluster of Hox genes is expressed in a proximal-distal gradient and may respond to these signals.

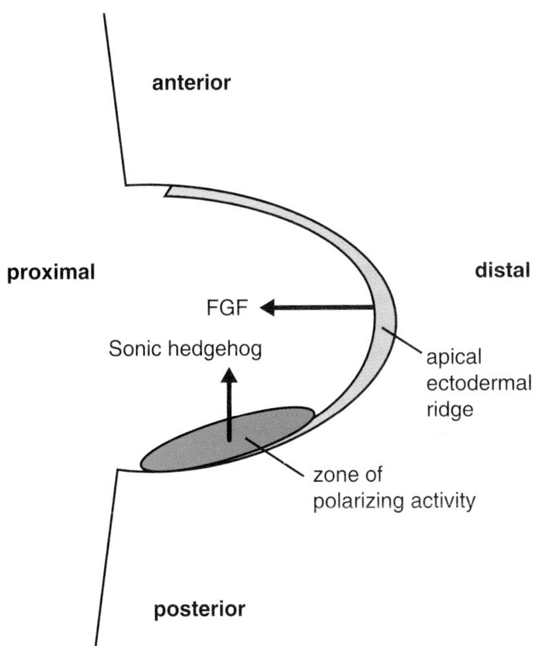

FIGURE 9-12 Diagram of limb bud, indicated apical ectodermal ridge, which determines proximal-distal axis, and zone of polarizing activity, which determines posterior-anterior axis. (Redrawn by permission from Tabin C. The initiation of the limb bud: growth factors, *Hox* genes, and retinoids. Cell 1995;80:671–674.)

Retinoic acid, an analog of vitamin A, is another substance involved in the regulation of limb development. It is present in an anterior-posterior gradient both along the axis of the embryo and in the flank prior to the formation of the limb. Retinoic acid appears to be important for establishing the location of the future ZPA in the prelimb flank. Retinoic acid enters the cell and binds to a receptor in the nucleus, which in turn binds to the promoter region of certain genes. Hox genes located toward the 3′ end of the cluster have greater sensitivity to induction by retinoic acid than those at the 5′ end, which may explain, in part, the anterior-posterior gradient of expression of these genes.

Retinoic acid sometimes is used as a medication to treat skin disorders such as acne. One form is administered systemically, as a pill. This preparation has been shown to be highly teratogenic, causing limb, nervous system, and cardiac malformations. Presumably, high levels of retinoic acid present in the embryo when the mother takes this drug upset normal morphogenetic signals. A topically administered form of retinoic acid is not absorbed in large quantities and does not appear to be teratogenic.

Jean is now 7 years old, and she is in second grade. She is enrolled in the public school, in a class with both hearing and deaf children. There is a sign language interpreter in the classroom. Jean is doing very well in school. Karen has had one other child, who has normal hearing. She is still in contact with Eric, who is remarried. Eric has had another child who is deaf and has Waardenburg's syndrome.

9.6 Human Dysmorphology

Approximately 3% of all pregnancies end with the birth of a child with a birth defect. Usually, these are isolated defects, such as cleft lip or a neural tube defect; the child is otherwise healthy, although he or

Perspective

Living with Waardenburg's Syndrome

KRISTIN PUORRO

I was born the first of two children with Waardenburg's syndrome in my family. Two years later, my younger brother was born and then, at age 9 months, he was determined to have been profoundly deaf from birth. It was soon discovered that his deafness was due to Waardenburg's syndrome and that, I too, had the syndrome. When I was born, my pediatrician suspected that I may be deaf because of my characteristics of the syndrome.

Because I am hearing, living with the syndrome has not been that difficult. Of course, I had to learn sign language as a child, but I can't remember learning it; it was just something I grew up knowing. As for the characteristics of the syndrome, those that are most apparent in me are a patch of white skin on my ankle and a wide space between my eyes and, at the age of 24, I am prematurely gray. I started turning gray when I was 14. No one likes to look different, nor does one like to be told that he or she is different, especially as an adolescent. However, for me, my symptoms are more cosmetic than debilitating or limiting. Later on in life, we realized that my father, too, had the syndrome but, like me, his characteristics were not severe, so no one ever thought to recognize them as being part of this syndrome. He has the wide space between his eyes, and one of his eyes is half brown and half blue, the other one being entirely brown. Both my brother and I have deep blue eyes. My father started turning gray by the time he was 30, but whether this is part of the syndrome or just premature graying, I am not sure.

However, for my brother, living with the syndrome has been more difficult. Apart from the deafness, the most prominent characteristics in him are a white forelock and the wide space between his eyes. When his deafness was first diagnosed, he saw numerous specialists—everyone from neurologists and cardiologists to psychiatrists and geneticists. Unfortunately, these doctors were insensitive to the feelings of my parents, who, at a young age, were told simply that their child would never be like other children. Most doctors who were encountered by my parents focused on the things that my brother would *never* be able to do instead of focusing on the things he *would* be able to do. They told my parents that he would never hear like us, he would never talk like us, and he would never attend the same schools as other children. They did not understand the gravity of his deafness. They understood the genetics of it and the physiologic aspects of not hearing, and they could attach a syndrome to it, but once my parents and brother left the examining room or doctor's office, the implications of my brother's deafness would become apparent only to my family. My parents had a wealth of information about his deafness

she may suffer serious consequences from the presence of the malformation. Most of these malformations occur sporadically and are believed to have a multifactorial etiology. Others are determined by single genes or chromosomal abnormalities.

Aside from isolated malformations, some babies are born with a complex of multiple congenital anomalies. The pioneer in the study of these anomalies, David Smith, referred to this study as *dysmorphology*.

Smith distinguished **malformations** from **deformations** and **disruptions**. A *malformation* is the result of abnormal development of tissue. Developmental mechanisms somehow are interfered with, and the tissue does not form properly. *Deformation* is defined as the distortion of a normally formed tissue by extrinsic pressure. An example is asymmetry of the skull due to pressure from a benign uterine tumor called a *fibroid*. *Disruption* is the damage of a normally formed tissue. For example, a tear in the amniotic cavity can trap a limb, amputating part of the extremity. Disruptions tend to be asymmetric, whereas malformations are more often (but not always) symmetric.

Various patterns of malformations are seen. Some compose **syndromes**, such as Down syndrome and the syndrome that runs in our family but, when it came down to everyday living, they were very much uninformed. It did not matter that the deafness was due to Waardenburg's syndrome; attaching a name to it did not erase the fact that my brother was deaf. They needed to make decisions that would affect my brother for the rest of his life. My parents were given all the medical information and statistical data about the syndrome, but what they needed to know and what would be more important was how to bring up a deaf child in a hearing world.

Growing up, my brother and I were constantly being stared at for our signing or for his white hair, which was apparent from birth. Even today, people will stare or ask us why we have white hair. Surprisingly, a lot of times the staring is from grown adults. My mother recalls a time when a woman did not want her child to play with my brother because he was different; apparently she thought her child could catch deafness. I think one of the most difficult things about being handicapped or being "different" is not being accepted by some people, who can often be extremely ignorant. As a child and adolescent, I found myself constantly defending my brother or myself against others who were making fun of us or who were not accepting of us.

I have also found that a lot of doctors have never heard of Waardenburg's syndrome, so that when they find out that we have it, they want to look at our white hair, our eyes, and our white skin. This quest for knowledge is understandable, and I think it is important that more physicians be made aware of Waardenburg's syndrome, but sometimes we felt like a specimen as opposed to a patient.

There also is the misconception that deaf people are less intelligent than hearing people. We have encountered people, even those in the medical profession, who refer to deaf people as the "deaf and dumb." This phrase is archaic and is analogous to a racial slur. My brother attends college just like anyone else; he can read, he can study, he can understand, he just can't hear. Many people have pity on him when they find out that he is deaf. Emotionally and socially, my brother is very fortunate in that he has not let his deafness stop him from doing or thinking what he wants.

The gene for Waardenburg's syndrome was discovered recently. This may have implications for my future. There is a strong possibility that I may have a deaf child someday. If prenatal diagnosis were to become available in the future, I don't think I would want any kind of testing. For me, having a deaf child would not be a devastating experience. In this respect, there is a positive side to my experiences with Waardenburg's syndrome. However, I need to think about my future spouse. I will need to discuss my syndrome and explain the risk factors involved with it to whomever I may marry. I hope he will be accepting of the possibility of having a child with Waardenburg's syndrome. This is an obstacle for me, but it is something I will need to face, as I am sure most people with inherited syndromes need to do.

or Waardenburg's syndrome. These are sets of congenital defects that are the consequence of some defined, ultimate cause. Down syndrome results from having an extra copy of chromosome 21. Waardenburg's syndrome is caused by a Pax3 or micro-ophthalmia gene mutation. Syndromes can also result from exposure to teratogenic agents such as retinoic acid. Whatever the cause—abnormality of one gene or a group of genes or exposure to a teratogen—the outcome is that multiple developing systems are perturbed in a reproducible way.

A second pattern of malformations is called a **sequence.** A sequence generally arises as the consequence of a single primary event—for example, underdevelopment of the lower jaw in the Pierre-Robin sequence. Other anomalies that comprise the sequence are secondary effects of the primary malformation. In the Pierre-Robin sequence, the tongue is too large for the small mouth and interferes with closure of the palate, resulting in cleft palate. Here, cleft palate is not a primary malformation but a secondary consequence of underdevelopment of the jaw.

The third pattern is referred to as an **association.** It has been observed that particular sets of congenital anomalies tend to occur together more often than expected due to chance. These do not comprise syndromes; the causes are unknown, and many associations are etiologically heterogeneous. Indeed, the majority occur sporadically. One example is the VACTERL association, the major components of which are *v*ertebral anomalies, *a*nal atresia, *c*ardiac anomalies, *t*racheo-*e*sophageal fistula, *r*enal anomalies, and *l*imb defects (Figure 9-13). Various combinations of these features occur in different babies, reflecting the nonrandom association of these malformations. It is not known why these associations occur. They may reflect processes that have some molecular or morphologic event in common or processes that all occur at the same time in development.

The medical practice of dysmorphology is partly science and partly art. A dysmorphologist examines for both subtle and major anomalies, looking for recognizable patterns. Correctly establishing a diagnosis can help one to provide anticipatory guidance: What does the future hold? Is the child likely to have additional anomalies that are not obvious on cursory examination? It also permits accurate genetic counseling. Sometimes the recurrence risk will be high; other times the family may be surprised to learn that recurrence is unlikely.

Until recently, the molecular basis of syndromes of abnormal development was largely unknown. This vastly limited the tools available for diagnosis. The identification of genes involved in both normal and abnormal development by positional cloning or the candidate gene approach gradually is changing this picture. It is likely that our knowledge of the cause of congenital anomalies—and concomitantly our ability to diagnose these and possibly prevent them—will increase dramatically in the years to come.

Case Study

Part I

The pregnancy has proceeded uneventfully right through the time of delivery. Jane and Albert were offered ultrasound examination during the pregnancy but did not choose to do this, as they would not terminate a pregnancy regardless of the outcome. Labor begins spontaneously, but the baby's head fails to engage in the pelvis, necessitating cesarean section. It is obvious at the moment of birth that there are problems. The baby, a girl, has a large head and short neck and chest, with markedly bowed lower legs. Although she breaths spontaneously in the delivery room, it is immediately apparent that she is having respiratory distress. She is intubated in the delivery room and brought to the special care nursery. Jane and Albert barely have time to see her. Within hours, the baby is transported to a nearby academic medical center. Jane and Albert are confused and frightened.

It is increasingly common to offer ultrasound examination as a component of routine prenatal care. In addition to helping to recognize fetal malformations, ultrasonography provides accurate assessment of gestational age. Many couples find that having information about their baby's health prenatally helps them to plan, even if they do not choose termination of pregnancy in the event of fetal problems. The stressful environment of the delivery room is not a good place in which first to learn that a baby is in trouble. This baby apparently has multiple congenital anomalies, including orthopedic problems

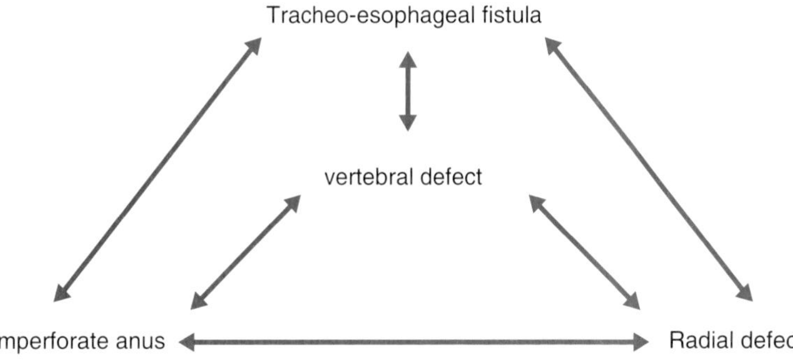

FIGURE 9-13 Diagram indicating association of tracheo-esophageal fistula, imperforate anus, vertebral defects, and radial defects. These malformations tend to occur together more often than might be expected due to chance. (Redrawn by permission from Quan L, Smith DW. The VATER association—vertebral defects, anal atresia, t-e fistula with esophageal atresia, radial and renal dysplasia: a spectrum of associated defects. J Pediatr 1973;82:104.)

and respiratory distress. Such problems require assessment and care outside the purview of a community hospital, necessitating transfer to an intensive care nursery in a tertiary medical center.

Part II

A geneticist is called to see the baby soon after her arrival at the medical center. On examination, she is found to have a very short thorax and neck. Her head circumference is 41.5 cm (much greater than the ninety-fifth percentile), and she has a flat facial profile, low nasal bridge, and bilateral epicanthal folds. Her ears are low-set and posteriorly rotated. She has a cleft palate and small lower jaw. There is marked shortening of both lower limbs, with pretibial skin dimples (Figure 9-14). The upper limbs appear to be normal. Genital examination reveals enlarged and pigmented labia majora and cliteromegaly (see Figure 9-14). A vaginal opening is present.

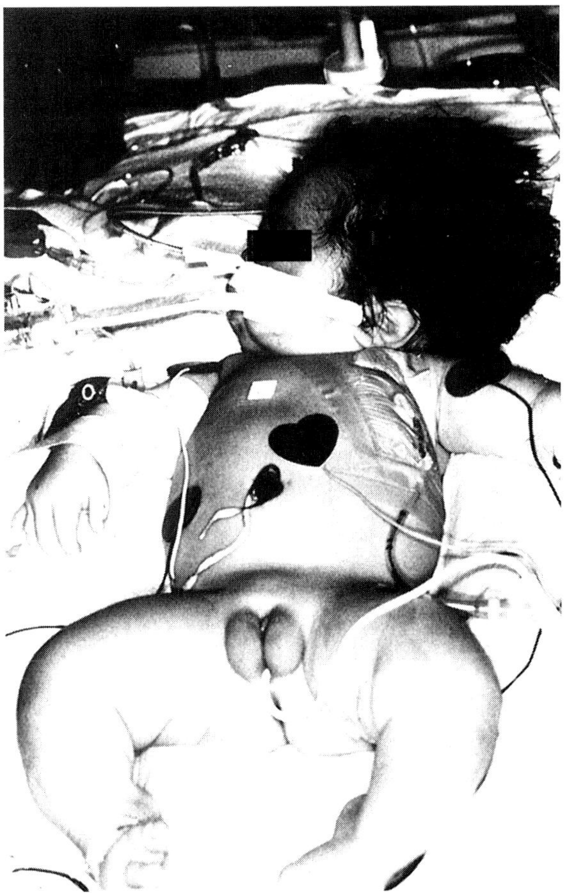

FIGURE 9-14 Photograph of child showing pretibial dimples (best seen in right leg) and hypertrophic labia majora. (Courtesy of Dr. Kathryn North, Children's Hospital, Boston.)

The physical findings are indicative of a skeletal dysplasia—that is, a congenital disorder of bone formation. The baby has a large head but small neck, thorax, and lower limbs. Although the skull is large, the bones of the face are small, leading to the low nasal bridge and flat facial profile. It is likely that the large head is responsible for the baby's inability to traverse the birth canal, leading to the cesarean section. Low-set, posteriorly rotated ears, epicanthal folds (extra folds of skin at the inner canthi of the eyes), cleft palate, and small lower jaw (micrognathia) are features found in a large number of syndromes of multiple congenital anomalies. The genital findings are not necessarily abnormal, and there is no doubt that the phenotype is female. The lower-limb anomalies are striking owing to the bowed tibias and skin dimples.

Part III

The geneticist suggests that a number of investigations be performed. These include x-rays of the entire skeleton (Figure 9-15), head and abdominal ultrasonography, and chromosomal analysis. Albert meets with the neonatologist later that evening, who explains that the baby is very sick but that no definite diagnosis has been made so far. The plan has been to provide maximum support as further information is collected. The orthopedic service has been called to see the baby, but has not sent someone by yet. Albert tells the care team that he and Jane have decided to name the baby Lisa.

The x-rays reveal hypoplastic scapulae, curving of the femurs and tibiae, dislocation of the hips, and abnormal cervical vertebrae. There are 11 ribs bilaterally. Head and renal ultrasound examinations are normal. A uterus is seen by pelvic ultrasonography. It is likely that the short thorax accounts, at least in part, for the respiratory distress, although the baby might also have tracheomalacia.

There are a large number of skeletal dysplasia syndromes that may present at birth. They are classified by clinical and radiologic features. One of the most common and easily recognized is achondroplasia. Affected babies have a large head but have short

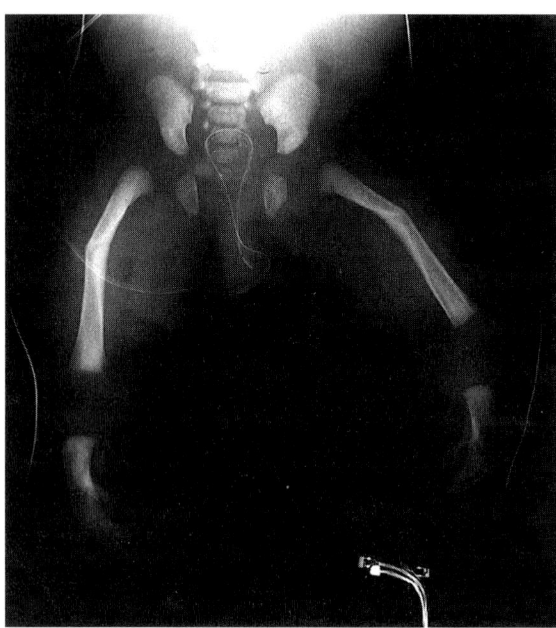

FIGURE 9-15 X-ray showing dysplastic femurs, tibias, and fibulas.

arms and legs. There are usually no other congenital anomalies or respiratory problems. Thanatophoric dysplasia has been found to be allelic due to achondroplasia, both being due to mutations in the gene FGFR3, one of a family of fibroblast growth factor receptor genes. Thanatophoric dysplasia is characterized by extremely short, bowed limbs, large head with hydrocephalus, and small thorax. Unlike achondroplasia, it is lethal in the newborn period.

Chondrodysplasia punctata includes a set of disorders in which there is stippling of the epiphyses, the cartilaginous growth plates. Other congenital skeletal dysplasias have characteristic radiographic features. In diastrophic dysplasia, there are broad metaphyses and the ears become swollen in infancy. In metatrophic dysplasia, the vertebral bodies are markedly flattened, and limbs are deformed. Kneist's dysplasia consists of flat facies, myopia, flattened vertebrae, and clublike metaphyses.

This baby's features do not fit these syndromes but fit very well with a rare disorder called *campomelic dysplasia*. The bowed tibiae with pretibial skin dimples are pathognomonic. The facial features, cleft palate, large head size, hypoplastic scapulae, 11 ribs, and respiratory insufficiency also are compatible with this diagnosis.

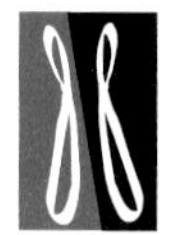

Part IV

Lisa is now almost 24 hours old. The geneticist meets with Albert and explains that Lisa has campomelic dysplasia. It is explained that this disorder is most often lethal in the early days of life, due to respiratory insufficiency. On the other hand, many children have been known to survive for long periods, although some require continued respiratory support. Among long-term survivors, prognosis for intellectual function is variable, with cognitive problems noted in some. A family history is obtained, revealing no prior history of similar problems. This is Jane and Albert's first child. Jane had one previous miscarriage at 8 weeks' gestation. Jane and Albert are not related to one another. The geneticist notes that the syndrome is usually sporadic, although recurrence in families has been reported in rare instances, suggesting possible recessive inheritance.

The major medical complication associated with campomelic dysplasia is respiratory insufficiency. Some babies are stillborn and are unable to make any initial respiratory efforts. The respiratory problems are attributable to a combination of factors, including small trachea and larynx, short thorax, and small lower jaw with airway obstruction by the tongue. Virtually all these babies require intensive ventilatory support at birth. Those who survive may need continued ventilation and tracheostomy.

The prognosis for central nervous system function has been difficult to assess, mostly because of major respiratory problems at birth; there are few long-term survivors. Some infants with this disorder have central nervous system malformations, including absence of the olfactory bulbs, dilated ventricles, and abnormalities of neuronal migration. However, there is relatively little experience with intellectual outcome of infants who were provided aggressive respiratory support from the time of birth.

In addition to the skeletal and cerebral malformations, there can be anomalies involving other organ systems. These include renal anomalies, congenital heart defects, and gastrointestinal malformations. Finally, a proportion of phenotypic females with campomelic dysplasia are found to have a 46,XY karyotype and have partial or complete absence of testes, with varying degrees of abnormality of the external genitalia.

The vast majority of cases of campomelic dysplasia occur sporadically, without prior family history. The existence of families with recurrence in siblings has suggested possible autosomal recessive transmission in some cases. Some consanguineous families have been reported, although overall the segregation ratio has been lower than the expected 25%. The genetics of camptomelic dysplasia were explained only after the responsible gene was identified.

Part V

Jane comes in with Albert the next day, and both speak with the neonatology staff and with the geneticist. They have decided to continue aggressive treatment. Lisa is currently still ventilator-dependent, having failed one attempt at extubation. The chromosomal analysis is reported later that day and provides another setback: Lisa is found to have the 46,XY karyotype. The geneticist explains that gender reversal has been reported before in camptomelic dysplasia. Eventually, Lisa will require exploratory surgery to remove gonadal tissue to protect her from possible tumor formation, assuming that she survives her respiratory problems. The geneticist also notes that further literature review has revealed that the gene responsible for camptomelic dysplasia has been discovered just in the past few weeks. Studies have revealed that affected individuals generally have mutations in just one allele, indicating that the trait is dominant and so most affected individuals represent new mutations. Blood is drawn from Lisa and sent to a laboratory that is involved in research on the camptomelic dysplasia gene.

Females with camptomelic dysplasia generally have normal genitalia, but males may have genital phenotypes ranging from normal male through female-appearing external genitalia. Internal genitalia can be either male or female in XY infants with camptomelic dysplasia and, finally, gonads can be either testes or undifferentiated gonads. Understanding the basis for these abnormalities requires a review of normal sexual differentiation. Herein also lies the key to discovering the gene responsible for camptomelic dysplasia.

Differentiation of the internal and external genitalia begins with the development of the gonad into either an ovary or a testis (Figure 9-16). Up to 8 weeks of gestation, the gonad, which contains germ cells, is undifferentiated and can become either an ovary or a testis. After that time, a testis forms in the presence of a Y chromosome, and an ovary forms if a Y is absent. The specific gene on the Y chromosome has been identified and is referred to as *SRY*. SRY appears to be a transcription factor; its binding to other genes sets into motion a cascade of events that lead to differentiation of the testes.

If a testis forms, Sertoli cells in the testis secrete a peptide hormone, müllerian inhibiting substance, that leads to regression of the müllerian ducts. Testosterone produced by the testis leads to growth of the wolffian ducts into epididymes, vas deferens, and seminal vesicles. The external genitalia also are stimulated by testosterone to grow into a penis and scrotum.

In the absence of SRY, an ovary forms instead of a testis. The müllerian ducts become fallopian tubes, uterus, and part of the vagina. The wolffian ducts degenerate due to lack of androgen stimulation, and the external genitalia remain in the female form.

Disorders of sexual differentiation can be explained in terms of this basic process. Presence of SRY generally results in formation of a testis. In some instances, SRY is deleted from or mutated on an otherwise normal Y chromosome, resulting in an XY female; other times, SRY is present due to translocation to a chromosome other than Y, leading to an XX male. Y chromosome abnormalities resulting in mosaicism for an XY cell line and a 45,X cell line can lead to phenotypes ranging from normal male, if testes form bilaterally, to Turner's syndrome if the 45,X cell line predominates in the gonads. In some cases, a normal testis may form on one side and a "streak" gonad on the other if one side has a Y chromosome and the other only 45,X cells. Müllerian inhibiting substance functions locally, so there will be a fallopian tube and uterus only on the side with 45,X cells.

The external genitalia depend on androgen stimulation for normal formation. If testes are not present, or do not function normally, the external genitalia will be female. Conversely, if there is exposure to androgens in a 46,XX fetus with ovaries, the external genitalia will appear male, although there will be no gonads in the scrotum. This occurs in females with congenital adrenal hyperplasia, in which deficiency of an enzyme required for adrenal hormone synthesis results in excessive androgen synthesis due to shunting of precursors proximal to the enzyme block into the androgen biosynthetic pathway. Testosterone binds to a cytoplasmic receptor that is encoded by a gene on the X chromosome. Deficiency of the receptor leads to androgen insensitivity, associated with normal testes and no müllerian structures but female external genitalia.

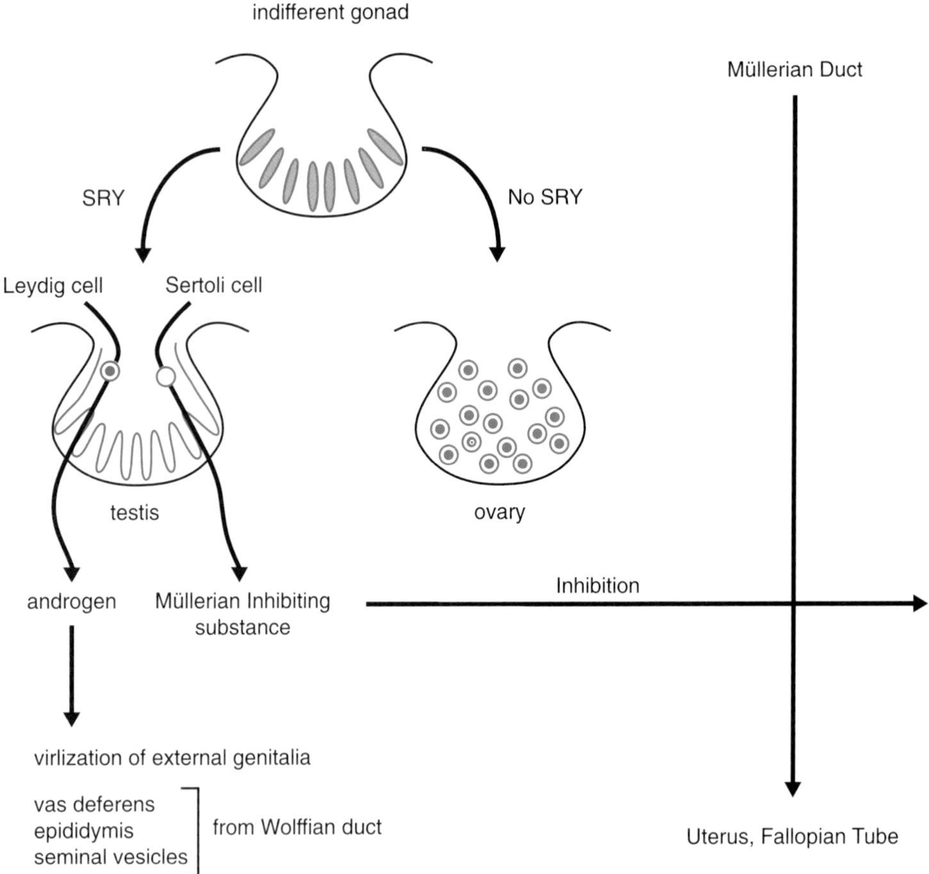

FIGURE 9-16 Pathway of sexual differentiation.

The gene responsible for camptomelic dysplasia was identified through a combination of positional cloning and candidate gene approaches, using knowledge of the association with gender reversal to provide a clue to the nature of the candidate gene. The probable location of the gene was determined by the finding of several individuals with camptomelic dysplasia who had chromosome rearrangements involving region 17q24.1–q25.1. In December 1994, two groups reported the isolation of a gene, referred to as *SOX9,* from the region of the translocations and found mutations in individuals with camptomelic dysplasia who did not have chromosomal abnormalities.

SOX9 is a gene that has a region of homology with SRY. The region is a stretch of 80 amino acids called the *high-mobility group domain*, which is believed to bind to DNA and stimulate transcription. Genes that include this domain are called *SOX genes* (for *S*RY b*ox*). SOX9 was mapped to a region of mouse chromosome 11 that is homologous to human 17q24, making it a candidate for campomelic dysplasia. The two groups independently cloned translocation breakpoints in different individuals with campomelic dysplasia and found that SOX9 mapped to the region. Inactivating mutations were found in SOX9 in others with campomelic dysplasia.

SOX9 is believed to participate in gonad formation, probably as part of the cascade of activated genes. Why some XY individuals with mutations experience gender reversal and others do not is not understood, but this may reflect functional differences in different parts of the protein. Those with gender reversal have dysplastic gonads rather than normal testes. These gonads do not produce müllerian inhibiting substance, so fallopian tubes and

uterus are present. There is deficient androgen production, however, so virilization of the external genitalia is incomplete. Dysplastic gonads can lead to malignant gonadoblastoma, and so they are surgically removed before adolescence.

The SOX9 gene is also active in developing bone, consistent with the phenotype of campomelic dysplasia. Affected individuals are heterozygous for SOX9 mutations, indicating that the disorder is dominant, not recessive. In one case, however, the mother of an affected child was found to be mosaic for a SOX9 mutation. This probably explains instances of recurrence in families. There is no evidence for autosomal recessive transmission of camptomelic dysplasia.

Part VI

Jane and Albert have had more education in genetics during the past few weeks than ever during their years in school. They have agonized over the medical decisions regarding Lisa's care, not wanting to put her through unnecessary suffering if her prognosis is indeed poor. Because no central nervous system malformations have been noted by magnetic resonance imaging of the brain, and the literature indicates a range of cognitive outcomes, they are inclined to continue active support. They are encouraged when they meet with another family with a 3-year-old child with campomelic dysplasia, who is on a ventilator at home and is doing fairly well. They decide to continue ventilatory support, and a tracheostomy is placed. After 1 month, Jane and Albert take Lisa home.

This case illustrates a number of issues commonly encountered in genetic counseling. First, the problems are complex and usually are well outside the knowledge of lay individuals. The parents have never heard of the syndrome and may have little or no understanding of the principles of mendelian inheritance, let alone molecular genetics. Counseling is complicated further by the emotional trauma of having a sick child, for whom life-and-death decisions need to be made. Also, the counseling itself may change as new information is gained through research.

It is often helpful to introduce a family to others who have firsthand experience with a genetic syndrome. Although specific manifestations and severity can differ from individual to individual, such contact provides at least a glimpse of what life may be like for a person with the disorder in question. The long-term prognosis for children with camptomelic dysplasia remains in question, largely due to limited experience.

REVIEW QUESTIONS

1. Individuals with the X-linked recessive disorder testicular feminization have a mutation in the androgen receptor that renders them insensitive to the hormone. These individuals have an XY sex chromosome constitution.
   a. Would you expect them to have testes?
   b. Would you expect them to have a uterus?
   c. Would you expect them to have male external genitalia?
2. Exposure to certain teratogens in the early weeks of pregnancy leads to miscarriage rather than congenital anomalies, whereas later exposure produces specific patterns of congenital anomalies. Why might this be so?
3. A child is born with holoprosencephaly (a brain defect in which the two hemispheres have failed to separate, leading to a single nonsegmented cortex with severe developmental impairment). The child has normal chromosomes, and there is no family history of a similar problem. The mother had suffered from a urinary tract infection in the eighth month of pregnancy for which she took an antibiotic. Can you reassure her about her concerns that this may have caused her child's problems?
4. An infant is found to have congenital heart disease and tracheo-esophageal fistula. Does the presence of these two distinct anomalies indicate that the problems have a genetic basis?

CASE STUDY

Foster JW, Dominguez-Steglich M, Guioli S, et al. Campomelic dysplasia and autosomal sex reversal caused by mutations in an SRY-related gene. Nature 1994;372:525–530.

Throughout this
considered mair
individual and of th
Duchenne's muscul
drome—are extreme
uals throughout the
Sachs disease, cystic
occur with relativel
prevalent in some pc
ers are very common
morbidity, at least i
often multifactorial
diabetes.

In this chapter,
from a different pers
Although genes act
families, the forces t
act at the level of pc
can have important
ics. First, we can un
appear to be singled

Houston CS, Opitz JN camptomelic synd 1983;15:3–28.

Kwok C, Weller PA, (the gene responsil autosomal sex rev 1995;57:1028–1(

10.1 Thalassemia

1975

Gino is a 12-year-old boy who lives in Cagliari, Sardinia. When he was 6 months old, Gino was having difficulty feeding and was irritable and listless. He also was noted to be pale, and his abdomen was protuberant. A blood test confirmed his physician's impression that Gino had thalassemia. Gino has been receiving blood transfusions every 4 weeks since he was 18 months old. This has improved his energy level and growth. His spleen was removed when he was 6, and he has required antibiotic treatment for infections ever since. Over the past several years, Gino's major problems have been diabetes mellitus and severe heart failure. He has required hospitalization several times in the past 2 years for management of these problems. His family realizes that probably Gino will not survive much longer. He has normal intelligence, although he is far behind in school as a result of his illness. He is well aware of the seriousness of his medical condition. Recently, Gino has been started on an experimental therapy that requires injection of a new drug. It is too soon to know whether this will help with his problems and prolong his life.

Thalassemia is a form of chronic anemia due to an inherited deficiency in the production of one of the chains of hemoglobin. Hemoglobin, the oxygen-carrying molecule of the red blood cell (RBC), consists of two pairs of globin chains, each of which binds one heme molecule (see Chapter 1). The major form of hemoglobin in the adult, hemoglobin A, is a tetramer of two alpha-globin-chains and two beta-chains. There are other betalike chains that can bind to the alpha-chains and are expressed at different times in life (Figure 10-1). Epsilon-chains are produced during early embryonic development and gamma-chains during fetal life. These bind with alpha-globin to produce embryonic and fetal hemoglobins, respectively. Beta-chain synthesis begins during fetal development, but beta-globin does not become predominant until after birth. An additional betalike chain, delta, is made in small quantities after birth and comprises a minor species of hemoglobin called A_2.

Individuals with beta-thalassemia have deficient production of beta-chains, whereas those with alpha-thalassemia have deficient alpha-chain production. The pathophysiology of beta-thalassemia (Figure 10-2) begins with chronic anemia. There may be no problems in the early months of life while fetal hemoglobin ($\alpha_2\gamma_2$) is made. As fetal hemoglobin production declines, however, the anemia begins. This leads to poor growth, lack of energy, and irritability.

The production of RBCs is regulated by the body's need for oxygen. In children with thalassemia, RBC production is stimulated because of chronic tissue hypoxia. The bone marrow expands to increase its output of RBCs, which leads to weakening of the long bones and enlargement of bones in the skull that normally do not contribute to hematopoiesis. Fractures and a characteristic facial deformity result. Hematopoiesis also occurs in the liver, leading to hepatomegaly. The extra production of RBCs, of course, is futile, as the newly made erythrocytes also are deficient in beta-globin. Alpha-chains unite with the few delta- or gamma-chains present, but most precipitate in the cell. These RBCs are removed from circulation by the spleen. The spleen enlarges owing to extramedullary hematopoiesis as well as trapping of defective RBCs. Massive enlargement of the spleen necessitates splenectomy, which in turn leads to susceptibility to bacterial infection.

Left untreated, beta-thalassemia is fatal due to the effects of anemia, bone destruction, and rupture of the spleen. The obvious treatment is to replace RBCs by transfusion. Transfusion of intact RBCs corrects the anemia; bone marrow spaces return to normal size, and extramedullary hematopoiesis ceases. Energy and growth resume normally, and quality of life can be vastly improved. Unfortunately, the treatment is not without adverse effects. There is a risk of infection, particularly with hepatitis virus. More significant, though, is the problem of chronic iron overload. Transfused RBCs have a limited life span and, when they degenerate, the iron in heme must be disposed of. The body's capacity to excrete iron is overwhelmed rapidly, and iron accumulates in tissues, where it is toxic, leading to cirrhosis of the liver, diabetes mellitus, and heart failure. The treatment of thalassemia by transfusion, then, is ultimately as bad as the disease.

In the mid-1970s, an approach was developed to deal with iron overload, based on the use of a chelating agent, deferoxamine. Deferoxamine binds iron and renders it soluble, so that it can be excreted

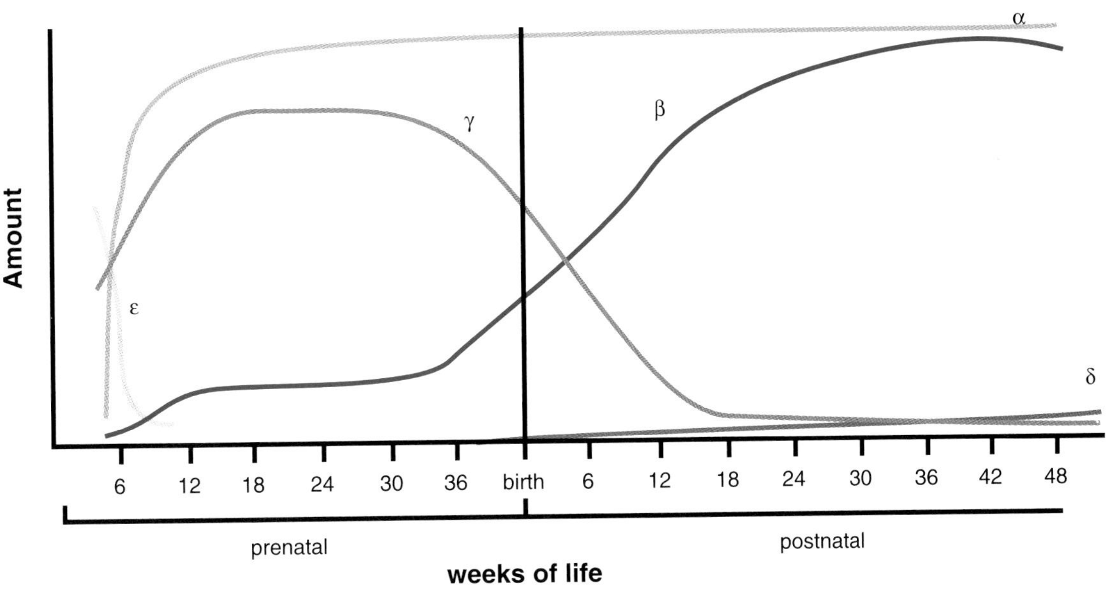

$\alpha_2\varepsilon_2$ or ε_4	embryonic hemoglobin
$\alpha_2\gamma_2$	fetal hemoglobin
$\alpha_2\beta_2$	hemoglobin A
$\alpha_2\delta_2$	hemoglobin A_2

FIGURE 10-1 Synthesis of globins in fetal and adult life. Alpha-globin synthesis begins in early fetal life and continues through adult life. The epsilon-chain is made for a short time in the early embryo and contributes to embryonic hemoglobin ($\alpha_2\varepsilon_2$). The gamma-chain is made for a period during fetal life, complexing with alpha-globin to make fetal hemoglobin ($\alpha_2\gamma_2$). Beta-globin synthesis begins in the fetus but achieves predominance after birth. This is the constituent of adult hemoglobin, hemoglobin A ($\alpha_2\beta_2$). A minor adult hemoglobin species, A_2, consists of $\alpha_2\delta_2$.

in the urine. The drug is not absorbed through the gastrointestinal tract and so must be administered by subcutaneous infusion. This is administered overnight using a pump connected to a needle placed under the skin.

With regular transfusions and deferoxamine, a person with thalassemia can live a productive life. Activity and energy level can be normal. The therapy, though, is very expensive—tens of thousands of dollars annually. This can strain the finances of a family and of the community. Moreover, thalassemia is particularly prevalent in some areas of the world, many of which can ill afford the burden of this disease on their health care systems.

10.2 Population Frequency of Thalassemia

1978

Gino died earlier this year of congestive heart failure. Although his family had known for a long time of the seriousness of his medical condition, they still were shaken by the loss. The experience has caused also concern among relatives about their chances of having children with thalassemia. In particular, Gino's cousin Rosa was recently married, and she and her husband Antonio are thinking about starting a family. Rosa asks her doctor about her risks of being a thalassemia carrier and about whether Antonio might be a carrier.

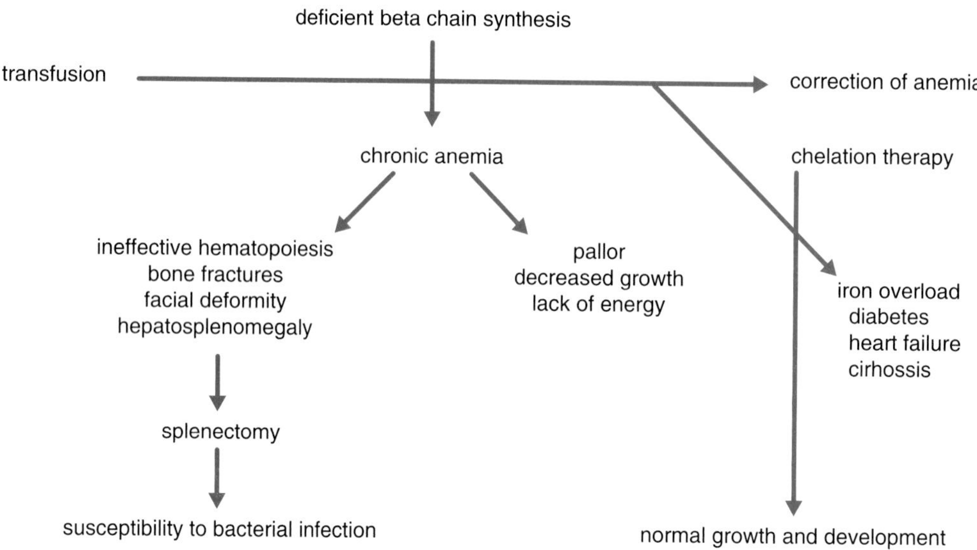

FIGURE 10-2 Pathophysiology of thalassemia.

Thalassemia is not uniformly distributed around the world. Prevalence is highest in the Mediterranean region, North-Central Africa, the Middle East, India, and Southeast Asia. In Sardinia, where the frequency of the disease was 1 in 213 persons in the mid-1970s, virtually everyone knew somebody—a friend or a relative—who was affected. Thalassemia and its complications consumed a large fraction of funds for health care on the island and were among the most common reasons for hospitalization.

Thalassemia is inherited as an autosomal recessive trait; hence, both parents must be carriers to have an affected child. In a population where the disorder is prevalent, such as Sardinia, there is a substantial chance for a person to be a carrier. How can the risk be measured? One way would be to screen the population for carrier status. Unlike many autosomal recessive disorders, carrier testing for thalassemia is possible by fairly simple means. Beta-thalassemia carriers have mild anemia, which is clinically silent but can be detected by blood testing. They also have elevated levels of hemoglobin A_2. Although such testing became the basis for population screening for clinical purposes, as will be described shortly, it was not necessary to go to this extent to estimate the carrier frequency. That could be done easily using the principles of population genetics.

Already in Chapter 3, we encountered the equation used to calculate allele frequencies, the Hardy-Weinberg equation, To review, consider a gene locus with two alleles, *A* and *a*, *A* being dominant to *a*. The frequency of allele *A* will be designated by the variable *p* and that of *a* by the variable *q*. If all alleles are either *A* or *a*, then $p + q = 1$. Hardy and Weinberg showed that in a very large population with random mating, the frequency of genotype *AA* is p^2, *Aa* is $2pq$, and *aa* is q^2. Chapter 3 presented a conceptual view of the derivation of the Hardy-Weinberg equation. A more rigorous derivation is provided on page 317.

The Hardy-Weinberg formulation also predicts that the gene frequencies will remain stable from generation to generation, provided that there is no mutation of *A* to *a* or of *a* to *A*, and no migration of individuals to or from the population, random mating, and no selection. This may seem an obvious result: under the assumptions of an "ideal" Hardy-Weinberg population, the alleles *A* and *a* have nowhere to go; they cannot leave or enter the population by migration, be lost by infertility, change by mutation, or dwindle by chance, so their stability is assured.

The Hardy-Weinberg equation is especially useful in the calculation of carrier frequencies for autosomal recessive traits. Direct determination of carrier frequency is difficult or impossible because carriers generally do not express the trait. Disease frequency, however, is relatively easy to determine as affected individuals usually come to medical attention. We have seen (in Chapter 3) how the Hardy-Weinberg equation is used to calculate the frequency of carriers for a cystic fibrosis allele. Use of the Hardy-Weinberg formula in assessing carrier frequency might seem to be suspect in populations that do not obey the assumptions of no mutation, selection, and so on. In most large populations, however, the fit is close enough for reasonable approximation. The biggest violation is usually that homozygous individuals are at significant disadvantage in mating or may be unable to mate entirely. For most rare disorders, though, this has a small impact on calculation of carrier frequency because the vast majority of heterozygous individuals arise from matings of heterozygotes, not from matings of *AA* individuals with the rare *aa*s.

On Sardinia in 1975, 1 of every 213 individuals was affected with thalassemia. This is a frequency of $0.0047 = q^2$. Therefore $q = 0.0685$ and $p = 1 - q = 0.9315$. The carrier frequency is $2pq = 2(0.0685)(0.9315) = 0.01276$, or 12.76%. Matings between thalassemia carriers on Sardinia indeed were not uncommon in the 1970s.

Derivation of the Hardy-Weinberg Equation

Start with a population in which the frequency of $AA = x$, $Aa = y$, $aa = z$ (that is, the frequencies of the three genotypes are found to have values of x, y, and z, where $x + y + z = 1$).

At this point, the frequency of $A = p = x + \frac{1}{2}y$ and the frequency of $a = q = z + \frac{1}{2}y$. We will show that, after one generation of random mating, the frequency of $AA = p^2$, $Aa = 2pq$, and $aa = q^2$.

First, let's consider all possible matings:

Mating Type	Frequency	Outcome
$AA \times AA$	x^2	All AA
$AA \times Aa$	$2xy$	½ AA, ½ Aa
$AA \times aa$	$2xz$	All Aa
$Aa \times Aa$	y^2	¼ AA, ½ Aa, ¼ aa
$Aa \times aa$	$2yz$	½ Aa, ½ aa
$aa \times aa$	z^2	All aa

A mating type such as $AA \times Aa$ can occur in two ways: The male can be *AA* and the female *Aa*, or *vice versa*; hence the frequency is $2xy$ (not xy).

Now let's tally the three genotypes in the next generation:

$$AA = x^2 + \tfrac{1}{2}(2xy) + \tfrac{1}{4}(y^2) = (x + \tfrac{1}{2}y)^2 = p^2$$

$$Aa = \tfrac{1}{2}(2xy) + 2xz + \tfrac{1}{2}(y^2) + \tfrac{1}{2}(2yz) = 2(x + \tfrac{1}{2}y)(z + \tfrac{1}{2}y) = 2pq$$

$$aa = \tfrac{1}{4}(y^2) + \tfrac{1}{2}(2yz) + z^2 = (z + \tfrac{1}{2}y)^2 = q^2$$

10.3 Population Screening for Thalassemia

1978

Rosa and Antonio are referred to a thalassemia screening program that has begun in Southern Sardinia (Figure 10-3). The two meet with a counselor, fill in a questionnaire about their health and family history, and are told about thalassemia and the screening test. Because there is a history of thalassemia in Rosa's family, she is tested first and is found to be a carrier. Antonio then is tested. The results for both Rosa and Antonio are shown in Figure 10-4.

In 1977, a pilot program was launched to screen for thalassemia carriers in Southern Sardinia. Carriers can be easily identified by an accurate and inexpensive blood test, showing reduced RBC volume and relatively increased hemoglobin A_2 compared with A (because of a reduction in $\alpha_2\beta_2$ production with normal $\alpha_2\delta_2$). A public education program was initiated using town meetings, posters, pamphlets, and advertising via radio, television, and newspapers. Couples planning a pregnancy were targeted, and one member of each couple was tested first, with the partner tested only if the first partner tested positively. If both partners were found to be carriers, prenatal diagnosis was offered.

1979

Both Rosa and Antonio are found to be thalassemia carriers. They are counseled that they have a 25% risk of having a child with beta-thalassemia. Rosa is now 18 weeks pregnant. A sample of fetal blood is

FIGURE 10-3 Poster publicizing thalassemia screening program in Sardinia. (Courtesy of Dr. Antonio Cao.)

drawn under ultrasound guidance, and the fetal blood cells are incubated with [^{3}H]-leucine. The newly synthesized globin is analyzed for the alpha-beta ratio. The ratio is found to be abnormal, indicating that the fetus is affected with thalassemia. After agonizing over their decision, Rosa and Antonio decide to terminate the pregnancy.

In the 1970s, the only way to diagnose beta-thalassemia prenatally was to draw fetal blood from the placenta and to measure the production of beta-globin. Diagnosis was complicated by the fact that very little beta-chain is made early in pregnancy, at which time the majority of globin is fetal hemoglobin ($\alpha_2\gamma_2$). Beta-globin production was determined by incubation of fetal blood in the presence of tritiated leucine. Beta- and gamma-globin were separated by electrophoresis and quantified by measuring incorporation of radioactivity.

During the first 3 years of screening, 4057 individuals were tested. The testing was voluntary and included approximately two-thirds of the population of childbearing age in the region. Six hundred twenty-two thalassemia carriers were detected (for a carrier frequency of 13.6%, which included a few carriers for globin disorders other than beta-thalassemia). An additional 2402 carriers were detected on the basis of a known family history of beta-thalassemia. A total of 694 couples were found to be at risk of having an affected child, and 177 pregnancies were monitored by prenatal diagnosis. The rate of acceptance of prenatal testing changed dramatically during the 3 years of pilot testing, increasing from approximately 73% in 1977 to nearly 93% in 1980. Among the 177 tested pregnancies, 5 resulted in miscarriage due to fetal hemorrhage or premature labor. Forty-two fetuses were found to be affected with thalassemia, and 39 were electively terminated. The frequency of beta-thalassemia in this region of Sardinia declined from 1 in 213 persons in 1976 to 1 in 290 in 1978.

At around the same time, population screening for thalassemia also was initiated on the island of Cyprus, with similar success. It is interesting, though, to compare these programs with two other population screens for recessive disorders begun in the United States in the 1970s—one for sickle cell anemia and one for Tay-Sachs disease.

Sickle cell carrier screening began shakily in the United States in the early 1970s. The test used at first did not distinguish sickle cell carriers from those with the disease and, unfortunately, the same was true for much of the educational material disseminated at the

	Mean Cell Volume (fl)	% HbA_2	Electrophoresis
Rosa	65	6.7	A-F-A_2
Antonio	68	7.1	A-F-A_2
Normal Range	80-110	1.5 — 3.5	A-A_2

FIGURE 10-4 Results of thalassemia screening in Rosa and Antonio. (*A* = hemoglobin A; *F* = fetal hemoglobin; A_2 = hemoglobin A_2.)

time. Therefore, those who tested positively were, in many cases, subjected to discrimination (e.g., denial of employment or health or life insurance, or raising of insurance premiums). Furthermore, when the screening began, prenatal diagnosis of sickle cell disease was not possible. This vastly limited the options of those found to be at risk of having an affected child. The screening program was poorly organized, overly ambitious, and racially divisive.

Several changes have occurred over the years that have led to a reappraisal of sickle cell screening. First, it has been found that early identification of sickle cell disease is important as affected children are at high risk of developing life-threatening infections. Newborn screening now is offered widely and is performed on all newborns, not just African-Americans. Although the carrier frequency for sickle cell anemia is highest among blacks (approximately 10%), as many as 7% of those affected with the disorder are not black. Another development is the advent of methods of prenatal diagnosis based on DNA technology. This provides the option of prenatal diagnosis to couples found to be at risk. A recent pilot study of carrier screening for hemoglobinopathies in New York has demonstrated the wide acceptance of a program in which the testing is readily available and adequate and effective counseling and education are provided.

Screening for carriers of Tay-Sachs disease began in many countries around the world in the early 1970s. Nearly 1 in 30 Jewish individuals of Eastern European (Ashkenazi) background is a carrier for the infantile form of Tay-Sachs disease, which is a tragically progressive lethal disorder. Screening has targeted this population with an aggressive program of public education, rigorous quality control of laboratory testing, and ready availability of genetic counseling. Prenatal diagnosis of Tay-Sachs disease is possible with a reliable biochemical test or, more recently, with a DNA-based test. For ultraorthodox Jewish groups in which abortion is not acceptable, confidential screening has been used to avoid marriages between carriers. From 1971 through 1992, almost 1 million individuals had been tested and more than 36,000 carriers detected. Prenatal testing was done for more than 2400 pregnancies at risk. Prior to screening, approximately 60 new Tay-Sachs disease cases were identified around the world each year. Currently, the rate has fallen to three to five cases per year, comparable to the frequency in non-Jewish couples.

These experiences, along with the experience in newborn screening for metabolic disorders discussed in Chapter 1, highlight the factors that determine the success of a population screening program. First, the screening must be based on a test that is reliable, accurate, cost-effective, and readily available. Second, the screening must be organized in a manner that is appropriate to the customs of the target population. Third, there must be options available to those found to be at risk that are acceptable in the population. Finally, the goal of screening must be the improvement of the health of individuals, not the improvement of the gene pool of the population. These lessons will likely be increasingly important as the genetic basis for common diseases comes to light and mass screening for other genetic traits is contemplated.

10.4 Globin Gene Mutations

1983

Rosa's second pregnancy was in 1980 and was found to be unaffected, resulting in the birth of a healthy boy, Paolo. A third pregnancy in 1982 also resulted in a healthy child, a girl named Maria. Now Rosa is pregnant again. This time, the pregnancy is monitored by amniocentesis, and a DNA test is used (Figure 10-5).

By 1987, the frequency of beta-thalassemia in Sardinia had fallen to approximately 1:1100, despite the cumbersome method of prenatal diagnosis used in the early years. Because beta-globin is produced only by RBCs, prenatal diagnosis by amniocentesis was not possible. It remained to develop a DNA-based test to simplify the approach to prenatal diagnosis.

Beta-globin was one of the first human genes to be cloned. Nearly 95% of mRNA in reticulocytes corresponds with globin message, so physical isolation of globin mRNA was relatively easy. This material was used to make cDNA, which was radioactively labeled to identify cDNA clones from a reticulocyte cDNA library, as well as clones from a genomic library. The alpha- and beta-globin genes have similar structure, indicating origin from a common ancestral sequence. The genes each consist of three exons and two introns. Two alpha-genes are

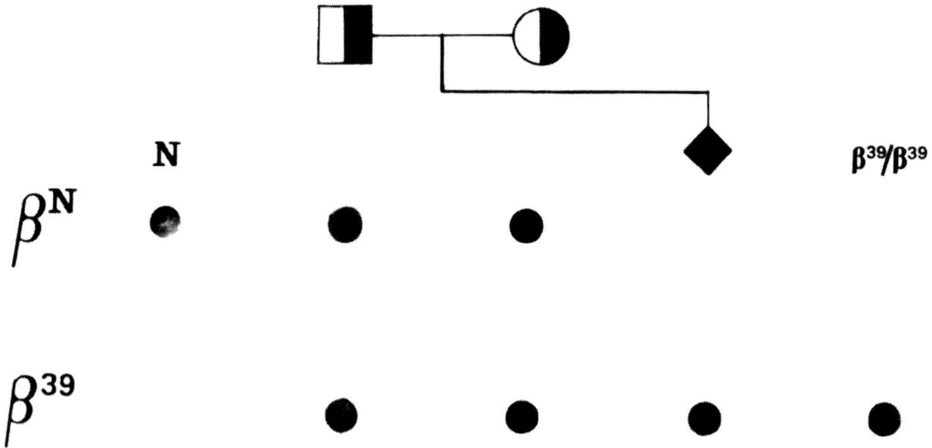

FIGURE 10-5 Dot-blot test for beta^{39} mutation. The beta^{39} region is amplified by polymerase chain reaction, and either normal (top) or mutant (bottom) oligonucleotide is hybridized with the sequence. An individual homozygous for the wild-type allele hybridizes only with the wild-type oligonucleotide (*N* at left). Both heterozygous parents hybridize with both sequences. The fetus, and a beta^{39} homozygote, hybridizes only with the mutant oligonucleotide. (Courtesy of Dr. B Handelin.)

tandemly duplicated on chromosome 16 (Figure 10-6), located alongside an alphalike gene, zeta-globin, and three pseudogenes, ψα and ψζ. These pseudogenes are homologous to alpha-globin but lack some feature essential for expression and therefore are not transcribed. The betalike genes are clustered on chromosome 11, organized from 5′ to 3′ in the order of their expression during development.

Characterization of globin gene mutations has revealed the diversity of ways in which a genetic change can alter the function of a gene product and also provides a basis for molecular diagnosis. The most prevalent beta-globin mutation in Sardinia was identified in 1981. Southern blot analysis with a beta-globin cDNA did not reveal gross alterations of the gene in DNA samples from Sardinians with

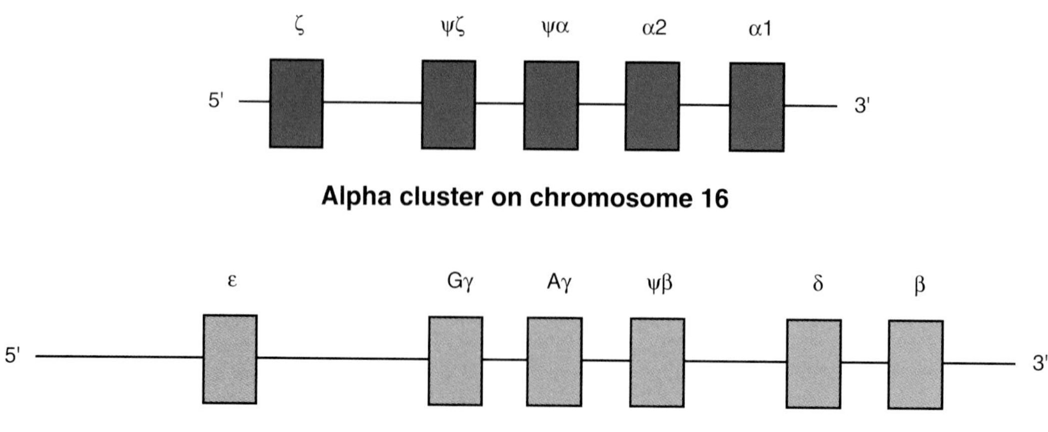

FIGURE 10-6 Map of alpha- and beta-globin clusters on chromosomes 16 and 11, respectively. There are two alpha-genes, designated *α1* and *α2*. On chromosome 11, there are two gamma genes that differ at a single amino acid (glycine or alanine), designated *Gγ* and *Aγ*.

beta-thalassemia. Reticulocyte mRNA was isolated from affected individuals and translated into protein in a cell-free translation system (Figure 10-7). This system contains ribosomes, tRNAs, and enzymes necessary to synthesize protein from mRNA. Although RNA from unaffected individuals resulted in beta-globin production, no product resulted with RNA from a person from Sardinia with thalassemia. This indicated that the mutation interfered with protein synthesis. Hypothesizing that a stop mutation (i.e., a mutation that changes an amino acid codon to a stop codon) might be responsible, a special tRNA called a *suppressor tRNA* was added to the translation system. This tRNA inserts a serine amino acid when a UAG stop codon is encountered, and it allowed beta-globin production to occur with mRNA from a Sardinian with beta-thalassemia. A mutant beta-globin cDNA then was cloned and sequenced, and it was found that codon 39 was changed from CAG, encoding glutamine, to UAG, a stop codon.

The codon 39 mutation results in a form of thalassemia referred to as *beta^0*, because no beta-globin is made. A person who is homozygous for this mutation produces a 39 amino acid beta-globin peptide (full-length beta-globin is 146 amino acids) that is degraded in the cell. More than 95% of Sardinians with beta-thalassemia have this mutation. Another rare mutation was found in approximately 2% of the carriers, this one a frameshift at codon 6 that also results in a truncated protein. A low proportion of Sardinians with thalassemia have a form referred to as *beta$^+$*, in which a small amount of beta-globin is produced. Two mutations were found to be associated with this phenotype in different individuals. Both interfere with mRNA splicing, one at nucleotide 110 in intron 1 and one at nucleotide 745 in intron 2.

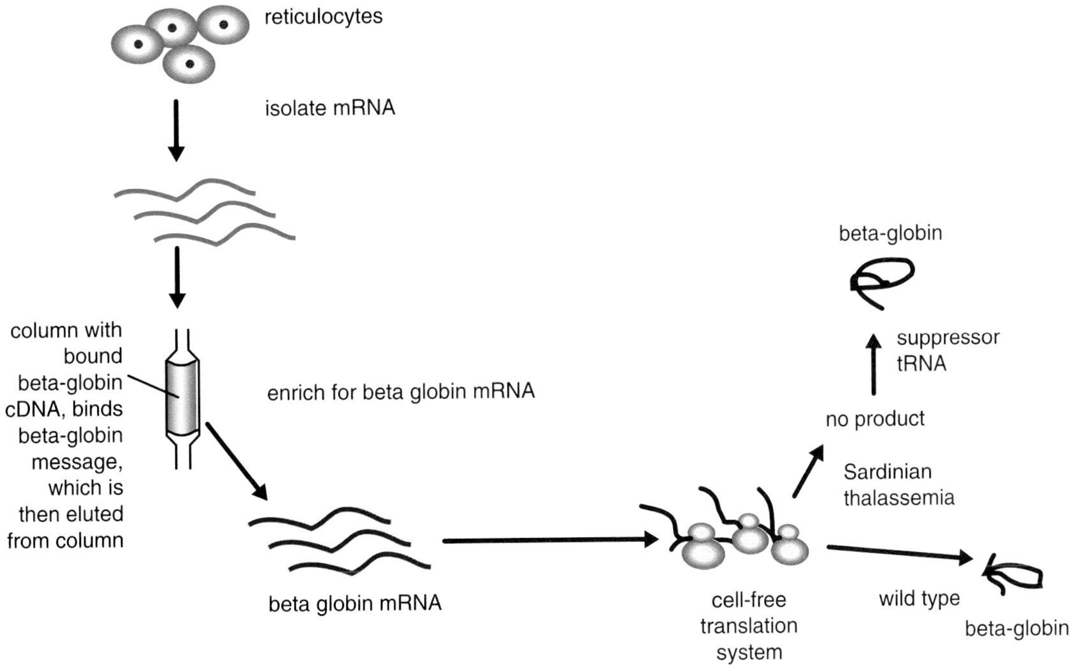

FIGURE 10-7 Cell-free translation assay for beta-globin mutation. Reticulocyte mRNA was isolated from individuals with thalassemia or normal controls and passed through a column to which beta-globin cDNA was bound. Reticulocyte beta-globin RNA binds to the column and then is eluted. The RNA is bound in vitro with ribosomes and the proteins necessary for translation into protein. RNA from unaffected individuals results in production of beta-globin, whereas RNA from Sardinians with thalassemia does not. If, however, a mutant tRNA is used that inserts a serine where a UAG stop codon is encountered (called *suppressor tRNA*, because it suppresses a stop mutation), functional protein is made. This indicates that the likely mutation is the occurrence of a stop codon in the coding sequence. (Data from Trecartin RF, et al. Beta zero thalassemia in Sardinia is caused by a nonsense mutation. J Clin Invest 1981;68:1012–1016.)

These four mutations account for more than 98% of thalassemia mutations in the Sardinian population.

Both Rosa and Antonio are found to carry the beta[39] mutation. This same assay is used to test amniotic fluid cells from Rosa's next pregnancy, and the fetus is found to be a carrier. The pregnancy is continued, and a healthy baby girl, Gabriella, is delivered.

Study of globin mutations around the world has revealed a similar story. In each population, a relatively small number of mutations accounts for the majority of affected individuals, yet these mutations differ widely from one region to another. The entire process of gene expression, from initiation of transcription at the promoter, to RNA splicing, to polyA addition, are vulnerable to mutation (Figure 10-8). Missense and nonsense mutations and gene deletions have been found.

Mutations responsible for alpha-thalassemia are likewise diverse. Gene deletions, though, are relatively common in individuals with alpha-thalassemia (Figure 10-9). Both alpha-globin genes on chromosome 16 are expressed, and either or both of these genes may be deleted. Deletion of both copies produces a "– –" allele, found almost exclusively in Southeast Asia. Deletion of a single alpha-gene, the –α allele, is found in Asia, Africa, and the Mediterranean region. Homozygotes for the – – allele, – –/– – individuals, make no alpha-globin, which is a lethal condition known as *hydrops fetalis*. Compound heterozygotes with the genotype – –/–α have a severe form of alpha-thalassemia known as *hemoglobin H disease* (hemoglobin H is a tetramer of beta-chains that precipitates in RBCs). The genotypes – –/αα or –α/–α produce the phenotype of alpha-thalassemia trait, which is a mild anemia. Those with –α/αα are silent carriers.

In both alpha- and beta-thalassemias, the diversity of mutations tends to be limited within a population, yet mutations vary widely between populations. The forces that maintain this distribution of mutations illustrate two important principles of population genetics, which we will now explore.

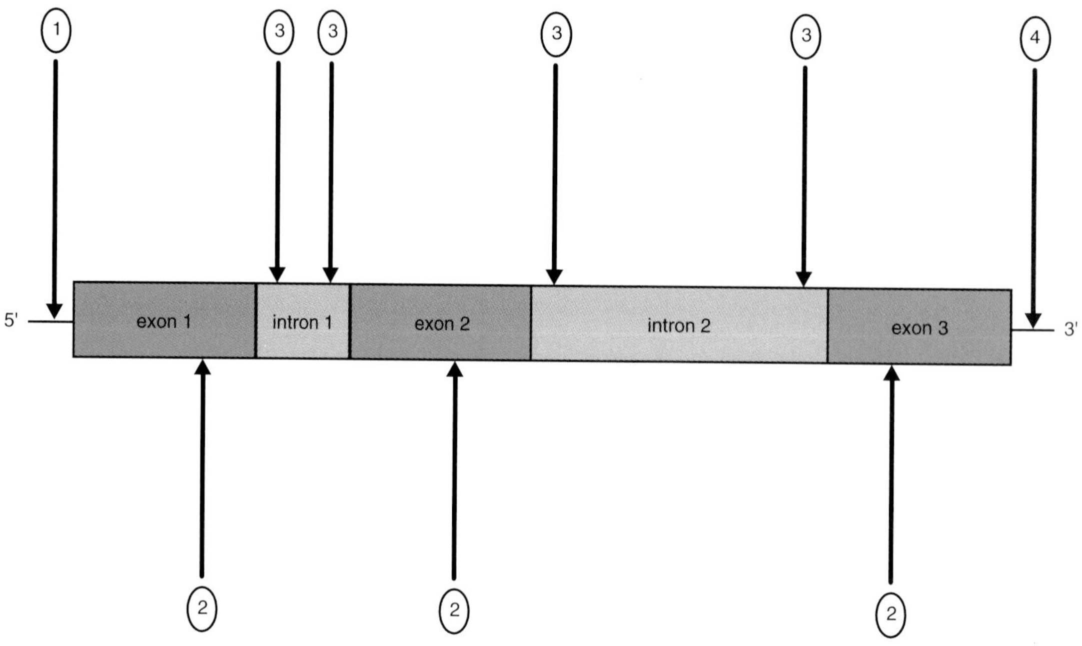

FIGURE 10-8 Sites of mutation in the beta-globin gene. (*1*) Promoter mutations lead to decreased level of transcription. (*2*) Point mutations within exons cause amino acid substitutions or cessation of translation (stop mutations). (*3*) Intron mutations affect the fidelity of splicing. (*4*) PolyA addition site mutations leads to decreased mRNA stability.

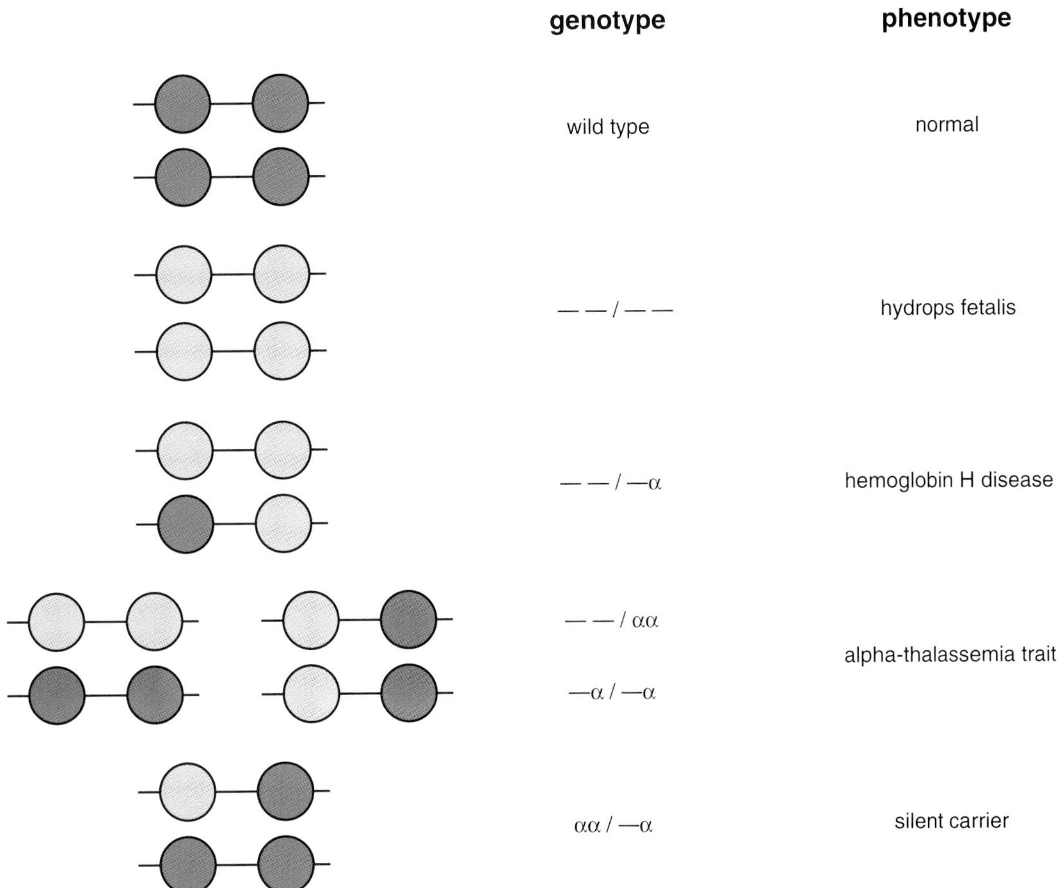

FIGURE 10-9 Alpha-globin gene deletions. There are normally four alpha-globin genes, two on each chromosome. Deletion of all four results in a lethal condition, hydrops fetalis. Deletion of three genes leads to a severe anemia, hemoglobin H disease. Deletion of two alpha-globin genes results in carrier status, alpha-thalassemia trait. Deletion of a single alpha-globin is clinically silent.

10.5 Genetic Polymorphism

In 1949, the geneticist J.B.S. Haldane pointed out that the distribution of thalassemia parallels that of malaria (Figure 10-10), and proposed that malaria might constitute a selective force that could maintain thalassemia alleles in the population. Malaria is a parasitic infection of RBCs and at least one form, falciparum, often is fatal without treatment. How can a parasitic infection influence the worldwide distribution of a genetic trait?

To answer this, let us first consider the simple situation of a pair of alleles, *A* and *a*, with *A* dominant to *a*. Suppose that the frequency of both *A* and *a* is 0.5. If the population obeys the assumptions of the Hardy-Weinberg equation, we would expect these frequencies to remain stable over time. Approximately 25% of individuals would be *AA*, 50% *Aa*, and 25% *aa*.

Now, let's change things. Suppose that *aa* individuals suddenly are rendered unable to reproduce. It doesn't matter how this happens: The *aa*s might die before reproductive age or be healthy but sterile. What matters is that they do not contribute to the gene pool of the next generation. Geneticists refer to this as a lethal trait, but it may be only the germ cells that die.

One might expect that this change would lead to a gradual loss of *a* alleles and a corresponding increase in the proportion of *A*. In the first generation after imposition of selection, 25% of individuals

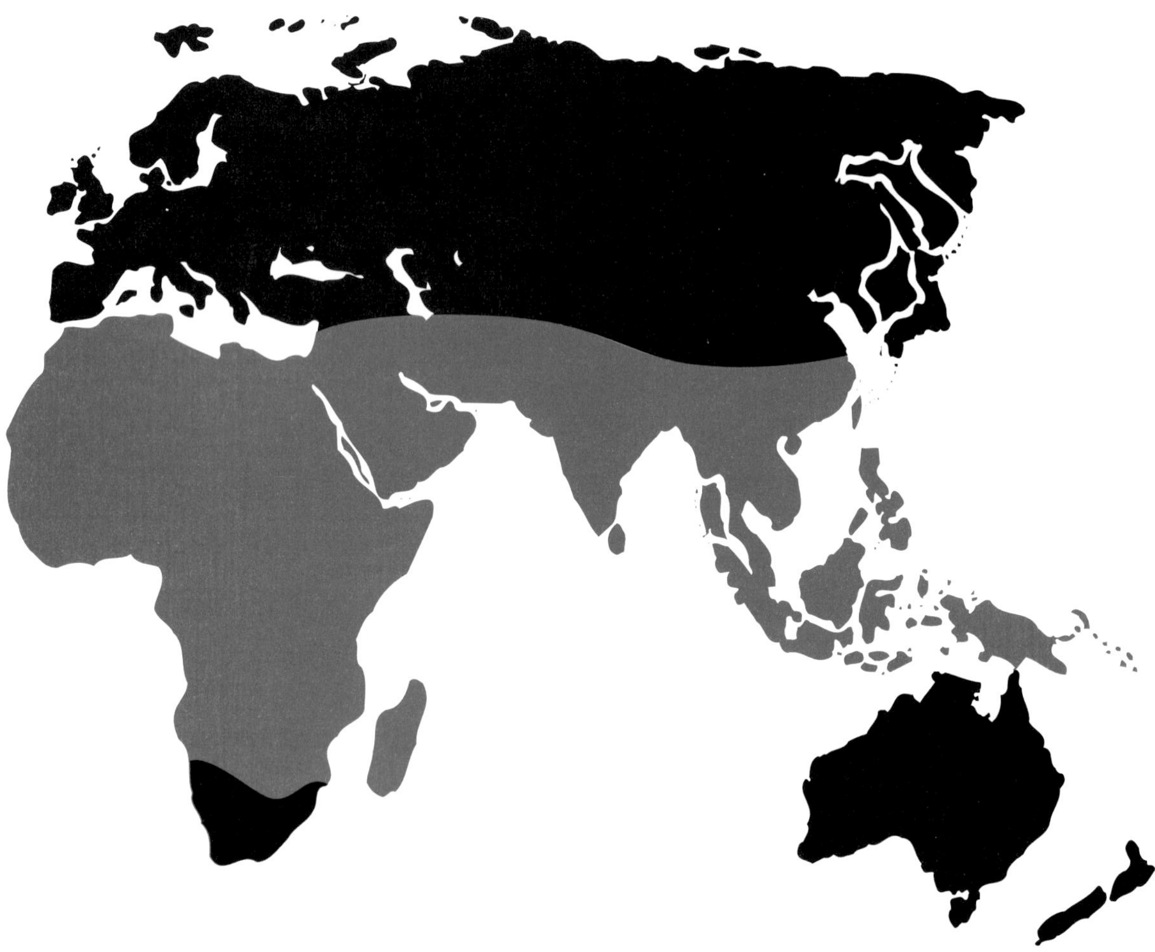

FIGURE 10-10 Worldwide distribution of thalassemia, paralleling the distribution of malaria (light shading).

are *aa* and are lost to the gene pool. Among the survivors, the 25% *AA* and 50% *Aa*, two-thirds of the alleles are *A* and one-third are *a*, so the frequency of *a* diminishes from 0.5 to 0.33. In the next generation, some *aa* homozygotes are produced through matings between heterozygotes, and the frequency of *a* will decrease again due to the inability of *aa* individuals to produce offspring. As the frequency of *a* diminishes, however, the proportion of heterozygotes also diminishes, so, with each generation there are fewer and fewer *aa* offspring to be subjected to selection (Figure 10-11). Therefore, the rate of decrease of the frequency of *a* slows. A graph of the frequency of *a* as a function of time shows an exponential decline that approaches zero asymptotically (Figure 10-12). In an infinitely large population, *a* will approach, but will never reach, zero. In a real population, the day may come when there are very few *Aa* individuals who, by chance, happen not to produce any offspring with the *a* allele. At that point, the frequency of *a* becomes 0, and *a* is said to be **extinguished** and *A* **fixed**.

This extreme scenario can be softened if *aa* homozygotes are able to produce some offspring but do so less efficiently than *AA* or *Aa* individuals. The *aa*s are said to be *selected against*, or to have reduced **reproductive fitness**. The rate of decrease of the frequency of *a* will be slower, but the frequency will decrease nevertheless.

It is easy to find examples of genetic traits that are subject to selection: In fact, most genetic disorders we have considered in this book are selected against to some degree, by reduction in fertility, poor health, or death before reproductive years. Something, then,

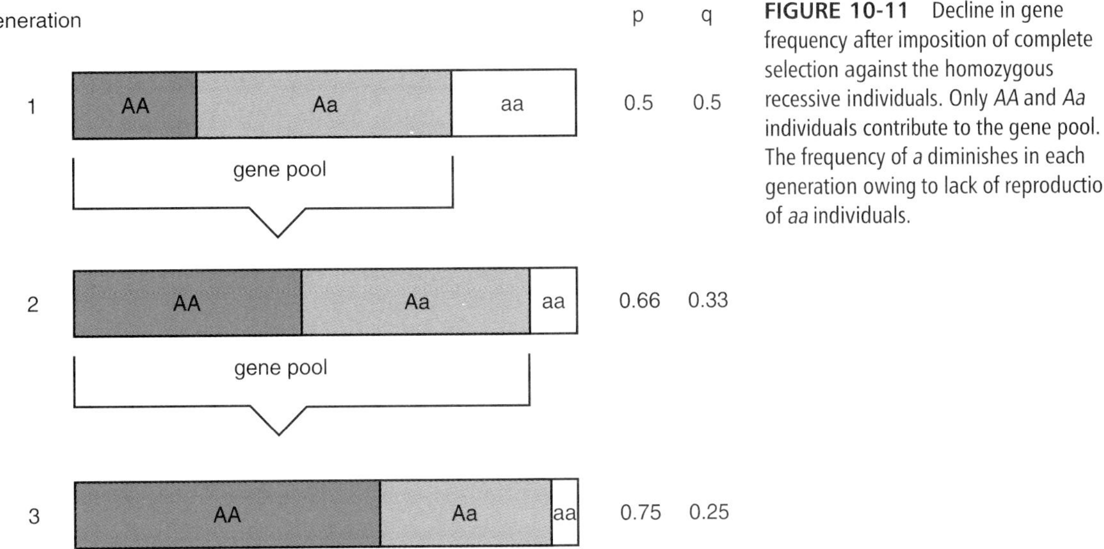

FIGURE 10-11 Decline in gene frequency after imposition of complete selection against the homozygous recessive individuals. Only *AA* and *Aa* individuals contribute to the gene pool. The frequency of *a* diminishes in each generation owing to lack of reproduction of *aa* individuals.

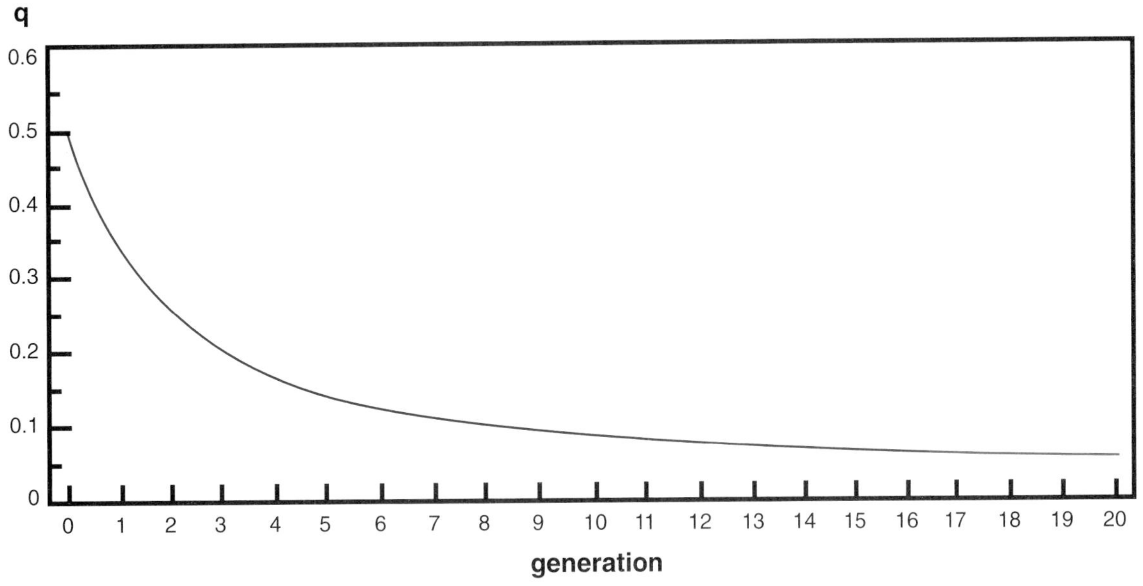

FIGURE 10-12 Graph of q as a function of generations after the imposition of complete selection, assuming that, at the start, $p = q = 0.5$. Note that the frequency decreases quickly at first, as there are many *aa* individuals. As the frequency of *aa* individuals declines, however, the rate of decrease in q also declines.

must maintain these traits in the population despite their selective disadvantage.

Consider Duchenne's muscular dystrophy, an X-linked recessive trait that is lethal in males. In a given generation, the proportion of affected males is q and of heterozygous females is $2pq$, which is nearly equal to $2q$ because the disease gene is quite rare and thus p equals approximately 1 (see page 329). The proportion of mutant alleles in each generation is therefore $q/(q + 2q) = \frac{1}{3}$. The frequency of the mutant allele would dwindle gradually to near 0 unless something replenishes it, and this indeed happens in the form of new mutation. In each generation, there is a probability designated by the variable μ that a given copy of the dystrophin gene will mutate on any X chromosome. A point will be reached where this rate of new mutation exactly balances the loss of alleles in each generation, at which point the gene frequency will be stable. This can be expressed by the equation $\mu = q/3$. At equilibrium, then, the value of q will equal 3μ. We can measure q, the disease frequency, which is approximately 1 in 3000 (0.000333). This allows us to estimate that the mutation rate for this locus, or μ, is 1.1×10^{-4} per gamete per generation. The wide diversity of dystrophin mutations found in different affected individuals accords with this notion that a substantial proportion (one-third) of cases arise owing to fresh mutation.

Similar arguments apply for other disorders in which new mutation counterbalances loss of alleles due to selection. In neurofibromatosis, which occurs with a frequency of 1 in 4000 persons, for example, approximately half of the cases in the world arise by new mutation. This predicts a gene frequency of 1 in 8000 and a mutation rate of 1 in 16,000. Table 10-1 lists mutation rates for a number of human genetic disorders.

Genetic disorders that have a high frequency of new mutation tend to occur throughout the world and do not display a racial, ethnic, or regional predilection. This is true for Duchenne's muscular dystrophy and neurofibromatosis and reflects the fact that new mutations occur at random. Clearly, this is different from the situation of the globin disorders, with their relatively narrow geographic distribution.

TABLE 10-1 Estimated mutation rates for a number of human genetic disorders

Disorder	Mutation Rate
Achondroplasia	1×10^{-5}
Retinoblastoma	6×10^{-6}
Neurofibromatosis	1×10^{-4}
Hemophilia A	5×10^{-5}
Marfan's syndrome	5×10^{-6}

Note: Rates are expressed as mutations per gamete per generation.
Reproduced by permission from Vogel F, Rathenberg R. Spontaneous mutation in man. Adv Human Genet 1975;5:223–318.

Also, disorders with a high rate of new mutation tend to be genetically heterogeneous: A wide range of different mutations account for the disorder in different people. Again, this is true for Duchenne's muscular dystrophy and neurofibromatosis, but it is not true for the globin disorders. Although different globin mutations are characteristic of different populations, within a population the range of mutations can be quite restricted, as is the case for the beta[39] mutation in Sardinia.

The worldwide distribution of globin mutations reflects the action of two forces that mold gene frequencies. The first is referred to as **balanced polymorphism**, the second the **founder effect**. If we look back in history, most, if not all, globin mutations would be expected to have been lethal. In the days before antibiotics, transfusions, and deferoxamine, few persons with sickle cell disease or thalassemia were likely to be capable of reproduction. The rare globin mutant allele should have been gradually extinguished due to selection. At the same time, however, falciparum malaria was also often lethal. It turns out, though, that carriers of a globin mutation are relatively resistant to malaria. Homozygous normal individuals were highly susceptible to malaria, and the homozygotes for globin mutations died of anemia. This left the heterozygotes as the fittest individuals in the population. Each generation, some of their offspring would die of malaria and some would die of anemia. During an outbreak of malaria in some region, large numbers might die, leaving only a few founding individuals who might be carriers for some specific globin mutation, making that mutation prevalent in their offspring. In this new population, eventually an equilibrium would be reached, such that the loss of wild-type alleles due to malaria would balance the loss of globin mutant alleles due to anemia (Figure 10-13). The globin mutation would be retained in the population for as long as the counterbalancing selective pressures are maintained.

In Chapter 1, *polymorphism* was defined as the occurrence of two or more alleles at a locus that each have a frequency of at least 1%. The case of the globin genes represents a balanced polymorphism, in which the loss of a mutant allele due to selection is balanced by the loss of the wild-type allele due to selection of a different kind. This phenomenon of balanced polymorphism—also known as **heterozygote advantage**—is well-known in plants and other animals.

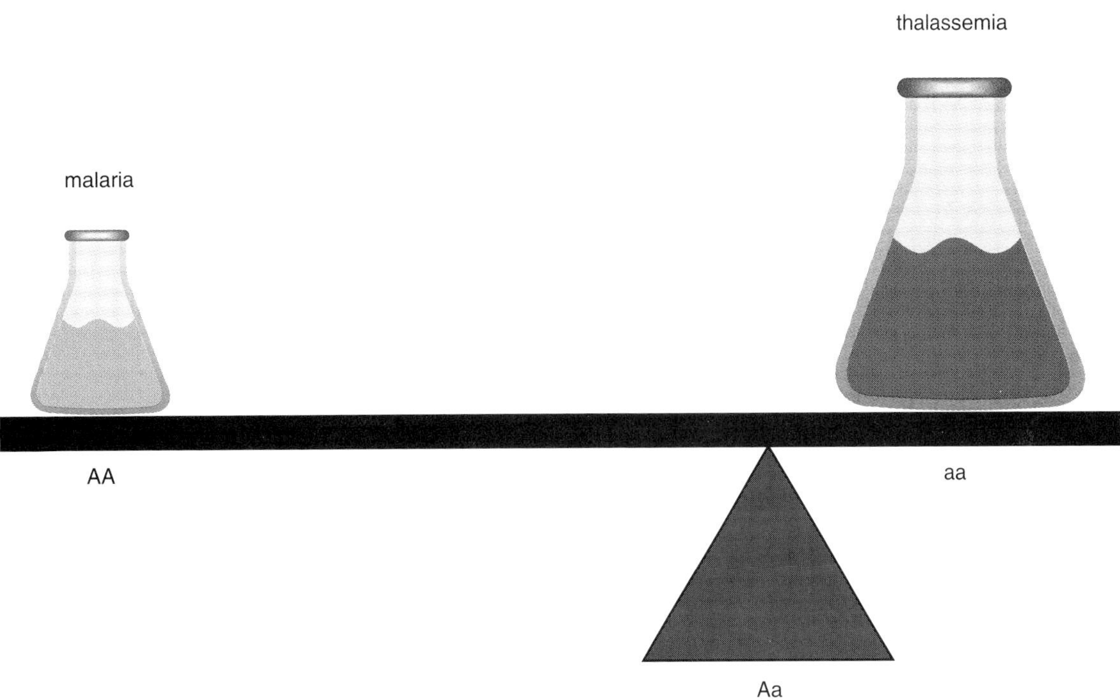

FIGURE 10-13 Concept of balanced polymorphism. Both the homozygous wild-type and homozygous mutant individuals are subjected to selection by malaria and anemia, respectively. The *Aa* individuals have the greatest reproductive fitness and serve as a reservoir for the *a* allele.

The globin-malaria system is the best documented case in humans, although other examples undoubtedly exist.

Heterozygote advantage may explain why globin mutations are prevalent in the malaria belt but not why different types of mutations are found in different regions. This is explained by the founder effect. The Hardy-Weinberg equation assumes that random breeding occurs in a very large population. Even if the frequency of an allele is low, some matings will occur that involve heterozygotes or homozygotes for the allele to maintain it in the population. If the population is very small, however, such matings might not occur, and the frequency of the allele would fall even if it were not subjected to selection (Figure 10-14). This is referred to as **genetic drift**.

The populations of Sardinia, Central Africa, or Southeast Asia are not small, of course, but, within a region, a population may well have been subjected to a "bottleneck," perhaps due to an outbreak of malaria. If a large part of the breeding population succumbed to malaria and, among the few survivors, there was one individual with a specific globin mutation, that mutation would become more prevalent in the new population that formed in the region after the bottleneck (Figure 10-15). This illustrates the concept of founder effect.

A striking example of a founder effect has been demonstrated among French Canadians with tyrosinemia type I, in which a single mutation in the enzyme fumarylacetoacetate hydrolase accounts for most cases. This disorder is an inborn error of tyrosine metabolism that leads to liver damage caused by buildup of toxic metabolites. It is rare around the world, but in the Saguenay–Lac St. John region of Quebec it affects 1 of every 1846 newborns, predicting a carrier frequency of 0.045, or nearly 1 in 22. DNA testing has revealed that one mutation—a splicing mutation in intron 12—occurs in 90% of carriers in this region and therefore occurs in homozygous form in approximately 81% of affected individuals. In contrast, this mutation was found in only 28% of tyrosinemia carriers elsewhere in the world. It is presumed that the mutation was introduced into the Saguenay–Lac St. John region by a founder individual sometime in the past several

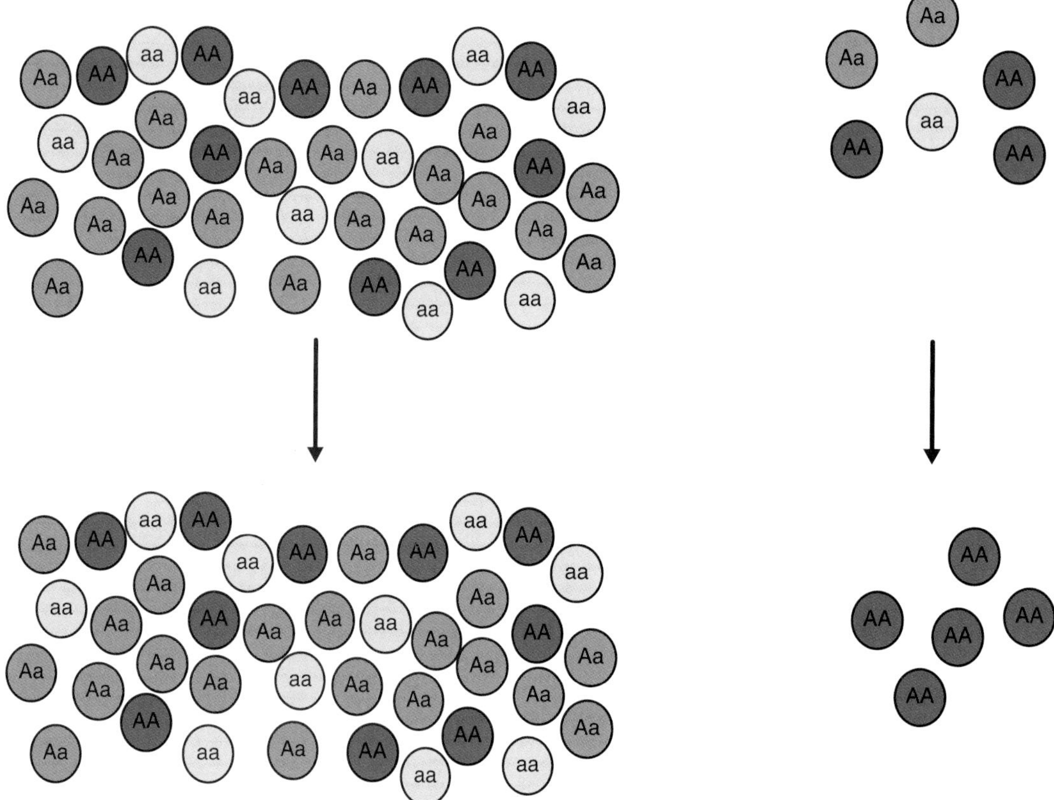

FIGURE 10-14 Concept of genetic drift. In a large population with random mating, large fluctuations of gene frequency are unlikely. In a small population, however, gene frequency can change dramatically from one generation to the next, if, for example, only the *AA* individuals participate in mating by chance.

hundred years, and the relative isolation of the region has led to a high frequency of the allele due to the founder effect.

The founder effect has also been invoked to explain other examples of high prevalence of genetic disorders in specific populations, such as cystic fibrosis in Northern Europeans or Tay-Sachs disease in Eastern European Jews, but here it cannot be the whole story. Both cystic fibrosis and Tay-Sachs disease display genetic heterogeneity, even within the populations of highest disease prevalence. The ΔF508 mutation accounts for only 70% of cystic fibrosis mutations in Europe, and at least three mutations are found among Ashkenazi carriers of Tay-Sachs disease. As for the globin mutations, it has been suggested that heterozygote advantage may, in part, explain the unusual geographic distribution of these mutations.

The selective force responsible for the maintenance of the cystic fibrosis mutations may have taken the form of another infectious disease, cholera. Cholera, an infection of the intestine by the bacteria *Vibrio cholerae*, causes death by inducing massive secretory diarrhea, leading to fluid and electrolyte depletion. Cholera toxin increases intracellular levels of adenosine 3′:5′-cyclic phosphate (cAMP), which in turn leads to secretion of chloride through the cAMP-regulated chloride channel, which is the cystic fibrosis transmembrane conductance regulator (CFTR). Could mutation of the CFTR gene reduce this chloride and fluid loss and hence protect against the effects of cholera toxin? The hypothesis has been tested using a mouse model for cystic fibrosis created by knockout of the murine CFTR gene. Homozygous normal mice, heterozygotes, and homozygous affected mice were tested for intestinal

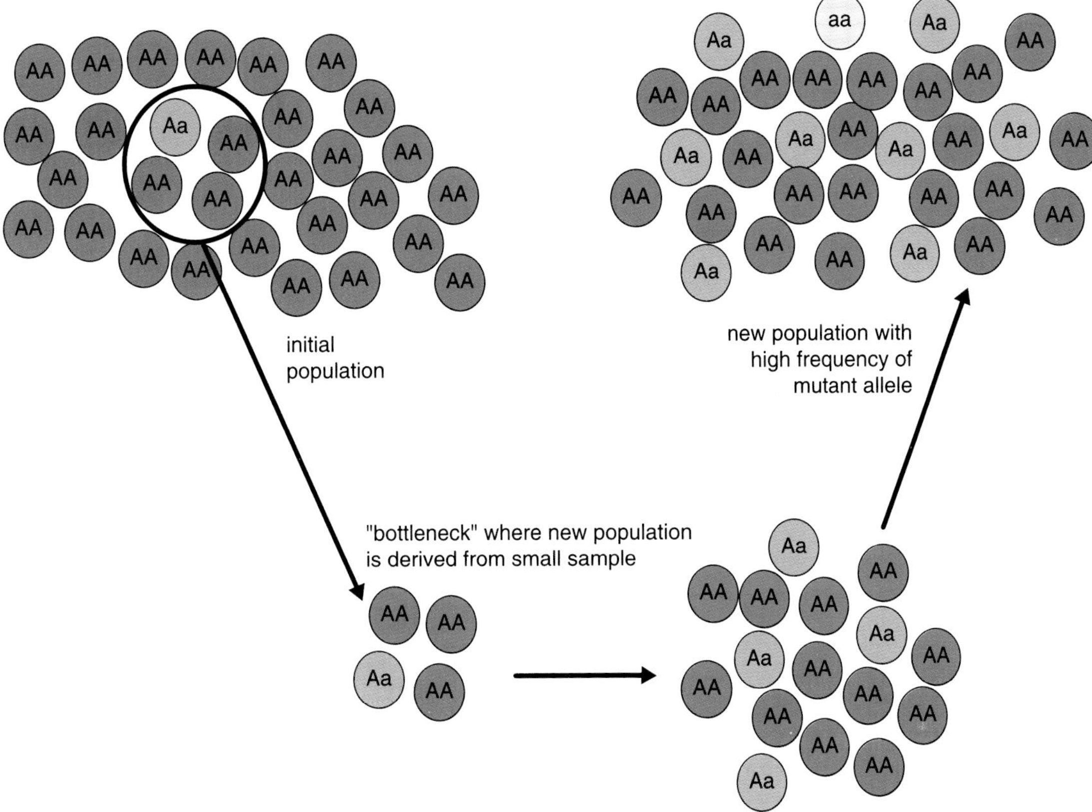

FIGURE 10-15 Founder effect. The frequency of the *a* allele is low in the initial population, but a small subset, in which one individual is *Aa*, is removed from the large population and founds a new population. The frequency of *a* is markedly higher in this new population, due to its relatively high frequency in the founders.

fluid secretion after ingestion of purified cholera toxin. Fluid loss was highest in the wild-type animals, lowest in the mutants, and intermediate in the carriers, supporting the potential selective advantage of the carrier state (Figure 10-16).

It is likely that selective forces have acted on other genetic traits in human history, although in most cases these forces have not been identified. Traits that are now considered deleterious may well have resulted in a selective advantage to heterozygotes at some point in history. If the same selective forces acted independently in different populations, there may be different mutations at the same locus that were subject to selection, accounting for genetic heterogeneity between populations and a founder effect within a population. These phenomena—heterozygote advantage, founder effect, and genetic heterogeneity—do not act in isolation but together represent some of the major forces that mold gene frequencies in populations.

Hardy-Weinberg Equation for X-Linked Recessive Trait

For an X-linked recessive gene, males are hemizygous; the frequency of males with the genotype *A* is simply the frequency of the *A* allele, or p. Likewise, the frequency of males with the *a* genotype is q. Females can be *AA*, *Aa*, or *aa*, with the usual frequencies of p^2, $2pq$, and q^2, respectively.

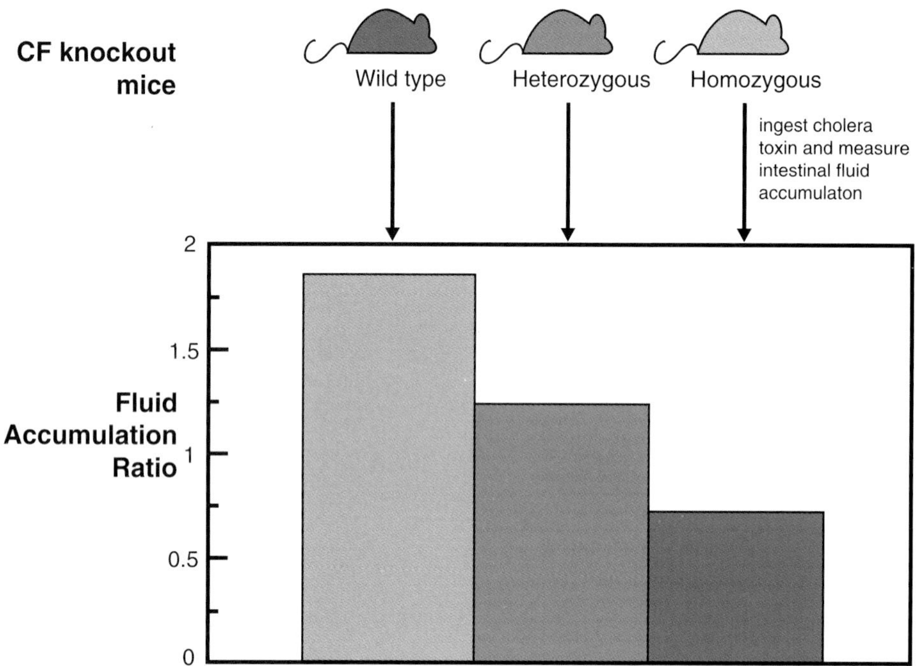

FIGURE 10-16 Mice with cystic fibrosis (*CF*) knockout created by transgenic technology are exposed to cholera toxin. It was found that the cystic fibrosis mutation exerted a protective effect, reducing the amount of diarrhea (measured as fluid accumulation ratio). (Reproduced by permission from Gabriel SE, et al. Cystic fibrosis heterozygote resistance to cholera toxin in the cystic fibrosis mouse model. Science 1994;266:107–109.)

10.6 New Approaches to Treatment of Thalassemia

1994

Rosa's youngest sister, Maria, recently was married, and she and her partner, Alberto, are screening for thalassemia carrier status. While they are waiting for the test results, they ask the genetic counselor what is new in treatment for thalassemia.

The use of chelation therapy has vastly improved the clinical outcome for persons with beta-thalassemia. Most of the long-term complications of the disease are due to iron overload, but early institution of deferoxamine treatment and faithful compliance has been shown to prevent the accumulation of iron and its consequences. Using the end point of cardiac disease, one recent study has shown that 91% of patients treated with deferoxamine were free of heart disease after 15 years; survival in this study correlated closely with serum ferritin measurements, which reflects stored iron. Transfusion and chelation therapy is not without drawbacks, however. There is risk of developing hepatitis, although donor blood now is screened for this virus. The cost of treatment can be high, more than $30,000 per year in 1991. Moreover, treatment must be continued throughout life, requiring overnight administration of deferoxamine and transfusions as often as every 3 weeks. Investigators have therefore sought alternative means of treatment.

Replacement of hematopoietic cells with new cells that do not have mutant beta-globin genes would provide definitive treatment of thalassemia. This can be accomplished by bone marrow transplantation. In one study, bone marrow transplantation was done using human leukocyte antigen–identical donor marrow for patients with beta-thalassemia who were already suffering from the effects of iron overload. Disease-free survival for 3 years or more was 80%. Another study showed 85% disease-free survival when patients were treated prior to the appearance of signs of iron overload. These are encouraging results, but the treatment also comes at a price, both medical and financial. The bone marrow transplant procedure involves major discomfort, a small proportion of recipients suffer complications or rejection, and some die (7 of 89 in a recent study). The cost of the procedure in the United States was more than $175,000 in 1991, less costly than a lifetime of transfusion and chelation therapy but still very expensive.

A long-term goal might be to replace the defective beta-globin gene in bone marrow cells from an affected individual and to transplant these cells back into that person's marrow. This would avoid the risks of graft rejection or rejection of the recipient by the graft (graft-versus-host disease). Beta-globin cDNA has been inserted into hematopoietic stem cells using retroviral vectors. The challenge has been to obtain adequate quantities of beta-globin synthesis, which has not yet been achieved. Gene therapy for beta-thalassemia is therefore not yet possible.

An alternative form of treatment is suggested by the existence of a cluster of betalike genes. Individuals with beta-thalassemia do not suffer anemia during fetal life because of the production of fetal hemoglobin, which uses the gamma-chains instead of beta-chains. Gamma-chain production falls to very low levels soon after birth. The genes, of course, still are present, but somehow they are permanently inactivated. Can they be reactivated in individuals with beta-globin mutations? Although fetal hemoglobin has greater oxygen affinity than adult hemoglobin and is best suited to oxygen transport in the oxygen-deprived fetal environment, it might function well enough in the adult to ameliorate many of the effects of beta-thalassemia.

The drug azacytidine has been found to stimulate gamma-globin production. Two individuals with severe complications of beta-thalassemia were able to stop transfusions within weeks of starting azacitidine. Unfortunately, the drug also causes neutropenia, which has limited its clinical usefulness. In addition, it may be carcinogenic. Another agent that has a similar ability to stimulate fetal hemoglobin production but is associated with fewer side effects is the fatty acid butyrate. It was observed that the switch from gamma- to beta-globin production is delayed in infants of diabetic mothers who have high levels of γ-amino-*n*-butyric acid in their plasma. In one study of three individuals with sickle cell anemia and three with beta-thalassemia treated by intravenous infusion of butyrate, fetal globin synthesis increased from 6% to 45% with minimal side effects. Less encouraging results were obtained, however, in another study. Recently hydroxyurea has been used to stimulate fetal hemoglobin production with better results.

The question sometimes is raised: What will be the impact on the gene pool of allowing the survival of individuals with deleterious genetic traits? If individuals with beta-thalassemia are able to reproduce, will the thalassemia trait become more prevalent, making a significant proportion of the population dependent on medical therapy? We can look at this from the perspective of the Hardy-Weinberg equation. If the carrier frequency of beta-thalassemia in Sardinia is approximately 13% and there is no selection against the beta-thalassemia allele, we would expect this frequency to remain stable over time. We can assume, though, that there is no longer selection against the wild-type allele—that is, that few individuals in Sardinia now die of malaria. The frequency of thalassemia, of course, has declined due to population screening and prenatal diagnosis. We can calculate how long it would take to halve the gene frequency if all pregnancies with affected fetuses are terminated (see below).

We see that it would take 14½ generations (perhaps 300 years) of population screening to reduce the gene frequency from nearly 6% to 3%, and hence the carrier frequency from 12% to 6%. Another 300 years would be required to halve the frequency yet again. Gene frequencies change very slowly, even if total selection against a genetic trait is instituted. Changing population frequency of a genetic trait is therefore not a realistic goal and may not be ethical. A more feasible and defensible goal of population screening is to improve the health of individuals in the population, not to improve the gene pool of the society.

Maria is found to be a thalassemia carrier, but Alberto is not. The partners are informed that they are not at risk of having a child with thalassemia.

Change in Frequency of Thalassemia After Complete Selection

At the start, the frequency of thalassemia is 1:213, so $p^2 = 0.00469$ and $p = 0.0685$; therefore $q = 0.9314$ and $2pq = 0.1276$. Let's look at what happens if suddenly all

Perspective

Living with a Child with Beta-Thalassemia

LILIA AND RUDI VISCOMI

When I first met my wife, Lilia, I was 19 years old, and she was 17. During our 5-year courtship, we both realized that we were carriers of the thalassemia trait; ironically, during our teenage years our carrier status was diagnosed by the same hematologist. I remember being told by this physician that thalassemia was a very common disease in people of Mediterranean descent. He also said that if I were to marry someone with thalassemia trait, our children would have a 50/50 chance of survival.

When we decided to get married, we sought advice from our family physician. At our session, we made him aware that we were both carriers of the thalassemia trait. Being unfamiliar with this disease, he looked up the word *thalassemia* in a medical dictionary and read to us its meaning. He stated that the chance of an offspring being born with the disease was very slim and, therefore, we were reassured and told not to worry about it.

During my wife's pregnancy with our first child, she informed her obstetrician of our medical condition. The obstetrician promised that he would speak to a hematologist and that at her next visit he would let her know the findings. In fact, at that visit the obstetrician stated that he had spoken with the hematologist and led us to believe that there was no cause for concern.

When our son Michael was born in March 1983, the pediatrician that we chose was the first physician to be concerned with our being thalassemia trait carriers. By chance, he knew about thalassemia because he had a patient in his care who had thalassemia. The pediatrician began to test our son's blood every 2 weeks, and at each test result, his hemoglobin level was lower than the previous test.

At 6 months of age, the physician asked us to bring our son to a hematology specialist. Shortly thereafter, our son was determined to have thalassemia. Michael's problems began at 7 months of age with his first transfusion. Transfusions were as frequent as two times per month; due to this frequency, it had become very difficult for medical personnel to find veins in his tiny arms in which to place an IV. Over the course of his very young life, this had required an emergency cut-down procedure and a

affected individuals are unable to reproduce (e.g., because affected fetuses are detected prenatally and the pregnancies are terminated). Call the gene frequency of the mutant allele at the start q_0 and one generation after selection q_1. What is the value of q_1? The only source of individuals with the mutant allele are heterozygotes, and the total population consists of homozygous wild type and heterozygotes. Therefore:

$$q_1 = 2p_0q_0/2(p_0^2 + 2p_0q_0)$$

[Note that the denominator is $2(p_0^2 + 2p_0q_0)$ because each individual contributes two alleles.]

We can reduce this by factoring out p and using the equation $p = 1 - q$:

$$q_1 = q_0/(1 + q_0).$$

If we look at q after another generation of selection, q_2, by the same reasoning we come up with:

$$q_2 = q_1/(1 + q_1).$$

Now we can express q_2 in terms of our starting point, q_0:

$$q_2 = \frac{q_0/(1 + q_0)}{1 + q_0}$$

central venous line. Our biggest concern during this time was the possibility of our son contracting the AIDS virus owing to the frequent transfusions.

Transfusions have helped Michael to live an almost normal life. However, the same treatment that is helping him also has its side effects. With each transfusion, Michael receives a large amount of iron, and there is the risk of iron overload, which could damage major organs. To prevent this, Michael needs to be hooked up to a small infusion pump that injects Desferal (an iron chelator) into his body for 12 hours every night. Michael began to use his pump at 2 years of age; at that time, it was very difficult to insert a needle in a child's body on a nightly basis. Not only was it difficult for us to do as parents, but he would be resistant to this process, as one can imagine. As a preteen, Michael injects himself, yet at times he becomes resistant to this routine, and compliance could become a problem as he gets older.

There is always a risk in receiving blood transfusions. We worry about AIDS and other blood-related diseases. Michael has received a diagnosis of hepatitis C.

We have been encouraging Michael to talk to his friends about his disease. As parents, we have become very active in the community and have participated in educational fund-raisers for thalassemia. However, this has resulted in some disappointments. Some of Michael's friends, due to their parents' misinterpretation of the disease, are not allowed to play with Michael for fear that they will contract the disease. Often, he has been called *virus boy*. Then there is the astronomical financial burden: Michael's medical care costs our insurance carrier more than $70,000 annually. With a lifetime maximum of $1 million, Michael's insurance coverage will soon run out.

Despite all the frustration, we have been able to find comfort through the Cooley's Anemia Foundation (*Cooley's anemia* is another name for thalassemia). We are very active in this organization. We take comfort, too, in many family members and friends who have been supportive. Also, we have learned to cope with this nuisance. Our concerns remain the risks associated with transfusions and the fact that Michael, as a young adult, may become noncompliant with his treatment. At the same time, we have faith in those institutions that are tirelessly conducting research and experimenting with genetic transplantation and butyrate for a possible cure. After all, there has been a lot of progress with respect to the understanding of this disease in just a few short years.

This reduces to $q_2 = q_0/(1 + 2\,q_0)$.

Generally, the value of q after n generations of selection is:

$$q_n = q_0/(1 + n\,q_0)$$

So how long does it take to halve the gene frequency; that is, for what value of n is $q_n/q_0 = ½$?

$$q_n / q_0 = 1/(1 + n\,q_0)$$

Therefore, if $q_n/q_0 = ½$, $1 + n\,q_0 = 2$, and $n = 1/q_0$. For beta-thalassemia on Sardinia, this means that the gene frequency would be halved in 1/0.0685 generations, or approximately 14½ generations.

Case Study

Part I

Margie and Charles are planning to be married. They are second cousins (Margie's paternal grandfather and Charles's maternal grandfather are brothers), and their families are very unhappy about the union, mainly for fear that the couple will have children with genetic problems. Margie and Charles decide to speak with a genetic counselor to learn more about the risks.

It is common lore that matings between close relatives are at increased risk of having offspring with genetic problems. There is a genetic basis for this belief, which is illustrated in the pedigree for Margie's and Charles's family shown in Figure 10-17. Because they share a common set of great-grandparents, there is a possibility that both Margie and Charles will be heterozygous carriers for some rare recessive allele that, if homozygous in an offspring, would cause a genetic disorder. Here it is assumed that the great-grandfather was the carrier.

We can calculate the risk that a child would be homozygous for some allele by virtue of inheriting it from each parent. Such a child is said to have alleles that are *identical by descent*—that is, he or she is not only homozygous, but the two alleles are exactly the same because they were derived from a common ancestor. Consider the hypothetical pedigree in Figure 10-18.

Assume that the child, V-1, inherits the recessive allele from his mother, IV-1. We want to know the probability that he will also inherit the same allele from his father, IV-2. IV-1 could have inherited *a* from either of her parents; the probability that she inherited it from III-2 is ½. Likewise, III-2 could have inherited it from either of his parents; the likelihood that it came from II-2 is also ½. If II-2 carries the gene, he had to have received it from one of his parents, and the chance that his sibling II-3 also is a carrier is ½. The chance for II-3 to pass the gene to III-3 is ½, and for it to be passed in turn to IV-2 is

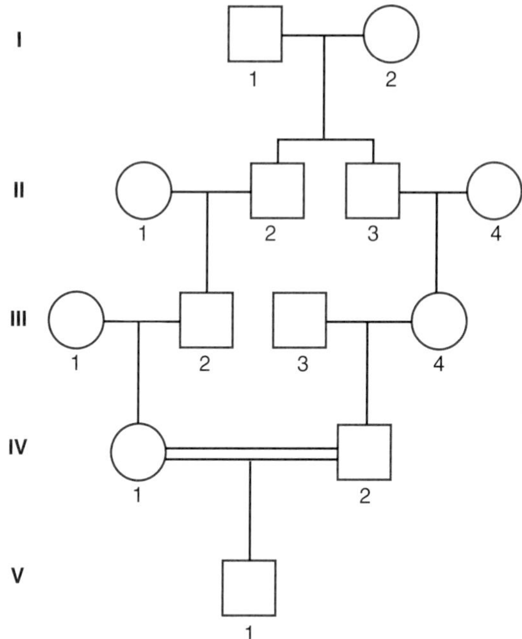

FIGURE 10-18 Hypothetical pedigree illustrating alleles that are identical by descent.

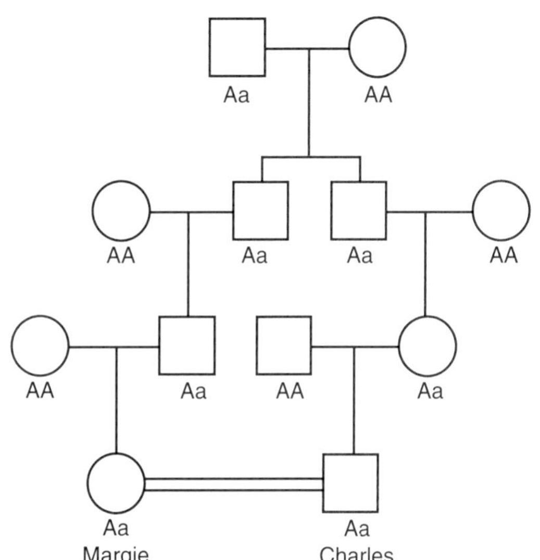

FIGURE 10-17 Pedigree for Margie's and Charles's family.

again ½. Finally, the chance for IV-2 to transmit the gene to V-1 is also ½. Considering the entire path from IV-1 up through the great-grandparents and then down to V-1, with each step representing an independent event, we get a total probability of $\frac{1}{2} \times \frac{1}{2} \times \frac{1}{2} \times \frac{1}{2} \times \frac{1}{2} \times \frac{1}{2} = \frac{1}{64}$.

Put differently, the child of a pair of second cousins has 1 chance in 64 of having alleles that are identical by descent at any locus, or is identical by descent at approximately 1/64 of his or her genetic loci. For the vast majority of genes, such homozygosity does no harm. There is a chance, however, that one of the great-grandparents happened to carry an allele that, if homozygous, would cause a genetic disorder. Such an allele can be passed from generation to generation in heterozygous form without being noticed. If, however, a consanguineous mating occurs, a homozygous offspring might be born.

How likely is it that a child will be homozygous for a deleterious allele? That depends on how many such deleterious alleles one of the great-grandparents carried. It has been estimated that all persons carry a few alleles that, if homozygous, would cause a lethal phenotype. These alleles are likely to be different from person to person, so normally such lethal phenotypes are rare, requiring the chance union of two carriers for the same lethal allele. A consanguineous mating increases the risk that the lethal alleles carried by the parents will be the same. We can quantify this risk as follows:

Suppose that every person carries one recessive lethal allele. The chance that the great-grandfather passes this allele to both his great-granddaughter and his great-grandson, and then that they each pass it to a child, is 1 in 256. The same can be said for the great-grandmother, giving a total risk of $\frac{1}{256} + \frac{1}{256}$ or $\frac{1}{128}$, a little less than 1%. The risk goes up, though, if the number of lethal equivalents per individual in the population is assumed to be higher.

Clearly, then, there is some basis for concern about a consanguineous mating. How is this approached by the genetic counselor?

Part II

The genetic counselor learns that Margie and Charles are both in good health. They are of French Canadian ancestry. Margie has a cousin (her father's sister's child) with cystic fibrosis. Otherwise, there is no family history of genetic disorders or congenital malformations.

Before worrying about theoretic risks applying to consanguinity, the counselor must consider more tangible risk factors. First, is there a family history of any inherited condition? In this case, there is cystic fibrosis in Margie's cousin. The pedigree can be redrawn with this new information, as shown in Figure 10-19.

What is the risk of cystic fibrosis in a child of Margie and Charles? Margie's father's sister must be a cystic fibrosis carrier, so Margie's father has a 50% risk of also being a carrier. Margie herself, therefore, has a 25% risk of being a carrier. Her actual risk is a little higher than this, since there is a slight chance (about 4%) that Margie's mother is a cystic fibrosis carrier, but this only adds an additional 2% risk to Margie, and we will not consider that in this calculation.

The calculation of Charles's risk is more complicated. We must consider for him the chance that he inherited one of the cystic fibrosis alleles present in the affected relative, and, in addition, the chance that he has inherited a different cystic fibrosis allele. The chance that he inherits the same allele that Margie is at risk of carrying is 1/16. This is calculated as follows: Although one of Margie's father's parents must be a cystic fibrosis carrier, probably only one is. The chance that the carrier is Margie's father's father is then 1/2. If he is a carrier, the risk of his brother being a carrier is also 1/2. The risk that the cystic fibrosis gene is passed on to Charles's mother is 1/2, and to Charles another 1/2. Hence the total risk to Charles is 1/16 to carry one of the alleles that affects Margie's cousin.

In considering Charles's risk of being a carrier we have to consider four scenarios:

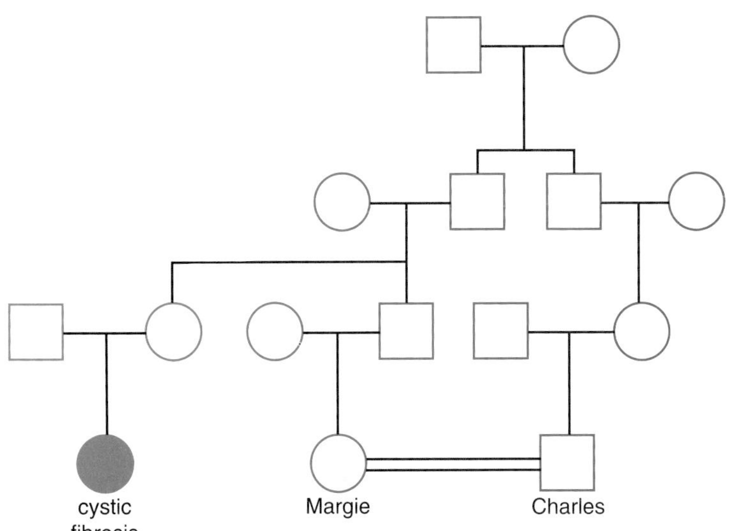

FIGURE 10-19 Redrawn pedigree for Margie's and Charles's family.

1. Charles could have received a wildtype cystic fibrosis allele from both his parents. The probability of his inheriting a wildtype allele from his father is p, the gene frequency. The probability of inheriting a wildtype allele from his mother is (15/16)p, that is the product of not inheriting the allele known to be present in the family (1-1/16 = 15/16) times the probability that his mother carries a wildtype allele, which is p. Therefore this scenario has a probability of $15/16p^2$.
2. Charles could have inherited the cystic fibrosis mutation that affects Margie's cousin from his mother and a wildtype allele from his father. The former has a probability of 1/16, the latter of p, so this scenario has a probability of (1/16)(p).
3. Charles could have inherited a cystic fibrosis allele from his mother that is not the same as one that affects Margie's cousin, and inherited a wildtype allele from his father. The former has a risk of (15/16)(q), i.e., the probability of not inheriting the allele in Margie's cousin (15/16) times the risk that the allele he inherits from his mother is a cystic fibrosis allele (q). The latter has a risk of p, as we have already seen, so this scenario has a total risk of (15/16)(pq).

Overall, the risk that Charles is a carrier is the sum of scenarios 2–4 divided by all four possibilities:

$$\frac{1/16p + 15/16(pq) + 15/16(pq)}{15/16(p^2) + 1/16p + 15/16(pq) + 15/16(pq)}$$

factoring out a p/16, and substituting 1-q for p, the equation reduces to:

$$\frac{1 + 30q}{16 + 15q}$$

If q = 0.02, this gives a risk of 0.0982. The overall risk to Margie and Charles, then is 1/4(0.0982)1/4 = 0.0061, about 1 in 163.

The second concern is to consider the ethnic background of the couple and determine whether this places them at risk for any specific condition. For a couple of French Canadian ancestry, one disorder to consider is Tay-Sachs disease. The frequency of carrier status for Tay-Sachs disease in French Canadians is approximately 1 in 50, half the frequency in Ashkenazi Jews but twice the frequency in the general population. With or without consanguinity, this disorder should be discussed in counseling Margie and Charles, and carrier testing should be offered.

Part III

Medical records are obtained from Margie's cousin with cystic fibrosis and the cousin is found to have the genotype ΔF508/621 + 1. Margie and Charles are tested for these CF mutations, and 30 others, and both are negative. In addition, they are tested for hexosaminidase A activity. Charles is found to be a Tay-Sachs carrier and Margie is not. They are told that they are at very low risk of having a child with cystic fibrosis, and that they are not at risk of having a child with Tay-Sachs disease. They are provided empiric counseling that their risk of having a child with a birth defect or genetic disorder is 1% to 2% above the background frequency of 3% in the general population. The family is somewhat reassured, and Margie and Charles are married.

The cystic fibrosis mutation testing indicates that Margie and Charles have a very low risk of having a child with cystic fibrosis. Since only Charles is a Tay-Sachs carrier, their offspring are not at risk of this disorder. In the absence of quantifiable risks of known genetic disorders, the basis for counseling of consanguineous couples is empiric data. These data have been gathered from population surveys of children with congenital anomalies and mendelian genetic disorders. Overall, such surveys indicate a very slight increase in the risk of genetic disorders in the offspring of consanguineous couples. It is estimated that nearly 3% of all pregnancies result in the birth of a child with a birth defect of some kind. The risk to parents who are first cousins adds approximately 3% to that baseline risk, for a total risk of 6% or so. For parents who are second cousins, the empiric data suggest a risk of 1% to 2% above baseline. This is less risk than many expect. Of course, it accounts only for birth defects that are readily evident early in life, and it does not include problems that cause miscarriage.

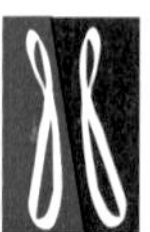

Part IV

Margie is now pregnant with her first child. She asks about prenatal testing. She is offered triple screening for alpha-fetoprotein, unconjugated estriol, and human chorionic gonadotropin. Her risk of having a child with a chromosomal anomaly is estimated to be 1 in 890, taking into account the results of testing and her age (which is 26). An ultrasound examination is normal.

Even if this pregnancy is at slightly greater risk for genetic problems than most, the risk does not predict any specific disorder. Therefore, there is no specific means of prenatal diagnosis available. Ultrasound screening can be done to look for evidence of major structural anomalies, but this has limited sensitivity. Prenatal care otherwise is the same as routine care.

It has become common to offer a set of blood tests, often called the *triple test*, as part of regular prenatal care (see Chapter 5). The values in this pregnancy indicate a risk of Down syndrome of 1 in 890, which is below the threshold of 1 in 270 used commonly as a cutoff at which genetic amniocentesis is recommended.

Part V

Margie's first child, a girl named Theresa, is now 3 years old and is in good health. For Margie's second pregnancy, the triple screen again indicates a low risk of Down syndrome, and ultrasound examination is normal. Margie gives birth at term to an 8-pound boy. He is found to have a cleft lip and palate but no other congenital anomalies. They name him Louis.

Cleft lip is a congenital anomaly wherein the tissue planes that form the philtrum—the skin just above the lip—fail to fuse in early development. Cleft lip can occur in isolation, in combination with cleft palate, or as part of a syndrome of multiple congenital anomalies. Clinically, the major significance of cleft lip is cosmetic. There is also a problem with forming a seal around a nipple, requiring the use of a special nipple during infancy. Isolated cleft lip usually is surgically repaired in the early weeks of life. It is not associated with other health problems.

From the genetic point of view, cleft lip with or without cleft palate, when present without other congenital malformations, is a classic multifactorial trait. As such, it is believed that a combination of both genetic and nongenetic factors contribute to the etiology. As we have learned from discussion of multifactorial inheritance, this implies that both parents have contributed genetic "liability" to development of the trait. Because in this case the parents are related, one might expect that they share some genetic liability just as they may share recessive alleles in common. It might be expected, therefore, that the risk of multifactorial traits would be increased slightly in the offspring of consanguineous parents, just as the risk of recessive traits is elevated. It is difficult to quantify the degree of this increased risk, and it is likely to be included in the empiric risk of birth defects quoted to the family. Of course, it is impossible to prove that Louis's cleft lip actually is attributable to the fact that his parents are second cousins. The family should be reassured that this is a relatively common congenital anomaly that occurs in the absence of consanguinity.

Part VI

Theresa is now 22 years old, and Louis is 19. Louis had surgical repair of his cleft in the first year of life, and he has done well, both physically and cognitively. Theresa is now planning to be married, and her partner, who is unrelated, is of Italian ancestry. The couple asks whether the fact that Theresa's parents are second cousins will increase the risk to children of Theresa and her partner.

The risks owing to parental consanguinity apply to the offspring of the consanguineous couple and are due to the possibility that identical deleterious genes will be passed to an offspring from both parents. When the child of a consanguineous couple seeks counseling, however, the fact that his or her parents were related is of no consequence. What matters is whether that person's partner is also a relative; if not, the couple is at no greater risk of passing on genetic problems to offspring than is anyone in the general population.

REVIEW QUESTIONS

1. Although mutations of the globin gene tend to occur commonly in various parts of the world where malaria is prevalent, the specific globin mutations tend to be region-specific. Explain this in terms of population genetic theory.
2. Ataxia telangiectasia (AT) is an autosomal recessive disorder characterized by progressive ataxia (unsteadiness), telangiectasia (dilated skin capillaries), immunodeficiency, and predisposition to lymphomas. The frequency of

AT in the population is approximately 1 in 40,000 individuals.

a. It has been suggested that carriers of the AT mutant allele might also have increased risk of malignancy. What is the carrier frequency of the AT gene?

b. If this hypothesis were correct, and 80% of AT carriers died of cancer between ages 50 and 70 years, what would this selection do to the frequency of the AT gene?

3. Consider an autosomal recessive disorder with a population frequency of 1/10,000 in individuals of African descent. What would be the approximate risk that a child of first cousins would have alleles identical by descent for this disorder assuming that only one of that child's common grandparents was of African descent.

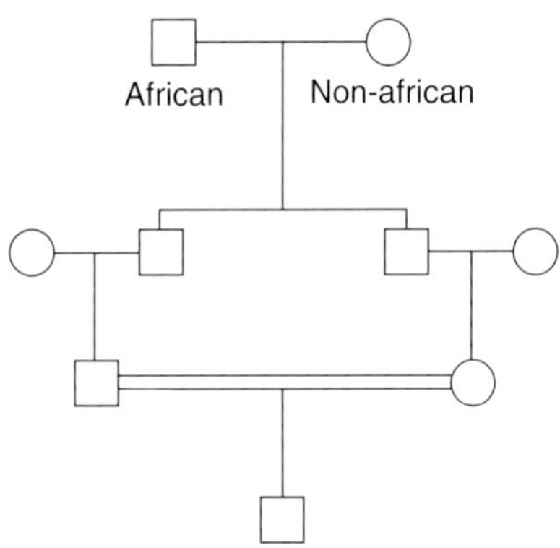

4. A woman has a child with Duchenne's muscular dystrophy. She has no other children, no brothers, and no prior history of muscular dystrophy. What is the likelihood that she is a carrier for Duchenne's muscular dystrophy, as opposed to her son's being a new mutation?

FURTHER READING

10.3 Population Screening for Thalassemia

Cao A, et al. Prevention of homozygous β-thalassemia by carrier screening and prenatal diagnosis in Sardinia. Am J Hum Genet 1981;33:592–605.

Kaback M, et al. Tay-Sachs disease—carrier screening, prenatal diagnosis, and the molecular era. JAMA 1993;270:2307–2315.

Loader S, et al. Prenatal screening for hemoglobinopathies: II. Evaluation of counseling. Am J Hum Genet 1991;48:447–451.

Rowley PT, et al. Prenatal screening for hemoglobinopathies: I. A prospective regional trial. Am J Hum Genet 1991;48:439–446.

Rowley PT, et al. Prenatal screening for hemoglobinopathies: III. Applicability of the health belief model. Am J Hum Genet 1991;48:452–459.

10.5 Genetic Polymorphism

Grompe M, et al. A single mutation of the fumarylacetoacetate hydrolase gene in French Canadians with hereditary tyrosinemia type I. N Engl J Med 1994;331:353–357.

10.6 New Approaches to Treatment of Thalassemia

Lowrey CH, Nienhuis AW. Brief report: treatment with azacitidine of patients with end-stage β-thalassemia. N Engl J Med 1993;329:845–848.

Lucarelli G, et al. Bone marrow transplantation in adult thalassemia. Blood 1992;80:1379–1381.

Lucarelli G, et al. Marrow transplantation in patients with thalassemia responsive to iron chelation therapy. N Engl J Med 1993;329:840–844.

Oliveri NF, et al. Survival in medically treated patients with homozygous (-thalassemia. N Engl J Med 1994;331:574–578.

Perrine SP, et al. A short-term trial of butyrate to stimulate fetal-globin-gene expression in the β-globin disorders. N Engl J Med 1993;328:81–86.

Sher GD, Ginder GD, Little J, et al. Extended therapy with intravenous arginine butyrate in patients with β-hemoglobinopathies. N Engl J Med 1995;332: 1606–1610.

CASE STUDY

Young, Ian D. *Introduction to Risk Calculation in Genetic Counseling*. Oxford: Oxford University Press, 1991.

Answers to Review Questions

Chapter 1

1. Risk = (2/3)(1/50)(1/4) = 1/300

2. (a) The cDNA hybridizes with exons that are present on different restriction fragments, due to the presence of introns. (b) A number of possible explanations exist. There may be an insertion of material into one restriction fragment, or a polymorphism might exist, leading to an allele of slightly larger size for one part of the cDNA. There could also be a deletion that removes one restriction site and juxtaposes another. (c) Possibly this is a pathogenic mutation, but not necessarily. It would be important to show, by testing unaffected individuals, that the change is not a polymorphism and to show that it segregates with the disease in the family.

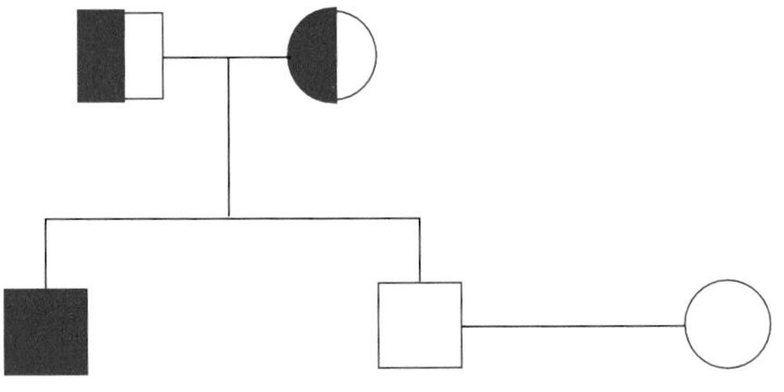

3. (1) Homozygous wild-type; (2) homozygous mutant; (3) heterozygous; (4) either failure of PCR reaction or some other mutation or deletion that causes both oligonucleotides not to hybridize.

4. A frameshift, which leads to stop codon, is likely to occur, causing truncation of the protein.

Chapter 2

1. A missense mutation will result in 50% of fibrillin having an abnormal structure, which will severely disrupt interactions between fibrillin molecules. A frameshift near the 5' end will result in lack of expression of that allele. Total amount of fibrillin may be reduced, but what remains will have normal structure.

2. There may be multiple restriction enzyme recognition sites within the cloned region. Alternatively, multiple homologous sequences may exist at various sites in the genome. Unlike the case for a cDNA probe, intron-exon structure does not explain this result, as a genomic clone will contain both introns and exons.

3. The child is most likely affected due to new mutation, as both parents are unaffected. The parents could have another affected child, though, if one of them is a germ-line mosaic. The sister, however, appears not to have inherited the gene mutation. She would be at risk of having an affected child only due to an independent mutation, which is very rare.

4. Severe dominant phenotypes often interfere with reproduction. Most cases, therefore, arise due to new mutation. Recessives, on the other hand, are transmitted from asymptomatic carrier parents, so the level of severity of the disorder does not influence the likelihood of having an affected child.

Chapter 3

1. Tom has inherited the *B* allele from his mother, whereas it is the *A* allele that is in coupling with the polycystic kidney disease gene. Barring recombination, Tom probably is not affected.

2. Recombination has occurred between the marker gene and the disease gene, suggesting that the marker gene is not the disease gene itself.

3. (a) The fetus probably is a carrier. (b) Genetic heterogeneity would cause a diagnostic error if the disease gene in this family is not linked to the marker used.

4. (a) The mutation is a deletion, indicated by the fact that I-2 did not transmit her 5-kb allele to II-1. Apparently I-2 is hemizygous, having one 5-kb allele but no allele on the other chromosome. (b) II-2 is heterozygous, having inherited her mother's normal 5-kb allele.

Chapter 4

1. (a) Autosomal dominant: nonpenetrance or new mutation. (b) Autosomal recessive: both parents are carriers. (c) X-linked: mother is a carrier.

2. The daughter could have Turner's syndrome (single X chromosome), some other X chromosome abnormality, or nonrandom X chromosome inactivation.

3. She probably is not a carrier. Her mother received the *B* allele from her mother, who is a carrier, and she passed the *A* allele to the consultand. She would be a carrier only if a recombination event occurred. Note that her uncle has Duchenne's muscular dystrophy yet has the *A* allele. This indicates that a recombination event occurred either from his mother to him or from his mother to his sister. This recombination, however, does not influence the diagnosis for the consultand.

4. A single exon deletion would be inferred from the absence of a single band on the multiplex PCR gel. It is

possible, however, that a point mutation at the PCR primer binding site has occurred, which would give the same result. This possibility can be excluded by confirming the deletion by Southern analysis. A multiexon deletion, however, cannot be explained in this way.

Chapter 5

1. (a) What is being sought is translocation between chromosome 21 and another chromosome, most commonly 13, 14, 15, or 22. This would convey risk of Down syndrome to an offspring of this couple. (b) This balanced translocation confers risk of unbalanced products in offspring. Prenatal diagnosis by amniocentesis or chorionic villus sampling should be offered.

2. (a) Nondisjunction has occurred in the embryo. The embryo may have started as normal, and a trisomy 15 cell line may have occurred in chorionic villi. Alternatively, the embryo may have started out with trisomy 15 but subsequently lost one copy of 15 in some tissues. (b) The finding of Prader-Willi syndrome suggests that the second alternative probably is correct: The embryo initially had trisomy 15. This would have been lethal, but loss of the paternal copy occurred, restoring disomy but causing Prader-Willi syndrome due to uniparental disomy.

3. (a) The balanced translocation led to the development of gametes with unbalanced chromosomes. (b) Most likely, the NF1 gene has been split by the translocation. One end of the gene likely remains on chromosome 17 and the other end on chromosome 1. FISH probes from the two ends of the gene should reveal this.

4. Presumably, the degree of physical involvement is related to the proportion of cells with a 45,X karyotype in critical tissues. There is a bias of ascertainment wherein women in whom the diagnosis is made postnatally are more likely to be severely affected and hence come to medical attention. Many of the fetuses recognized by prenatal diagnosis would not have manifested an obvious phenotype, at least not early in life. Some may go on to develop signs of Turner's syndrome as they grow older, though others will not.

Chapter 6

1. (a) Genetic liability toward a multifactorial trait generally is believed to arise from both sides of the family. There is no reason to believe that one or the other parent contributed disproportionately. (b) The threshold for expression is apparently higher in females than males, so if a female is affected, there is higher liability toward the trait in the family, making recurrence risk higher. (c) An affected child would more likely be male. Even though the family has higher liability, based on their having an affected female, recurrence risk would still be higher for the gender with the lower threshold.

2. Folic acid should be taken prior to conception to protect against neural tube defects. Neural tube closure is complete by 4 weeks' gestation.

3. The risk would be higher if both parents are affected, because each would contribute liability toward the trait in a child.

4. The nondisjuction event must have occurred postzygotically, after separation of the embryos.

Chapter 7

1. (a) Individuals at risk are I-2, II-2, -3, -4, -5, and III-1, -2, and -3, as this is a mitochondrial disorder. (b) He is not at risk of passing mutant mitochondria on to his children.

2. She still may carry the mutant mitochondria in her germ line. Her mother is likely to have carried the mutation and passed at least some mutant mitochondria on to her daughter.

3. The patient has Kearns-Sayre disease and is heteroplasmic for a mitochondrial DNA deletion.

4. It is possible that the child has an autosomal recessively inherited mitochondrial disorder.

Chapter 8

1. The gene defect leads to an increased rate of chromosome rearrangement and mutation. Some of these changes result in loss of recessive oncogenes or activation of dominant oncogenes, leading to malignancy.

2. (a) There is loss of heterozygosity in the tumor, suggesting that the NF2 gene acts as a tumor suppressor. (b) The mother's (normal) allele would be expected to be lost from the tumor.

3. The retinoblastoma gene acts as a tumor suppressor for many tumor types. Loss of this gene may contribute to progression of various types of malignancy, although it probably is not the rate-limiting step toward tumor formation that it is in retinoblastoma. In these tumors, loss of Rb function is an acquired event for both alleles.

4. Each B cell produces a single type of antibody. A tumor arises clonally from a single cell and hence expresses just one type of antibody.

Chapter 9

1. (a) The presence of a Y chromosome indicates that testes would be present. (b) If testes are present, müllerian inhibiting substance would be made, leading to absence of a uterus. (c) Androgen effect is required for virilization of the external genitalia. If there is androgen insensitivity, no virilization would occur.

2. The first few weeks of embryonic development is a major period of cell division and organogenesis. Teratogen exposure during this period often leads to such profound disruption as to cause miscarriage.

3. Major brain development occurs in the early weeks of pregnancy. Any exposure in the eighth month of pregnancy would have occurred too late to cause holoprosencephaly.

4. Congenital heart disease and tracheo-esophageal fistula tend to occur in association. The etiology, however, varies from one individual to another. This association is not necessarily indicative of a genetic cause.

Chapter 10

1. The high prevalence of globin mutations in malaria-prone regions is an example of a balanced polymorphism. The tendency for mutations to be more or less region-specific probably is due to a founder effect.

2. (a) If the disease frequency is 1 in 40,000, the gene frequency is 1 in 200 and the carrier frequency is 1 in 100. (b) It would have no effect because the selection occurs after the reproductive age.

3. The great-grandfather's risk of being a carrier is calculated by: $q^2 = \frac{1}{10,000}$ therefore $q = \frac{1}{100}$ and $2q = \frac{1}{50}$. The risk that the child is identical by descent for one of great-grandfather's alleles is $\frac{1}{64}$. The risk, therefore, that the child is identical by descent for the mutant allele is $(\frac{1}{50})(\frac{1}{64}) = \frac{1}{3200}$

4. Two-thirds of women who have sporadically affected sons are carriers of Duchenne's muscular dystrophy.

Glossary

acrocentric placement of centromere near one end of a chromosome

adenine one of four bases of DNA; abbreviated "A"; pairs with thymine or uracil

age-dependent penetrance increasing likelihood of manifesting signs or symptoms of a genetic disorder with increasing age

allele specific form of a gene

alpha fetoprotein (AFP) protein secreted during fetal life; high AFP is indicative of congenital anomalies such as open neural tube defects; low AFP can be a sign of Down syndrome

alternative splicing different patterns of exon splicing of a transcript, resulting in production of peptides that differ in amino acid sequence

Alu sequence one of a class of intermediate repeated DNA sequences, concentrated within coding regions of the genome

amino acid chemical building block of a protein, consisting of an amine group, a carboxyl group, and one of 20 specific functional groups

amino acid substitution mutation that leads to production of a peptide with a different amino acid at one site

amniocentesis method of prenatal diagnostic testing in which a sample of amniotic fluid is withdrawn for analysis

amniotic fluid fluid that bathes the fetus within the amniotic cavity, consisting largely of fetal urine with skin, bladder, and amnion cells

aneuploidy nonintegral multiple of the haploid chromosome set due to one or more missing or extra chromosomes

anonymous marker segment of cloned DNA of unknown function used as a reference point for DNA mapping

anticipation phenomenon whereby a genetic disorder becomes more severe from one generation to the next; characteristic of triplet repeat expansion disorders

anticodon tRNA sequence that recognizes codon in mRNA to insert appropriate amino acid into the growing peptide

antisense oligonucleotide sequence of DNA that is complementary to part of an mRNA, used to specifically inhibit expression of that gene

apoptosis programmed cell death

association group of congenital anomalies seen together more often than expected by chance

autocrine growth control production of growth factor by the same cell that responds to that factor

autoradiography technique of visualizing the presence of a radioactively labeled substance by exposure of silver grains in photographic emulsion

autosomal dominant mode of genetic transmission wherein a single mutant allele is sufficient to produce a phenotype, carried on a nonsex chromosome

autosome nonsex chromosome

bacteriophage virus that infects bacterial cells

balanced polymorphism genetic variant that is maintained at relatively high frequency in a population due to selection against homozygotes for the wild-type or variant sequence, with fitness being highest in heterozygotes

balanced translocation exchange of segments between chromosomes so that no genetic material is lost or gained

Barr body condensed copy of inactivated X chromosome in interphase nucleus

B cell antibody-producing lymphocyte

bias of ascertainment selection of individuals for study by virtue of their having a trait that thereby inflates the estimate of the frequency of that trait in the study population

blastomere pleuripotent embryonic cell during the first few divisions following fertilization

candidate gene gene thought to be involved in a specific genetic trait or disorder on the basis of mapping information or physiological evidence

carboxyl terminal last amino acid in a protein (in order of assembly)

carrier individual who has a mutant gene (usually used to describe individual who does not manifest signs of the trait)
cascade testing mode of screening for carriers of a genetic trait by testing relatives of an individual known to be affected
C banding chromosome staining technique that produces dark staining at centromeres
cDNA DNA copy of RNA made using enzyme reverse transcriptase
cDNA library collection of cloned cDNA sequences, representing portion of sequences transcribed in the cells or tissue of origin
centimorgan unit of genetic distance corresponding with 1% recombination
centromere site of attachment of spindle fibers to the chromosome representing the last point of separation of replicated chromatids
centromeric heterochromatin highly condensed chromatin near the centromere consisting of repeated DNA
chimera individual containing two genetically distinct cell lines resulting from separate fertilization events
chorionic villus sampling method of prenatal diagnosis in which fetal placenta is sampled either transabdominally or transcervically
chromatid one of two replicated arms of a chromosome
chromatin DNA with associated proteins
chromosome structure in cell on which genes are located, consisting of a highly compacted stretch of DNA with associated proteins
chromosome abnormality clinically significant change of chromosome number or structure
chromosome banding means of staining chromosomes to elicit characteristic and specific patterns to aid chromosome identification
chromosome jumping means of DNA cloning in which one cloned segment is used to identify another located a long distance away on the same chromosome
chromosome painting means of staining a chromosome based on hybridization with fluorescent-labeled DNA sequences specific to that chromosome
clone group of cells derived from a common progenitor; in recombinant DNA denotes a purified sequence of DNA
codon triplet of bases that encode a specific amino acid
cofactor substance that participates, along with an enzyme, in a chemical reaction
colchicine chemical that disrupts the mitotic spindle; used to collect cells at metaphase for chromosome analysis
complementarity occurrence of sequence of bases in DNA or RNA that will form stable double helix by pairing of A to T (or U) and G to C
compound heterozygote individual with two different mutant alleles at a locus
congenital anomaly structural abnormality present at birth
consanguinity blood relationship between a couple
conservative change substitution of one amino acid for another with similar chemical properties, with little or no effect on the structure and function of the protein
contig set of overlapping clones of DNA covering a large region
cosmid cloning vector consisting of plasmid with sequences that allow packaging in a lambda phage head; used for cloning large segments of DNA
couple testing approach to carrier screening, in which a couple are tested together, and are offered prenatal testing only if both are found to be carriers
coupling situation in which two alleles at linked genes are present on the same chromosome
CpG dinucleotides adjacent pair of nucleotides, with 5′-C-G-3′; the C in this location may be methylated
CpG islands region in which there are many CpG dinucleotides, often near the 5′ end of a gene
critical region part of a chromosome believed to contain genes involved in a specific phenotype
crossover consequence of genetic recombination
cryptic donor or acceptor sequence within an intron that can serve as a splice donor or acceptor if the usual site is disrupted by mutation
CVS acronym for chorionic villus sampling
cytogenetic pertaining to chromosomes
cytosine one of four bases of DNA; abbreviated "C"; pairs with guanine

D (diversity) segment component of immunoglobulin heavy chain gene
deformation alteration of developing structure in embryo or fetus due to extrinsic pressure

deletion mode of mutation due to loss of one or more bases of DNA, or to loss of large chromosomal region

denaturation separation of double-stranded nucleic acid into single strands

dideoxy sequencing mode of DNA sequencing using 2′ deoxynucleotides

dinucleotide repeat stretch of DNA containing a pair of bases (often C–A) repeated many times; repeat number may be polymorphic

diploid having two copies of each chromosome (except sex chromosomes)

disomic having two copies of a specific chromosome

disruption alteration of developing structure in embryo or fetus due to destruction of tissue

dizygotic twins twins resulting from separate fertilization events

DNA acronym for deoxyribonucleic acid, the chemical basis of heredity

dominant allele that exerts its phenotypic effect whether present in heterozygous or homozygous state

dosage compensation mechanism of equalizing X-linked gene expression in males and females by suppression of one X in females

Down syndrome complex of congenital anomalies resulting from an extra copy of chromosome 21

duplication mode of mutation due to repetition of one or more bases of DNA, or of a chromosomal region

dysmorphology study of structural abnormalities of human development

electrophoresis separation of chemical substances by differential migration in an electric field

endoplasmic reticulum subcellular structure representing site of protein synthesis

enzyme protein that catalyzes a specific chemical reaction

ethidium bromide fluorescent dye that intercalates into DNA; used to stain DNA

exon segment of gene that encodes amino acid sequence of protein; adjacent exons are separated by introns, which are spliced out during RNA processing

exon skipping mutation that alters pattern of splicing and results in splicing out of an exon

expressivity range of phenotypic variability of a genetic trait in a population

extinction elimination of a genetic trait from a population due to selection or genetic drift

fertilization union of sperm and egg leading to initiation of embryonic development

fibroblast cell type prevalent in connective tissue

FISH acronym for fluorescence in situ hybridization

fixation establishment of an allele as the sole allele at a given locus in a population due to extinction of other alleles

fluorescence in situ hybridization means of localization of cloned segment of DNA on chromosome by binding of complementary DNA and visualization by fluorescence microscopy

founder effect prevalence of a specific allele in a population due to its presence in one of the original members of the population and tendency of members of the population to be relatively inbred

fragile X syndrome genetic disorder associated with the expansion of a triplet repeat at the FMR 1 locus; associated with X-linked mental retardation

frameshift mutation that disrupts the sequence so that the reading frame is altered

fraternal twins nonidentical twins

G-banding mode of chromosome staining by Giemsa resulting in characteristic patterns of light and dark bands along chromosome

gene walking cloning overlapping segments of DNA to isolate large region starting from single cloned site

genetic counseling means of communication by medical professionals to an individual or family to educate them about natural history and genetics of a clinical disorder, along with available options for testing and treatment

genetic drift fluctuation of frequency of an allele in a small population from generation to generation due to statistical variation

genetic heterogeneity occurrence of multiple alleles at a gene locus or multiple loci that result in a similar phenotype

genetic linkage proximity of a set of gene loci on the same chromosome

genome collection of genes in an organism

genomic library collection of cloned DNA fragments from random sites in the genome

genotype set of specific alleles at a gene locus in an individual

germ cell sperm or egg cell

germ-line mosaicism occurrence of two or more cell lines derived from a single fertilization event but differing by presence or absence of one or more mutant alleles

guanosine one of four bases of DNA; abbreviated "G"; pairs with cytosine

haploid having only one complete set of chromosomes

haplotype set of alleles at a group of linked genes together on a specific chromosome

Hardy-Weinberg equilibrium mathematical statement of the relation between allele frequencies and frequencies of corresponding genotypes in a population

heavy chain one of two types of chains associated with immunoglobulin molecules

hemizygosity presence of one copy of a gene at a given locus instead of two; applies to X-linked genes in males, or any gene whose homologous copy has been deleted

heterochromatin chromatin that remains highly condensed during interphase and usually contains DNA that is genetically inactive

heteroduplex double-stranded molecule derived from two similar, but not identical DNA or RNA sequences

heteroplasmy occurrence of two or more populations of genetically distinct mitochondrial DNAs in a cell

heterozygote advantage selective advantage of heterozygous individuals over homozygotes; results in balanced polymorphism

heterozygous having two different alleles at a gene locus

histones basic proteins that form a complex with DNA in chromosome

HLA acronym for human lymphocyte antigen region on chromosome 6

homeobox DNA sequence that encodes a 60 amino acid region that is common to homeotic genes

homeotic genes genes involved in the control of development, discovered originally in *Drosophila* mutants, that lead to alteration in structures forming in specific body segments

homologous chromosomes pair of chromosomes, one inherited from each parent

homozygous having a pair of identical alleles for a particular gene

housekeeping genes genes that are expressed in a wide variety of cell types and are involved in common, basic mechanisms of cell physiology

hox abbreviation for homeobox

hybridization formation of double helix from complementary strands of DNA or RNA derived from different sources

hypermutability genetic trait that leads to high rate of mutation during DNA replication; characteristic of some neoplastic cells

identical twins monozygotic twins

imprinting differential expression of maternally and paternally derived genes

in situ hybridization technique of identifying chromosome region that contains sequence complementary with a cloned segment of DNA

informative mating mating in which one partner is heterozygous such that the polymorphic marker and alleles can be distinguished from those contributed by partner to offspring; used in genetic linkage analysis

informativeness likelihood of informative mating for particular genetic polymorphism

insertional mutation mutation resulting from insertion of one or more bases into DNA sequence

intron segment of noncoding DNA between exons, spliced out during RNA processing

inversion mutation due to reverse orientation of a segment of DNA

J (joining) segments part of sequence contributing to formation of immunoglobulin heavy chain

Kb abbreviation for kilobase

kilobase one thousand bases

lethal trait genetic trait that renders an individual unable to reproduce

liability tendency toward expression of a multifactorial trait, consisting of a combination of genetic and nongenetic factors

library set of cloned nucleic acid segments (genomic or cDNA)

light chain one of two chains of immunoglobulin molecule
LINE sequence type of repeated DNA, concentrated in noncoding regions
linkage proximity of a set of gene loci on the same chromosome
linkage disequilibrium nonrandom association of alleles at linked loci
linkage equilibrium random association of alleles at linked loci
liposome lipid bilayer used to transfer substance, such as cloned gene, across cell membrane
locus site of specific DNA sequence on chromosome
lod score logarithm of odds ratio of likelihood of data given specified value of recombination compared with random segregation; often abbreviated z
long-range restriction map map of sites of cutting of restriction endonucleases over a large region of DNA
Lyon hypothesis scheme of dosage compensation involving inactivation of one X chromosome in every cell in females; formulated by Mary Lyon
lysogeny stable incorporation of bacterial virus into bacterial genome
lysosomal storage disease clinical disorder due to absence of activity of a specific lysosomal enzyme, leading to build-up of substrate in lysosome

malformation abnormality of the formation of a fetal structure
marker gene polymorphic DNA sequence used in linkage mapping
maternal transmission characteristic of mitochondrial genetic traits, passed from a mother to all her offspring
maximum likelihood estimate value of recombination fraction (θ) at which peak lod score is obtained
Mb abbreviation for megabase
megabase one million bases
meiosis process of reduction division in germ line, leading to formation of haploid germ cells
Mendelian trait genetic trait that follows patterns of simple Mendelian inheritance
metacentric presence of centromere at center of chromosome
metaphase stage of cell division when chromosomes are aligned at the center of the cell prior to separation
metastasis spread of neoplastic cells from their site or origin to remote sites
methylation addition of methyl groups to cytosine bases (usually at CpG dinucleotides)
microsatellite polymorphisms DNA sequence variants due to different numbers of repeats of a simple sequence
missense mutation base change in the coding sequence of a protein that leads to amino acid substitution
mitochondrial DNA circular double-stranded DNA within mitochondrion that encodes 13 mitochondrial proteins, transfer and ribosomal RNAs
mitochondrion structure in cell involved in aerobic metabolism
mitosis process of cell division
mitotic spindle structure in dividing cell that pulls chromatids to opposite poles to make daughter cells
molecular diagnostic testing identification of specific nucleic acid sequences for medical diagnosis
monosomy presence of one rather than two copies of a specific chromosome in an individual
monozygotic twins genetically identical twins resulting from single fertilization event
mosaicism occurrence of two or more genetically distinct cell lines derived from a common progenitor
mRNA processed gene transcript ready for translation into protein
Müllerian duct embryonic structure that gives rise to uterus and Fallopian tubes
multifactorial inheritance traits determined by a combination of multiple genetic and/or nongenetic factors
multiplex PCR simultaneous amplification of multiple sequences in a single reaction using multiple sets of primers
mutation change in the sequence of DNA at a genetic locus

neoplasm clone of cells released from normal controls of growth
nondisjunction failure of proper chromosome segregation, leading to both copies of a chromosome (or both chromatids) going to the same daughter cell
nonhistone protein protein associated with DNA, not one of the basic histone proteins

nonpaternity finding that stated father is not the biological father of a child
nonpenetrance the absence of phenotype in a person known to carry a specific mutant gene
nonsense mutation mutation that changes a codon for an amino acid to a stop codon
NOR staining method of staining nucleolus-organizer region of acrocentric chromosomes
northern blot method of identification of RNA separated by electrophoresis, blotted onto membrane, and hybridized with labeled nucleic acid
nucleolus organizer regions sites on acrocentric chromosomes containing ribosomal DNA
nucleosome knob-like structure consisting of approximately 140 base pairs of DNA and associated histones, forming a structural unit of chromatin
nucleotide base of DNA or RNA

oligonucleotide segment of several bases of DNA or RNA
oncogene gene that confers some neoplastic properties on a cell, usually derived by activation of a proto-oncogene
open reading frame region of cDNA sequence starting with an AUG codon that initiates protein synthesis and including a region that encodes protein, ending at a stop codon
oogonial cell immature egg cell, prior to meiosis

p arm short arm of chromosome
P1 phage vector cloning vector derived from P1 bacteriophage that can accommodate large inserts of foreign DNA
paired box DNA-binding domain of set of gene products involved in development, consisting of 128 amino acids
paracentric inversion inversion of chromosome region not involving centromere
pathogenic mutation mutation responsible for a genetic disorder
Pax abbreviation for paired box
PCR acronym for polymerase chain reaction
pedigree diagram of family using standard symbols
penetrance expression of genetic trait in an individual with mutant genotype
pericentric inversion inversion of chromosome region involving centromere
phage lambda bacterial virus used as a cloning vector
phenotype physical manifestations resulting from specific genotype
phytohemagglutinin substance derived from kidney beans that stimulates division of T-cells; used to culture cells for chromosomal analysis
plasmid circular double-stranded DNA capable of autonomous replication in bacterial cells; used as a cloning vector
pleiotropy diverse physical characteristics resulting from a single genetic trait
point mutation single base change of DNA sequence
polyacrylamide gel electrophoresis method for high-resolution separation of nucleic acids or proteins
polygenic inheritance traits determined by two or more separate genes
polymerase chain reaction means of amplification of a DNA sequence by multiple cycles of replication starting from the pair of primers that flank the sequence
polymorphism occurrence of at least two alleles at a locus each having a frequency of at least 1%
population genetics study of the factors that influence the frequency of genetic traits in a population
positional cloning means of cloning a gene based on its location in the genome
preimplantation diagnosis genetic diagnosis from early embryo, prior to implantation in uterus
premutation sequence variation that predisposes to mutation; usually moderate expansion of a triplet repeat that leads to further expansion, such as in fragile X syndrome
prenatal diagnosis diagnosis of a medical problem in embryo or fetus
primer oligonucleotide used as the point of initiation of DNA synthesis
proband individual in family who brings family to medical attention
probe cloned nucleic acid sequence used to identify homologous sequence by nucleic acid hybridization
prokaryote primitive microorganism lacking cell nucleus
prophase stage of mitosis when nuclear membrane disappears and chromosomes condense
proposita female proband
propositus male proband

proto-oncogene cellular gene that, when appropriately altered, becomes an oncogene
pseudoautosomal gene located on both the X and Y chromosomes, and hence segregating as an autosomal locus
pseudogene DNA sequence with substantial homology to a gene, but not encoding protein
pseudomosaic chromosomally abnormal cell line found in cultured prenatal sample that is believed to have arisen during the culture process
pulsed field gel electrophoresis method for separating DNA fragments of large size
purine chemical structure of adenine and guanosine
pyrimidine chemical structure of thymine, cytosine, and uracil

q arm long arm of a chromosome
Q-banding method of chromosome banding using the fluorescent dye quinacrine
quinacrine fluorescent dye used to elicit Q banding

random segregation independent segregation of nonlinked genes to gametes
R-banding method of chromosome staining that produces bands that are the reverse of G-bands
reading frame set of triplet codons in a gene that encode the protein
recessive allele that exerts its phenotypic effect only if present in homozygous state
recessive oncogene gene that, when homozygously mutated, confers some properties of the neoplastic state; tumor suppressor gene
reciprocal translocation exchange of segments between two or more chromosomes
recombination association of new set of alleles in coupling on a given chromosome due to crossing over in meiosis
recombination fraction (θ) probability of recombination between two genetic loci
renaturation reannealing of separated single strands of nucleic acid into a double helix
reproductive fitness relative ability of individuals with a specified genotype to reproduce
repulsion presence of specified alleles of linked loci on opposite homologous chromosomes
restriction endonuclease enzyme that specifically cuts double-stranded DNA at a defined base sequence
restriction fragment length polymorphism genetic polymorphism involving a base change that affects the ability of a specific restriction endonuclease to cut the site
reverse transcriptase enzyme present in retroviruses that copies RNA into DNA
RFLP acronym for restriction fragment length polymorphism
ribosome cellular structure involved in translation of mRNA into protein
RNA acronym for ribonucleic acid
Robertsonian translocation translocation involving fusion of the long arms of a pair of acrocentric chromosomes

selection impairment of reproductive fitness of individuals with a specific genotype
sex chromosome X or Y chromosome involved in sex determination
sex-linked gene gene present on X or Y chromosome
Sherman paradox phenomenon whereby fragile X syndrome is transmitted via nonmanifesting male carriers to daughters, who have affected offspring; due to males with premutations
single-stranded conformational polymorphism method for identification of DNA sequence harboring a gene mutation, based on altered mobility of single-stranded sequences in polyacrylamide gel due to sequence-specific conformational change
somatic cell hybridization fusion of somatic cells using Sendai virus or polyethylene glycol
somatic mosaicism presence of two or more genetically distinct cell lines in an individual, derived from a common progenitor
Southern blot method of identification of DNA fragments separated by electrophoresis, blotted onto membrane, and hybridized with labeled nucleic acid
SOX acronym for SRY box
splice acceptor sequence at 3′ border of intron
splice donor sequence at 5′ border of intron
splicing process of removal of introns and ligation of exons during processing of RNA
splicing mutations mutations that change the patterns of RNA splicing, usually by altering splice donor or acceptor sites
SRY gene on Y chromosome required for differentiation of the testes
SSCP acronym for single strand conformation polymorphism

stepwise testing approach to carrier screening in which one member of a couple is tested first, and the partner is tested only if the first individual is found to be a carrier

stop codon codon that leads to termination of translation

submetacentric location of a centromere between the middle of the chromosome and one end, producing a short and long arm

suppressor tRNA altered tRNA that inserts an amino acid at a stop codon

syndrome set of reproducible clinical features due to a common underlying mechanism

T-banding method of staining chromosome ends

T cell lymphocyte involved in modulation of the immune response or cytotoxic activity

telomere specific DNA structure at the ends of chromosomes

temperature sensitive mutation that renders a protein unstable at specific temperature

teratogen substance that interferes with normal embryonic development

theta (θ) abbreviation for recombination fraction

threshold model theory of multifactorial inheritance stating that a trait occurs when liability exceeds a threshold

thymine one of four bases of DNA; abbreviated "T"; pairs with adenine

transfection introduction of segment of foreign DNA into a cell

transformed cell cell capable of unlimited number of rounds of replication

transgene cloned gene inserted into foreign genome

transgenic mouse mouse into which a foreign gene has been inserted into germ line

transition mutation that substitutes a pyrimidine for a pyrimidine (e.g., G to A or A to G), or a purine for a purine (e.g., T to C or C to T)

translation process of production of protein from mRNA

translocation exchange of segments between chromosomes

transversion mutation that substitutes a purine for a pyrimidine (e.g., T to G, etc.) or vice versa

triple test measurement of levels of alpha fetoprotein, human chorionic gonadotrophin, and unconjugated estriol as a screen for Down syndrome

triplet repeat expansion type of mutation in which a gene segment containing multiple copies of a triplet of bases is expanded in length

trisomy presence of three copies of a chromosome rather than two

tRNA RNA molecule that carries a specific amino acid and recognizes the corresponding codon, inserting that amino acid into a growing peptide

tumor suppressor gene see recessive oncogene

two-hit hypothesis hypothesis formulated by A. Knudson, postulating that malignant transformation occurs following a two-step process

ultrasound mode of imaging using high-frequency sound waves, used to visualize a developing fetus for prenatal diagnosis

uniparental disomy inheritance of both copies of a chromosome from the same parent

uracil base that substitutes for thymine in RNA and pairs with adenine; abbreviated "U"

variable (V) segments segment of DNA encoding the antigen-recognition site of an immunoglobulin heavy or light chain

vector DNA sequence that conveys an inserted segment into a cell

wild type most common allele in a population at a particular gene locus

Wolffian duct embryonic structure that gives rise to epididymis vas deferens, and seminal vesicle

X chromosome one of the sex chromosomes

X inactivation center (Xic) region on X chromosome at which X inactivation is initiated

Xist gene on X chromosome believed to be involved in the initiation of X inactivation

Y chromosome one of the sex chromosomes

yeast artificial chromosome (YAC) cloning vector that allows replication of inserted DNA in yeast cells, allowing cloning of very large segments

zoo blot Southern blot containing DNA from multiple species, used to detect species conservation

zygote sperm or egg cell

Index

Note: Page numbers followed by an f *indicate figures; those followed by a* t *indicate tables.*